Cephalopods Present and Past: New Insights and Fresh Perspectives

Cephalopods Present and Past: New Insights and Fresh Perspectives

Edited by

Neil H. Landman
Division of Paleontology (Invertebrates)
American Museum of Natural History
New York, NY, USA

Richard Arnold Davis
Department of Biology
College of Mount St. Joseph
Cincinnati, OH, USA

Royal H. Mapes
Department of Geological Sciences
Ohio University
Athens, OH, USA

Springer

A C.I.P. Catalogue record for this book is available from the Library of Congress.

ISBN 978-1-4020-6461-6
ISBN 978-1-4020-6806-5 (e-book)

Published by Springer,
P.O. Box 17, 3300 AA Dordrecht, The Netherlands.

www.springer.com

Cover illustration: Reconstruction of the life cycle of *Manticoceras*, depicting the orientations of the aperture of four representative growth stages. Figure by Christian Klug, Universität Zürich.

Printed on acid-free paper

Preface

Cephalopods are diverse, highly developed molluscs capable of swimming and jet propulsion. These animals are an important component of present-day marine ecosystems throughout the world and comprise approximately 900 species. They also have an extraordinary fossil record, extending back to the Cambrian Period, with as many as 10,000 extinct species. Throughout their long history, they have experienced spectacular radiations and near-total extinctions. Because of their superb fossil record, they also serve as ideal index fossils to subdivide geologic time. This book touches on many of these themes, and it treats both fossil and present-day cephalopods. The chapters are outgrowths of presentations at the Sixth International Symposium "Cephalopods – Present and Past," at the University of Arkansas in Fayetteville, September 16–19, 2004. The Symposium was organized principally by Walter L. Manger of the Department of Geology, University of Arkansas. The editors gratefully acknowledge Walter for his terrific job in putting together this symposium and for making it such an intellectual, and social, success. Other publications related to this Symposium include the abstract volume, assembled by W. L. Manger, and two field-trip guidebooks, one written by W. L. Manger, and the other by R. H. Mapes.

Because this symposium was held in North America, it honored four cephalopod workers from this continent: William A. Cobban (US Geological Survey, Denver, Colorado), Brian F. Glenister (University of Iowa, Iowa City, Iowa), William M. Furnish (University of Iowa, Iowa City, Iowa), and Gerd E. G. Westermann (McMaster University, Hamilton, Ontario). These four workers are giants in their fields, and through their research on the biology, systematics, and biostratigraphy of fossil cephalopods, they have enormously expanded our understanding of these animals and the history of planet Earth. This volume is dedicated to them – in recognition of their phenomenal accomplishments.

This volume contains 20 chapters covering a wide range of topics about both fossil and present-day cephalopods. We have grouped these chapters into three sections, although we recognize that many of the subjects overlap:

- Phylogeny and Systematics (Chapters 1–7)
- Morphology of Soft and Hard Tissues (Chapters 8–14)
- Biogeography, Biostratigraphy, Ecology, and Taphonomy (Chapters 15–20)

Within each section, ammonoids are treated first, followed by coleoids, in order of geologic time.

Every chapter was examined by at least two outside reviewers, and their suggestions and other comments, together with those of the editors, were forwarded to the authors. The reviewers made many helpful suggestions; this resulted in substantially improving the quality of the manuscripts. In addition, authors were encouraged to follow General Recommendation 10 of the International Code of Zoological Nomenclature, which suggests that the author and date of every taxon in a publication be cited at least once in that publication.

The editors extend their sincere thanks to the following people who reviewed the manuscripts: Emily G. Allen (Bryn Mawr College, Bryn Mawr, Pennsylvania), Roland Anderson (Seattle Aquarium, Seattle, Washington), R. Thomas Becker (WWU, Geologisch-Paläontologisches Institut, Münster, Germany), Hugo Bucher (Paläontologisches Institut und Museum der Universität Zürich, Zürich, Switzerland), Antonio G. Checa (Universidad de Granada, Granada, Spain), William A. Cobban (US Geological Survey, Denver, Colorado), Régis Chirat (Université Claude Bernard Lyon 1, Villeurbanne Cedex, France), Larisa A. Doguzhaeva (Palaeontological Institute of the Russian Academy of Sciences, Moscow, Russia), Jean-Louis Dommergues (Centre des Sciences de la Terre, Université de Bourgogne, Dijon, France), Desmond T. Donovan (University College London, London, United Kingdom), Dirk Fuchs (Freie Universität Berlin, Berlin, Germany), Roger A. Hewitt (Leigh-on-Sea, Essex, United Kingdom), W. James Kennedy (University Museum, Oxford, United Kingdom), William T. Kirchgasser (SUNY, Potsdam, New York), Christian Klug (Paläontologisches Institut und Museum der Universität Zürich, Zürich, Switzerland), Dieter Korn (Museum für Naturkunde der Humboldt-Universität zu Berlin, Berlin, Germany), Cyprian Kulicki (Polska Akademia Nauk, Warsaw, Poland), Neal L. Larson (Black Hills Museum of Natural History, Hill City, South Dakota), George R. McGhee (Rutgers University, New Brunswick, New Jersey), Lisa K. Meeks (Exxon Mobil Development Company, Houston, Texas), Pascal Neige (Centre des Sciences de la Terre, Université de Bourgogne, Dijon, France), W. Bruce Saunders (Bryn Mawr College, Bryn Mawr, Pennsylvania), Dolf Seilacher (Yale University, New Haven, Connecticut), Kazushige Tanabe (University of Tokyo, Tokyo, Japan), Janet R. Voight (The Field Museum, Chicago, Illinois), Frank Weise (Freie Universität Berlin, Berlin, Germany), Wolfgang Weitschat (Geologische-Paläontologisches Institut und Museum der Universität Hamburg, Hamburg, Germany), Gerd E. G. Westermann (Hamilton, Ontario, Canada), and Margaret M. Yacobucci (Bowling Green State University, Bowling Green, Ohio).

The editors also thank Susan M. Klofak, Kathy B. Sarg, Steve Thurston, and Stephanie Crooms (American Museum of Natural History) for help in working with the manuscripts (proofing, mailing, word processing, and scanning images), and Judith Terpos (Springer) for guidance in putting the book together.

Neil H. Landman
New York, New York
Richard Arnold Davis
Cincinnati, Ohio
Royal H. Mapes
Athens, Ohio

Contents

Chapter 3 • Conch Form Analysis, Variability, Morphological Disparity, and Mode of Life of the Frasnian (Late Devonian) Ammonoid *Manticoceras* from Coumiac (Montagne Noire, France)
Dieter Korn and Christian Klug

Chapter 4 • GONIAT – The Current State of the Paleontological Database System on Paleozoic Ammonoids
Jürgen Kullmann

Chapter 5 • Ornamental Polymorphism in *Placenticeras kaffrarium* (Ammonoidea; Upper Cretaceous of India): Evolutionary Implications
Tapas K. Gangopadhyay and Subhendu Bardhan

Chapter 9 • Cameral Membranes in Carboniferous and Permian Goniatites: Description and Relationship to Pseudosutures
Kristin Polizzotto, Neil H. Landman, and Royal H. Mapes

Chapter 10 • Soft-tissue Attachment of Middle Triassic Ceratitida from Germany
Christian Klug, Michael Montenari, Hartmut Schulz, and Max Urlichs

Chapter 11 • The Preservation of Body Tissues, Shell, and Mandibles in the Ceratitid Ammonoid *Austrotrachyceras* (Late Triassic), Austria
Larisa A. Doguzhaeva, Royal H. Mapes, Herbert Summesberger, and Harry Mutvei

Chapter 12 • Connecting Ring Ultrastructure in the Jurassic Ammonoid *Quenstedtoceras* with Discussion on Mode of Life of Ammonoids
Harry Mutvei and Elena Dunca

List of Contributors

Emily G. Allen
Department of Geology
Bryn Mawr College, Bryn Mawr
PA 19010, USA
E-mail: egallen@brynmawr.edu

Subhendu Bardhan
Department of Geological Sciences
Jadavpur University
Kolkata 700032, India
E-mail: s_bardhan01@yahoo.co.uk

William A. Cobban
70 Estes Street Lakewood
CO 80226, USA

Larisa A. Doguzhaeva
Paleontological Institute of the Russian Academy of Sciences
St. Profsoyuznaya 123
Moscow 117997, Russia
E-mail: ldoguzhaeva@rambler.ru

Elena Dunca
Department of Palaeozoology
Swedish Museum of Natural History
S-10405, Stockholm, Sweden
E-mail: elena.dunca@nrm.se

Theo Engeser
Institut für Geologische Wissenschaften
Fachrichtung Paläontologie
Freie Universität Berlin
Malteserstrasse 74-100
D-12249 Berlin, Germany

Dirk Fuchs
Institut für Geologische Wissenschaften
Fachrichtung Paläontologie
Freie Universität Berlin
Malteserstrasse 74-100
D-12249 Berlin, Germany
E-mail: drig@zedat.fu-berlin.de

Tapas K. Gangopadhyay
Department of Geology
Bengal Engineering and Science University Shibpur
Howrah 711103, India
E-mail: tapasgeolbe@yahoo.com

Helmut Keupp
Institut für Geologische Wissenschaften
Fachrichtung Paläontologie
Freie Universität Berlin
Malteserstrasse 74-100
D-12249 Berlin, Germany

Susan M. Klofak
Division of Paleontology (Invertebrates)
American Museum of Natural History
79th Street and Central Park West

New York, NY 10024
and Department of Biology
City College of the City University
of New York
Convent Avenue and 138th Street
New York, NY 10031, USA
E-mail: klofak@amnh.org

Christian Klug
Paläontologisches Institut
und Museum, Universität Zürich
Karl Schmid-Strasse 4
CH-8006 Zürich, Switzerland
E-mail: chklug@pim.uzh.ch

Dieter Korn
Museum für Naturkunde der
Humboldt-Universität zu Berlin
Invalidenstrasse 43, D-10115 Berlin
Germany
E-mail: dieter.korn@museum.hu-berlin.de

Jürgen Kullmann
Institut für Geowissenschaften
Universität Tübingen, Sigwartstrasse 10
D-72076 Tübingen, Germany
E-mail: juergen.kullmann@
uni-tuebingen.de

Neil H. Landman
Division of Paleontology (Invertebrates)
American Museum of Natural History
79th Street and Central Park West
New York, NY 10024, USA
E-mail: landman@amnh.org

Neal Larson
Black Hills Institute of Geological
Research
PO Box 643
117 Main Street
Hill City SD 57745 and
Black Hills Museum of Natural History
Hill City SD 57745, USA
E-mail: ammoniteguy@bhigr.com

A. Lauster
Environmental Sciences, Alaska
Pacific University
Anchorage, AK 99508, USA

Royal H. Mapes
Department of Geological Sciences
316 Clippinger Laboratories
Ohio University, Athens, OH 45701
USA
E-mail: mapes@ohio.edu

Vasilij Mitta
Russian Academy of Sciences
Paleontological Institute
Profsojuznaja 123, Moscow 117997
Russia
E-mail: vmitta@mail.ru

Michael Montenari
Earth Sciences and Geography
Keele University
William Smith Building
Newcastle (Staffordshire)
ST5 5BG United Kingdom

Sebastien Moyne
UMR CNRS 5561 Biogéosciences
6 bd Gabriel, F-21000 Dijon, France
E-mail: sebastien.moyne@u-bourgogne.fr

Harry Mutvei
Department of Palaeozoology
Swedish Museum of Natural History
S-10405, Stockholm, Sweden
E-mail: harry.mutvei@nrm.se

Pascal Neige
UMR CNRS 5561 Biogéosciences
6 bd Gabriel, F-21000 Dijon
France
E-mail: Pascal.neige@u-bourgogne.fr

Svetlana Nikolaeva
Paleontological Institute

Russian Academy of Sciences
Profsoyuznaya 123, Moscow 117997
Russia
E-mail: 44svnikol@mtu-net.ru;
s.nikolaeva@nhm.ac.uk

Alexei M. Orlov
Russian Federal Research Institute
of Fisheries and Oceanography
(VNIRO)
17, V. Krasnoselskaya
Moscow 107140, Russia
E-mail: orlov@vniro.ru

Kristin Polizzotto
Division of Paleontology (Invertebrates)
American Museum of Natural History
79th Street and Central Park West
New York, NY 10024 and
Department of Biological Sciences
Kingsborough Community College
2001 Oriental Boulevard
Brooklyn, NY 11235, USA
E-mail: kpolizzotto@
kingsborough.edu

Isabelle Rouget
UMR CNRS 5143, Case 104
T. 46-56, 5éme E
4 place Jussieu, F- 75252 Paris
Cedex 05, France
E-mail: rouget@ccr.jussieu.fr

Pinaki Roy
Department of Geological Sciences
Jadavpur University
Kolkata 700032, India

David Scheel
Environmental Sciences
Alaska Pacific University
Anchorage, AK 99508, USA
E-mail: dscheel@alaskapacific.edu

Hartmut Schulz
Institut für Geowissenschaften
Eberhard-Karls-Universität
Sigwartstrasse 10, D-72076 Tübingen
Germany

Sabyasachi Shome
Geological Survey of India
15A&B Kyd Street
Kolkata 700016, India

Herbert Summesberger
Naturhistorisches Museum Wien
A-1014, Wien, Burgring 7, Austria
E-mail: herbert.summesberger@
nhm-wein.ac.at

Max Urlichs
Staatliches Museum für Naturkunde
Rosenstein 1
D-70191 Stuttgart, Germany
E-mail: urlichs.smns@naturkunde
museum-bw.de

T. L. S. Vincent
Environmental Sciences
Alaska Pacific University
Anchorage, AK 99508
USA

Kerstin Warnke
Freie Universität Berlin
FR Paläontologie
Malteserstrase 74-100, Haus D-12249
Berlin, Germany
E-mail: warnke@zedat.fu-berlin.de

Part I
Phylogeny and Systematics

Chapter 1
Phylogenetic Practices Among Scholars of Fossil Cephalopods, with Special Reference to Cladistics

Pascal Neige,[1] Isabelle Rouget,[2] and Sebastien Moyne[1]

[1]UMR CNRS 5561 Biogéosciences, 6 bd Gabriel, F-21000 Dijon, France, Pascal.neige@u-bourgogne.fr; sebastien.moyne@u-bourgogne.fr;

[2]UMR CNRS 5143, Case 104, T. 46–56, 5ème E, 4 place Jussieu, F-75252 Paris Cedex 05, France, rouget@ccr.jussieu.fr

Keywords: cephalopods, cladistics, phylogeny, taxonomy

1 Introduction

One of the most popular activities among paleontologists is to attribute species names to fossil specimens and then to classify species in a hierarchical pattern: the so-called Linnaean classification. This taxonomic activity is vital, ensuring a large corpus of knowledge of past life across geological times. By-products are: the study of biodiversity through time, the discovery of some extraordinary events such as mass extinctions and major radiations, and the slicing of geological time into singular associations of fossils (known as biozones) to date sediments.

Within the last decade, methods for studying fossils for whatever purpose have been largely modified, especially under scientific pressure to adhere as closely as possible to quantified and reproducible approaches. Scholars of fossil cephalopods have contributed to this scientific revolution. Some were not merely following the movement; they were largely ahead of their time. This was particularly the case of David Raup in his work on morphometry (1967). New discoveries and methods of study have drastically increased our knowledge of past cephalopods in terms of diversity, taxonomy, paleobiogeography, ontogeny, dimorphism, mode of life, and so on. Surprisingly, it seems that the community of fossil cephalopod scholars as a whole has tended to bypass one

N. H. Landman et al. (eds.), *Cephalopods Present and Past: New Insights and Fresh Perspectives*, 3–14.

of the major changes and advances in biological and paleontological sciences: the cladistic approach and its implications for phylogeny and taxonomy. This approach is now widely used to reconstruct phylogenetic patterns, and has proved to be efficient when applied to present and past organisms of any kind (metazoans, plants, etc.). At a time when some biologists and paleontologists propose the abandonment of Linnaean taxonomy, and its replacement by the PhyloCode (see Cantino and de Queiroz, 2003; Laurin, 2004), scholars of fossil cephalopods have still not – to our way of thinking – clearly opened the debate concerning the respective merits of the different phylogenetic methods, and especially the interpretative power of cladistics.

In this study we present an in-depth study of phylogenetic practices among fossil cephalopod scholars, with particular emphasis on the use of cladistics. Reasons for such underuse of cladistics applied to fossil cephalopods will be briefly explained. This paper must be seen as a first step toward a larger debate concerning the choice of phylogenetic method within our favorite fossil group.

2 Sampling Phylogenetic Practices: Review of Paleontological Literature from 1985 to 2003

Paleontological literature is explored here from 1985 up to 2003. The year 1985 corresponds to the organization in Tübingen (Germany) of the 2nd International Cephalopod Symposium. The first, in York (England), was entirely devoted to ammonoids, and it is generally considered that the second edition held in Tübingen counts as the first symposium dealing with various present and past cephalopod groups. Because our purpose is to evaluate phylogenetic practices among fossil cephalopods as a whole, we decided that the Tübingen symposium acts as a starting date. The year 2003 is the last complete year at the time of writing, and thus provides a full year's supply of journal volumes.

Two databases have been compiled for this period of publication. One is based on all regular volumes of five paleontological journals: *Geobios*, *Journal of Paleontology*, *Lethaia*, *Palaeontology*, and *Paleobiology*. These journals have been chosen for the following reasons: they are peer reviewed for the complete range of years we are working on; they have no taxonomic or stratigraphic restrictions (i.e., they are not dedicated to a particular taxonomic or stratigraphic field); all are well known and easily accessible in libraries all over the world. It is true that they do not cover the complete range of paleontological publications, but we believe that this sampling is a valid representation of the range. Our first database ensures precise monitoring of phylogenetic practices for fossil cephalopods, and also a point of comparison for other taxonomic groups. Our second database comes from the compilation of specific fossil cephalopod literature during the same period of time. We have selected only proceedings that followed symposiums and were subject to a peer-review process. We believe that this will ensure the best monitoring of specialized fossil cephalopod literature, and that these collective contributions act as landmarks for fossil cephalopod knowledge. Selected proceedings are: *Proceedings of the 2nd International Cephalopod Symposium* (Wiedmann and Kullmann, 1988), then the 3rd (Elmi et al., 1993), the 4th (Oloriz and Rodriguez-Tovar 1999), and the 5th (Summesberger et al., 2002). We also added the *Proceedings of the International Symposium, Coleoid Cephalopods Through Time* (Warnke et al., 2003) held in Berlin (Germany) in 2002, which gathered together present and past studies on coleoid cephalopods.

2.1 Regular Paleontological Publications

Papers with an explicit taxonomic or phylogenetic section (i.e., with a taxonomic list, a taxonomic treatment, or a new taxon name) have been counted for the period of publication studied (Table1.1). Counting only such papers reduces the sample to those more or less linked to a phylogenetic perspective; 3,031 publications devoted to various taxa, and with an explicit taxonomic section, have been published for the period studied and the journals analyzed. Among them, 440 have a cladistic section (a cladistic section is recognized here if a cladogram or equivalent parenthetic notation appears in the publication, even if it is not based on a parsimony analysis). The relative cladistic contribution is thus 14.5% (number of papers with cladistic section divided by number of papers with explicit taxonomic section).

Fluctuations in taxonomic and cladistic publications over time are quite different (Fig. 1.1). In both cases, we note an increase in the number of publications. This was tested using Spearman's nonparametric Rank Correlation Test (Table 1.2) (see Swan and Sandilands, 1995), the equivalent of Pearson's Classic Correlation Test but applied to not normally distributed variables, which is the case here. Results indicate that both taxonomic and cladistic publications increase in number over time. However, the increase in the number of cladistic publications is clearly much more marked (Fig. 1.1). A second test was performed on the percentage of cladistic publications to test for the net increase in the cladistic approach (to eliminate the increase in cladistic

Table 1.1 *Number of publications in regular paleontological literature (1985–2003). "Taxonomy" refers to publications with an explicit taxonomic section, "Cladistics" to publications with a cladogram (or a parenthetic notation) constructed with or without parsimony analysis (see text).*

Title	Taxonomy	Cladistics (%)
Geobios	575	36 (6.26)
Journal of Paleontology	1,501	154 (10.26)
Lethaia	112	41 (36.61)
Palaeontology	741	140 (18.89)
Paleobiology	102	69 (67.65)
Total	3,031	440 (14.52)

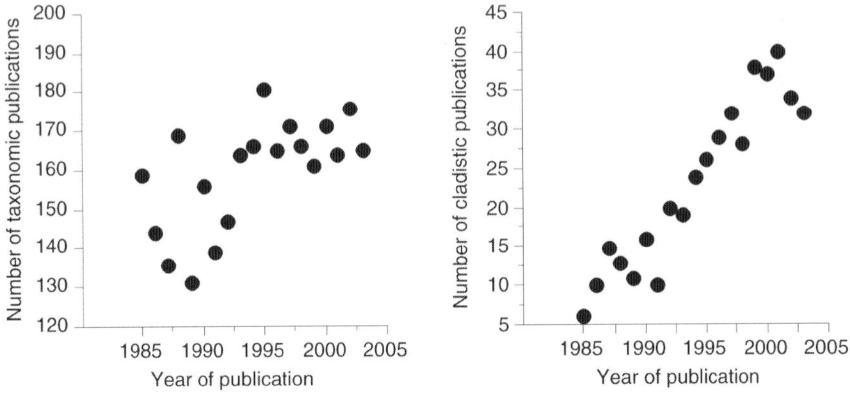

Fig. 1.1 *Number of publications over time from five regular paleontological journals (see text).*

Table 1.2 *Nonparametric Spearman's Rank Correlation Test for various data sets (see text). rs: Spearman's Rank Correlation Coefficient, NS: test nonsignificant, *: test significant at 95% confidence level, **: at 99% confidence level, ***: at 99.9% confidence level. Results are given with ex-aequo correction (see Swan and Sandilands, 1995).*

	r_s	p-values	
Year, Number of Taxonomic publications	0.60	0.011	*
Year, Number of Cladistic publications	0.93	<0.0001	***
Year, Percentage of Cladistic publications	0.94	<0.0001	***

publications due to the increase in the total number of publications). Results indicate that the cladistic approach is increasingly used in paleontology (Table 1.2).

To explore phylogenetic practices in detail, each of the 3,031 previous papers is attributed to the taxon with which it deals. Here, 12 major taxonomic entities are recognized: plants, corals and relatives, sponges and relatives, trilobites, arthropods (excluding trilobites), brachiopods and bryozoans, bivalves, gastropods, cephalopods, echinoderms, graptolites, and finally vertebrates. They may not correspond to identical taxonomic levels, yet each of them reflects a particular bauplan organization, reserved for a particular community of scholars. Results reflect the domination of the vertebrate community in the number of papers with taxonomic purpose published (Fig. 1.2A). The pecking order is then brachiopods and bryozoans, arthropods (excluding trilobites), echinoderms, and cephalopods in fifth rank, just before trilobites. In contrast, the situation is quite different for papers with cladistic purpose (Fig. 1.2A). Vertebrates are still dominant, but the ranking of other groups has drastically changed: second are echinoderms, then trilobites, brachiopods and bryozoans, and arthropods (except trilobites) in fifth position. Cephalopods are only ranked in eighth position, after other mollusk representatives (gastropods and bivalves, respectively). The proportion of cladistic papers per major taxonomic

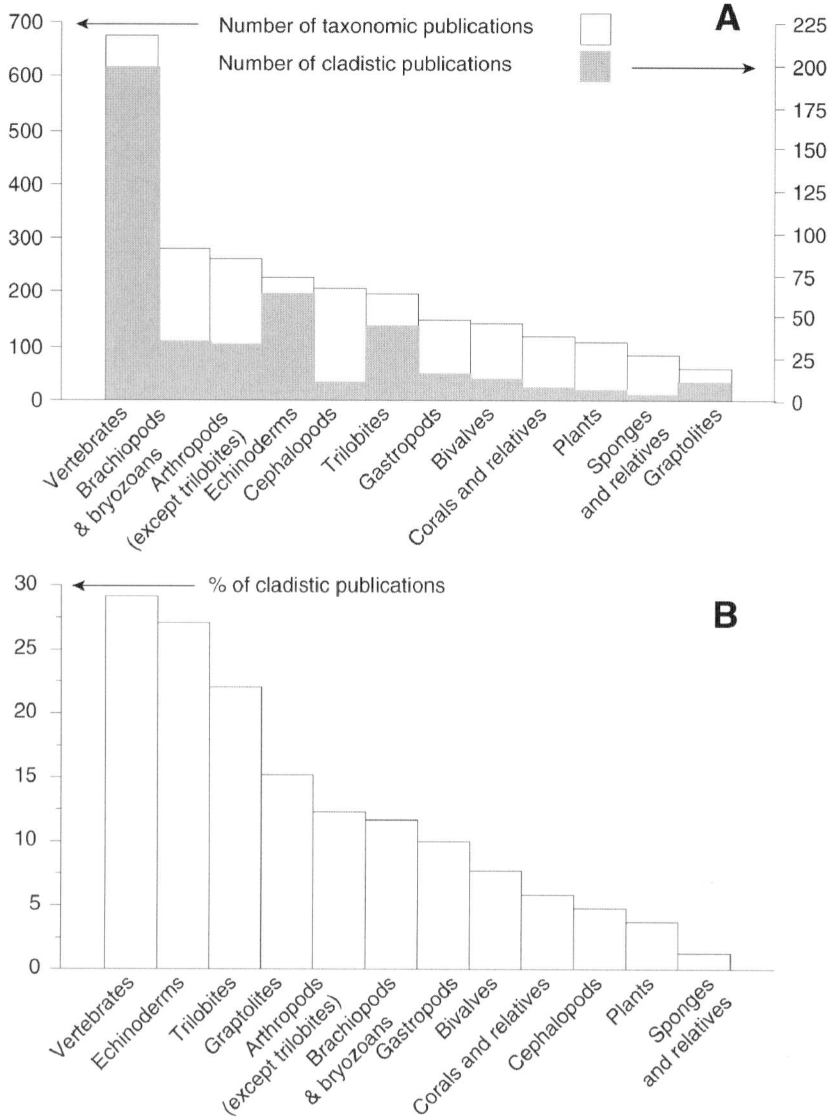

Fig. 1.2 *Comparative number of publications for twelve higher-level groupings (not of equal rank), over time, in five regular paleontological journals (see text). A: number of papers; B: percentage of cladistics papers.*

entity demonstrates the dramatic underuse of cladistics within the fossil cephalopod community (Fig. 1.2B). The ratio is less than 5%, and only sponges and relatives and plants get lower scores. The percentage of cladistic publications for fossil cephalopods varies from one journal to another (Table 1.3; see Appendix for detailed bibliographic references).

Table 1.3 *Number of publications for cephalopods only, in regular paleontological literature (1985–2003). "Taxonomy" refers to publications with an explicit taxonomic section, "Cladistics" to publications with a cladogram (or parenthetic notation) constructed with or without parsimony analysis (see text and Appendix).*

Title	Taxonomy	Cladistics (%)
Geobios	60	1 (1.67)
Journal of Paleontology	81	2 (2.47)
Lethaia	7	2 (28.60)
Palaeontology	56	3 (5.36)
Paleobiology	2	2 (100)
Total	206	10 (4.85)

Results for fossil cephalopods are very different from those for all taxa as a whole. In Fig. 1.3, there is quite clearly a negative correlation for the number of taxonomic papers, and no correlation for the number of cladistic papers. These observations are confirmed using Spearman's Rank Correlation Test (Table 1.4).

2.2 Specialized Fossil Cephalopod Literature

The exploration of specialized cephalopod literature confirms the previous results. For the five major international symposia dealing with fossil cephalopods (Table 1.5; see Appendix for detailed bibliographic references), only eight papers using the cladistic approach have been found (and two are based on molecular data for recent species). Interestingly, for cephalopods, the proportion of papers dealing with cladistics is clearly much higher here than in regular paleontological publications (16.98%

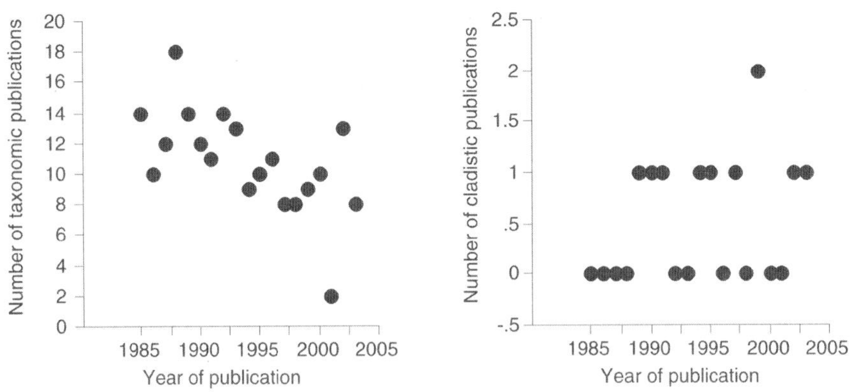

Fig. 1.3 *Number of publications over time, restricted to papers based on cephalopods, from five regular paleontological journals (scale very different from Fig. 1 and see text).*

Table 1.4 *Nonparametric Spearman's Rank Correlation Test for various data sets (see text). rs: Spearman's Rank Correlation Coefficient, NS: test nonsignificant, *: test significant at 95% confidence level, **: at 99% confidence level, ***: at 99.9% confidence level. Results are given with ex-aequo correction (see Swan and Sandilands, 1995).*

	r_s	p-values	
Year, Number of Taxonomic publications	−0.63	0.008	**
Year, Number of Cladistic publications	0.29	0.21	NS
Year, Percentage of Cladistic publications	0.36	0.13	NS

Table 1.5 *Number of publications in specialized fossil cephalopod literature (see text). "Cladistics" refers to publications with a cladogram (or parenthetic notation) constructed with or without parsimony analysis (see text and Appendix).*

	Number of publications	
Symposium	Taxonomy	Cladistics (%)
II Cephalopods, Tübingen (1985)	11	2 (18.18)
III Cephalopods, Lyon (1990)	13	0 (0)
IV Cephalopods, Granada (1999)	7	1 (14.29)
V Cephalopods, Vienna (2002)	15	2 (13.33)
Coleoid Ceph., Berlin (2003)	7	4 (57.14)
Total	53	9 (16.98)

versus 4.85%, respectively). This can be explained by the publication in specialized fossil cephalopod literature of a few papers using cladistics without any parsimony procedure, thus increasing the number of papers classed here under cladistics. Publishing this kind of analysis (cladistics but no parsimony) is clearly unusual in regular paleontological literature.

3 Discussion

As demonstrated here, in existing literature (up to 2003), the cladistic approach is only rarely applied to fossil cephalopods. Examples of resolved phylogenetic relationships using cladistics may be found in Landman (1989), Korn (1997), Yacobucci (1999), Monks (2000), Rouget (2002), and Moyne and Neige (2004). We believe that published papers using this approach reflect personal choice rather than accepted usage among cephalopod scholars. It is perfectly normal for cladistic analyses to start out as a rarity. For this still to be the case years after the first cladistic publication on fossil cephalopods, is however rather surprising. Paradoxically, no published paper can be found that properly demonstrates the inadequacy of cladistics when applied to fossil cephalopods: rejection of this method seems to be more a question of habit. Consider for example, the case of Landman's Red Ammonoid book (Landman et al., 1996), a landmark publication on ammonoids. It contains lengthy, detailed contributions on Paleozoic (Becker and Kullmann, 1996) and

Mesozoic (Page, 1996) ammonoids, respectively entitled *Paleozoic Ammonoids in Space and Time* and *Mesozoic Ammonoids in Space and Time*. Each of these chapters contains phylogenetic relationship hypotheses for major taxa (Figs. 1 and 5 in Becket and Kullman 1996, and Figs. 1 and 2 in Page 1996). Interestingly, both are included in a larger section entitled *Biostratigraphy and Biogeography*. In neither of these chapters, nor in any other in the Red Book, is there a discussion concerning the methods used to reconstruct phylogeny. In general, in a paper dealing with phylogenetic relationships among fossil cephalopods, the method used is implicitly considered to be well known by other scholars, and to use both morphologic and stratigraphic arguments. The main problem with this habit in our opinion is that scholars rely on morphology or stratigraphy in varying proportions, depending on context and other factors, but generally without explaining their choice. The consequence of such an absence of discussion within our scientific community is that there might easily be as many phylogenetic methods as scholars, which creates an unnecessary heterogeneity in phylogenetic hypotheses.

It is not easy to explain the reasons for such underuse of cladistic methods for fossil cephalopods, even more so in the absence of any clear debate on this point within our scientific community. However, we believe that five main factors prevent (or should we say unconsciously prevent?) a majority of scholars from using cladistics (see Rouget et al., 2004 for a similar discussion, restricted to ammonites).

- Factor 1: Cladistics is not needed because fossil cephalopod taxonomy is already perfect.

We strongly believe that no scholar with even minimal knowledge of systematics and phylogeny in fossil cephalopods could argue this point. On the contrary, many scholars point out the need for new phylogenetic investigations. For example, Donovan (1994: 1040) claims that it is very difficult, if not impossible at the present time, to write diagnoses of major taxa in Mesozoic ammonites.

- Factor 2: Fossil cephalopods do not display a sufficient number of characters to reconstruct phylogenetic relationships using cladistics.

To our point of view, this claim is simply wrong. Korn (1997) used 24 characters on Carboniferous ammonites, Monks (1999) used 27 characters on Cretaceous ammonites, and Moyne and Neige (2004) used 16 characters on Jurassic ammonites. Moreover, this argument is rarely used when reconstructing phylogenetic relationships if other methods than cladistics are used.

- Factor 3: Homeomorphies are too numerous to allow the use of cladistics.

This is probably the most valid argument against the use of cladistics, and we agree that homeomorphy implies more complex cladistic reconstructions. However, two points in favor of cladistics have to be noted. First, cases of complete homeomorphy are exceedingly rare, almost impossible, as some characters at least are likely to be different. For example, ornamental features may be identical for two independent taxa, but their suture line different. Second, detecting homeomorphy is one of the goals of cladistics. Thus, we strongly encourage scholars to test for homeomorphy using cladistics, rather than simply believe that homeomorphy would flaw their results if they used cladistics.

- Factor 4: The stratigraphic succession of fossils gives phylogenetic relationships, so that no other arguments are needed to resolve them. Thus for some, using cladistics based on morphological characters means abandoning a potentially useful data set based on stratigraphy.

Two points should be noted. First, attempts to reconstruct phylogenetic relationships using cladistics have led to results that are not in conflict with stratigraphic order (see Monks, 2000). In other words, cladistic order and stratigraphic order generally fit. Second, if we consider that both cladistics and stratigraphy may be of use in reconstructing phylogenies, then we must consider these two methods individually in order to compare results. If the results do not fit it means that one or other method, or both, must have given an invalid phylogenetic hypothesis. For example, observed stratigraphic succession may be erroneous because one of the taxa is in fact older, but has not yet been discovered at this older age. In contrast, selected morphological characters may have been misinterpreted during anatomical alignment. Our feeling is that if a mismatch exists between stratigraphic and cladistic hypotheses, such data must be analyzed and interpreted, but certainly not used as an argument to reject cladistics (for quantitative analyses, see Siddall, 1998; Wills, 1999). Some authors have proposed alternative procedures including both stratigraphy and morphology to resolve phylogenies (e.g., the stratocladistic approach, Fisher, 1994; Wagner, 1995). To our knowledge, these latter have never been applied to cephalopods.

- Factor 5: Rapid diversification of cephalopod species produces a phylogenetic pattern which is difficult to resolve using cladistics.

As Yacobucci (1999) demonstrated, radial evolution may occur for fossil cephalopods (i.e., branching events are concentrated on a single ancestral lineage during a brief period of time). The consequences of such an evolutionary process are that taxa have many autapomorphies and share few or no synapomorphies (which are necessary for phylogenetic reconstruction). This generates some "hard" polytomies where branching order is difficult or even impossible to resolve. Thus the resulting pattern may not be amenable to cladistic analysis. However, it has been suggested (Wagner and Erwin, 1995) that these hard polytomies reflect phylogenies, as did the example developed by Yacobucci (1999). This would mean that the phylogenetic pattern can be reconstructed. Anyway even if radial evolution produces phylogenetic patterns which are unresolvable using the cladistic approach, it is clear that the pattern will be no easier to resolve using stratigraphy, as radial evolution implies the simultaneous appearance of species in the fossil record.

Finally we believe that cladistic methods are perfectly tailored to the resolution of phylogenetic relationships for fossil cephalopods. We consider that using such methods pragmatically will be of great help in reassessing fossil cephalopod phylogeny and taxonomy (see Rulleau et al., 2003 for such an approach where various types of arguments have been compared: cladistic, paleogeographic, and stratigraphic). Moreover, the use of robust methods such as cladistics to reconstruct phylogenies is the only way to maintain fossil cephalopods as a model taxon to study evolutionary dynamics in time and space.

Acknowledgments

We thank J. L. Dommergues for many helpful exchanges on cladistics and ammonites, and the participants at the *VI International Cephalopod Symposium* held in Fayetteville (September 2004) for many interesting discussions, with conflicting points of view. Thanks to M. M. Yacobucci, E. G. Allen, and N. H. Landman for their valuable remarks during the review process. We thank Carmela Chateau for reviewing the English version.

Appendix. Bibliographic list of articles using the cladistic method included in the database (see text).

Note that this is not an exhaustive list of the cladistic approach applied to fossil cephalopods, but the list resulting from our sample of paleontological literature (see text). * = studies based on recent cephalopods only.

Bandel, K., and S. v. Boletzky. 1988. Features of development and functional morphology required in the reconstruction of early coleoid cephalopods. *In* J. Wiedmann, and J. Kullmann (editors), *Proceedings of the 2nd International Cephalopod Symposium, Cephalopods – Present and Past*, pp. 229–246. Stuttgart: E. Schweizerbart'sche Verlagsbuchhandlung.

Dommergues, J. L. 1994. The Jurassic ammonite *Coeloceras*: an atypical example of dimorphic progenesis elucidated by cladistics. *Lethaia* **27**: 143–152.

Donovan, D. T., L. A. Doguzhaeva, and H. Mutvei. 2003. Two pairs of fins in the late Jurassic coleoid *Trachyteuthis* from southern Germany. *In* K. Warnke, H. Keupp, and S. v. Boletzky (editors), *Coleoid Cephalopods Through Time. Berliner Paläobiologische Abhandlungen* **3**: 91–99.

Engeser, T., and K. Bandel. 1988. Phylogenetic classification of coleoid cephalopods. *In* J. Wiedmann, and J. Kullmann (editors), *Proceedings of the 2nd International Cephalopod Symposium, Cephalopods – Present and Past*, pp. 105–115. Stuttgart: E. Schweizerbart'sche Verlagsbuchhandlung.

Evans, D. H., and A. H. King. 1990. The affinities of early oncocerid nautiloids from the lower ordovician of Spitsbergen and Sweden. *Palaeontology* **33**: 623–630.

Fuchs, D., H. Keupp, and T. Engeser. 2003. New record of soft parts of *Munsterella scutellaris* Muenster, 1842 (Coleoidea) from the late Jurassic plattenkalks of Eichstätt and their significance for octobrachian relationships. *In* K. Warnke, H. Keupp, and S. v. Boletzky (editors), *Coleoid Cephalopods Through Time. Berliner Paläobiologische Abhandlungen* **3:** 101–111.

Haas, W. 2002. The evolutionary history of the eight-armed coleoidea. *In* H. Summesberger, K. Histon, and A. Daurer (editors), *Cephalopods – Present and Past. Abhandlungen der Geologischen Bundesanstalt* **57**: 341–351.

Haas, W. 2003. Trends in the evolution of the Decabrachia. *In* K. Warnke, H. Keupp, and S. v. Boletzky (editors), *Coleoid Cephalopods Through Time. Berliner Paläobiologische Abhandlungen* **3**: 113–129.

Harvey, A. W., R. Mooi, and T. M. Gosliner. 1999. Phylogenetic taxonomy and the status of *Allonautilus* Ward and Saunders, 1997. *Journal of Paleontology* **73**: 1214–1217.
Landman, N. H. 1989. Iterative progenesis in Upper Cretaceous ammonites. *Paleobiology* **15**: 95–117.
Landman, N. H., J. L. Dommergues, and D. Marchand. 1991. The complex nature of progenetic species – examples from Mesozoic ammonites. *Lethaia* **24**: 409–421.
Monks, N. 1999. Cladistic analysis of Albian Heteromorph ammonites. *Palaeontology* **42**: 907–925.
Monks, N. 2002. Cladistic analysis of a problematic ammonite group: the Hamitidae (Cretaceous, Albian–Turonian) and proposals for new cladistic terms. *Palaeontology* **45**: 689–707.
Rulleau, L., M. Bécaud, and P. Neige. 2003. Les ammonites traditionnellement regroupées dans la sous-famille des Bouleiceratinae (Hildoceratidae, Toarcien): aspects phylogénétiques, biogéographiques et systématiques. *Geobios* **36**: 317–348.
Ward, P. D., and W. B. Saunders. 1997. *Allonautilus*: a new genus of living nautiloid cephalopod and its bearing on phylogeny of the Nautilida. *Journal of Paleontology* **71**: 1054–1064.
*Warnke, K., R. Söller, D. Blohm, and U. Saint-Paul. 2002. Assessment of the phylogenetic relationship between *Octopus vulgaris* Cuvier, 1797 and *O. mimus* Gould 1852, using mitochondrial 16S rDNA in combination with morphological characters. *In* H. Summesberger, K. Histon, and A. Daurer (editors), *Cephalopods – Present and Past. Abhandlungen der Geologischen Bundesanstalt* **57:** 401–405.
*Warnke, K., J. Plötner, J. I. Santana, M. J. Rueda, and O. Llinas. 2003. Reflections on the phylogenetic position of spirula (cephalopoda): preliminary evidence from the 18S ribosomal RNA gene. *In* K. Warnke, H. Keupp, and S. v. Boletzky (editors), *Coleoid Cephalopods Through Time. Berliner Paläobiologische Abhandlungen* **3**: 253–260.
*Wray, C. G., N. H. Landman, W.B. Saunders, and J. Bonacum. 1995. Genetic divergence and geographic diversification in *Nautilus. Paleobiology* **21**: 220–228.
Yacobucci, M. M. 1999. Plasticity of developmental timing as the underlying cause of high speciation rates in ammonoids. *In* F. Oloriz, and F. J. Rodriguez-Tovar (editors), *Advancing Research on Living and Fossil Cephalopods*, pp. 59–76. New York: Kluwer Academic/ Plenum Press.

References

Becker, R. T., and J. Kullmann. 1996. Paleozoic ammonoids in space and time. *In* N. H. Landman, K. Tanabe, and R. A. Davis (editors), *Ammonoid Paleobiology*, pp. 711–753. New York: Plenum Press.
Cantino, P. D., and K. de Queiroz. 2003. PhyloCode: a phylogenetic code of biological nomenclature (http://www.ohiou.edu/phylocode/).
Donovan, D. T. 1994. History of classification of mesozoic ammonites. *Journal of the Geological Society* **151**: 1035–1040.
Elmi, S., C. Mangold, and Y. Alméras. 1993. 3ème Symposium International sur les Céphalopodes actuels et fossiles. *Geobios Mémoire spécial* **15**.
Fisher, D. C. 1994. Stratocladistics: morphological and temporal patterns and their relation to phylogenetic process. *In* L. Grande, and O. Rieppel (editors), *Interpreting the Hierarchy of Nature*, pp. 135–171. New York: Academic Press.
Korn, D. 1997. Evolution of the Goniatitaceae and Visean – Namurian biogeography. *Acta Palaeontologica Polonica* **42**: 177–199.
Landman, N. H. 1989. Iterative progenesis in upper Cretaceous ammonites. *Paleobiology* **15**: 95–117.
Landman, N. H., K. Tanabe, and R. A. Davis (editors). 1996. *Ammonoid Paleobiology*. New York: Plenum Press.

Laurin, M. (editor). 2004. *Abstract Volume of the First International Phylogenetic Nomenclature Meeting*, Paris.

Monks, N. 1999. Cladistic analysis of Albian Heteromorph ammonites. *Palaeontology* **42**: 907–925.

Monks, N. 2000. Functional morphology, ecology, and evolution of the Scaphitaceae Gill, 1871 (Cephalopoda). *Journal of Molluscan Studies* **66**: 205–216.

Moyne, S., and P. Neige. 2004. Cladistic analysis of the Middle Jurassic ammonite radiation. *Geological Magazine* **141**: 115–223.

Oloriz, F., and F. J. Rodriguez-Tovar (editors). 1999. *Advancing Research on Living and Fossil Cephalopods*. New York: Kluwer Academic/Plenum Press.

Page, K. 1996. Mesozoic ammonoids in space and time. *In* N. H. Landman, K. Tanabe, and R. A. Davis (editors), *Ammonoid Paleobiology*, pp. 755–794. New York: Plenum Press.

Raup, D. M. 1967. Geometric analysis of shell coiling: coiling in ammonoids. *Journal of Paleontology* **41**: 43–65.

Rouget, I. 2002. *Reconstruction phylogénétique chez les ammonites: confrontation des approches cladistiques et stratigraphiques. Le cas des Dayiceras (Ammonitina, Eodeoceratoidea)*, Unpublished Ph.D. thesis, University of Burgundy.

Rouget, I., P. Neige, and J. L. Dommergues. 2004. L'analyse phylogénétique chez les ammonites: état des lieux et perspectives. *Bulletin de la Société Géologique de France* **175**: 507–512.

Rulleau, L., M. Bécaud, and P. Neige, 2003. Les ammonites traditionnellement regroupées dans la sous-famille des Bouleiceratinae (Hildoceratidae, Toarcien): aspects phylogénétiques, biogéographiques et systématiques. *Geobios* **36**: 317–348.

Siddall, M. E. 1998. Stratigraphic fit to phylogenies: a proposed solution. *Cladistics* **14**: 201–208.

Summesberger, H., K. Histon, and A. Daurer (editors). 2002. *Cephalopods–Present and Past. Abhandlungen der Geologischen Bundesanstalt* **57**.

Swan, A. R. H., and M. Sandilands. 1995. *Introduction to Geological Data Analysis*. Oxford: Blackwell Science.

Wagner, P. J. 1995. Stratigraphic tests of cladistic hypotheses. *Paleobiology* **21**: 153–178.

Wagner, P. J., and D. H. Erwin. 1995. Phylogenetic patterns as tests of speciation models. *In* D. H. Erwin, and R. L. Anstey (editors), *New Approaches to Speciation in the Fossil Record*, pp. 87–122. New York: Columbia University Press.

Warnke, K., H. Keupp, and S. v. Boletzky (editors). 2003. *Coleoids Cephalopods Through Time. Berliner Paläobiologische Abhandlungen* **3**.

Wiedmann, J., and J. Kullmann. (editors).1988. *Proceedings of the 2nd International Symposium: Cephalopods – Present and Past*. Stuttgart: E. Schweizerbart'sche Verlagsbuchhandlung.

Wills, M. A. 1999. Congruence between phylogeny and stratigraphy: randomization tests and gap excess. *Systematic Biology* **48**: 559–580.

Yacobucci, M. M. 1999. Plasticity of developmental timing as the underlying cause of high speciation rates in ammonoids. *In* F. Oloriz, and F. J. Rodriguez-Tovar (editors), *Advancing Research on Living and Fossil Cephalopods*, pp. 59–76. New York: Kluwer Academic/ Plenum Press.

Chapter 2
Patterns of Embryonic Development in Early to Middle Devonian Ammonoids

Susan M. Klofak,[1] Neil H. Landman,[2] and Royal H. Mapes[3]

[1]Division of Paleontology (Invertebrates), American Museum of Natural History, 79th Street and Central Park West, New York, NY 10024, USA, and Department of Biology, City College of the City University of New York, Convent Avenue and 138th Street, New York, NY, 10031, USA, klofak@amnh.org;

[2]Division of Paleontology (Invertebrates), American Museum of Natural History, 79th Street and Central Park West, New York, NY 10024, USA, landman@amnh.org;

[3]Department of Geological Sciences, 316 Clippinger Laboratories, Ohio University, Athens, OH 45701–2979, USA, mapes@ohio.edu

Keywords: Devonian, ammonitella, Agoniatitidae, Mimagoniatitidae, Anarcestidae, New York, Morocco

1 Introduction

The basic structure of the ammonitella or embryonic shell of the Ammonoidea has been well documented (Sandberger, 1851; Branco, 1879–1880, 1880–1881; Schindewolf, 1933; Erben, 1960). The ammonitella begins with a small egg-shaped or spherical initial chamber (=protoconch) (for discussion of terms see Schindewolf, 1933; House, 1996; and Landman et al., 1996; see Landman et al., 1996 for additional references) and extends as a straight shaft or coiled tube (called the ammonitella coil by House, 1996: 168). In most species the ammonitella extends approximately one whorl ending in a primary constriction (Landman et al., 1996; Klofak et al., 1999). Internally, the siphuncle originates in the initial chamber as a rounded caecum and is attached to the shell wall by a prosiphon (Landman, 1988: Figs. 1, 2 and Klofak et al., 1999: Figs. 1, 2).

N. H. Landman et al. (eds.), *Cephalopods Present and Past: New Insights and Fresh Perspectives*, 15–56.

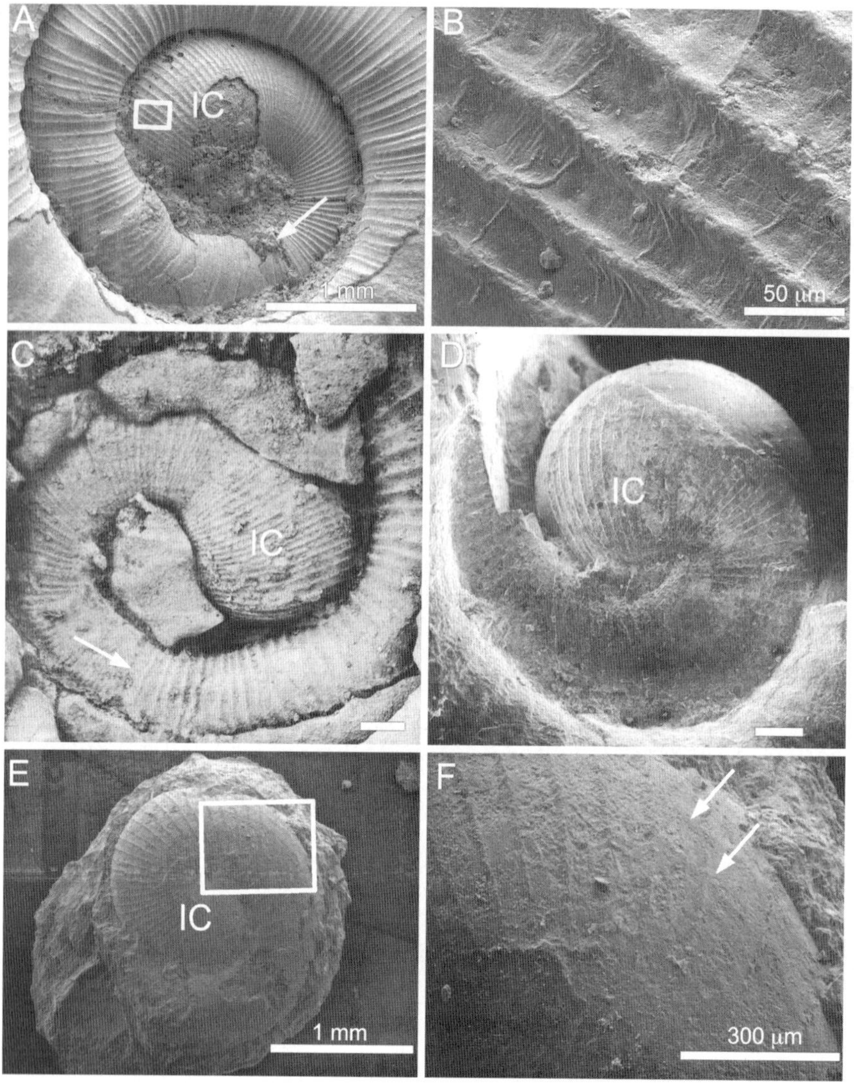

Fig. 2.1 *Ammonitellas in the families Mimagoniatidae and Agoniatitidae. (A, B)* Archanarcestes obesus *(AMNH 45370, Devonian, Morocco). (A) Lateral view of ammonitella. The initial chamber (IC) is visible and the end of the ammonitella is marked by breakage of the shell (arrow). Scale bar = 1.00 mm. (B) Close-up of transverse lirae on the initial chamber IC showing well-developed lirae with "wrinkle-like" creases stretched perpendicular between them. Area of photograph is marked by a box in 1A. Scale bar = 50.0 µm. (C)* Archanarcestes obesus *(AMNH 45374, Devonian, Morocco). Lateral view of ammonitella showing the initial chamber IC and primary constriction (arrow). Scale bar = 200 µm. (D)* Agoniatites vanuxemi *(NYSM 3545, Devonian, New York State). Lateral view of ammonitella. Scale bar = 200 µm. (E, F)* Fidelites fidelis *(AMNH 50417, Devonian, Morocco). (E) Lateral view of ammonitella showing the flattening on the ventral side of the ammonitella. Scale bar = 1.00 mm. (F) Close-up of transverse lirae showing faint, perpendicular "wrinkle-like" creases (arrows) between them. Area of close-up is indicated by a box on 1E. Scale bar = 300 µm.*

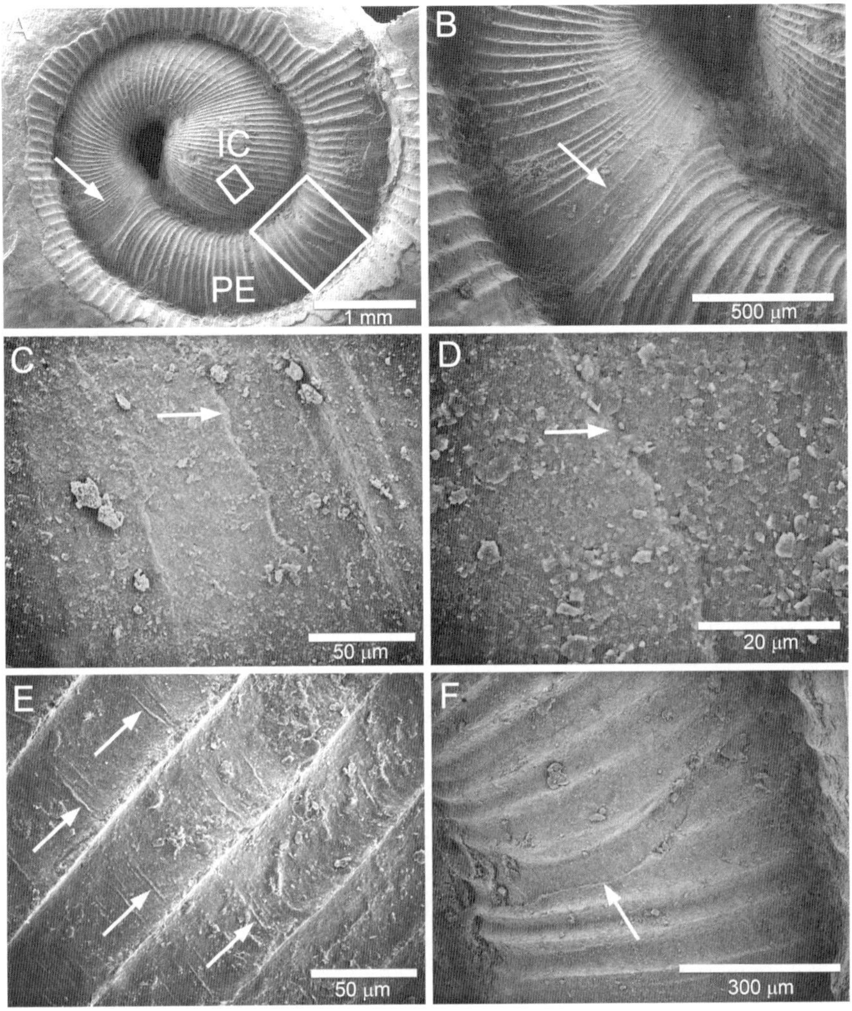

Fig. 2.2 Mimagoniatites fecundus *(AMNH 46645, Devonian, Morocco). (A) Lateral view of ammonitella showing initial chamber (IC), the apertural edge of ammonitella (arrow), and about one-half whorl of postembryonic shell (PE). Scale bar = 1.00 mm. (B) Close-up of aperture (arrow) showing the reduction in size and spacing of transverse lirae. Postembryonic shell is to the right. Scale bar = 500 μm. (C) Close-up from B, rotated approximately 45° to the right, showing the apertural edge of the ammonitella (arrow). The postembryonic shell can be seen emerging from beneath the ammonitella edge on the right. Scale bar = 50.0 μm. (D) Close-up of the apertural edge of the ammonitella from C (arrow). Scale bar = 20.0 μm. (E) Close-up of the transverse lirae on the initial chamber IC showing the "wrinkle-like" creases stretched perpendicularly between them (arrows). Area of photograph is indicated by the small box on the initial chamber IC in 2A. Scale bar = 50.0 μm. (F) Close-up of the postembryonic shell showing a healed break in the shell (arrow) which disrupted the production of the ornament. Area of close-up is indicated by box on the postembryonic shell in 2A. Scale bar = 300 μm.*

Most studies on ammonitellas are based on well-preserved Mesozoic specimens and it has been inferred that at least some of the same features could be extrapolated to the earliest Devonian forms (Schimansky, 1954; Erben et al., 1968; Kulicki, 1974, 1979; Drushschits et al., 1977; Bandel, 1982; Tanabe, 1989; Landman et al., 1996). The significance of these features and their implication for embryonic development has, however, been debated (see Klofak et al., 1999). Studies on Devonian taxa have demonstrated that many of the features defined for Mesozoic forms are present in their Devonian predecessors, for example, the prosiphon, caecum, and primary constriction (Klofak et al., 1999). There are critical differences, such as the ornament on the shell. In the advanced Paleozoic and Mesozoic ammonoids, the surface of the embryonic shell is either smooth or covered with fine tubercles (Tanabe et al., 1994; Landman et al., 1996; Sprey, 2002). In Devonian taxa the ammonitella is covered by transverse lirae (Clarke, 1899; Miller, 1938; Erben, 1960, 1964b; Clausen, 1969; House, 1996; Klofak et al., 1999). Another important difference occurs at the end of the ammonitella. In post-Devonian ammonoids the primary (or nepionic) constriction is accompanied by a thickening of the shell wall called the primary varix. Studies of well-preserved material have shown that this varix is absent in Devonian ammonoids (Klofak et al., 1999).

These differences have fueled much of the debate as to how the ammonitellas of Devonian ammonoids formed. Most models for post-Devonian ammonitellas call for a nonaccretionary mode of growth because these ammonitellas do not possess any ornament that might be interpreted as having formed by marginal accretion at the aperture. Devonian ammonoids, however, possess transverse lirae, which has suggested to some authors that these embryonic shells formed in an accretionary way (Erben, 1964b, 1966; Erben et al., 1968; Tanabe, 1989). The embryonic lirae are structurally different from those found on the postembryonic shell, however, which might suggest that the embryonic and postembryonic shell formed differently (Klofak et al., 1999). A similar situation has been described for Jurassic ammonites (Kulicki, 1974, 1979; Sprey, 2002) and Triassic ceratites (Landman et al., 2001). The authors in all of these studies describe tuberculate micro-ornament on both the embryonic and postembryonic shell. The morphology of the embryonic and postembryonic tubercles is different in each case, and, hence, it is likely that the two parts of the shell formed differently. Ultimately, the solution may lie in finding and examining shell microstructure, something we have not yet been able to do for Devonian ammonoids.

Previous descriptions of Devonian ammonitellas have been incorporated in numerous taxonomic descriptions (Miller, 1938; Erben, 1953, 1960, 1964b; Petter, 1959; House, 1962; Bogoslovsky, 1969; Chlupáč and Turek, 1983; Bensaïd, 1974; Göddertz, 1987, 1989; Wissner and Norris, 1991; Klug, 2001). Commonly, the size (diameter) and shape of the initial chamber and the degree of coiling of the ammonitella are given. Descriptions of lirae have generally been limited to noting their presence, largely due to the lack of well-preserved material. The exception is Erben (1964b) who described the pattern of the lirae in several Devonian taxa.

Our study entails an examination and quantification of the lirae of well-preserved ammonitellas from three families of Devonian ammonoids: the Mimagoniatitidae, the Anarcestidae, and the Agoniatitidae. The three families are all closely related phylogenetically, albeit of different taxonomic rank. Most workers have placed them within the same higher taxonomic group whether it be the superfamily Agoniatitacea (Miller, 1938), superfamily Anarcestaceae (Miller and Furnish, 1954; Petter, 1959; Erben, 1964b), suborder Agoniatitina (Bogoslovsky, 1969; House, 1981), order Anarcestida (Becker and House, 1994), or order Agoniatitida (Ruzhentsev, 1960, 1974; Becker and Kullmann, 1996; Korn, 2001; Korn and Klug, 2002). In most previously proposed phylogenies, the Mimagoniatitidae are viewed as ancestral to both the Agoniatitidae and the Anarcestidae, but generally more closely related to the Agoniatitidae. For example, within his superfamily Agoniatitacea, Miller (1938) placed the Mimagoniatitinae and Agoniatitinae in the family Agoniatitidae and the Anarcestinae in their own family, the Anarcestidae. More recently, the mimagoniatids and agoniatids were placed within the suborder Agoniatitina and the anarcestids were placed within the Anarcestina, both within the order Agoniatitida (Becker and Kullmann, 1996; Korn, 2001).

It is hoped that a study of the pattern of lirae spacing on the ammonitella may reveal data useful in defining both how the individual lirae as well as the ammonitella as a whole formed. By including data from three related families, the pattern of lirae spacing can then be compared among taxa. Differences that emerge may define useful taxonomic characters. They may also aid in our understanding of the early radiation of the Ammonoidea.

2 Material and Methods

The taxa used in this study are: *Archanarcestes obesus* (Erben, 1960) and *Mimagoniatites fecundus* Barrande, 1865, from the Mimagoniatitidae; *Agoniatites vanuxemi* (Hall, 1879) and *Fidelites fidelis* Barrande, 1865, from the Agoniatitidae; and *Latanarcestes* sp. from the Anarcestidae. Six of the specimens are from Morocco. The remaining specimen (*A. vanuxemi*) is from the Cherry Valley Limestone of New York State (see Clarke, 1899). Descriptions of Moroccan localities are given below. For a more detailed description of the geology and stratigraphy of these localities, see Becker and House (1994) and Klug (2000).

AMNH locality 3233: *Archanarcestes obesus* (AMNH 43374) from layer "4B" (Becker and House, 1994). Lower-Middle Devonian. East of Bou Tcharfine, near Erfoud, Morocco; latitude 31° 16′ 46″ N, longitude 3° 53′ 0.54″ W.

AMNH locality 3237: *Archanarcestes obesus* (AMNH 45370), Emsian, Devonian. Bou Tcharfine, near Erfoud, Morocco.

AMNH locality 3311: *Mimagoniatites fecundus* (AMNH 46645) and *Latanarcestes* sp. (AMNH 46646 and 50416). Emsian, Devonian. East of Bou Tcharfine, near Erfoud, Morocco; latitude 31° 22.51′ N, longitude 4° 4.38′ W.

AMNH locality 3312: *Fidelites fidelis* (AMNH 50417). Late Emsian, Devonian. Jebel Ouaoufilal, near Taouz, Tafilalt; latitude 30° 55.93′ N, longitude 4° 1.14′ W.

Specimens were examined using the scanning electron microscope (SEM). Four of these specimens have been previously illustrated (Klofak et al., 1999); three specimens are studied for the first time. AMNH 45374 described in Klofak et al. (1999) is revised here as *Archanarcestes obesus*.

Ornament was examined to confirm previous findings and expand the database. In addition to qualitative observations, measurements were collected on the distance between lirae on both the ammonitella and postembryonic shell. Measurements were taken from SEM photographs using the Quartz PCI Imaging Management System. Measurements were taken from three positions on the shell, wherever possible: ventral, midflank, and dorsal (Fig. 2.5a). The ventral and dorsal measurements were not taken directly on the venter or dorsum, but as close as possible in lateral view. The midflank measurements were taken along a line on the middle part of the flank. This represents the point of maximum whorl width. Measurements began at the apex of the shell and extended to the aperture of the embryonic shell. "Distance 1" is the distance between the first and second lirae measured between the crests of successive lirae (Fig. 2.5b); "distance 2" is the distance between the crests of the second and third lirae, and so forth. Measurements began at the closest measurable lira starting at the apex. Postembryonic lirae are designated in the same manner with the first postembryonic lirae distance designated "distance 1."

Specimens used in this study are reposited in the American Museum of Natural History, New York, New York (AMNH), and the New York State Museum, Albany, New York (NYSM).

3 Results

3.1 External Features of the Ammonitella

Ornament for all taxa consists of transverse lirae that cover the entire ammonitella except for a "bald spot" at the apex (Klofak et al., 1999). "Wrinkle-like" creases stretch between the lirae and are best defined on the initial chamber. These "wrinkle-like" creases are most strongly expressed in the Mimagoniatitidae (Figs. 2.1b, 2.2e), less strongly expressed in the Anarcestidae (Fig. 2.3b), and least strongly expressed in the Agoniatitidae (Fig. 2.1f).

The lirae are a relief feature and are not preserved on the steinkern as shown in *Latanarcestes* where the shell is broken away on part of the initial chamber (Fig. 2.3a, c, d; see also Klofak et al., 1999: Figs. 3c, 8b). The lirae on the embryonic shell differ from those on the postembryonic shell in that the embryonic lirae are symmetrical, while the postembryonic lirae are asymmetrical and more steeply sloped on the apical side (Klofak et al., 1999).

Just adapical of the primary constriction, the lirae become much weaker and more closely spaced (Figs. 2.2b, 2.3d, 2.4b). The edge of the ammonitella is marked by a primary constriction, with no evidence of an accompanying varix in any speci-

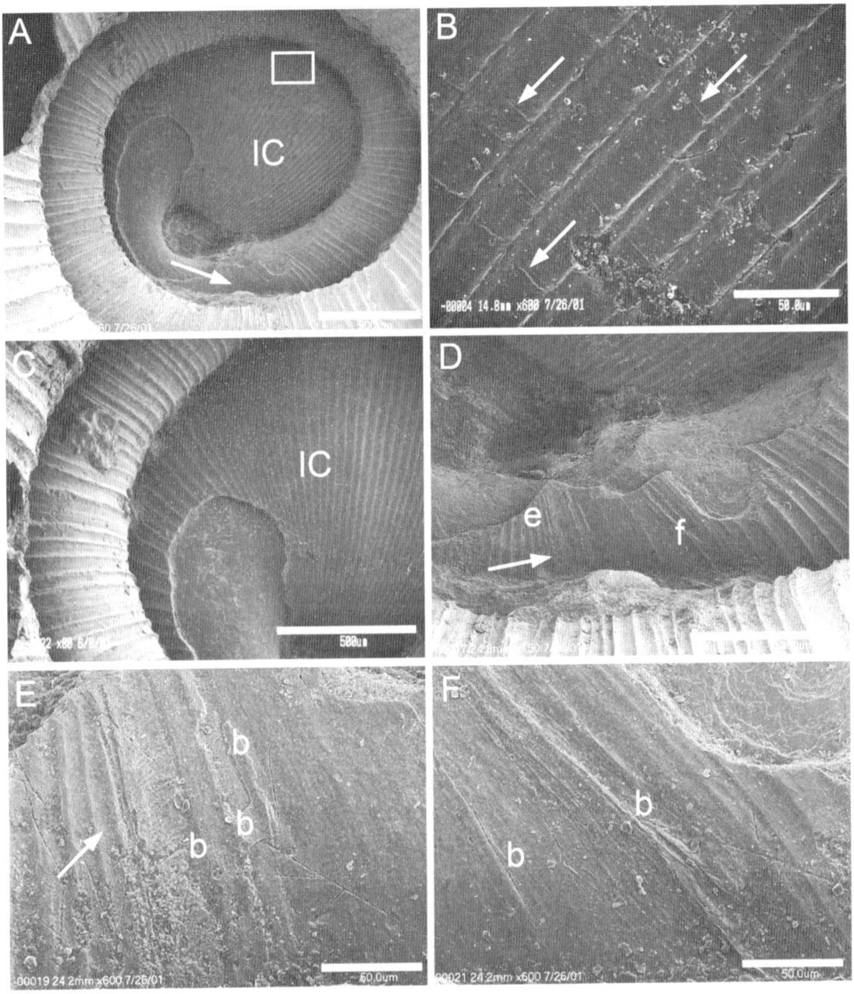

Fig. 2.3 Latanarcestes *sp. (AMNH 46646, Devonian, Morocco). (A) Lateral view of ammonitella showing initial chamber (IC) and part of apertural edge of ammonitella (arrow). Scale bar = 1.00 mm. (B) Close-up of transverse lirae on initial chamber IC showing perpendicular "wrinkle-like" creases (arrows) stretched between them. Area of close-up is indicated by a box on 3A. Scale bar = 50.0 µm. (C) Close-up near the end of the initial chamber IC where shell has broken away exposing the steinkern. Note that the surface of the steinkern is smooth, unaffected by the ornament on the shell. Scale bar = 500 µm. (D) Close-up of ammonitella showing the apertural edge of the ammonitella (arrow). Scale bar = 80.0 µm. (E) Close-up of the apertural edge of the ammonitella (arrow) showing the postembryonic shell emerging from beneath it and followed by a series of breaks (b) in the postembryonic shell, with new shell emerging from beneath the existing shell. Area of close-up marked by letter e in 3D. Scale bar = 50.0 µm. (F) Close-up of smooth area of postembryonic shell to the right of 3E showing a series of breaks (b) with new shell emerging from beneath the existing shell. Area of close-up is marked by letter f in 3D. Scale bar = 50.0 µm.*

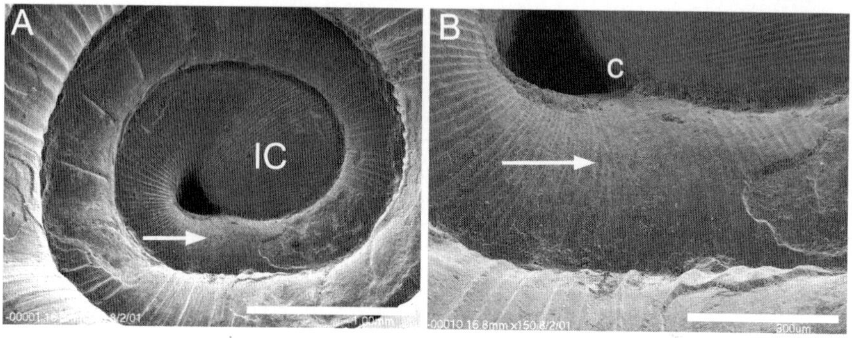

Fig. 2.4 Latanarcestes *sp. (AMNH 50416, Devonian, Morocco). (A) Lateral view of ammonitella showing initial chamber (IC), with visible constriction (arrow). Scale bar = 1.00 mm. (B) Close-up of apertural edge of ammonitella (arrow). Constriction is visible on dorsal edge (c). Scale bar = 300 μm.*

men examined (Figs. 2.2b, c, d, 2.3d, e, f, 2.4b). In previously studied specimens, the apertural edge of the ammonitella appears as a crease or break (Fig. 2.1a, c; see also Klofak et al., 1999). A perfectly preserved aperture is present in a specimen of *Archanarcestes obesus* (Fig. 2.2b, c, d) where the apertural edge parallels the ornament and forms a distinct line across the flank, slightly irregular in appearance. The postembryonic shell can be seen emerging from beneath the embryonic shell (Fig. 2.2c, d). This occurs in the area of smaller, finer lirae. In *Latanarcestes* sp., the area of small faint lirae is more extensive when compared to the specimens of *Archanarcestes* (Figs. 2.3d, 2.4b; see also Klofak et al., 1999: Fig. 8d). The apertural edge also appears in this area and, like *Archanarcestes,* is irregular in its course. Multiple breaks occur adoral of the apertural edge with successive shell layers emerging from beneath the previously secreted shell (Fig. 2.3e, f).

3.2 Lirae Spacing

The second part of our study is an analysis of lirae spacing in three of the earliest families of ammonoids: the Mimagoniatitidae, the Anarcestidae, and the Agoniatitidae. In total, seven specimens were studied; three mimagoniatids, two anarcestids, and two agoniatids.

The results are presented in Figs. 2.6–2.11. The horizontal axis is the lirae number beginning with the most adapical lira on the embryonic shell; in some graphs, the first lira measured is on the postembryonic shell. The vertical axis is the distance between lirae (see Fig. 2.5b), given in microns. Data are presented for the three measured positions: ventral, midflank, and dorsal (Fig. 2.5a). They are graphed together using different symbols for each position on the shell. Data are given in Appendices 1–11.

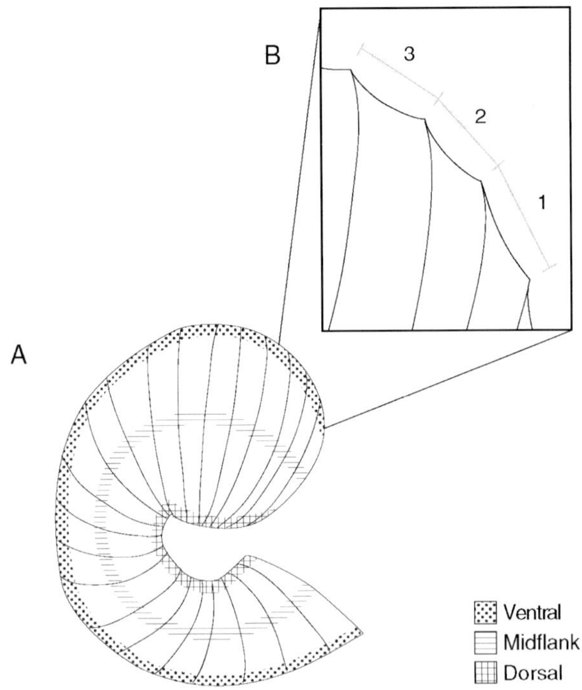

Fig. 2.5 *Sketch of the ammonitella of a primitive ammonoid showing the areas where the lirae spaces (distances between successive lirae) were measured. (A) The three areas are marked with different patterns: Ventral = dots; Midflank = horizontal lines; and Dorsal = cross hatch. (B) Close-up of ventral edge of ammonitella with lirae shown raised. The horizontal line (1, 2, 3) indicates how the distances were measured.*

The ratios of the ventral (V) and dorsal (D) lirae distance are also given, where possible. The horizontal axis is the lirae number and the vertical axis is the ratio (V/D).

Means and standard deviations were calculated for the distance between lirae using data from the midflank position. Data are given in Table 2.1. Because of the extent of the overlap of standard deviations, these data cannot be used to differentiate among the three taxonomic groups. The results of an F-test confirm that the means are significantly different at the 0.01 level.

3.2.1 Mimagoniatitidae

The three specimens from the Mimagoniatitidae show the same general pattern, although there is some variation (Figs. 2.6a, 2.7a, 2.8a). Two of the specimens are *Archanarcestes obesus* (AMNH 45374, Fig. 2.7, Appendix 2; AMNH 45370, Fig. 2.8, Appendix 1).

Table 2.1 *Means of the distance between successive lirae calculated at the mid flank positions. Measurements are given in microns.*

Family	N	Lirae spacing (μm) Mean ± SD	Range
MIMAGONIATIDAE			
AMNH 46645	76	45 ± 16	14–75
AMNH 45374	52	31 ± 9	9–49
AMNH 45370	55	28 ± 10	14–52
AGONIATITIDAE			
NYSM 3545	56	40 ± 8	55–25
AMNH 50417	29	47 ± 15	28–126
ANARCESTIDAE			
AMNH 46646	70	28 ± 10	8–70
AMNH 50416	94	18 ± 6	10–37

The third specimen is *Mimagoniatites fecundus* (AMNH 46645, Fig. 2.6, Appendix 4). As expected, based on simple geometry, the ventral distance between lirae is greater than the dorsal distance because the venter subscribes a greater arc. The mid-flank position contains measurements midway between the dorsum and venter.

There is a general increase in the distance between lirae over the initial chamber at the ventral, midflank, and dorsal positions. Then at approximately 5–10 lirae from the apex, there is a decrease in the distance. This is generally slight, but in one specimen (AMNH 45374, Fig. 2.7a), it is expressed very strongly. After this initial drop, there is a second increase, followed by an abrupt drop in lirae distance in all three specimens. This decrease marks the end of the initial chamber and occurs at about 25–30 lirae from the apex. Over the course of the ammonitella coil, there is a gentle increase in spacing followed by a decrease at the ventral, midflank, and dorsal positions. This decrease is abrupt and occurs at the end of the ammonitella. In all, there are 70–77 lirae on the ammonitella of the Mimagoniatitidae.

The spacing between lirae can fluctuate within the general trend. There appears to be no pattern to this fluctuation. The result is a graph with a very jagged appearance. These fluctuations are also not perfectly aligned among the three measured positions.

In two of the specimens (AMNH 46645, Fig. 2.6b, Appendix 5; AMNH 45374, Fig. 2.7b, Appendix 3), the postembryonic distance between lirae was measured. Again, the ventral distance was the widest, the dorsal distance was the narrowest, and the midflank distance was in between. There is a rapid increase in spacing from the first measurable postembryonic lira.

The ratio of the ventral distance/dorsal distance (V/D) on the embryonic shell was compared to that of the postembryonic shell (Figs. 2.6c, d, 2.7c, d). The most complete data set (AMNH 46645) shows that there is much greater correlation between the dorsal and ventral distances between lirae on the postembryonic shell than on the embryonic shell (Fig. 2.6c, d).

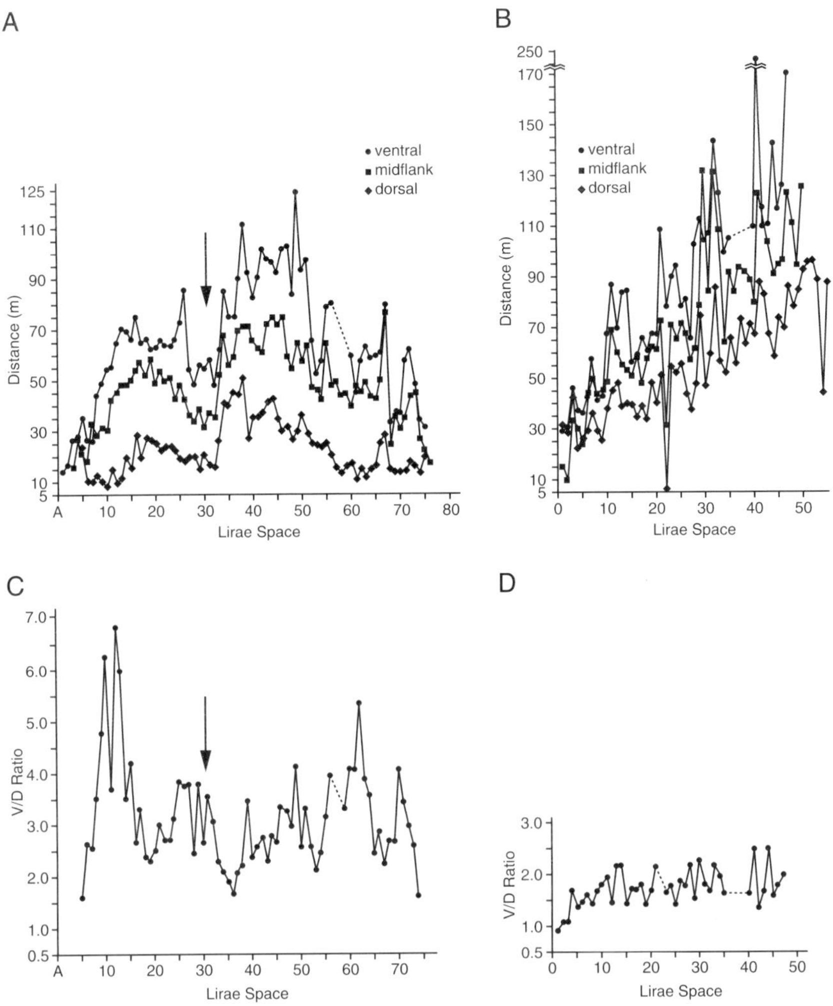

Fig. 2.6 Mimagoniatites fecundus *(AMNH 46645, Devonian, Morocco). (A) Lirae spacing on embryonic shell for ventral, midflank, and dorsal positions. Symbols given in graph. The X axis is the number of lirae space (see methods and materials for details). The Y axis is the measured distance between two lirae given in microns (μm). Data are given in Appendix 4. (B) Lirae spacing on postembryonic shell for ventral, midflank, and dorsal positions. The X axis is the number of lirae space. The Y axis is the measured distance between two lirae given in microns (μm). Data are given in Appendix 5. (C) Ratio of the ventral lirae space/dorsal lirae space (V/D) on the embryonic shell. Data are given in Appendix 4. (D) Ratio of the ventral lirae space/dorsal lirae space (V/D) on the postembryonic shell. Data are given in Appendix 5.*

3.2.2 Anarcestidae

Two specimens from the Anarcestidae were measured, both identified as *Latanarcestes* sp. (AMNH 46646 and AMNH 50416). The results are presented in

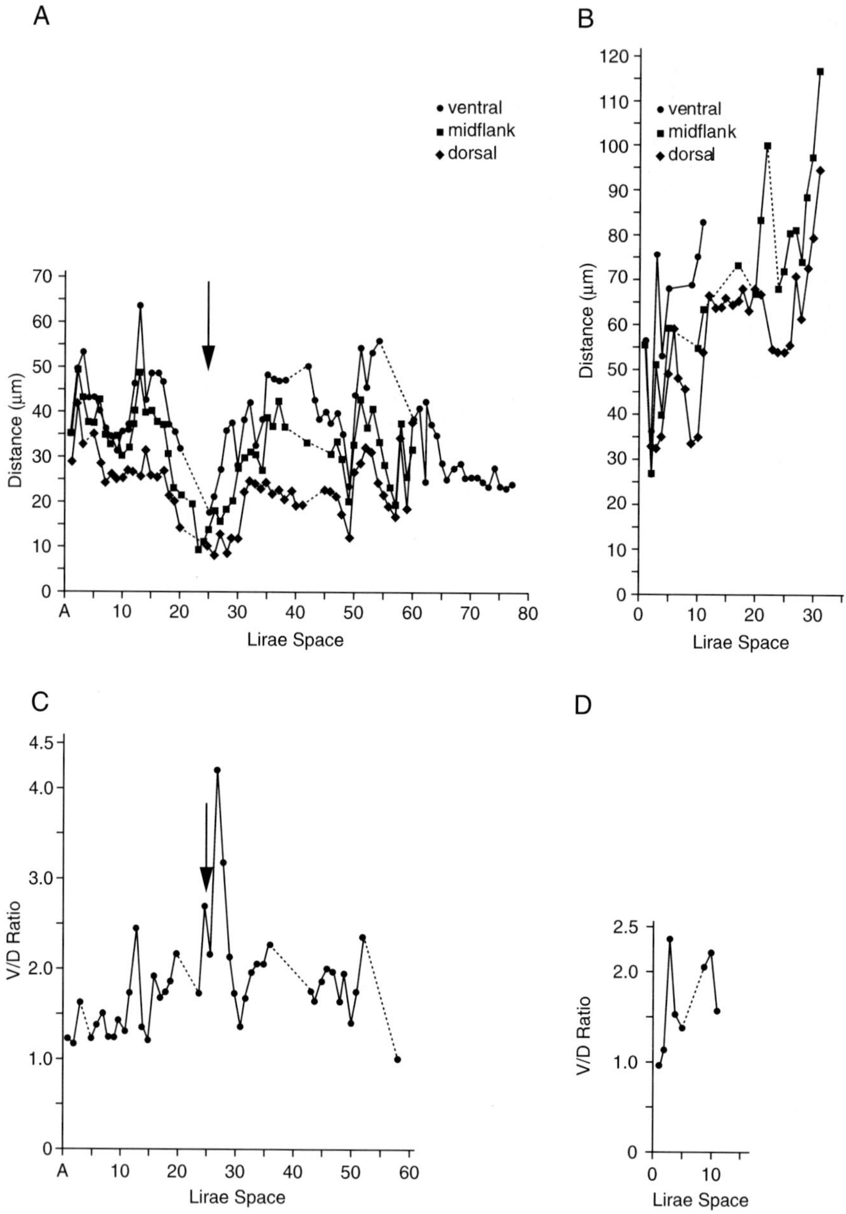

Fig. 2.7 Archanarcestes obesus *(AMNH 45374, Devonian, Morocco). (A) Lirae spacing on embryonic shell for ventral, midflank, and dorsal positions. Symbols given in graph. X axis is the number of lirae space. Y axis is the measured distance between two lirae given in microns (μm). Data are given in Appendix 2. (B) Lirae spacing on postembryonic shell for ventral, midflank, and dorsal positions. X axis is the number of lirae space. Y axis is the measured distance between two lirae given in microns (μm). Data are given in Appendix 3. (C) Ratio of the ventral lirae space/dorsal lirae space (V/D) on the embryonic shell. Data are given in Appendix 2. (D) Ratio of the ventral lirae space/dorsal lirae space (V/D) on the postembryonic shell. Data are given in Appendix 3.*

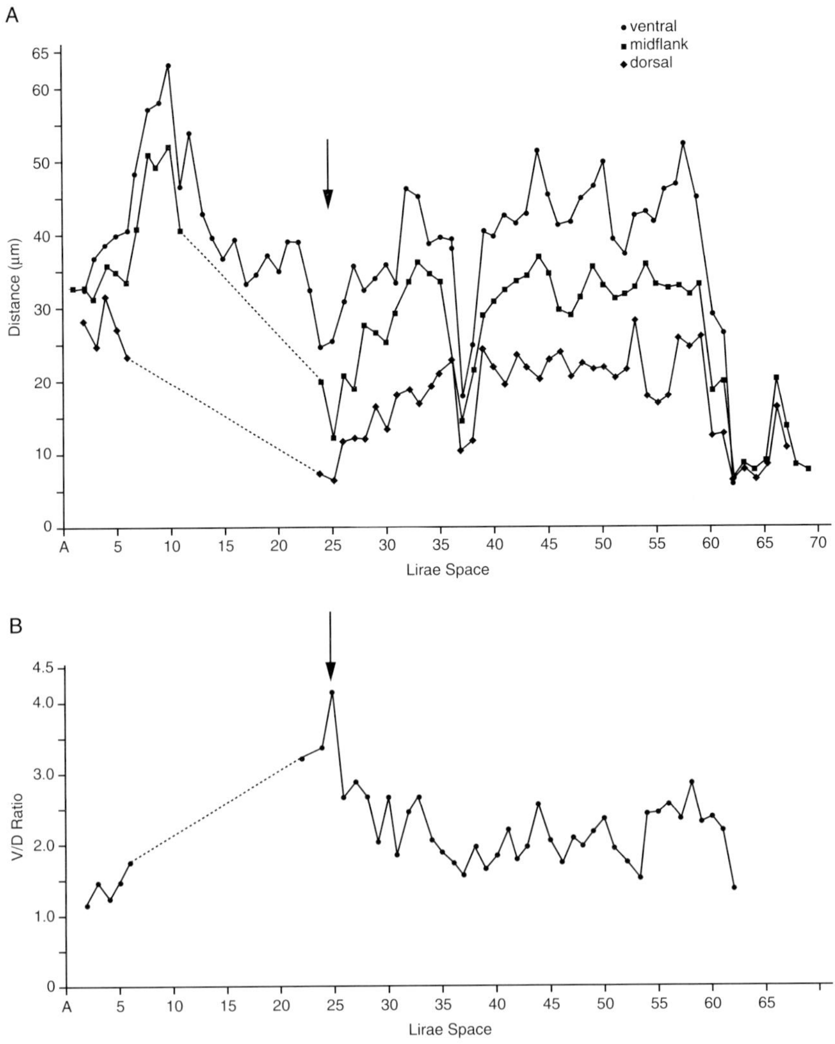

Fig. 2.8 Archanarcestes obesus *(AMNH 45370, Devonian, Morocco). (A) Lirae spacing on embryonic shell for ventral, midflank, and dorsal positions. Symbols given in graph. X axis is the number of lirae space. Y axis is the measured distance between two lirae given in microns (µm). (B) Ratio of the ventral lirae space/dorsal lirae space (V/D) on the embryonic shell. Data are given in Appendix 1.*

Figs. 2.9–2.10 and Appendices 6–9. Except for a small gap, the data for AMNH 50416 are complete across the ammonitella (Fig. 2.9a). AMNH 46646 has dorsal measurements for only the initial chamber (Fig. 2.10a). Thereafter, the ventral data are the most complete. On the initial chamber, as in the Mimagoniatitidae, the ventral distance is the widest, the dorsal the narrowest, and the midflank somewhere in between. The ventral distance increases, then decreases to the end of the initial

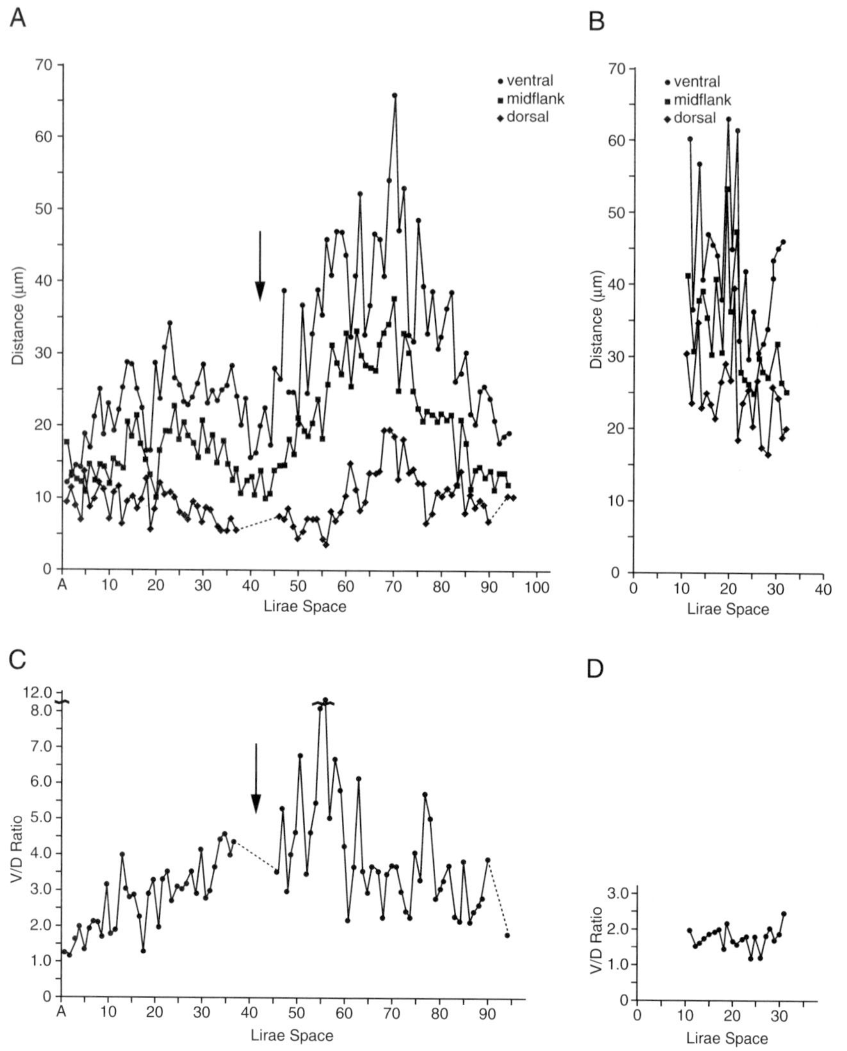

Fig. 2.9 Latanarcestes *sp. (AMNH 50416, Devonian, Morocco). (A) Lirae spacing on embryonic shell for ventral, midflank, and dorsal positions. Symbols given in graph. X axis is the number of lirae space. Y axis is the measured distance between two lirae given in microns (µm). Data are given in Appendix 8. (B) Lirae spacing on postembryonic shell for ventral, midflank, and dorsal positions. X axis is the number of lirae space. Y axis is the measured distance between two lirae given in microns (µm). Data are given in Appendix 9. (C) Ratio of the ventral lirae space/dorsal lirae space (V/D) on the embryonic shell. Data are given in Appendix 8. (D) Ratio of the ventral lirae space/dorsal lirae space (V/D) on the postembryonic shell. Data are given in Appendix 9.*

chamber, seen most strongly in AMNH 50416 (Fig. 2.9a). The pattern of midflank distance parallels that of the ventral distance. The dorsal distance, in contrast, decreases slowly from the apical end of the initial chamber to its end. A decrease in the distance between lirae marks the end of the initial chamber in the ventral and midflank data. There are 40–45 lirae on the initial chamber.

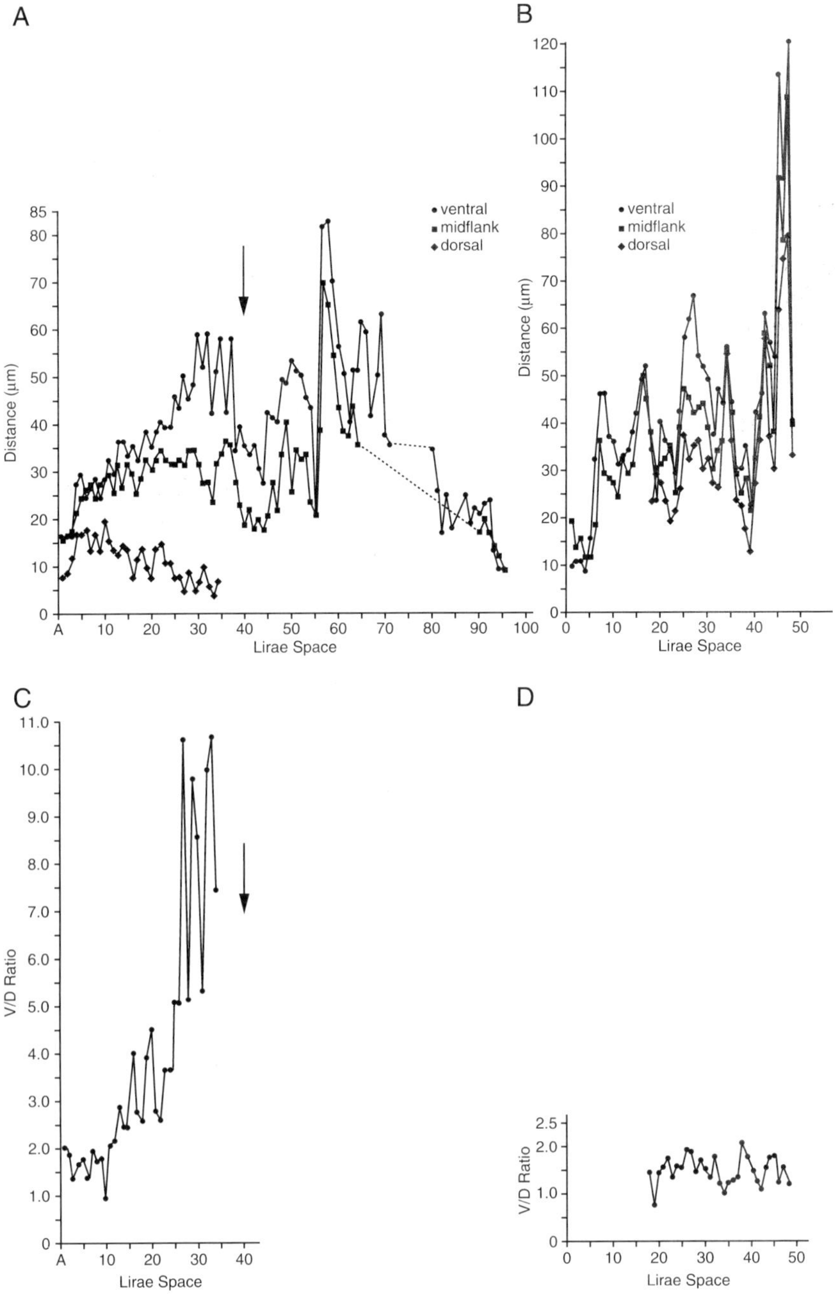

Fig. 2.10 Latanarcestes *sp. (AMNH 46646, Devonian, Morocco). (A) Lirae spacing on embryonic shell for ventral, midflank, and dorsal positions. Symbols given in graph. X axis is the number of lirae space. Y axis is the measured distance between two lirae given in microns (μm). Data are given in Appendix 6. (B) Lirae spacing on postembryonic shell for ventral, midflank, and dorsal positions. X axis is the number of lirae space. Y axis is the measured distance between two lirae given in microns (μm). Data are given in Appendix 7. (C) Ratio of the ventral lirae space/dorsal lirae space (V/D) on the embryonic shell. Data are given in Appendix 6. (D) Ratio of the ventral lirae space/dorsal lirae space (V/D) on the postembryonic shell. Data are given in Appendix 7.*

Immediately after the decrease in distance, there is an increase in distance on the ammonitella coil. This peaks at about 20 lirae from the end of the initial chamber. Adoral of this, the distance decreases on the ammonitella coil until the end of the ammonitella. This pattern is observed on the dorsal, midflank, and ventral positions (Figs. 2.9a, 2.10a). There are 95 lirae on the ammonitella of these Anarcestidae. The distances of the postembryonic lirae show no initial increase, just wide fluctuations (Figs. 2.9b, 2.10b). Only after about 40 lirae does the distance between the lirae begin to increase.

The V/D ratios again show a greater amount of fluctuation between successive lirae on the embryonic shell when compared to the postembryonic shell (Figs. 2.9c, d, 2.10c, d).

3.2.3 Agoniatitidae

Two specimens from the Agoniatitidae were studied, identified as *Agoniatites vanuxemi* (NYSM 3545) and *Fidelites fidelis* (AMNH 50417). The data are presented on Fig. 2.11 and Appendices 10–11. It was not possible to measure the distance between lirae on the postembryonic shell. The data for the two Agoniatitidae are less complete than those for the Mimagoniatitidae and the Anarcestidae. There are gaps in various places at all three measurement positions. In *F. fidelis*, the lirae can only be measured to just adoral of the initial chamber (Fig. 2.11b). Here again, the ventral distance is the widest, the dorsal the narrowest, and the midflank measurement lies somewhere in between. In both genera, after an initial increase in the distance between lirae (1–10) at all three positions, there is a slight decrease. Then, at the ventral and midflank positions, the distance increases rapidly to the end of the initial chamber. At the dorsal position, however, the distances vary widely, but remain essentially level. There are about 30–35 lirae on the initial chamber. There is a decrease at the end of the initial chamber, followed by a very rapid increase in the distance between lirae at the ventral and midflank position and a less rapid increase at the dorsal position. For the remainder of the ammonitella in *A. vanuxemi* (Fig. 2.11a), there is a rapid decrease in the distance followed by a more gradual increase, then a leveling off or a slight decrease. Data end just before the end of the ammonitella in both specimens. There are approximately 65 lirae on the ammonitella of the Agoniatitdae.

Because of gaps in the data, there are fewer data available for calculation of the V/D ratio. What is evident is that over the initial chamber, the ratio increases rapidly at about 20 lirae (V > > D) and reaches a maximum at this point (Fig. 2.11c, d).

4 Discussion

These specimens confirm previous findings on the nature of the ammonitella of early Devonian ammonoids (Klofak et al., 1999). The end of the ammonitella is marked by a reduction in both the spacing and size of the ornament, until a relatively smooth surface appears. A constriction is present with no accompanying

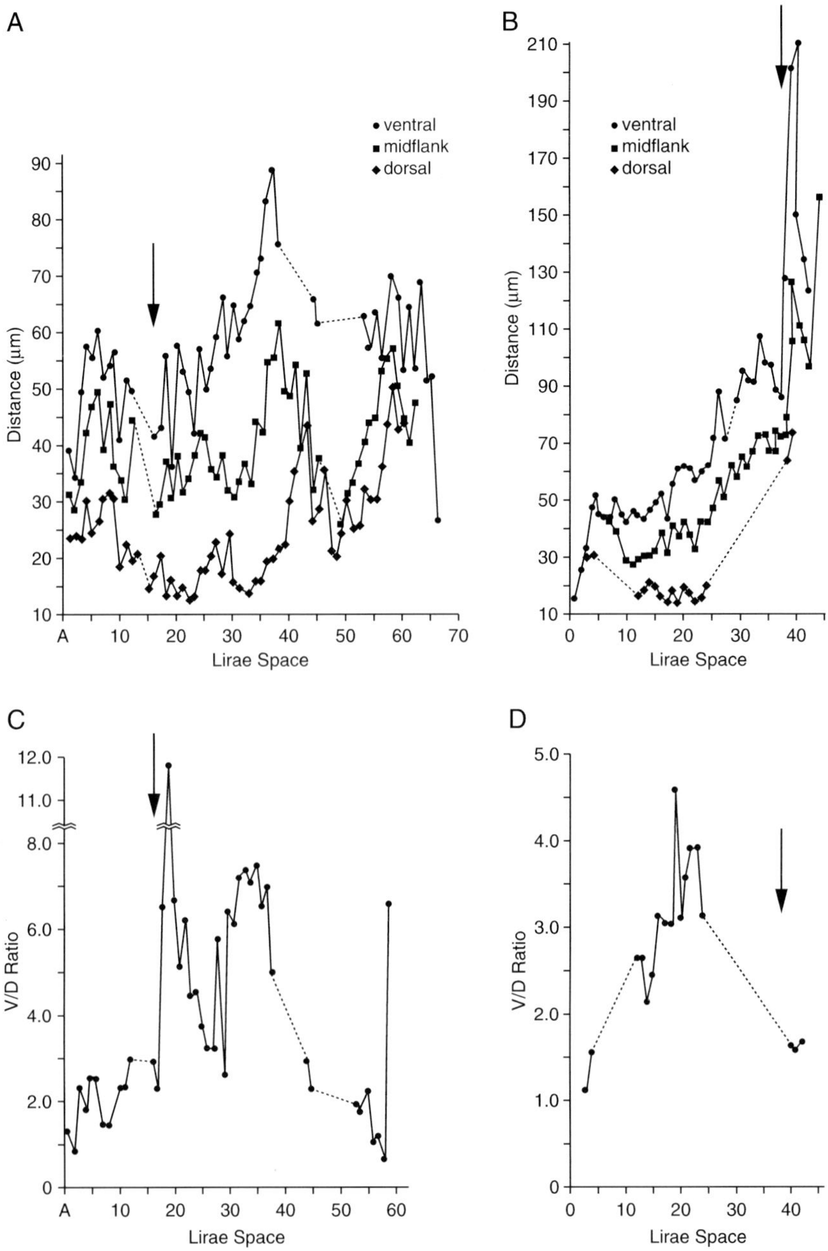

Fig. 2.11 *(A, C)* Agoniatites vanuxemi *(NYSM 3545, Devonian, New York State, USA) (A) Lirae spacing on embryonic shell for ventral, midflank, and dorsal positions. Symbols given in graph. X axis is the number of lirae space. Y axis is the measured distance between two lirae given in microns (µm). (C) Ratio of the ventral lirae space/dorsal lirae space (V/D) on the embryonic shell. Data are given in Appendix 10. (B, D)* Fidelites fidelis *(AMNH 50417, Devonian, Morocco). (B) Lirae spacing on embryonic shell for ventral, midflank, and dorsal positions. Symbols given in graph. X axis is the number of lirae space. Y axis is the measured distance between two lirae given in microns (µm). (D) Ratio of the ventral lirae space/dorsal lirae space (V/D) on the embryonic shell. Data are given in Appendix 11.*

varix. Previously, the absence of a primary varix was documented for the genus *Archanarcestes* in the family Mimagoniatitidae (see Klofak et al., 1999). A second genus in the Mimagoniatitidae demonstrates the same type of ammonitella aperture, that is, a constriction with no varix. Finally, we have also extended this apertural type to a second family, the Anarcestidae. It is known that a varix is present in the Tornoceratina (House, 1965), a phylogenetically more advanced taxon (Becker and Kullmann, 1996; Klug and Korn, 2002). To date, no varix has been found in phylogenetically more primitive ammonoids, increasing the likelihood that this may be the ancestral state for ammonoids. Bactritids also do not possess a varix, but the microstructure of their embryonic shells is different from that of these ammonitellas and may represent a different mode of formation. Bactritids are different enough that some workers have separated them from the Ammonoidea (Ruzhentshev, 1960, 1974; Erben, 1964a; Mapes, 1979; Doguzhaeva, 2002; Korn and Klug, 2002).

The postembryonic shell can be seen emerging from beneath the ammonitella edge of several specimens (AMNH 46646, Fig. 2.3f; AMNH 46645, Fig. 2.2c, d). In the specimen of *Latanarcestes,* the area just anterior to the ammonitella edge shows several breaks, with new shell emerging from below the previous surface (Fig. 2.3f). This is not dissimilar to the end of the embryonic shell in modern *Nautilus* where the thin apertural edge is prone to breakage at hatching (Arnold et al., 1987). It also supports an idea proposed by Kulicki (1974) that the primary varix in post-Devonian ammonoids evolved to prevent breakage during hatching.

It is also clear from the data collected, that while marking the end of a phase in embryonic development, the end of the initial chamber does not exhibit any discernable breaks in shell production (Fig. 2.3c). There is an obvious change in the shape of the shell (see "1. Wachstums-Änderung" of Erben, 1962, 1964b) and a reduction in the distance between lirae in all three taxa at the end of the intial chamber. In addition, Erben (1964b) noted that the ventral sinus appears at this point, perhaps marking the first appearance of the hyponome. This was part of his reason for interpreting this point as the end of embryonic development, though this interpretation has been disproved. Studies on *Nautilus* have shown that a functional hyponome is present early in embryonic development and its appearance is not correlated with hatching (Tanabe et al., 1991).

The original intent of this project was to discover if there were taxonomic differences in the distance between lirae in different taxonomic groups. It became apparent that there was enough variation across an individual ammonitella that any individual measurement or a mean measurement was not informative (Table 2.1). However, the pattern itself proved to be useful.

Based on the shell geometry of a coiled tube with radial ornament, one would expect the distance between any two lirae to be widest on the venter, narrowest on the dorsum, and somewhere in between on the midflank. Projecting a three dimensional object onto two dimensions may distort some measurements. However, the venter and dorsum are not affected because they are measured in the same plane. Distortion could affect the midflank position on the initial chamber but this is minimized because the midflank measurements always lie between the venter and dorsum. Therefore, the patterns in our data are real and not artifacts of the methodology.

There are differences in the distances between lirae on the initial chamber among the different ammonoid families. In the Mimagoniatitidae, the ventral and dorsal parts of the ammonitella show the same pattern (Figs. 2.6a, 2.7a, 2.8a). This is not the case for the Anarcestidae where, as the ventral distance between lirae increases across the initial chamber, the dorsal distance between lirae decreases (Figs. 2.9a, 2.10a). This is also true for the Agoniatitidae (Figs. 2.11a, b). The distance between lirae on the initial chamber of the Agoniatitidae also shows a rapid increase before the decrease marking the end of the initial chamber. These changes reflect subtle changes in the shape and symmetry of the nearly spherical initial chamber. In the Mimagoniatitidae, the arc of both the dorsum and venter are symmetrical, whereas in the Anarcestidae, the arc of the venter is circular but the dorsum is flattened.

With these data, it might be possible to use the change in the distance between lirae as an indicator of the end of the initial chamber. Currently, the initial chamber is defined based on the appearance of the first septum (proseptum) (Owen, 1878; Branco, 1879–1880; Hyatt, 1883; Schindewolf, 1933; Erben, 1960; Bandel, 1982; Landman et al., 1996). It makes better biological sense to define a developmental stage based on characters that formed or are present at that stage (change in distance between lirae) rather than ones added at a later stage (first septum). In other words, the end of the initial chamber would be the point where the distance between lirae shows a marked decrease.

The actual distance between lirae seems to vary widely across the ammonitella. The only constraint on lirae formation on the ammonitella is that the prescribed number of lirae is present by the end of embryonic development. This number is distinctive and taxonomically significant at a family level: 70–77 are present in the Mimagoniatitidae, 95 in the Anarcestidae, and an estimated 65 for the Agoniatitidae. In the Mimagoniatitidae and the Anarcestidae, the difference in number can be directly attributed to the different number of lirae formed on the initial chamber. There are 25–30 in the Mimagoniatitidae and 35–40 in the Anarcestidae, but approximately the same number occur on the ammonitella coil. For the Agoniatitidae, there seems to be a reduction in the number of lirae on the ammonitella coil. There are 30–35 lirae on the initial chamber, more than that in the Mimagoniatitidae, but only a total of 65 for the ammonitella, less than that in the Mimagoniatitidae. Both the Mimagoniatitidae and Anarcestidae have umbilical perforations; the Agoniatitidae do not. It is likely that the tighter coil of the agoniatitid ammonitellas led to a reduction in the size of the ammonitella coil and, hence, a reduction in the number of lirae. The initial chamber, however, remains large. Not only are the agoniatid ammonitellas more tightly coiled, but also the nature of the coiling appears different. The end of the initial chamber can only be defined based on the fold of the dorsal part of the initial chamber against the dorsal part of the ammonitella coil. The venter shows only a slight flattening, similar to the description given for some ammonitellas of Mesozoic ammonoids (Bandel, 1986; Landman et al., 1996). This is visible in the specimen of *Fidelites fidelis* (Fig. 2.1e).

A comparison of the V/D ratio of the embryonic shell and the beginning of the postembryonic shell in the Mimagoniatitidae and Anarcestidae shows the relationship between the dorsal and ventral parts of the shell. The V/D ratios for the postembryonic shells show that there is a closer correlation between the placement of the lirae on the dorsal and ventral sides of the postembryonic shell, than in the embryonic stage. This pattern is especially clear in Fig. 2.6c, d.

It is assumed that the embryo inside the egg case would have produced a shell in a constant environment. Upon hatching, the growth of the animal and production of shell would have been affected by many variables in the environment, such as temperature, salinity, food supply, predators, and water currents. For example, in Fig. 2.2f, less than a whorl after hatching, a small repaired break can be observed on the shell. Rather than be constrained by a more constant environment, the variability of the embryonic shell formation speaks of a high degree of developmental plasticity. This variability has already been documented by Erben (1950) where he observed several coiling shapes in *Mimagoniatites fecundus*. Such developmental plasticity likely played a role in the rapid radiation of the early ammonoids.

Several models for the development of the embryonic shell of ammonoids have been proposed (see Landman et al., 1996). These models feature opposing views. Either the ammonitella was calcified in its entirety (Bandel, 1982; Tanabe, 1989) or it was accreted as small increments at the aperture as in the adult shell (Erben, 1964b, 1966; Erben et al., 1968; Tanabe, 1989). The data presented here do not suggest a good fit with either of these views. The presence of transverse lirae has been used as evidence for an accretionary mode of formation (Erben, 1964b, 1966; Tanabe, 1989), with growth occurring at the apertural edge in small increments followed by hiatuses, producing growth lines (Bucher et al., 1996). The transverse lirae are not growth lines (Bucher et al., 2003) and no growth lines appear between lirae on the embryonic shell. The marginal accretionary mode of formation would likely result in a regular spacing as the mantle edge moved forward in small increments. The distance between any two consecutive lirae on the venter should be proportional to the distance between the same lirae on the dorsum. This is not what is observed for these primitive ammonoids. The V/D ratio shows wide fluctuations between consecutive lirae (Fig. 2.6c, d).

Other models for ammonitella formation suggest a nonaccretionary mode of formation (Bandel, 1982; Tanabe, 1989). The presence of the "wrinkle-like" creases between the transverse lirae suggest an organic component. The production of an organic nonmineralized shell eliminates the need for the lirae to be produced in an even fashion and the V/D ratio could then vary. This might also explain the differences in shape observed by Erben (1950) in the coiling of the ammonitella of *Mimagoniatites fecundus*. There is also no need for the entire nonmineralized shell to calcify simultaneously. Stepwise growth at the aperture has been observed in molluscs during the formation of large scale ornamental features (Vermeij, 1993). And similar models have been proposed for ammonoids (Checa, 1995; Bucher et al., 2003). While neither of these models produces ornament like that on these embryonic shells, it does suggest that ammonoids have the ability to produce the shell wall in a variety of ways, including growing short segments of organic material, and then calcifying it. This would also account for the fluctuations in the spacing observed if the segments produced were from lira to lira.

The models proposed for ammonitella formation are based, at least in part, on the embryonic development of extant taxonomic groups. Nonaccretionary models rely heavily on archaeogastropod developmental models (Bandel, 1982) and accretionary models on *Nautilus* development (Arnold et al., 1987). The embryonic shell of *Nautilus* grows by forward accretion at the aperture (Arnold et al., 1987). It produces a reticulate ornament with both longitudinal and transverse elements strongly developed. A similar ornament has been described in some Jurassic nautilids (Chirat and von Boletzky, 2003). This ornament is different from that described for Devonian ammonoids. The transverse elements in the nautilids appear as if imbricated and the longitudinal elements are not the "wrinkle-like" creases described here, but rather strongly developed longitudinal ridges that intersect the transverse lines. While examining related modern taxa is useful, the phylogenetic distance of these taxa must also be considered (Jacobs and Landman, 1993). The Nautiloidea lie some distance away from the Ammonoidea (Berthold and Engeser, 1987; Engeser, 1990, 1996). Furthermore, between the nautiloids and the three families in this study lie several taxonomic groups (Orthocerida, Bactritida, and several more basal ammonoid families, see Becker and Kullmann, 1996; Korn and Klug, 2002) several of which have distinctly different embryonic shells [see Doguzhaeva (1996a, b, 2002) for the bactritids and Ristedt (1968) for the orthocerids]. While data do exist for more basal ammonoid taxa (Sandberger, 1851; Schindewolf, 1933; Erben, 1950, 1960, 1962, 1964b, 1965; Chlupáč and Turek, 1983), they are scarce and the material is not as well preserved. Due to the open nature of their coiling, the inner whorls are often lost, and when they are preserved, they are often steinkerns. The external morphology of the shell is necessary for comparison. What is known is that members of one clade (the Agoniatitida) show remarkable variation at a very early developmental stage. Such developmental plasticity very early in ontogeny may have aided the rapid radiation of the ammonoids in the Devonian.

5 Conclusions

While still limited by a small number of specimens (seven), this study has added data to our understanding of the nature of and differences between three closely related families of early ammonoids. This study has confirmed a previous finding on the nature of the aperture of the ammonitella (Klofak et al., 1996). In both the Mimagoniatitidae and the Anarcestidae there is no primary varix. Furthermore, there is a suggestion that the absence of a varix produces a weak aperture, which is prone to breakage.

The Mimagoniatitidae, the Anarcestidae, and the Agoniatitidae all possess the same ornament, i.e., transverse lirae with longitudinal "wrinkle-like" creases stretched between them. However, the degree to which these "wrinkle-like" creases are expressed is different for the three families. The number of the lirae is also different for the three families. Most of the difference can be found on the initial chamber.

Other differences between these families can be seen in the pattern of the distances between the lirae on the ammonitella, especially on the initial chamber. In the Mimagoniatitidae, the changes in the spacing of the lirae follow the same pattern on both the ventral and dorsal sides of the shell. In the Anarcestidae and the Agoniatitidae, the pattern of change is different for the ventral and dorsal sides of the shell. In the Anarcestidae, where the ventral part of the shell shows an increase in the distance between lirae, the dorsal part of the shell shows no change or a slight decrease. And in the Agoniatitidae we see an increase in distance between the lirae on the ventral part of the shell, while there is a decrease on the dorsal part of the shell. The end of the initial chamber is marked, as in the Mimagoniatitidae and the Anarcestidae, by a decrease in the distance between lirae. This decrease provides a method of defining the limits of the initial chamber based on a character present at its formation, rather than a character added later in embryonic development as is currently used (the first suture). This makes better biological sense.

Acknowledgments

The authors would like to thank the following people: Kathy B. Sarg for helping with work on the SEM; Jacob Mey and Kevin Frischman for making sure everything worked right with the SEM; Steve Thurston for helping with photography and graphics; Ilya Temkin for translating Russian papers; and Stephanie Crooms, Bushra Hussaini, Kristin Polizzotto, and Yumiko Iwasaki for a variety of formatting and typing issues. Thanks are also extended to Michael House, Thomas Becker, Christian Klug, Bruno Fectay, Yumiko Iwasaki, and Kathy B. Sarg for assistance doing fieldwork in Morocco. Kazushige Tanabe, Roger Hewitt, and Christian Klug are thanked for editorial comments and Ed Landing for allowing us to examine New York material. This research was funded by the Norman D. Newell Fund for Invertebrate Paleontology at the AMNH and the Lerner-Gray Fund for Marine Research to S. M. Klofak at the AMNH.

Appendix 1

Lirae spacing (in microns) on the embryonic shell of *Archanarcestes obesus* (AMNH 45370).

Lirae space number	Ventral (V)	Midflank	Dorsal (D)	Ratio V/D
1		32.6		
2	32.6	32.4	28.0	1.2
3	36.8	31.0	24.8	1.5
4	38.6	35.6	31.6	1.2
5	40.0	34.6	27.0	1.5
6	40.6	33.2	23.4	1.7
7	48.2	40.6		
8	57.0	50.8		
9	58.0	49.0		
10	63.2	52.8		
11	46.4	40.4		
12	53.8			

(continued)

(continued)

Lirae space number	Ventral (V)	Midflank	Dorsal (D)	Ratio V/D
13	42.6			
14	39.6			
15	36.8			
16	39.4			
17	33.0			
18	34.6			
19	37.0			
20	34.8			
21	38.8			
22	38.8			
23	32.2			
24	24.4	19.8	7.2	3.4
25	25.6	12.4	6.2	4.1
26	30.8	20.6	11.6	2.7
27	35.4	18.6	12.0	2.9
28	32.2	27.4	12.0	2.7
29	33.8	26.4	16.8	2.0
30	35.6	25.0	13.4	2.7
31	33.0	29.0	18.0	1.8
32	46.0	33.2	18.8	2.5
33	45.0	35.0	17.0	2.7
34	38.4	33.2	19.0	2.0
35	39.2	32.2	20.8	1.9
36	39.0	37.0	22.6	1.7
37	17.8	14.2	11.4	1.6
38	24.6	21.2	12.6	2.0
39	40.2	28.6	24.2	1.7
40	39.6	30.6	21.6	1.8
41	42.2	32.0	19.2	2.2
42	41.2	33.4	23.2	1.8
43	42.6	34.0	21.8	2.0
44	51.2	36.8	20.0	2.6
45	45.2	34.6	22.4	2.0
46	41.0	29.2	23.8	1.7
47	41.4	28.8	20.6	2.0
48	44.6	31.0	22.4	2.0
49	46.2	35.4	21.2	2.2
50	49.8	32.8	21.6	2.3
51	39.0	31.0	20.4	1.9
52	37.0	31.6	21.2	1.8
53	42.4	32.4	28.0	1.5
54	43.0	35.8	17.8	2.4
55	41.4	33.0	16.8	2.5
56	46.0	32.4	17.8	2.6
57	46.4	32.8	19.6	2.4
58	51.6	31.8	18.2	2.8

(continued)

(continued)

Lirae space number	Ventral (V)	Midflank	Dorsal (D)	Ratio V/D
59	45.4	32.8	19.8	2.3
60	28.8	18.2	12.2	2.4
61	26.4	19.6	12.2	2.2
62	6.9	6.6	5.1	1.4
63	8.4	7.7		
64	7.5	6.1		
65	8.8	8.1		
66	20.3	16.1		
67	23.4	10.1		
68	8.4			
69	7.5			

Appendix 2

Lirae spacing (in microns) on the embryonic shell of *Archanarcestes obesus* (AMNH 45374).

Lirae space number	Ventral (V)	Midflank	Dorsal (D)	Ratio V/D
1	35.1	35.1	28.0	1.2
2	49.8	49.5	41.9	1.2
3	53.3	43.0	32.7	1.6
4	43.0	37.3		
5	43.1	37.1	35.0	1.2
6	40.1	42.5	28.9	1.4
7	35.2	34.8	24.1	1.5
8	33.4	32.5	26.2	1.3
9	31.3	34.7	25.1	1.3
10	35.7	30.1	25.2	1.4
11	35.9	32.0	27.2	1.3
12	46.2	40.2	26.8	1.7
13	63.4	49.6	25.9	2.5
14	42.6	39.9	31.7	1.3
15	48.4	40.0	25.9	1.2
16	48.4	37.8	25.2	1.9
17	46.5	37.0	27.3	1.7
18	37.0	29.8	21.3	1.7
19	35.5	23.2	19.3	1.8
20	30.9	21.8	14.3	2.2
22	19.6			
23	9.6			
24	11.2			
25	17.6	14.0	10.2	1.7
26	21.4	18.0	8.0	2.7
27	27.2	15.8	12.4	2.2

(continued)

(continued)

Lirae space number	Ventral (V)	Midflank	Dorsal (D)	Ratio V/D
28	36.0	18.2	8.6	4.2
29	37.6	20.4	11.8	3.2
30	28.0	27.4	11.8	2.1
31	38.4	30.0	22.2	1.7
32	42.2	31.2	24.8	1.7
33	32.8	30.4	24.0	1.4
34	38.2	26.8	23.0	1.7
35	48.3	38.6	24.6	2.0
36	47.6	36.6	23.2	2.1
37	47.0	42.4	22.8	2.1
38	47.2	36.4	20.6	2.3
39			22.4	
40			19.2	
41			19.2	
42	33.0			
43	42.6			
44	38.2			
45	40.0		23.0	1.7
46	37.4	30.6	22.8	1.6
47	39.8	33.4	21.4	1.9
48	35.0	29.2	17.4	2.0
49	23.6	20.0	12.0	2.0
50	43.8	32.8	26.4	1.7
51	54.4	42.4	28.4	1.9
52	45.6	36.2	32.2	1.4
53	53.2	40.6	31.0	1.7
54	56.8	33.2	24.4	2.3
55		28.2	22.0	
56		23.2	19.0	
57		19.6	16.8	
58		37.2	34.2	
59		25.6	18.6	
60	38.6	31.4	38.2	1.0
61	40.8			
62	24.2			
63	42.2			
64	37.2			
65	34.8			
66	28.6			
67	25.0			
68	27.6			
69	28.4			
70	25.4			
71	26.6			
72	25.6			
73	24.6			

(continued)

(continued)

Lirae space number	Ventral (V)	Midflank	Dorsal (D)	Ratio V/D
74	23.6			
75	27.8			
76	23.6			
77	23.0			
78	24.0			

Appendix 3

Lirae spacing (in microns) on the postembryonic shell of *Archanarcestes obesus* (AMNH 45374).

Lirae space number	Ventral (V)	Midflank	Dorsal (D)	Ratio V/D
1	54.8	56.0	56.0	1.0
2	36.0	26.2	32.4	1.1
3	75.0	50.4	32.0	2.3
4	52.6	39.2	34.6	1.5
5	67.4	58.6	48.8	1.4
6			59.0	
7			47.6	
8			45.2	
9	68.2		33.4	2.0
10	74.6	54.2	34.4	2.2
11	82.0	62.4	53.2	1.5
12			66.4	
13			63.4	
14			63.4	
15			65.6	
16			63.6	
17		72.6	64.8	
18			67.2	
19			62.6	
20		66.2	67.2	
21		82.4	66.2	
22		99.2	71.2	
23			54.0	
24		67.4	53.2	
25		71.2	53.2	
26		79.8	54.8	
27		80.4	70.0	
28		73.4	60.8	
29		87.8	71.8	
30		96.4	78.8	
31		115.8	93.8	

(continued)

Appendix 4

Lirae spacing (in microns) on the embryonic shell of *Mimagoniatites fecundus* (AMNH 46645).

Lirae space number	Ventral (V)	Midflank	Dorsal (D)	Ratio V/D
1		14.2		
2	16.6	17.4		
3	26.9	16.6		
4	28.4	27.7		
5	35.6	21.2	23.7	1.6
6	26.9	18.2	10.3	2.6
7	26.1	34.0	10.3	2.5
8	44.2	28.4	12.6	3.5
9	49.0	31.6	10.3	4.8
10	54.5	30.8	8.7	6.3
11	55.3	42.7	15.0	3.7
12	64.8	45.8	9.5	6.8
13	70.3	49.0	11.9	5.9
14	69.5	49.0	19.8	3.5
15	66.4	50.6	15.8	4.2
16	75.1	54.5	28.4	2.6
17	64.8	57.7	19.8	3.3
18	66.4	52.1	27.7	2.4
19	62.4	58.5	26.9	2.3
20	63.2	49.0	25.3	2.5
21	66.4	53.7	22.1	3.0
22	64.0	50.6	23.7	2.7
23	64.0	52.1	23.7	2.7
24	66.4	42.7	22.1	3.1
25	72.7	48.2	19.0	3.8
26	85.2	42.7	18.2	3.7
27	54.5	36.3	19.8	2.8
28	48.2	34.0	19.8	2.4
29	56.1	38.7	15.0	3.7
30	54.5	31.6	20.5	2.7
31	58.5	37.1	16.6	3.5
32	48.2	35.6	15.8	3.1
33	62.4	52.1	27.1	2.4
34	86.3	67.9	41.1	2.1
35	75.1	56.1	39.5	1.9
36	75.1	59.3	45.0	1.7
37	90.1	69.5	44.2	2.0
38	111.4	71.1	50.6	2.2
39	92.4	71.1	26.9	3.4
40	83.0	65.6	35.5	2.3
41	90.9	62.4	35.5	2.6
42	101.9	60.8	37.1	2.7

(continued)

(continued)

Lirae space number	Ventral (V)	Midflank	Dorsal (D)	Ratio V/D
43	98.0	71.9	41.1	2.3
44	96.4	74.3	42.7	2.8
45	92.4	71.1	34.8	2.7
46	101.9	74.3	30.0	3.4
47	102.7	59.3	31.6	3.3
48	83.7	54.5	26.9	2.9
49	124.0	64.8	30.0	4.1
50	93.2	57.7	36.3	2.6
51	97.2	64.0	29.2	3.3
52	65.6	47.4	25.3	2.6
53	52.1	46.6	24.5	2.1
54	57.7	42.7	23.7	2.4
55	79.0	64.8	25.3	3.1
56	80.6	48.2	20.5	3.9
57		49.8	15.8	
58		44.2	13.4	
59		45.0	17.4	
60	60.0	39.5	18.2	3.3
61	45.0	47.4	11.1	4.1
62	57.7	45.0	14.2	4.1
63	63.2	49.0	11.9	5.3
64	58.5	43.5	15.0	3.9
65	59.3	42.7	16.6	3.6
66	60.8	49.8	25.3	2.4
67	79.8	75.8	28.4	2.8
68	33.2	24.5	15.0	2.2
69	37.9	35.6	14.2	2.7
70	37.1	30.8	14.2	2.6
71	57.7	34.8	14.2	4.1
72	62.4	43.5	18.2	3.4
73	49.0	45.0	16.6	3.0
74	34.8	26.9	13.4	2.6
75	31.6	22.1	19.8	1.6
76		17.4		

Appendix 5

Lirae spacing (in microns) on the postembryonic shell of *Mimagoniatites fecundus* (AMNH 46645).

Lirae space number	Ventral (V)	Midflank	Dorsal (D)	Ratio V/D
1	29.2	15.0	31.6	0.9
2	30.8	9.5	28.4	1.1
3	46.6	3.2	42.7	1.1
4	37.1	30.0	22.1	1.7
5	36.3	23.7	26.1	1.4
6	42.7	44.2	29.2	1.5
7	57.7	49.0	36.3	1.6
8	41.1	43.5	29.2	1.4
9	42.7	45.0	25.3	1.7
10	67.2	48.2	37.9	1.8
11	86.9	68.7	45.0	1.9
12	69.5	60.0	48.2	1.4
13	83.7	55.3	38.7	2.2
14	84.5	52.9	39.5	2.1
15	56.1	50.6	39.5	1.4
16	59.3	58.5	34.8	1.7
17	65.6	48.2	38.7	1.7
18	60.8	57.7	34.0	1.8
19	67.9	62.4	48.2	1.4
20	67.2	61.6	40.3	1.7
21	108.2	72.7	51.4	1.7
22	78.2	31.6	6.3	12.4
23	90.1	71.1	54.5	1.7
24	94.0	65.6	52.9	1.8
25	78.2	71.1	55.3	1.4
26	80.6	67.2	43.5	1.9
27	65.6	56.9	37.1	1.8
28	102.7	61.6	47.4	2.2
29	113.0	78.2	74.3	1.5
30	104.3	131.1	46.6	2.2
31	106.7	83.7	59.3	1.8
32	143.0	129.6	85.3	1.7
33	122.5	108.2	56.9	2.2
34	99.5	64.0	52.1	1.9
35	105.1	91.6	65.6	1.6
36	83.7	56.1		
37	93.2	73.5		
38	91.6	64.0		
39	88.5	71.9		
40	109.8	79.8	67.9	1.6
41	216.5	122.5	87.7	2.5
42	109.8	116.9	83.0	1.3

(continued)

(continued)

Lirae space number	Ventral (V)	Midflank	Dorsal (D)	Ratio V/D
43	110.6	103.5	67.2	1.7
44	142.2	90.9	58.5	2.5
45	116.1	94.8	73.5	1.6
46	120.9	96.4	69.5	1.7
47	169.9	122.5	86.1	2.0
48	110.6	78.2		
49	93.2	84.5		
50	124.8	92.4		
51		95.6		
52		96.4		
53		88.5		
54		43.5		
55		87.7		

Appendix 6

Lirae spacing (in microns) on the embryonic shell of *Latanarcestes* sp. (AMNH 46646).

Lirae space number	Ventral (V)	Midflank	Dorsal (D)	Ratio V/D
1	15.8	16.8	7.9	2.0
2	16.8	16.8	8.9	1.9
3	17.8	16.8	11.9	1.4
4	27.7	21.7	16.8	1.7
5	29.6	24.7	16.8	1.8
6	24.7	26.7	17.8	1.4
7	26.7	27.7	13.8	1.9
8	28.7	24.7	16.8	1.7
9	24.7	27.7	13.8	1.8
10	28.7	28.7	19.8	1.5
11	32.6	29.6	15.8	2.1
12	29.6	25.7	13.8	2.1
13	36.6	31.6	12.8	2.9
14	36.6	26.7	14.8	2.5
15	33.6	31.6	13.8	2.4
16	35.6	29.6	8.9	4.0
17	32.6	25.7	11.9	2.8
18	35.6	28.7	13.8	2.6
19	38.5	32.6	9.9	3.9
20	35.6	30.6	7.9	4.5
21	38.5	32.6	13.8	2.8
22	40.5	34.6	14.8	2.6

(continued)

(continued)

Lirae space number	Ventral (V)	Midflank	Dorsal (D)	Ratio V/D
23	39.5	32.6	10.9	3.6
24	39.5	31.6	10.9	3.6
25	45.5	31.6	7.9	5.8
26	44.5	32.6	7.9	5.6
27	52.4	31.6	4.9	10.6
28	45.5	34.6	8.9	5.1
29	48.4	34.6	4.9	9.8
30	59.3	31.6	6.9	8.6
31	52.4	27.7	10.0	5.3
32	59.3	27.7	5.9	10.0
33	42.5	23.7	4.0	10.8
34	51.4	31.6	6.9	7.4
35	58.3	33.6		
36	42.5	36.6		
37	58.3	35.6		
38	34.6	27.7		
39	39.5	22.7		
40	35.6	18.8		
41	33.6	21.7		
42	35.6	17.8		
43	30.6	19.8		
44	27.7	17.8		
45	42.5	20.8		
46	41.5	27.7		
47	40.5	21.7		
48	49.4	33.6		
49	48.4	40.5		
50	53.4	25.7		
51	51.4	34.6		
52	50.4	32.6		
53	45.5	33.6		
54	43.5	23.7		
55	20.8	21.7		
56	38.5	38.5		
57	82.0	70.2		
58	83.0	65.2		
59	70.2	54.3		
60	56.3	43.5		
61	50.4	38.5		
62	40.5	37.5		
63	51.4	43.5		
64	51.4	35.6		
65	61.3			
66	59.3			
67	42.5			

(continued)

(continued)

Lirae space number	Ventral (V)	Midflank	Dorsal (D)	Ratio V/D
68	50.4			
69	62.2			
70	37.5			
71	35.6			
GAP				
80	34.6			
81	25.7			
82	16.8			
83	24.7			
84	17.8			
85				
86				
87	24.7			
88	18.8			
89	21.7			
90	20.8	16.8		
91	22.7	19.8		
92	23.7	16.8		
93	12.8	13.8		
94	8.9	11.9		
95	8.9	8.9		

Appendix 7

Lirae spacing (in microns) on the postembryonic shell of *Latanarcestes* sp. (AMNH 46646).

Lirae space number	Ventral (V)	Midflank	Dorsal (D)	Ratio V/D
1	10.0	19.8		
2	10.9	13.8		
3	10.9	15.8		
4	8.9	11.9		
5	15.8	11.9		
6	32.6	18.8		
7	46.4	36.6		
8	46.4	29.6		
9	37.5	28.7		
10	36.6	27.7		
11	31.6	24.7		
12	33.6	32.6		
13	34.6	29.6		
14	38.5	31.6		
15	42.5	42.5		
16	49.4	50.4		

(continued)

(continued)

Lirae space number	Ventral (V)	Midflank	Dorsal (D)	Ratio V/D
17	52.4	45.5		
18	34.6	38.5	23.7	1.5
19	23.7	30.6	29.6	0.8
20	40.5	31.6	27.7	1.5
21	36.6	32.6	23.7	1.5
22	34.5	35.6	19.8	1.8
23	29.6	25.7	21.7	1.4
24	42.5	39.5	26.7	1.6
25	58.3	47.4	37.5	1.6
26	62.2	45.5	32.6	1.9
27	67.2	42.5	35.6	1.9
28	54.3	43.5	36.6	1.5
29	52.4	44.5	30.6	1.7
30	49.4	39.5	32.6	1.5
31	37.5	30.6	27.7	1.4
32	47.4	34.6	26.7	1.7
33	44.5	36.6	36.6	1.2
34	55.3	56.3	55.3	1.0
35	44.5	42.5	36.6	1.2
36	30.6	29.6	23.7	1.3
37	30.6	25.7	22.7	1.4
38	36.6	28.7	17.8	2.1
39	22.7	21.7	12.8	1.8
40	42.5	32.6	27.7	1.5
41	46.4	41.5	36.6	1.3
42	63.2	58.3	59.3	1.1
43	57.3	52.4	37.5	1.5
44	54.3	38.5	30.6	1.7
45	114.6	91.9	64.2	1.8
46	91.9	29.0	75.1	1.2
47	121.5	109.7	80.0	1.5
48	40.5	40.5	33.6	1.2

Appendix 8

Lirae spacing (in microns) on the embryonic shell of *Latanarcestes* sp. (AMNH 50416).

Lirae space number	Ventral (V)	Midflank	Dorsal (D)	Ratio V/D
1	12.2	17.8	9.7	1.3
2	13.3	13.3	11.6	1.2
3	14.4	12.7	8.9	1.6
4	14.2	12.2	7.2	2.0
5	18.7	10.9	13.9	1.3

(continued)

(continued)

Lirae space number	Ventral (V)	Midflank	Dorsal (D)	Ratio V/D
6	16.9	14.8	8.8	1.9
7	21.3	12.6	10.0	2.1
8	25.3	14.8	12.4	2.0
9	18.9	14.6	11.2	1.7
10	23.1	12.0	7.3	3.2
11	19.1	15.4	10.9	1.8
12	22.3	14.9	11.8	1.9
13	25.3	14.2	6.4	4.0
14	28.8	20.5	9.5	3.0
15	28.6	18.5	10.2	2.8
16	25.2	21.4	8.8	2.9
17	22.5	17.7	10.0	2.3
18	16.5	15.3	12.8	1.3
19	16.5	13.3	5.7	2.9
20	28.8	10.5	8.8	3.3
21	23.7	16.9	12.2	1.9
22	30.9	19.2	10.4	3.0
23	34.1	19.2	10.7	3.2
24	26.7	22.9	10.1	2.6
25	25.6	18.0	8.2	3.1
26	23.3	20.4	7.7	3.0
27	22.9	18.3	7.0	3.3
28	24.0	17.4	9.5	2.5
29	25.8	15.4	8.8	2.9
30	28.5	20.5	6.9	4.1
31	23.1	16.3	8.7	2.8
32	24.7	18.6	8.3	3.0
33	23.3	14.9	6.4	3.6
34	24.7	18.0	5.6	4.4
35	25.4	14.8	5.6	4.5
36	28.2	12.5	7.1	4.0
37	24.2	14.0	5.6	4.3
38	20.3	10.7		
39	23.8	12.6		
40	15.6	12.9		
41	16.1	10.4		
42	19.7	13.9		
43	22.2	10.1		
44	17.3	10.7		
45	27.9	13.8		
46	26.5	14.3	7.6	3.5
47	38.7	14.4	7.3	5.3
48	24.9	18.3	8.4	3.0
49	24.7	16.1	6.2	4.0
50	20.3	21.3	4.4	4.6

(continued)

(continued)

Lirae space number	Ventral (V)	Midflank	Dorsal (D)	Ratio V/D
51	36.4	19.5	5.4	6.7
52	24.6	18.6	7.1	3.5
53	32.9	20.3	7.1	4.6
54	38.4	23.6	7.1	5.4
55	35.6	18.2	4.4	8.1
56	45.8	25.9	3.9	11.7
57	40.7	31.0	8.1	5.0
58	46.7	28.4	7.0	6.7
59	46.9	27.0	8.1	5.8
60	43.6	32.9	10.3	4.2
61	32.3	25.1	14.9	2.2
62	40.6	33.1	11.1	3.7
63	51.9	29.9	8.5	6.1
64	33.5	28.4	9.4	3.6
65	37.5	28.0	12.7	3.0
66	46.4	27.8	12.7	3.7
67	45.8	31.2	12.9	3.6
68	41.8	32.9	18.6	2.3
69	54.0	33.8	18.6	3.5
70	65.5	37.6	17.8	3.7
71	47.0	24.9	12.8	3.7
72	52.9	32.8	17.7	3.0
73	32.6	30.3	13.5	2.4
74	31.9	24.9	13.9	2.3
75	48.4	22.4	12.0	4.0
76	39.1	20.7	11.9	3.3
77	33.8	21.9	6.5	5.2
78	38.5	21.5	7.7	5.0
79	29.8	20.4	10.7	2.8
80	31.3	21.3	10.3	3.0
81	36.0	20.7	11.0	3.3
82	38.3	21.2	10.4	3.7
83	25.1	11.9	11.1	2.3
84	26.4	20.9	12.5	2.1
85	30.1	17.4	7.9	3.8
86	21.6	11.2	10.3	2.1
87	20.3	13.9	8.5	2.4
88	24.7	14.2	9.6	2.6
89	25.1	12.8	9.1	2.8
90	23.9	13.6	6.2	3.9
91	20.5	11.0		
92	16.6	13.4		
93	17.4	13.4		
94	18.0	11.9	10.2	1.8
95		10.2		

Appendix 9

Lirae spacing (in microns) on the postembryonic shell of *Latanarcestes* sp. (AMNH 50416).

Lirae space number	Ventral (V)	Midflank	Dorsal (D)	Ratio V/D
1	60.1	41.2	30.5	2.0
2	36.5	30.9	23.8	1.5
3	56.6	38.0	35.0	1.6
4	40.7	39.0	23.2	1.8
5	46.9	35.5	25.1	1.9
6	45.6	30.9	23.8	1.9
7	43.3	41.2	21.7	2.0
8	38.2	30.9	26.8	1.4
9	63.0	52.5	29.1	2.2
10	45.2	36.5	27.1	1.7
11	61.5	47.5	39.9	1.5
12	32.6	28.1	18.9	1.7
13	42.1	27.2	23.8	1.8
14	29.9	26.8	25.5	1.2
15	36.6	25.2	20.6	1.8
16	30.7	30.1	26.7	1.2
17	32.0	28.3	17.8	1.8
18	34.1	27.5	17.0	2.0
19	43.7	41.3	26.1	1.7
20	45.2	32.0	24.6	1.9
21	46.3	26.9	19.2	2.4
22		25.5	20.1	

Appendix 10

Lirae spacing (in microns) on the embryonic shell of *Agoniatites vanuxemi* (NYSM 3545).

Lirae space number	Ventral (V)	Midflank	Dorsal (D)	Ratio V/D
1	39.2	31.3	23.8	1.7
2	42.1	28.3	23.9	1.4
3	49.5	33.1	23.1	2.1
4	57.5	42.3	30.3	1.9
5	55.6	46.7	24.6	2.3
6	59.4	49.1	26.6	2.2
7	52.1	39.3	30.3	1.7
8	53.9	47.1	31.5	1.2
9	56.1	36.1	30.5	1.8
10	40.7	33.8	18.9	2.2
11	51.2	30.1	22.4	2.3

(continued)

(continued)

Lirae space number	Ventral (V)	Midflank	Dorsal (D)	Ratio V/D
12	49.6	44.4	19.8	2.5
13		20.8		
14				
15			14.4	
16	41.4	27.4	16.9	2.5
17	43.0	29.4	20.3	2.1
18	56.9	37.0	13.3	4.2
19	36.0	30.5	16.1	5.9
20	57.5	37.9	13.3	4.3
21	52.8	31.7	14.8	3.6
22	49.3	33.9	12.1	4.1
23	42.0	38.1	13.0	3.2
24	57.0	42.1	17.5	3.3
25	50.0	41.4	17.4	2.9
26	53.4	35.8	20.3	2.6
27	59.1	34.2	22.7	2.6
28	66.1	38.2	17.0	3.9
29	55.9	31.8	24.3	2.3
30	64.6	30.7	15.4	4.2
31	58.7	33.4	14.4	4.1
32	61.9	36.8	13.5	4.6
33	64.4	33.0	13.8	4.7
34	70.5	43.9	15.6	4.5
35	73.0	48.1	15.5	4.7
36	83.1	54.8	19.6	4.2
37	89.0	55.3	19.9	4.5
38	76.5	61.4	21.9	3.5
39	49.6	22.3		
40	48.8	30.1		
41	55.3	35.6		
42	39.4	38.9		
43	52.5	43.5		
44	65.9	31.8	26.8	2.5
45	61.7	37.6	28.5	2.2
46		35.9		
47		21.6		
48		20.7		
49	25.4	24.4		
50	31.3	30.1		
51	33.3	25.0		
52	36.6	25.4		
53	63.0	40.5	32.1	2.0
54	57.2	43.7	30.4	1.9
55	63.9	44.7	30.2	2.1
56	55.9	52.5	36.2	1.5

(continued)

(continued)

Lirae space number	Ventral (V)	Midflank	Dorsal (D)	Ratio V/D
57	70.3	56.0	43.8	1.6
58	66.1	57.9	50.2	1.3
59	53.4	51.2	42.9	
60	64.5	44.2		
61	53.5	40.6		
62	68.7	47.0		
63	51.3			
64	52.0			
65	26.5			

Appendix 11

Lirae spacing (in microns) on the embryonic shell of *Fidelites fidelis* (AMNH 50417).

Lirae Space number	Ventral (V)	Midflank	Dorsal (D)	Ratio V/D
1	15.2			
2	25.5			
3	33.7		29.9	1.1
4	47.9		30.7	1.6
5	45.3			
6	44.1			
7	44.1	42.7		
8	50.9	39.4		
9	45.3			
10	42.8	28.9		
11	46.2	27.4		
12	45.0	29.3	16.9	2.7
13	48.5	30.6	18.4	2.6
14	46.9	30.8	21.6	2.2
15	48.2	32.4	19.6	2.5
16	52.7	38.8	16.8	3.1
17	43.6	31.5	14.2	3.1
18	56.0	41.4	18.4	3.0
19	61.2	37.4	13.3	4.6
20	62.3	42.9	19.9	3.1
21	61.4	37.7	17.1	3.6
22	57.3	32.9	14.6	3.9
23	60.7	42.3	15.7	3.9
24	62.6	42.3	20.0	3.1
25	72.4	47.2		
26	89.0	57.4		
27	71.7	50.8		

(continued)

(continued)

Lirae space number	Ventral (V)	Midflank	Dorsal (D)	Ratio V/D
28		62.0		
29	85.0	58.0		
30	95.4	65.5		
31	92.0	62.0		
32	91.1	66.0		
33	107.3	72.8		
34	99.2	73.4		
35	97.6	67.6		
36	89.1	74.6		
37	86.6	72.6		
38	137.6	73.1		
39	202.0	126.4		
40	210.4	120.3		
41	134.1	107.0	84.8	1.6
42	123.5	93.5	74.2	1.7
43	156.6			

References

Arnold, J. M., N. H. Landman, and H. Mutvei. 1987. Development of the embryonic shell of *Nautilus*. *In* W. B. Saunders, and N. H. Landman (editors), *Nautilus – the Biology and Paleobiology of a Living Fossil*, pp. 373–400. New York: Plenum Press.

Bandel, K. 1982. Morphologie und Bildung der früh ontogenetischen Gehäuse bei conchiferen Mollusken. *Facies* **7**: 1–198.

Bandel, K. 1986. The ammonitella: a model of formation with the aid of the embryonic shell of archaeogastropods. *Lethaia* **19**: 171–180.

Barrande, J. 1865. Systême Silurien du centre de la Bohême. I. Vol. II. Céphalopodes. Praha and Paris.

Becker, R. T., and M. R. House. 1994. International Devonian goniatite zonation, Emsian to Givetian, with new records from Morocco. *Courier Forschungs-institut Senckenberg, Frankfurt am Main* **169**: 79–135.

Becker, R. T., and J. Kullmann. 1996. Paleozoic ammonoids in space and time. *In* N. H. Landman, K. Tanabe, and R. A. Davis (editors), *Ammonoid Paleobiology*, pp. 711–753. New York: Plenum Press.

Bensaïd, M. 1974. Étude sur des Goniatites à la limite du Dévonien moyen et supérieur du Sud marocain. *Notes du Service Géologique du Maroc* **36**(264): 81–140.

Berthold, T., and T. Engeser. 1987. Phylogenetic analysis and systematization of the Cephalopoda (Mollusca). *Verhandlungen Naturwissenscaftlichen vereins in Hamburg* (*NF*) **29**: 187–220.

Branco, W. 1879–80. Beiträge zur Entwicklungsgeschichte der fossilen Cephalopoden, Theile I. *Palaeontographica* **26**: 19–50, pls. 4–13.

Branco, W. 1880–81. Beiträge zur Entwicklungensgeschichte der fossilen Cephalopoden, Theile II. *Palaeontographica* **27**: 17–81, pls. 3–17.

Bogoslovsky, B. I. 1969. Devonskie Ammonoidei. I. Agoniatity. *Trudy Paleontologicheskogo Institita Akademiya Nauk SSSR* **124**: 341pp.

Bucher, H., N. H. Landman, S. M. Klofak, and J. Guex. 1996. Modes and rate of growth in ammonoids. *In* N. H. Landman, K. Tanabe, and R. A. Davis (editors), *Ammonoid Paleobiology*, pp. 407–461. New York: Plenum Press.

Bucher, H., R. Chirat, and J. Guex. 2003. Morphogenetic origin of radial lirae and mode of shell growth in *Calliphylloceras* (Jurassic Ammonoidea). *Eclogae geologicae Helvetiae* **96**: 495–502.

Checa, A. 1995. A model for the morphogenesis of ribs in ammonites inferred from associated microsculptures. *Palaeontology* **37**: 863–888.

Chirat, R., and S. v. Boletzky. 2003. Morphogenetic significance of the conchal furrow in nautiloids: evidence from early embryonic shell development of Jurassic Nautilida. *Lethaia* **36**: 161–170.

Chlupáč, I., and V. Turek. 1983. Devonian goniatites from the Barrandian area, Czechoslovakia. *Edice Rozpravy Ustrední_hoústavu geologického, svazek* **46**: 159pp. [Translated by I. Chlupáčand H. Zárubová]

Clarke, J. M. 1899. Notes on the early stages of certain goniatites. *Annual Report of the New York State Geologist 16* (*Annual Report of the New York State Museum*) **50**(2): 163–169.

Clausen, C. D. 1969. Oberdevonische Cephalopoden aus dem Rheinischen Schiefergebirge, II Gephuroceratidae, Beloceratidae. *Palaeontographica Abteilungen A* **132**: 95–178.

Doguzhaeva, L. A. 1996a. Microstructure of juvenile shells of the Permian *Hemibactrites* sp. (Cephalopoda: Bactritoidea). *Doklady Biological Sciences* **349**(2): 275–279.

Doguzhaeva, L. A. 1996b. Shell ultrastructure of the early Permian bactritella and ammonitella, and its phylogenetic implication. *Jost Wiedmann Symposium. Cretaceous Stratigraphy, Paleobiogeology and Paleobiogeography Abstracts*, pp. 19–25.

Doguzhaeva, L. A. 2002. Adolescent bactritoid, orthoceroid, ammonoid and coleoid shells from the upper Carboniferous and lower Permian of the South Urals. *In* H. Summesberger, K. Histon, and A. Daurer (editors), *Cephalopods – Present and Past. Abhandlungen Geologische Bundesanstalt* **57**: 9–55.

Drushschits, V. V., L. A. Doguzhayeva, and I. A. Mikhaylova. 1977. The structure of the ammonitella and the direct development of ammonites. *Paleontological Journal* **2**: 57–69.

Engeser, T. 1990. Major events in cephalopod evolution. *In* P. D. Taylor, and G. P. Larwood (editors), *Major Evolutionary Radiations. Systematics Association Special Volume* **42**: 119–138, Oxford: Clarendon Press.

Engeser, T. 1996. The position of the Ammonoidea within the Cephalopoda. *In* N. H. Landman, K. Tanabe, and R. A. Davis (editors), *Ammonoid Paleobiology*, pp. 3–19. New York: Plenum Press.

Erben, H. K. 1950. Bemerkungen zu Anomalien mancher Anfangswindungen von *Mimagoniatites fecundus* (Barr.) *Neues Jahrbuch für Geologie und Paläontologie, Monathefte*, pp. 25–32.

Erben, H. K. 1953. Goniatitacea (Ceph.) aus dem Unterdevon und Unterem Mitteldevon. *Neues Jahrbuch für Geologie und Paläontologie Abhandlungen* **98**: 175–225.

Erben, H. K. 1960. Primitive Ammonoidea aus dem Unterdevon Frankreichs und Deutschlands. *Neues Jahrbuch für Geologie und Paläontologie Abhandlungen* **110**(1): 1–128.

Erben, H. K. 1962. Über böhmische und türkische Vertreter von *Anetoceras* (Ammon., Unterdevon). *Paläontologische Zeitschrift* **36**(1/2): 14–27.

Erben, H. K. 1964a. Bactritoidea. *In* R. C. Moore (editor), *Treatise on Invertebrate Paleontology, Part K, Mollusca 4, Cephalopoda: K491–505*. Lawrence, Kansas, and New York: The University of Kansas Press and the Geological Society of America.

Erben, H. K. 1964b. Die Evolution der ältesten Ammonoidea, Lieferung 1. *Neues Jahrbuch für Mineralogie, Geologie, Paläontologie Abhandlungen* **120**(2): 107–212.

Erben, H. K. 1965. Die Evolution der ältesten Ammonoidea (Lieferung II). *Neues Jahrbuch für Mineralogie, Geologie Paläontologie Abhandlungen* **122**(3): 275–313.

Erben, H. K. 1966. Über den Ursprung der Ammonoidea. *Biological Review* **41**: 641–658.

Erben, H. K., G. Flajs, and A. Siehl 1968. Ammonoids: early ontogeny of ultramicroscopical shell structure. *Nature* **219**(5152): 396–398.

Göddertz, B. 1987. Devonische Goniatiten aus SW-Algerien und ihre stratigraphische Einordnung in die conodonten-Abfolge. *Palaeontographica Abteilung A* **197**: 127–220.

Göddertz, B. 1989. Unterdevonische Hercynische Goniatiten aus Deutschland, Frankreich und der Türkei. *Palaeontographica Abteilungen A* **208**(1–3): 61–89.

House, M. R. 1962. Observations on the ammonoid succession of the North American Devonian. *Journal of Paleontology* **36**(2): 247–284.

House, M. R. 1965. A Study on the Tornoceratidae: the succession of *Tornoceras* and related genera in the North American Devonian. *Philosophical Transactions of the Royal Society London Series B* **250**: 79–130.

House, M. R. 1981. On the origin, classification and evolution of the early Ammonoidea. *In* M. R. House, and J. R. Senior (editors), *The Ammonoidea. Systematic Association Special Volume* **18**: 3–36, New York: Academic Press.

House, M. R. 1996. Juvenile goniatite survival strategies following Devonian extinction events. *In* M. B. Hart (editor), *Biotic Recovery from Mass Extinction Events. Geological Society Special Publication* **102**: 163–185.

Hyatt, A. 1883. Fossil cephalopods in the Museum of Comparative Zoology. *American Association for the Advancement of Science Proceedings* **32**: 323–361.

Jacobs, D. K., and N. H. Landman. 1993. *Nautilus* – a poor model for the function and behavior of ammonoids? *Lethaia* **26**: 101–111.

Klofak, S. M., N. H. Landman, and R. H. Mapes. 1999. Embryonic development of primitive ammonoids and the monophyly of the Ammonoidea. *In* F. Oloriz, and F. J. Rodriguez-Tovar (editors), *Advancing Research on Living and Fossil Cephalopods*, pp. 23–45. New York: Plenum Press.

Klug, C. 2001. Early Emsian ammonoids from the eastern Anti-Atlas (Morocco) and their succession. *Paläontologische Zeitschrift* **74**(4): 479–515.

Klug, C., D. Korn, and A. Reisdorf. 2000. Ammonoid and conodont stratigraphy of the late Emsian to early Eifelian (Devonian) at the Jebel Ouaoufilal (near Taouz, Tafilalt, Morocco). *Travaux de l'Institut Scientifique, Rabat, Série Géologie et Géographie Physique* **20**: 45–56.

Korn, D. 2001. Morphometric evolution and phylogeny of Palaeozoic ammonoids. Early and Middle Devonian. *Acta Geologica Polonica* **5**(3): 193–215.

Korn D., and C. Klug. 2002. Ammoneae Devonicae. *In* W. Riegraf (editor), *Fossilium Catalogus I: Animalia* **138**: 1–375, Leiden: Backhuys.

Kulicki, C. 1974. Remarks on the embryogeny and postembryonal development of ammonites. *Acta Palaeontologica Polonica* **19**(2): 201–224.

Kulicki, C. 1979. The ammonite shell: its structure, development and biological significance. *Palaeontologia Polonica* **39**: 96–141.

Landman, N. H. 1988. Heterochrony in ammonites. *In* M. L. McKinney (editor), *Heterochrony in Evolution, pp.* 159–182. New York: Plenum Press.

Landman, N. H., K. Tanabe, and Y. Shigeta. 1996. Ammonoid embryonic development. *In* N. H. Landman, K. Tanabe, and R. A. Davis (editors), *Ammonoid Paleobiology, pp.* 343–405. New York: Plenum Press.

Landman, N. H., F. Bizzarini, K. Tanabe, R. H. Mapes, and C. Kulicki. 2001. Micro-ornament on the embryonic shells of Triassic ceratites (Ammonoidea). *American Malacological Bulletin* **16**(1/2): 1–12.

Mapes, R. H. 1979. Carboniferous and Permian Bactritoidea (Cephalopoda) in North America. *The University of Kansas Paleontological Contributions Article* **64**: 1–75.

Miller, A. K. 1938. Devonian ammonoids of America. *Geological Society of America Special Papers Number* **14**: 262pp.

Miller, A. K., and W. M. Furnish. 1954. The classification of the Paleozoic ammonoids. *Journal of Paleontology* **28**(5): 685–692.

Owen, C. B. 1878. On the relative positions to their constructors of the chambered shells of Cephalopods. *Proceedings of the Zoological Society of London* **1878**: 955–975.

Petter, G. 1959. Goniatites Dévoniennes du Sahara. *Publications du Service de la Carte Géologique de l'Algérie (Nouvelle Séries) Paléontologie Mémoire* **2**: 313pp.

Ristedt, H., 1968. Zur Revision der Orthoceratidae. *Akademie der Wissenschaften und der Literatur Mainz Abhandlungen der Mathematisch-Naturwissenschaftlichen Klasse* **1968**: 211–287.

Ruzhentsev, V. E. 1960. Printsipy sistematiki, sistema i filogeniya paleozoyskikh ammonoidey. *Trudy Paleontologicheskogo Instituta Akademiya Nauk SSSR* **83**: 331pp. [Principles of systematics, classification and phylogeny of Paleozoic ammonoids].

Ruzhentsev, V. E. 1974. Suborder Ammonoidea; general section. *In* V. E. Ruzhentsev (editor), *Fundamentals of Paleontology, Mollusca-Cephalopoda I* **5**: 371–503. Jerusalem: Kater Publishing House. [Translated by Israel Program for Scientific Translation].

Sandberger, G. 1851. Beobachtungen über mehrere schwierige Puncte der Organisation der Goniatiten. *Jahrbücher des Vereins für Naturkunde im Herzogthum Nassau* **7**: 292–304.

Schimansky, V. N. 1954. Pryamye nautiloidei i baktritoidei sakmarskogo artinskogo yarusov Yuzhnogo Urala (Straight nautiloids and bactritoids from the Sakmarian and Artinskian stages of the southern Urals). *Trudy Paleontologicheskii Institut Akademiia Nauk SSSR* **44**: 1–156.

Schindewolf, O. H. 1933. Vergleichende Morphologie und Phylogenie der Anfangskammern tetrabrachiater Cephalopoden – eine Studie über Herkunft, Stammesentwicklung und System der niederen Ammonoideen. *Abhandlungen preussischen Geologischen Landesanstalt, Berlin. Neue folge* **148**: 1–115.

Sprey, A. 2002. Tuberculate micro-ornament on the juvenile shell of Middle Jurassic ammonoids. *Lethaia* **34**: 31–35.

Tanabe, K. 1989. Endocochliate embryo model in the Mesozoic Ammonitida. *Historical Biology* **2**: 183–196.

Tanabe, K., N. H. Landman, and R. H. Mapes. 1994. Early shell features of some Late Paleozoic ammonoids and their systematic implications. *Transactions and Proceedings of the Palaeontological Society of Japan, New Series* **173**: 384–400.

Tanabe, K., J. Tsukahara, Y. Fukuda, and Y. Fukada. 1991. Histology of a living *Nautilus* embryo: preliminary observations. *Journal of Cephalopod Biology* **2**(1): 13–22.

Vermeij, G. J. 1993. *A Natural History of Sea Shells*. Princeton, New Jersey: Princeton University Press.

Wissner, U. F. G., and A. W. Norris. 1991. Middle Devonian goniatites from the Dunedin and Besa River Formations of Northeastern British Columbia. *Contributions to Paleontology, Geological Survey of Canada Bulletin* **412**: 45–79.

Chapter 3
Conch Form Analysis, Variability, Morphological Disparity, and Mode of Life of the Frasnian (Late Devonian) Ammonoid *Manticoceras* from Coumiac (Montagne Noire, France)

Dieter Korn[1] and Christian Klug[2]

[1]Museum für Naturkunde der Humboldt-Universität zu Berlin, Invalidenstraβe 43, D-10115 Berlin, Germany, dieter.korn@museum.hu-berlin.de;

[2]Paläontologisches Institut und Museum, Universität Zürich, Karl Schmid-Strasse 4, CH-8006 Zürich, Switzerland, chklug@pim.uzh.ch

Keywords: ammonoids, *Manticoceras*, morphological disparity, mode of life, Devonian

1 Introduction

Intraspecific variation of conch shape has been documented for many Mesozoic ammonoids (e.g., Aguirre-Urreta, 1998; Bhaumik et al., 1993; Dagys and Weitschat, 1993a, b; Mitta, 1990; Tanabe, 1993), while Paleozoic ammonoids are commonly regarded as expressing much less variability. Traditionally, species concepts which

N. H. Landman et al. (eds.), *Cephalopods Present and Past: New Insights and Fresh Perspectives*, 57–85.

were applied on Devonian ammonoids were strictly typological and based on the holotype and few paratypes. The investigation of larger populations was only rarely achieved (cf. Dzik, 1985 for Mesozoic ammonoids). However, it is urgently required to apply new methods such as population studies for a better understanding of phylogenetic relationships, intraspecific variability, and spatial biodiversity.

During the last 20 years, the Montagne Noire emerged as one of the most important regions in which Devonian sedimentary rocks are exposed. Detailed studies of the fossiliferous sections of pelagic sediments documented the completeness of the successions. This resulted in the designation of three global stratotypes, for the Givetian-Frasnian boundary at Col du Puech de la Suque (Klapper et al., 1987), for the Frasnian-Famennian boundary at Coumiac (Klapper et al., 1993), and for the Devonian-Carboniferous boundary at La Serre (Paproth et al., 1991).

The composition of ammonoid assemblages from immediately below and above the Frasnian-Famennian boundary in the abandoned quarries at Coumiac (Fig. 3.1) was already documented by House et al. (1985) and Becker and House (1993). They showed that ammonoid diversity decreased dramatically at the global Kellwasser event at the end of the Frasnian stage (Fig. 3.2).

The entire suborder Gephuroceratina, whose species dominated the Frasnian assemblages, disappeared at the top of the upper Kellwasser Horizon. After an earliest Famennian interval with an extremely sparse ammonoid fauna, the Tornoceratina experienced a rapid diversification.

Ammonoids from lower in the section, i.e., from intensely red-colored nodular limestones of Frasnian age, which have been commercially exploited in the

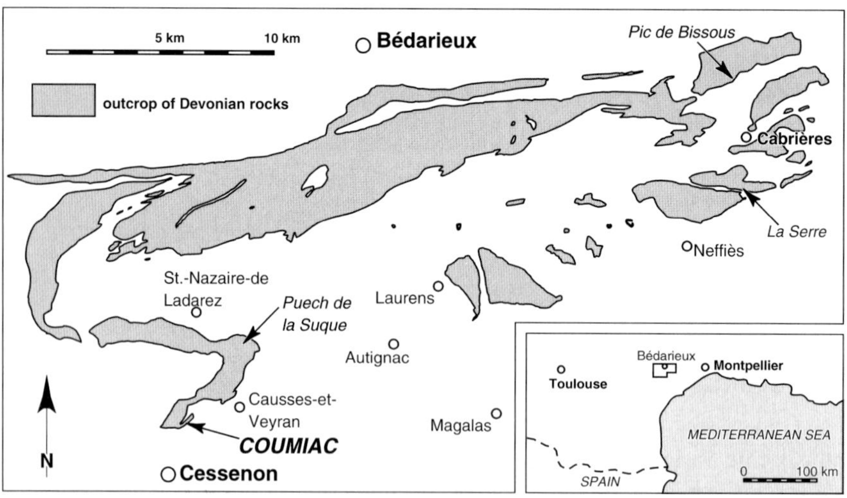

Fig. 3.1 *Map of the Montagne Noire (southern France) with outcrops of Devonian rocks and the ammonoid-bearing locality at Coumiac.*

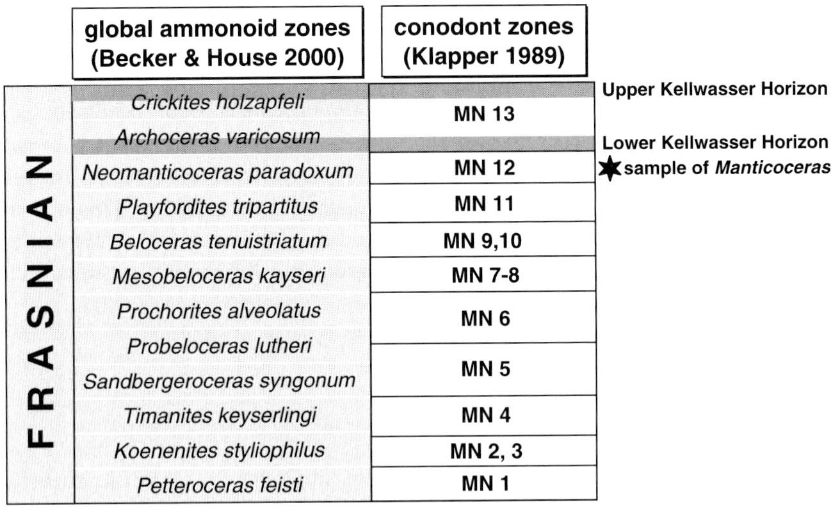

Fig. 3.2 *Ammonoid stratigraphy and regional conodont stratigraphy of the Frasnian (Late Devonian). The asterisk marks the age of the* Manticoceras*-bearing horizon in Coumiac.*

Coumiac quarry, are known since the end of the 19th century. Böhm (1935) figured a few specimens of *Manticoceras* and *Beloceras*, all of which are stored under the catalogue number UMDK in the Institut des Sciences de l'Evolution Montpellier (ISEM). In addition to these specimens, a suite of more than 100 specimens, assembled by the private collector J. Albeille, is available for study. Most likely, these specimens came from a single horizon. This collection is of considerably high value due to the current inability to extract such large numbers of specimens from the abandoned quarry.

The specimens belonging to *Manticoceras* display a remarkable variability in the unfolding of conch parameters in their ontogenetic development, and hence assignment to distinct species is extremely difficult or even impossible. This study is thus mainly focused on the range of morphological variability, rather than on taxonomic descriptions.

Intraspecific coiling variability among manticoceratid ammonoids has been known for a long time. It was explicitly noticed by Clarke (1899), but subsequent authors paid relatively little attention to this phenomenon when they described assemblages from various regions in a kind of parataxonomy. Perhaps this is one reason for the immense number of species which were attributed to *Manticoceras*. Until now, at least 85 species names are available for ammonoids that belong to *Manticoceras* and closely related genera, all of which are assembled in the family Gephuroceratidae (for an overview, see Korn and Klug, 2002). These were described in numerous monographs, of which those by Bogoslovsky (1958, 1969), Clarke (1899), Clausen (1969), Glenister (1958), Miller (1938), Petter (1959), and Wedekind (1913, 1918) provide an overview of the diversity of the Gephuroceratidae.

If one reduces the morphological range of the genus *Manticoceras* s. str., i.e., by separation of the oxyconic forms as *Carinoceras*, the more globose as *Sphaeromanticoceras*, etc. (as done by Becker and House, 1993), there still remain more than 40 species names. Considering that the ornament of all these species is very similar and that all have similar suture lines, species discrimination is mainly possible by using conch parameters. As we will show, these parameters are very plastic and uncover major problems in manticoceratid systematics. This can not only be demonstrated for the material from Coumiac, but also for supplementary samples of manticoceratids from the Rhenish Mountains, the Eifel Mountains, etc.

2 Material

From the Frasnian-Famennian stratotype section at Coumiac, 1.5 km northeast of Cessenon, Montagne Noire (Fig. 3.1), approximately 100 specimens of the genera *Manticoceras*, *Neomanticoceras*, and *Beloceras* are available for study. Apart from some specimens collected by Böhm, the majority were assembled by Albeille and donated to the ISEM (material listed below with the prefix UMDK). All were extracted from an intense red micritic limestone that is rich in buchiolid bivalves. Conodont samples were taken from carbonate material of large body chambers. This material yielded conodonts of the conodont Zone MN12 of Klapper (1989) which points to a position immediately below the lower Kellwasser Horizon (determinations by R. Feist, Montpellier, and Z. Belka, Poznan): *Ancyrodella curvata*, *Ancyrodella nodosa*, *Ancyrognathus triangularis*, *Icriodus symmetricus*, *Icriodus alternatus alternatus*, *Palmatolepis winchelli*, *Polygnathus alatus*, *Polygnathus* sp., and *Polygnathus webbi*.

The investigated manticoceratid specimens (Fig. 3.3) range, in their conch diameters, from 13 to 170 mm. The largest specimen is still fully septate and indicates that the maximum diameter of the conch can be estimated to be approximately 300 mm. Almost 40 of the 90 manticoceratid specimens were sectioned with the following size ranges represented:

- up to 30 mm –10 specimens
- 30 to 60 mm – 17 specimens
- 60 to 90 mm – 9 specimens
- more than 90 mm – 3 specimens

In many specimens, the inner whorls are well preserved and allowed the production of high-precision cross sections, meeting the largest diameter of the protoconch. From these cross sections, acetate peels were taken and enlarged by photographic and scanning magnification (×10). On the basis of these templates, precise computer images were drawn, of which eleven are figured in Fig. 3.4.

A supplementary sample consisting of 20 specimens from the eastern margin of the Rhenish Mountains was studied. The material came from localities such as

Fig. 3.3 Manticoceras *sp. from Coumiac (Montagne Noire), specimen UMDK 226 (coll. Albeille), x 1.5.*

Oberscheld and is stored in the Museum für Naturkunde, Berlin (MB.C. numbers). The specimens were already prepared by Jessen (a doctorate student of O. H. Schindewolf) who started a research program on the ammonoids from Büdesheim (Eifel) but never finished.

Additionally, the cross sections illustrated by Bogoslovsky (1969) and Clausen (1969) were redrawn and incorporated in the investigation. Thus, more than 70 cross sections were analyzed biometrically.

3 Conch Parameters

From the computer drawings of the cross sections, three of the basic conch parameters can immediately be obtained for each half volution (Fig. 3.5):

- conch diameter (dm)
- whorl width (ww)
- whorl height (wh)

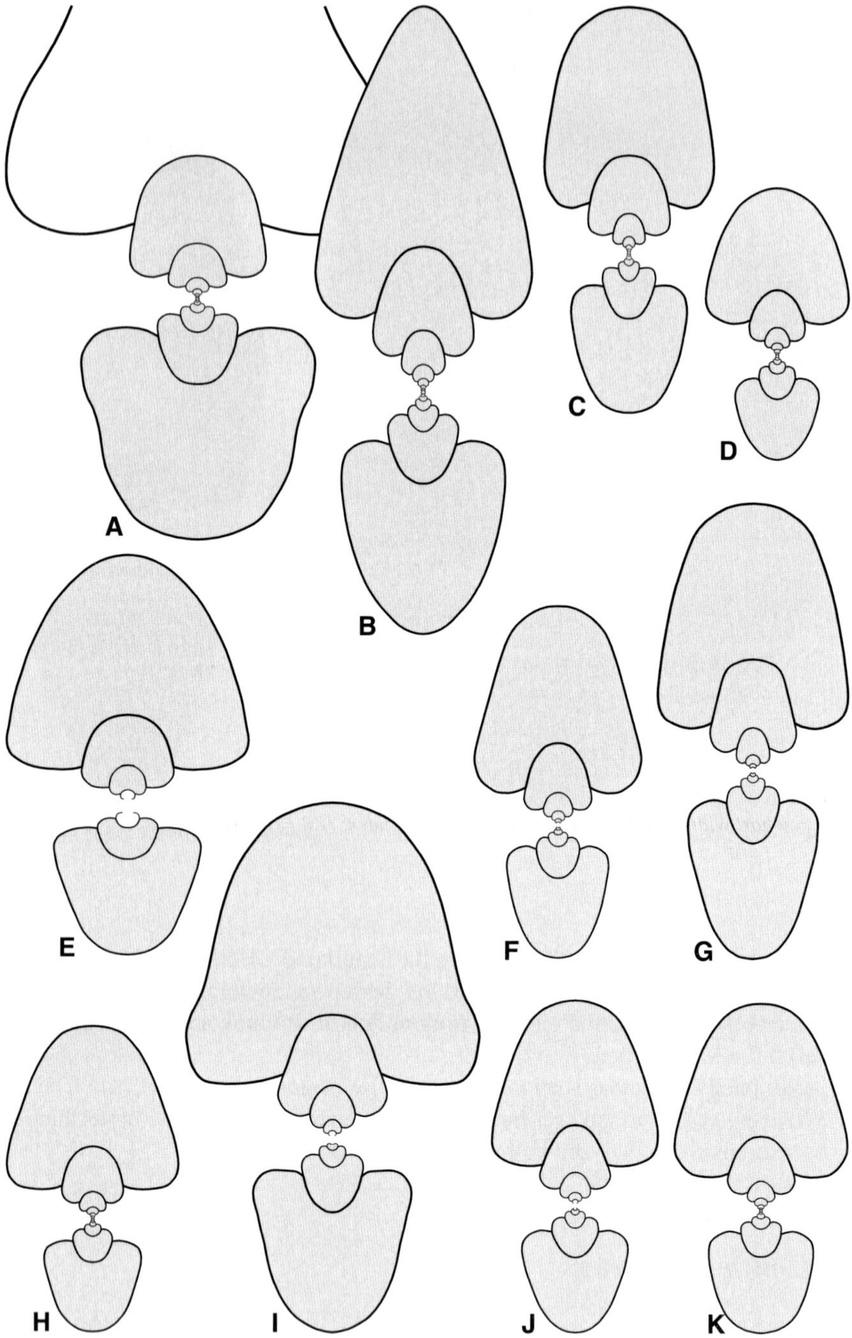

Fig. 3.4 *Selection of eleven cross sections of* Manticoceras *specimens from Coumiac (Montagne Noire); all x 0.75. A – specimen UMDK 218; B – specimen UMDK 219; C – specimen UMDK 275; D – specimen UMDK 202; E – specimen UMDK 236; F – specimen UMDK 232; G – specimen UMDK 281; H – specimen UMDK 276; I – specimen UMDK 220; J – specimen UMDK 235; K – specimen UMDK 282.*

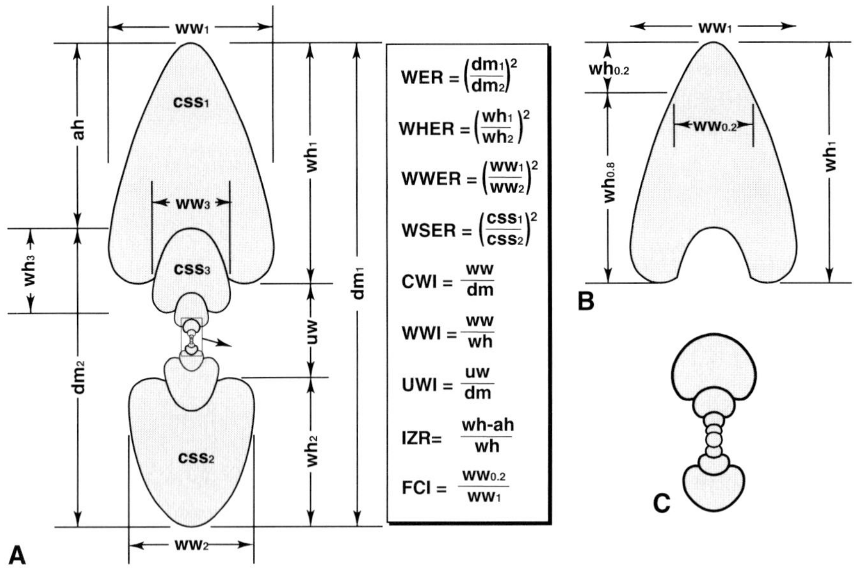

Fig. 3.5 *The conch parameters and ratios used in the text. A – conch parameters in a cross section of a* Manticoceras *specimen; B – parameters for calculating the flank convergence index (FCI); C – enlarged inner whorls showing the large protoconch (x 3) (dm = conch diameter; ww = whorl width; wh = whorl height; uw = umbilical width; ah = apertural height; css = cross-section surface; WER = whorl expansion rate; WHER = whorl height expansion rate; WWER = whorl width expansion rate; WSER = whorl surface expansion rate; CWI = conch width index; WWI = whorl width index; UWI = umbilical width index; IZR = imprint zone rate; FCI = flank convergence index).*

Using these basic parameters, secondary parameters can easily be computed:

- umbilical width (uw) = $dm_1 – wh_1 – wh_2$
- apertural height (ah) = $dm_1 – dm_2$
- imprint zone width (iz) = wh_1–ah or $wh_1 – (dm_1 – dm_2)$

Expansion rates (growth rates) and conch proportions were calculated in the following way by using the three basic conch parameters:

- whorl expansion rate (WER) = $(dm_1/dm_2)^2$ or $[dm_1/(dm_1 – ah)]^2$
- whorl height expansion rate ($WHER_1$) = $(wh_1/wh_2)^2$; $WHER_2 = wh_1/wh_3$
- whorl width expansion rate ($WWER_1$) = $(ww_1/ww_2)^2$; $WWER_2 = ww_1/ww_3$
- whorl surface expansion rate (WSER) = css_1/css_2
- umbilical width index (UWI) = uw/dm_1 or $(dm_1 – wh_1 – wh_2)/dm_1$
- imprint zone rate (IZR) = $wh_1 – ah/wh_1$ or $(wh_1 – (dm_1 – dm_2))/wh_1$
- conch width index (CWI) = ww_1/dm_1
- whorl width index (WWI) = ww_1/wh_1

To describe the shapes of the whorl cross sections, a further index was calculated:

- Flank convergence index (FCI) = The ratio of the maximal whorl width and the width at 0.80 of the whorl height, measured from the umbilicus. An FCI of 1 would mean nonconverging (i.e., parallel) flanks.

4 Conch of *Manticoceras*

Conchs of *Manticoceras* possess a number of more or less characteristic features (Fig. 3.5):

(1) The protoconch has a diameter of approximately 0.6 to 0.7 mm.
(2) The protoconch is often wider than the first volution.
(3) The inner two or three whorls (i.e., up to 5 mm diameter) are widely umbilicate (UWI = 0.30 to 0.45) with almost circular whorl cross sections. Above 5 mm, the UWI decreases constantly, and at 80 mm diameter, the umbilicus has a width of 0.12 to 0.20 of the conch diameter.
(4) The rate in which the whorls embrace the preceding whorls (IZR) increases slowly during ontogeny from 0.10 to 0.30. This trend is closely connected with the relative closure of the umbilicus.
(5) Adult whorls display three characters which show a wide range of plasticity and will be discussed in the following pages.
   – a more or less trapezoidal whorl cross section
   – moderately fast to fast expanding whorls (WER = 2.25–2.75)
   – a remarkably eccentric coiling

4.1 Whorl Expansion

The whorl expansion rate is the measure for the coiling of the ammonoid conch and describes the outline of the whorl spiral. Among the characters of the manticoceratid conch, it is one of the most plastic, which reflects ontogeny (i.e., eccentricity of the whorl coil) as well as variability within the population (Fig. 3.6A, B).

Typical for manticoceratids is the relatively low WER in juveniles. After a slight initial juvenile decrease, the WER reaches approximately 2.0 at 3–5 mm conch diameter. In this interval, the variability within the material is relatively low.

In the third or fourth whorl, the specimens from Coumiac show a rapid acceleration of whorl expansion, which usually reaches a maximum of 2.5–2.75 in the fourth or fifth whorl (20–50 mm conch diameter). Thereafter, the WER stagnates or decreases in the adult whorls. In this respect, all manticoceratids reflect the morphological development of their agoniatitid ancestors (see Klug, 2001; Korn, 2001; Korn and Klug, 2001).

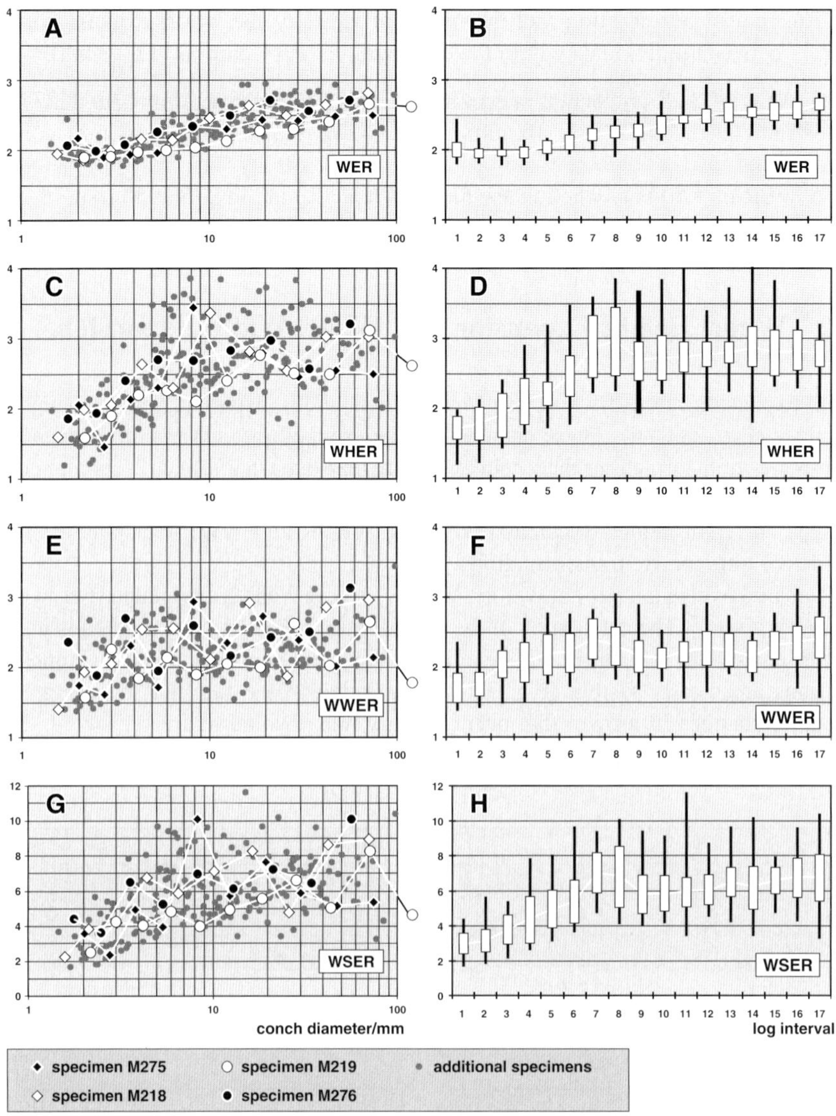

Fig. 3.6 *Ontogenetic trajectories (A, C, E, G) and intraspecific variability (B, D, F, H) in* Manticoceras *from Coumiac. The bivariate diagrams (A, C, E, G) show the ontogenetic development of the whorl expansion rate (WER = $(dm_1/dm_2)^2$), whorl height expansion rate (WHER = $(wh_1/wh_2)^2$), whorl width expansion rate (WWER = $(ww_1/ww_2)^2$), and whorl surface expansion rate (WSER = css_1/css_2); highlighted are the specimens UMDK 218, 219, 275, and 276. The box-and-whiskers diagrams (B, D, F, H) show the ontogenetic variations of intraspecific variability of the major conch ratios. Bold lines refer to the total range, boxes refer to the extension of the middle two quartiles, and white lines refer to the median value. Growth stages between 1 mm and 100 mm conch diameter were subdivided into 17 logarithmic intervals.*

The rate of acceleration and the maximal value of the whorl expansion rate differs between individuals within the community from Coumiac (Fig. 3.6A):

- A WER of 2.25 is reached in specimen UMDK276 at 6 mm, in UMDK275 at 8 mm, in UMDK218 at 10 mm, and in UMDK219 only at 16 mm conch diameter.
- A WER of 2.50 is reached in specimen 276 at 13 mm, in 275 at 50 mm, in 218 at 12 mm, and in 219 only at 50–60 mm conch diameter.

4.2 Whorl Height Expansion and Whorl Width Expansion

Both expansion rates of whorl height and whorl width display drastic ontogenetic changes (Fig. 3.6C–F). These changes can be grouped into (1) major ontogenetic trends, i.e., transformations that occur over more than two whorls, and (2) spontaneous changes, i.e., changes that occur only sporadically and are not part of continuous trends. Different modes of calculation of the WHER and WWER display the two modes of ontogenetic transformation:

Squared expansion per half whorl ($WHER_1$ and $WWER_1$), calculated $(wh_1/wh_2)^2$ and $(ww_1/ww_2)^2$. The calculation of the WHER and WWER by this method is much more sensitive to short-term ontogenetic changes, and hence spontaneous transformations are clearly visible (Fig. 3.7). This is due to the fact that by using this method, changes in growth that occur in only half a volution are registered. The

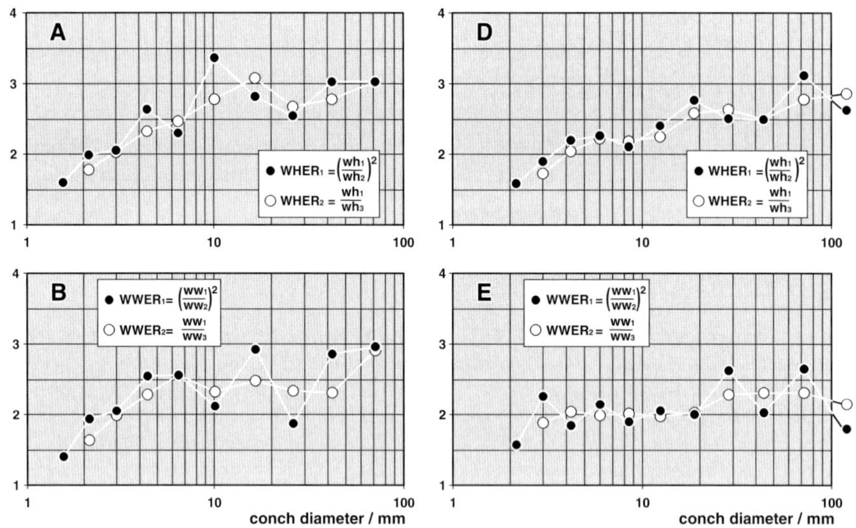

Fig. 3.7 *Ontogenetic trajectories of the expansion rates of whorl height (WHER) and whorl width (WWER) in four representative specimens of* Manticoceras *from Coumiac. A – specimen UMDK 218; B – specimen UMDK 219; C – specimen UMDK 275; D – specimen UMDK 276.*

oscillating curves mean that, after a sudden acceleration of the coiling rate, stagnation is expressed by a drawback.

Truly calculated expansion per whorl ($WHER_2$ and $WWER_2$), calculated wh_1/wh_3 and ww_1/ww_3. If the WHER and WWER are calculated directly per whorl by the quotient of the whorl height and whorl width of one whorl and the preceding whorl 360° earlier, a smoothed curve is visible. Spontaneous excursions of the growth trajectories are ruled out here, but longer-termed changes are more clearly expressed.

Two specimens (218 and 219) are chosen for illustration to show different degrees in which the expansion rates computed by the two methods are expressed (Fig. 3.7). Specimen 218 shows, both in the WHER and WWER, much more pronounced sudden changes, which are largely uncoupled.

The total plot of all calculated values of the $WHER_1$ displays a remarkable skewing (Fig. 3.6C). A general ontogenetic trend toward an increase of the $WHER_2$ is visible up to a conch diameter of approximately 10 mm, and thereafter, stagnation can be recognized. The variability within this character is highest in the growth interval with the strongest increase, and lower in later stages.

In the $WWER_1$ plot, a less pronounced ontogenetic trend is observable, and variability is also less conspicuous (Fig. 3.6E, F). An increase up to 10 mm conch diameter is also visible, followed by stagnation in later ontogeny.

It is interesting that the adult development of the WHER and WWER cannot be predicted when the early ontogeny is observed. Sudden changes particularly in the WWER occur especially in late ontogenetic stages of many specimens. Consequently, it is very difficult or even impossible to clearly distinguish between putative species within the population. The plot of all the sectioned specimens (Fig. 3.6C, E) shows that from specimens smaller than 40 mm, the adult morphology cannot be predicted. Thus, these parameters cannot be used to distinguish between species of *Manticoceras* based on small specimens.

4.3 Whorl Surface Expansion

The surface expansion rate is directly dependent on the WHER and WWER, and consequently, the trends observed in those characters can also be seen in the WSER plot (Fig. 3.6G). A prominent skewing is visible, but the box-and-whiskers diagram (Fig. 3.6H) illustrates an ontogenetic increase up to 10 mm conch diameter, followed by a reduction and a slight late ontogenetic increase.

The growth trajectories of the four selected specimens display striking differences. The curves of all four specimens perform, at different sizes, remarkable spontaneous fluctuations, visible in specimen 275 at approximately 8 mm conch diameter, in 218 at 40 mm, in 276 at 60 mm, and in specimen 219 at 70 mm.

The expansion rate of the whorl cross-section surface shows drastic ontogenetic changes in *Manticoceras*. In juvenile and preadult stages, only a slow increase in this rate occurs. Usually beginning with a value of 3.0 at 2–3 mm conch diameter, there is a more or less rapid increase, and a preadult maximum of WSER 6–8 is reached between 5 and 30 mm conch diameter (Fig. 3.6G, H). This maximum is followed by stagnation (specimens 219, 276) or a decrease (218, 275).

It is mainly the change of the whorl height that is responsible for the changes in the WSER. In the four specimens, the curve of the WHER largely parallels the WSER curve. WWER and WHER are seldom positively correlated. The WHER typically shows a more rapid preadult acceleration.

4.4 Whorl Cross-Section Shape and Flank Convergence Index

After a juvenile stage with an almost circular whorl cross section, an ontogenetic development toward a subtrapezoidal shape takes place during the fourth and fifth whorls. The differences between the specimens and their variability can best be illustrated by the four specimens UMDK 218, 219, 275, and 276 (Figs. 3.6G, 3.8).

Specimens 218 and 275 have similar whorl cross sections between 40 and 50 mm conch diameter. In later whorls, however, specimen 218 enlarged the whorl width remarkably, whereas specimen 275 conserved the preadult outline. Specimen 218 developed a broad umbilical rim while 219 became lenticular with a suboxyconic venter in the adult stage, and the cross section of 276 grew subtriangular.

The calculation of the convergence of the flanks led to the surprising conclusion that this character is rather uninformative for the analysis of the sample. An inconspicuous ontogenetic change occurs in this character, with a decrease from approximately 0.80 in the initial stage to an average value of 0.70 at 7 mm conch diameter (Fig. 3.9C, D).

Later in ontogeny, this mean value is 0.65 with only minor variability. The only exception is the last whorl of specimen 219, which becomes strikingly triangular and thus has a data point outside of the main range.

Traditionally, separation of *Manticoceras* species was performed by using the outline of the whorl cross section. *M. intumescens*, for instance, is defined by its subparallel flanks (as in UMDK275), and *M. cordatum* possesses converging flanks and a narrower venter (as in specimen 219 up to 50 mm conch diameter). In the material from Coumiac, the traditional subdivision can hardly be applied; all the morphologies are connected by intermediate forms. As shown in Figs. 3.4 and 3.6–9, no clearly definable boundary occurs even between the most striking cross-section shapes. These cross sections shown in Fig. 4 demonstrate that even conchs of adult individuals within the investigated population do not allow clear distinction of morphologies, and adult conchs with similar geometries may possess inner whorls that differ remarkably in their morphology.

The results uncover the major difficulties in the systematics of the genus *Manticoceras*. The question to be asked is, how many of the 40 erected "species" are in fact justified? According to the study of the material from Coumiac it is obvious that a taxonomy which is based on the principal conch parameters (relative thickness of the whorls, shape of the whorl cross section, umbilical width) does not lead to reasonable results.

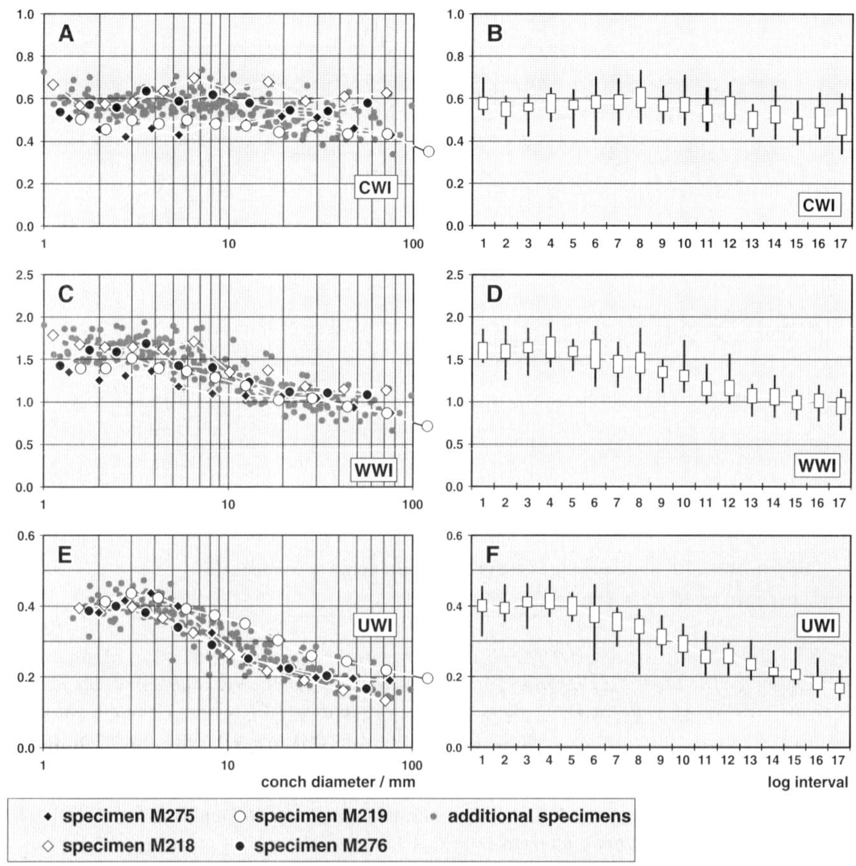

Fig. 3.8 *Ontogenetic trajectories (A, C, E) and intraspecific variability (B, D, F) in* Manticoceras *from Coumiac. The bivariate diagrams (A, C, E) show the ontogenetic development of the conch width index (CWI = ww/dm), whorl width index (WWI = ww/wh), and umbilical width index (UWI = uw/dm); highlighted are the specimens UMDK 218, 219, 275, and 276. The box-and-whiskers diagrams (B, D, F) show the ontogenetic variations of intraspecific variability of the major conch ratios. Bold lines refer to the total range, boxes refer to the extension of the middle two quartiles, and white lines refer to the median value. Growth stages between 1 mm and 100 mm conch diameter were subdivided into 17 logarithmic intervals.*

5 Comparisons with Other Samples of *Manticoceras*

5.1 Oberscheld

The random sample comprising 18 specimens from the eastern margin of the Rhenish Mountains displays an even wider range of conch morphologies than the sample from Coumiac. Represented are discoidal to pachyconic specimens with parallel or converging flanks (Fig. 3.10).

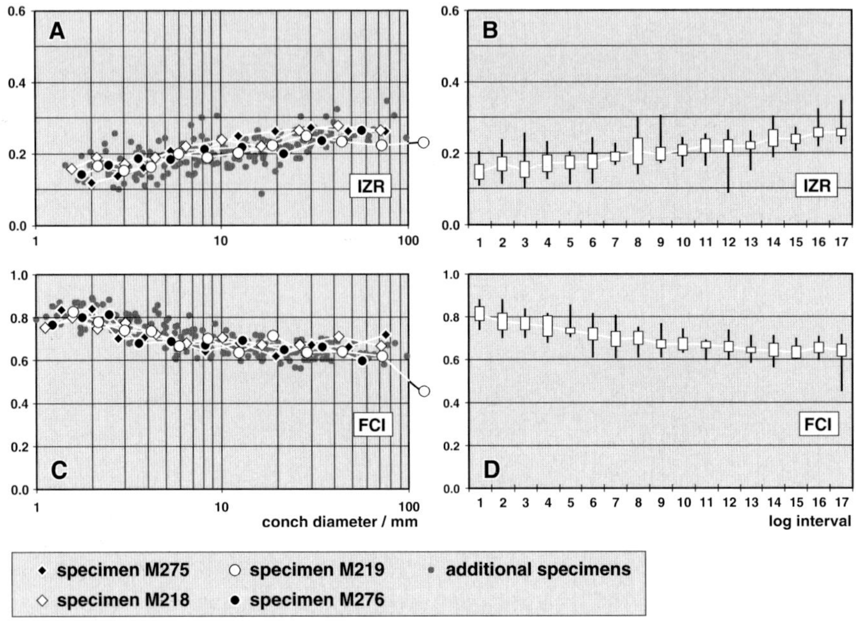

Fig. 3.9 *Ontogenetic trajectories (A, C) and intraspecific variability (B, D) in* Manticoceras *from Coumiac. The bivariate diagrams (A, C) show the ontogenetic development of the imprint zone rate (IZR = wh_1–ah/wh_1) and flank convergence index (FCI = $ww_{0.2}$/ww_1); highlighted are the specimens UMDK 218, 219, 275, and 276. The box-and-whiskers diagrams (B, D) show the ontogenetic variations of intraspecific variability of the major conch ratios. Bold lines refer to the total range, boxes refer to the extension of the middle two quartiles, and white lines refer to the median value. Growth stages between 1 mm and 100 mm conch diameter were subdivided into 17 logarithmic intervals.*

In contrast to the Coumiac sample, shapes with flattened and almost parallel-sided flanks are common. The material is stored in the Museum für Naturkunde, Berlin, under the catalogue numbers MB.C.3789 – MB.C.3806.

Of the four cross sections shown in Fig. 3.10, specimen MB.C.3791 can be attributed to *Timanoceras*, a genus that contains manticoceratids with crescent-shaped whorl sections. It differs from the others by its high conch width index (Fig. 3.10C). Specimens MB.C.3789 and MB.C.3790 cannot be separated in terms of morphometrics, and specimen MB.C.3792 differs by its flat conch form from the other three. However, Fig. 3.10C demonstrates that these three specimens do not represent separate ontogenetic traits; they are connected by intermediate specimens. Differences in other characters are not obvious. Accordingly, separation of distinct species within *Manticoceras* may be impossible on the basis of morphometrics. Because the suture lines as well as the ornament do not differ much, it appears unreasonable to retain all of the existing 40 species of *Manticoceras*.

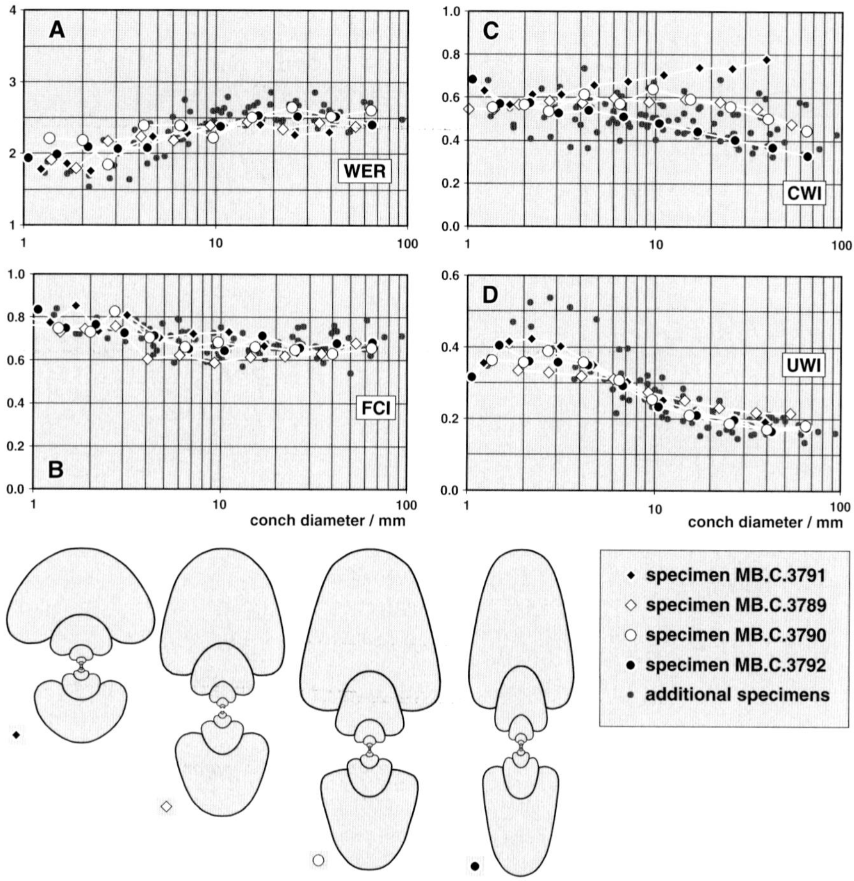

Fig. 3.10 *Bivariate diagrams showing the ontogenetic development of the whorl expansion rate (WER; in A), flank convergence index (FCI; in B), conch width index (CWI = ww/dm; in C), and umbilical width index (UWI = uw/dm; in D) of sectioned specimens of* Manticoceras *from the eastern Rhenish Mountains. Highlighted are the four specimens MB.C.3789–MB.C.3792; cross sections x 0.75.*

5.2 Büdesheim

Of the cross sections of manticoceratids from Büdesheim (Eifel) that were published by Clausen (1969), four characteristic individuals are highlighted in Fig. 3.11. The minute pyritized material comes from dark shales, and the question arose whether these specimens, which are less than 20 mm in diameter, portray similar ontogenetic trajectories to the normal-sized material from other localities.

The manticoceratids from Büdesheim have conchs with ontogenetically accelerated coiling rate (WER; Fig. 3.11A). The figures also show an ontogenetic reduction

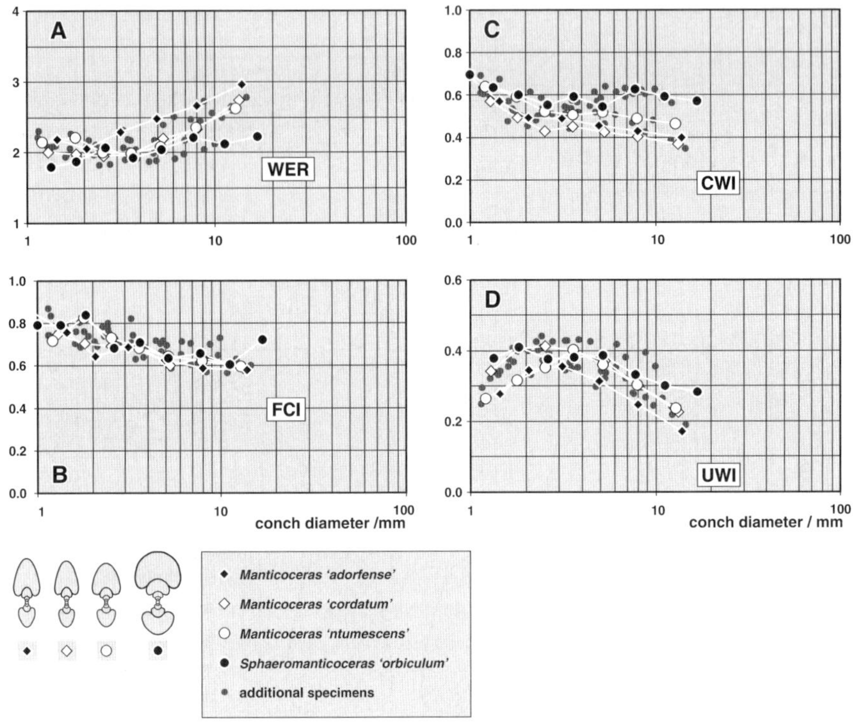

Fig. 3.11 *Bivariate diagrams showing the ontogenetic development of the whorl expansion rate (WER; in A), flank convergence index (FCI; in B), conch width index (CWI = ww/dm; in C), and umbilical width index (UWI = uw/dm; in D) of sectioned specimens of* Manticoceras *from Büdesheim (Eifel). Cross sections after Clausen (1969), x 0.75.*

of the conch width index (CWI; Fig. 3.11C) and a narrowing of the umbilicus (UWI; Fig. 3.11D).

Particularly Fig. 3.11C shows that the species of *Sphaeromanticoceras* can clearly be separated by their stouter conch, whereas the other three specimens are hardly separable. In the flank conversion index (Fig. 3.11B), they do not differ at all. In the whorl expansion rate (Fig. 3.11A) and the umbilical width index (Fig. 3.11D), only minor differences between the three can be recognized. Furthermore, intermediate forms are present and connect the morphologies of the three specimens.

In the three *Manticoceras* species, the whorl expansion rate experiences a remarkable juvenile increase (Fig. 3.11A). In the specimen of *Manticoceras adorfense*, a WER of 2.50 is reached already at 6 mm conch diameter, and in the specimens of *Manticoceras cordatum* and *Manticoceras intumescens*, the value exceeds 2.50 at 10 mm diameter. This means that the acceleration of the coiling rate is more pronounced in the Büdesheim material, which suggests an earlier maturity in these forms. Remarkably the material from Büdesheim shows the same ontogenetic allometry as the large limestone specimens. It can therefore be concluded that the minute forms represent adult populations.

5.3 Russia

Large specimens of manticoceratids from the Timan, the South Urals, and the Altay Mountains were sectioned by Bogoslovsky (1969). These few specimens display the same principal conch geometry and drastic ontogenetic changes (Fig. 3.12) as the material from Coumiac.

The analysis of five of the cross sections illustrated by Bogoslovsky (1969) supplements the results derived from the material from Coumiac and Oberscheld.

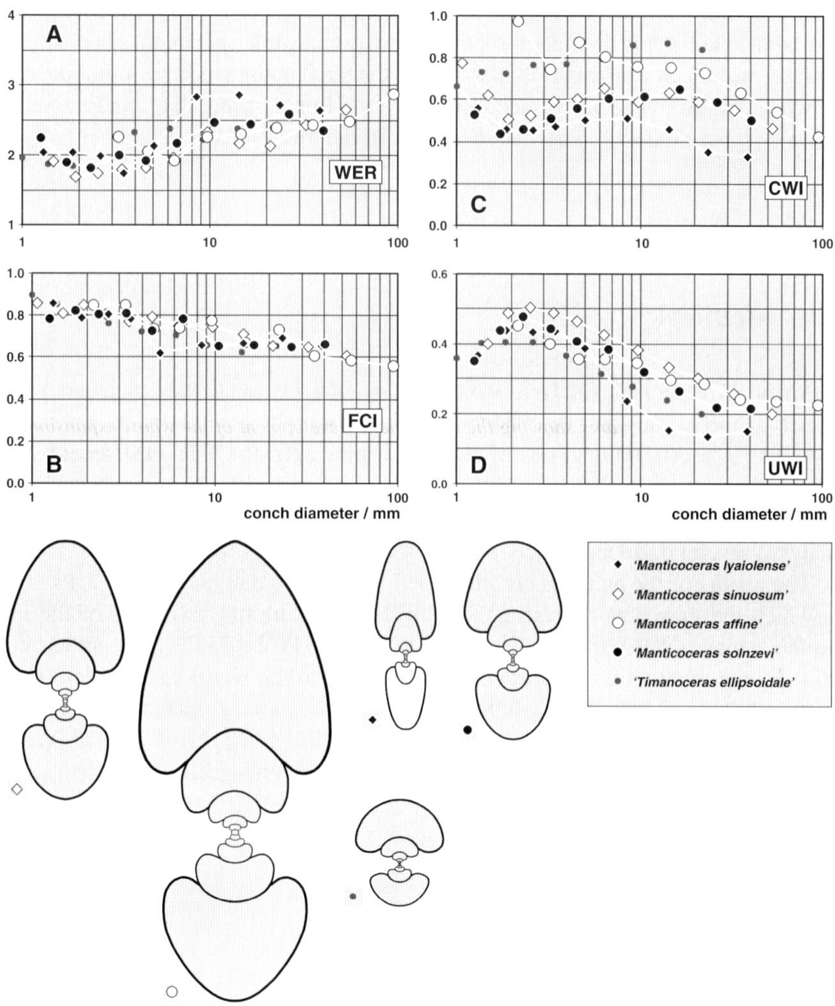

Fig. 3.12. *Bivariate diagrams showing the ontogenetic development of the whorl expansion rate (WER; in A), flank convergence index (FCI; in B), conch width index (CWI = ww/dm; in C), and umbilical width index (UWI = uw/dm; in D) of sectioned specimens of manticoceratid ammonoids from the Timan, the South Urals, and the Altay Mountains. Cross sections after Bogoslovsky (1969), x 0.75.*

Similar conch morphologies are represented, and the specimens of *Timanoceras ellipsoidale* and *Sphaeromanticoceras affine* are clearly distinguished by their stout conchs. It is practically impossible to separate the two species *Manticoceras sinuosum* and *M. solnzevi* in terms of morphometrics.

5.4 Morocco

Although only a few specimens were cut and measured, the extensive variability could also be found in manticoceratids from Morocco. In the eastern Anti-Atlas of Morocco, *Manticoceras* and its closest relatives locally occur in rock-forming numbers and attain giant diameters of up to 500 mm. Adult specimens display adult characters such as a terminal thickening of the shell and a slight terminal decrease in whorl height. The cross section may be almost triangular, compressed or depressed and all conch parameters vary as in the faunas discussed above.

6 PCA Analysis

6.1 Methods

A Principal Component Analysis was carried out with respect to the conch parameters, namely whorl expansion rate, whorl width index, umbilical width index, and imprint zone rate. Almost 500 data sets of material from Coumiac (237), Oberscheld (125), Büdesheim (80, after Clausen, 1969), and Russian material (48, after Bogoslovsky, 1969) compose the total *Manticoceras* morphospace (Fig. 3.13). On the basis of this analysis, several palaeobiological aspects of *Manticoceras* can be discussed.

The results of the analysis are illustrated in bivariate diagrams (Figs. 3.13, 3.14), in which the grey dots represent the total data set. The first two axes of the PCA explain a total of 83% of the variance (PC1 – 59.8%; PC2 – 23.2%). Variations along PC1 depend mainly on the whorl expansion rate and the whorl width index (Factor loading: WER = 0.875, WWI = 0.967, UWI = 0.737, IZR = –0.387), and PC2 is more associated with the shape of the body chamber (IZR = 0.901, UWI = 0.317, WER = –0.006, WWI = –0.124). For illustration of ontogenetic changes, the total data set was subdivided in six discrete size groups (Fig. 3.13D–I). In Fig. 3.14, distinct specimens of the four regions are highlighted to show the ontogenetic trajectories of chosen individuals.

6.2 Ontogenetic Trajectories

During ontogeny, a remarkable migration of the specimens across the morphospace can be observed, reflecting the transformation of the conch morphology from widely umbilicate to more involute, etc. The position of the centroids for distinct

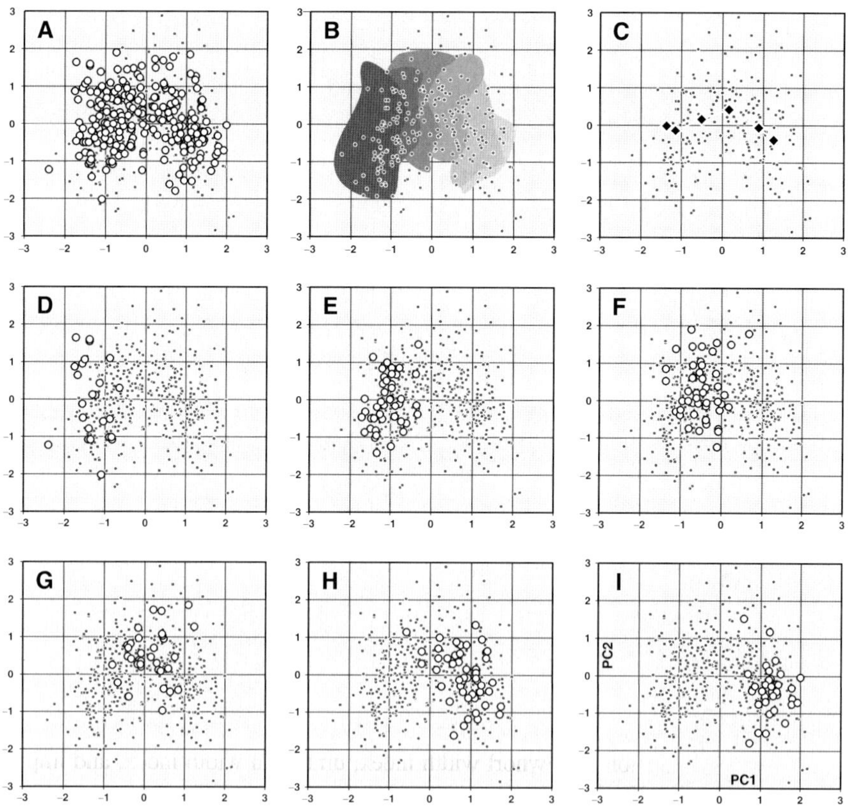

Fig. 3.13 *Principal components analysis of manticoceratid ammonoids, shown are PC1 (x axis) and PC2 (y axis). A – Morphospace of the complete data set; the specimens of* Manticoceras *from Coumiac are highlighted (circles). B – The migration of ontogenetic stages across the morphospace; light grey for juvenile specimens, dark grey for adult specimens. C – The migration of the centroid of the specimens from Coumiac across the morphospace. D–I – Morphospace occupation of different growth stages (relevant growth stages are shown as circles; other growth stages are shown as dots) of* Manticoceras *from Coumiac: D – specimens >40 mm dm, E – specimens from 20–40 mm dm, F – specimens from 10–20 mm dm, G – specimens from 6–10 mm dm, H – specimens from 3–6 mm dm, I – specimens < 3 mm dm.*

growth intervals display a slightly discontinuous morphological change with a more rapid character unfolding in juveniles.

For a closer view, the samples from Coumiac, Oberscheld, and Büdesheim were separately analyzed (Fig. 3.14). This shows that the Coumiac sample displays a rapid early ontogenetic modification of conch morphology, but a stagnation in stages larger than 20 mm conch diameter. The centroids of the samples 20–40 mm and > 40 mm differ only slightly. A similar picture can be seen in the material from Oberscheld.

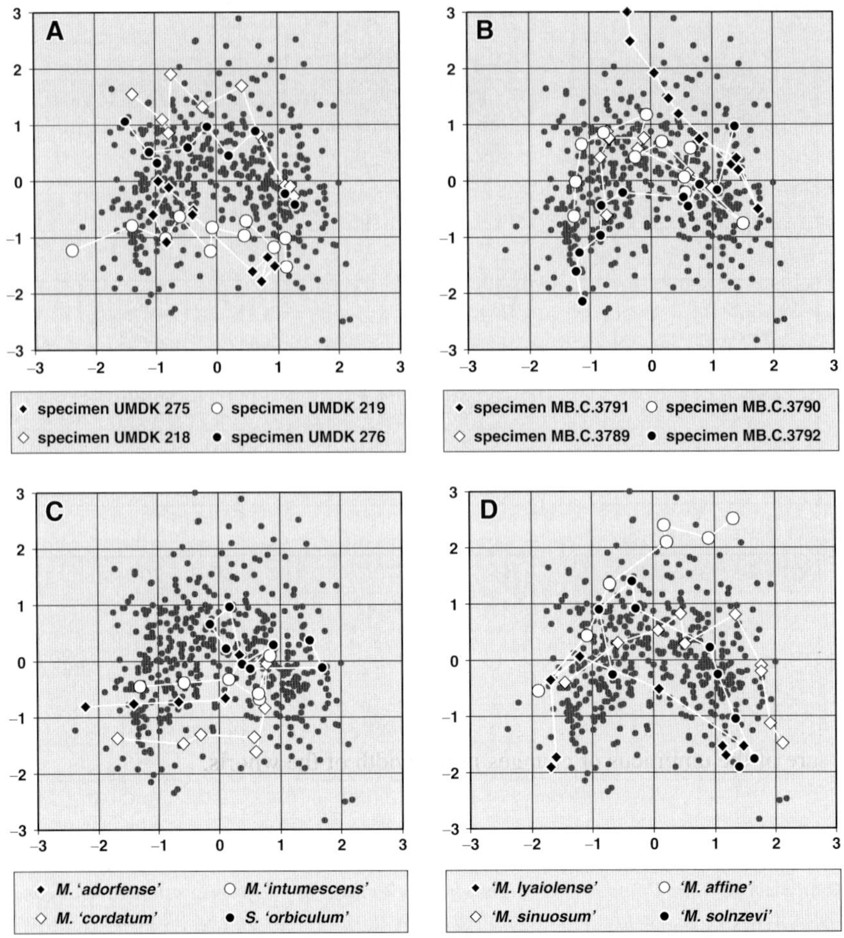

Fig. 3.14 *Principal components analysis of manticoceratid ammonoids and the migration of ontogenetic stages across the morphospace, shown are PC1 (x axis) and PC2 (y axis). Young specimens are on the right and older specimens on the left. A –* Manticoceras *from Coumiac; B – manticoceratids from the eastern Rhenish Mountains; C – manticoceratids from Büdesheim; D – manticoceratids from the Timan, South Urals, and Altay Mountains.*

The Büdesheim material differs markedly from the other two. The earliest juvenile morphospace of the Coumiac and Oberscheld material is not occupied by the specimens from Büdesheim, and the later ontogeny is strongly accelerated. Specimens of the 6–10 mm size range are morphologically similar to those from Coumiac and Oberscheld, but the specimens from Büdesheim (Fig. 3.14C) reach a conch geometry at 10–20 mm conch diameter that is only characteristic for specimens larger than 40 mm in the other two samples (Fig. 3.14A, B, D).

7 Orientation of the Aperture in *Manticoceras*

It is difficult to extract unambiguous information on the organism's mode of life from the ammonoid conch alone. Nevertheless, the conch provides some information, e.g., if the aperture was pointed directly upward or if it was oriented at an oblique angle as in Recent nautilids, if the living animal was a poor swimmer but able to reach the seafloor easily with its arms (as Recent *Nautilus* does) or whether it was a comparatively good swimmer. The orientation of the aperture of *Manticoceras* throughout its life is discussed subsequently.

Raup (1967) and Okamoto (1996) have demonstrated that the body chamber length (BCL) in ammonoids is correlated with the whorl expansion rate (WER). The nonlinear correlation shows that conchs with low WER possess long body chambers, and vice versa. This model is mainly applicable to forms which do not show remarkable ontogenetic changes of coiling. Eccentrically coiled conchs are much more difficult to compute, and only approximations can be produced so far. In order to obtain more precise results, other expansion rates (such as that of the whorl width) should be taken into account, because they also influence the volume of the soft body and consequently its length. The surface expansion rate can be regarded as a summary of all expansion rates and will be discussed below.

Trueman (1941), Raup (1967), and Saunders and Shapiro (1986) showed that particularly the length of the body chamber controls the orientation of the organism with its conch in the water column, i.e., the position of the aperture. Again, it is difficult to apply this method to conchs with ontogenetic modifications, such as the closure of the umbilicus or changes in the width of the whorls.

Manticoceras conchs are remarkable for their eccentric coiling and rather sudden ontogenetic changes of some of the whorl parameters, such as the whorl width. The accretion rate of shell material at the body chamber, which is the main factor in the accumulation of weight that has to be buoyed by the phragmocone, can only be calculated when the growth of several conch features is considered:

WER: The measure for the coiling of the ammonoid conch; it is practically identical with the expansion rate of the aperture.

WHER: The whorl height expansion rate is closely related to the WER. An increase of the WHER can be caused by enlargement of the aperture height (and thus also an increase of the WER), by the closure of the umbilicus (without increase of WER), or can be driven by both.

WWER: In *Manticoceras*, the ontogenetic development of the whorl width is largely uncoupled from the whorl height and the aperture height.

WSER: The whorl surface expansion rate depends on several other measures. A higher WSER can be caused by an increase of the aperture height (WER), the whorl height (WHER), and the whorl width (WWER), but also by reduction of the umbilical width and reduction of the imprint zone rate (IZR). All these parameters are easily calculable, in contrast to the WSER which requires some technical effort. Ontogenetic changes of these conch parameters influence the body chamber length. Additionally, the shape of the whorl cross section (circular, rectangular, or subtriangular) determines the whorl cross-section surface.

As shown diagrammatically in Fig. 3.6, the curves of the WSER, WHER, and WWER of the *Manticoceras* conch run almost parallel, meaning that the factor of the WHER and WWER leads to an extremely close approximation of the WSER. It can also be seen that the WER curve does not significantly deviate from the WSER curve. Despite the nonparallel development of whorl height and whorl width, no dramatic errors in the calculation of the BCL can be expected. Consequently, either WER or even better, WHER together with WWER, can be used as a good approximation for the WSER.

In the following reconstruction of a living ammonoid (Figs. 3.15, 3.16), the orientation of the aperture is based on the WER, which is regarded as the cardinal character. As Klug (2001) showed, the computed body chamber length (and consequently the orientation within the water column) changed markedly in the unusual Early Devonian ammonoid *Rherisites*. A similar assumption can be made for *Manticoceras*, although the extent of allometric growth is less striking than in *Rherisites*. Early

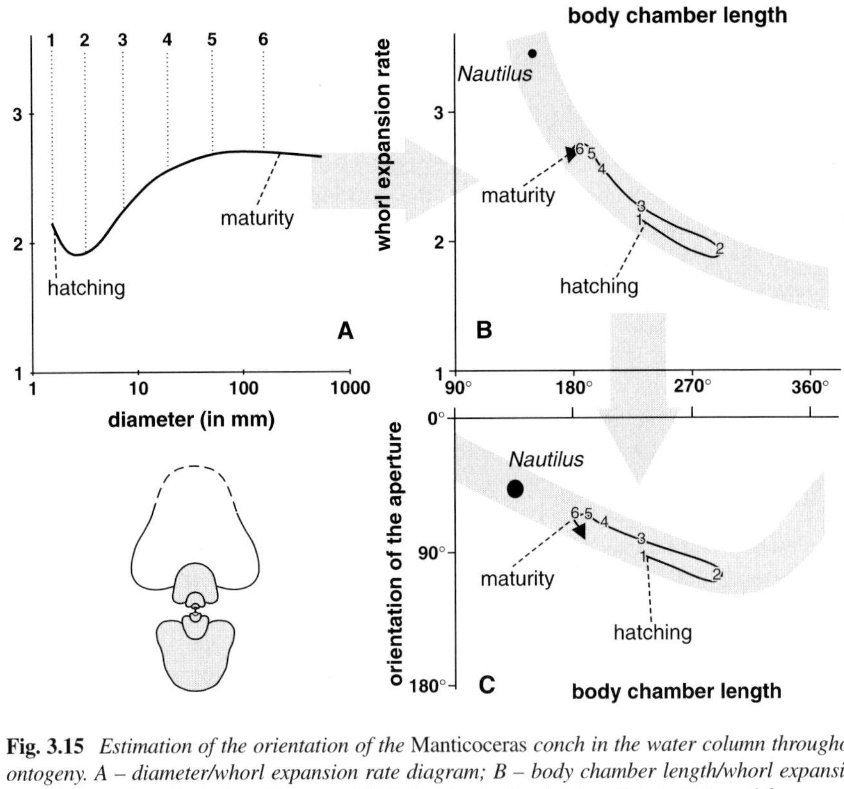

Fig. 3.15 *Estimation of the orientation of the* Manticoceras *conch in the water column throughout ontogeny. A – diameter/whorl expansion rate diagram; B – body chamber length/whorl expansion rate diagram (modified after Okamoto, 1996); C – body chamber length/orientation of the aperture diagram (modified after Saunders and Shapiro, 1986) with values of* Manticoceras. *The width of the curves in the latter two diagrams (right) schematically indicates the variability of the correlation between the parameters. This is caused by the poorly known variability of size, position, and density of the soft body and of the mandibles and cameral fluids (Klug, 2001). The numbers on the thin black line indicate the growth stages with the numbers of the whorls.*

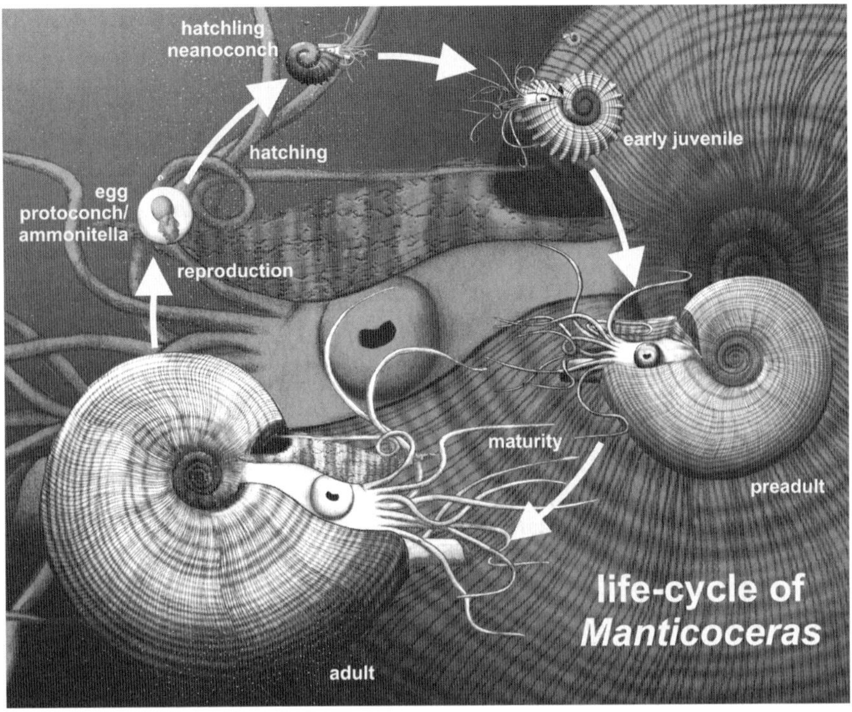

Fig. 3.16 *Reconstruction of the life cycle of* Manticoceras, *depicting the orientations of the aperture of four representative growth stages. The reconstruction of the soft part morphology is largely speculative, except for the position of the eye and the hyponome. The presence of a hood (as in Nautilus) might be indicated by rare finds of the black layer in situ (Keupp, 2000; Klug et al., 2004) which perfectly fits into the aperture. Natural diameters are: egg - 2mm, hatchling - 2mm, juvenile - 6mm, preadult - 50mm, adult - 400mm.*

juveniles of *Manticoceras* with a WER of 2.0 may have body chamber lengths of approximately 270°. Later in ontogeny, when the WER increased to 2.75–3.00, the BCL is reduced to 170°–180°. Using the diagram of Saunders and Shapiro (1986), the orientation of the aperture would change from approximately 90° (i.e., horizontal position) in early juveniles to 70° in preadults and 60° in adults.

Nevertheless, the effect of the jaws (Trauth, 1935; Clausen, 1969) on the orientation of the conch remains unclear, because their material and their position within the body chamber are unknown.

8 Life Cycle of *Manticoceras*

Because of their large maximal conch size, the adult body chambers of *Manticoceras* could have housed an enormous number of eggs. Presuming that ammonoids hatched during or after the formation of the primary constriction (at an ammonitella size of less than 2 mm) and that Recent cephalopod eggs grow during the development of the embryo (compare the coleoid *Illex*; Sakai et al., 1998), it appears reasonable to infer an initial egg size of manticoceratids of 1–2 mm, which increased to a size of 2–3 mm before hatching.

Presupposing an egg size of 2 mm before egg deposition in combination with a BCL of 180° in *Manticoceras*, a conch diameter of 400 mm (as measured in specimens from Morocco and the USA), and a portion of 20% of the body chamber volume filled by the gonads (c. 1,500 cm^3), 35,000 eggs would easily fit into this space (compared to possibly 5,000 in *Exopinacites*, Klug, 2001). In Recent coleoids, the number of eggs varies between 25 and 300,000 (Mangold, 1987). In contrast, Recent *Nautilus* produces just very few eggs at a time. Based on the strongly differing sizes of the embryonic shell in *Nautilus* and ammonoids, however, a much smaller egg size is very probable for ammonoids.

Most ammonoids probably were "r-strategists" which corresponds to a type III survivorship curve (Landman, 1985; Landman et al., 1996; Klug, 2001). Many Recent molluscs with similar survivorship curves have planktonic offspring. Consequently, the oblique orientation of the adult conchs of *Manticoceras* can be interpreted as a trade-off with respect to the large amount of eggs which could be carried by the adult females.

The WER of *Manticoceras* decreases clearly in the late embryonic and earliest juvenile conch (at 1–1.5 mm). During this phase, the primary constriction was formed and the ammonoid probably hatched. Among hatchlings, the WER values are around 2 mm; they had a BCL of approximately 250° to 300°. Consequently, the aperture was in an almost horizontal position. This trend is similar in the Agoniatitaceae and Mimagoniatitaceae, reflecting their close phylogenetic relationship (Korn, 2001). According to Jacobs and Chamberlain (1996), moderate to high-speed swimming in juvenile ammonoids was not very effective in terms of energy consumption.

With increasing size, transportation became more effective at higher speeds (Jacobs and Chamberlain, 1996), as in manticoceratids which grew to gigantic sizes (e.g., *Carinoceras*, see Miller, 1938). The high degree of the apertural angle in late juveniles and preadults (70°) supports the hypothesis that *Manticoceras* was capable of relatively high swimming velocities at these ontogenetic stages. It is unclear whether the downward tilt of the adult aperture to 60° had a significant influence on the maximum swimming speed. Adult manticoceratids certainly were still capable of performing relatively rapid movements in a horizontal direction, although they might have been slower than younger individuals. Furthermore, the energetic cost of transportation decreased only slightly with growing size at conch diameters of over 10 cm (Jacobs and Chamberlain, 1996). Therefore, manticoceratids were probably fairly good swimmers for most of their life. In combination with their possibly enormous amount of eggs, their swimming abilities may be the reason for their global distribution and an explanation for mass occurrences (e.g., in the Tafilalt, New York, Altay, Western Australia, Rhenish Massif). It appears probable that the late preadult and adult conch geometry gave mature specimens an ample choice of spawning sites.

The effect of the great variability in conch geometry among the specimens of the faunule on the orientation of their conchs was moderate. At conch diameters of 40–50 mm, the WER ranges from 2.3 to 3.0, implying BCLs of approximately 170°–210°. Consequently, the apertural angle amounted to 65° to 85°. Other factors such as shell thickness and the position of the jaw apparatus might also have influenced the orientation and could have compensated for (or intensified) this variability in orientation. Nevertheless, the manticoceratids are still likely to have been moderate to good swimmers because their hyponomes (location inferred from the ventral part of the aperture) were still positioned in approximately the same level as the centers of mass or above.

But why could *Phoenixites* (which also had a horizontally aligned aperture; cf. Sprey, 2002) survive into the Famennian while *Manticoceras* did not? One of the main differences between the two ammonoid genera is the larger conch size of the latter. Tentatively, it might imply a longer timespan until manticoceratids achieved maturity which could have been a disadvantage in a rapidly changing environment (Buggisch, 1991; House, 1993; Joachimsky and Buggisch, 1996).

9 Toward a Reconstruction of the *Manticoceras* Animal

Little is known about the external morphology of the soft parts of the Ammonoidea so far. Some internal details of the soft tissues can be inferred from attachment structures (e.g., Mutvei, 1964; Doguzhaeva and Mutvei, 1991, 1996; Richter, 2002). Additionally, very few remains of the intestines and the gills have been reported (Lehmann, 1971; Weitschat, 1986; Doguzhaeva and Mutvei, 1991, 1996).

The number of arms is still disputed and for various taxa, six (two arm pairs reduced, i.e., derived from the ten-arm state), eight or ten (like in Recent coleoids and embryonic *Nautilus*), or many arms (as in *Nautilus*) have been suggested by various ammonoid experts (compare Jacobs and Landman, 1993). In the reconstruction presented herein, ten arms are shown. This is still largely speculative, highlighting the absence of true evidence for the actual state of this character (number of arms). There is, however, a vague "phylogenetic bracket," although not entirely extant. Early *Nautilus* embryos have eight to ten arms and, on the other side of the ammonoids within the "family tree" of ammonoids, the belemnites have also been shown to have ten arms (Reitner and Urlichs, 1983; for more references see Westermann, 1996). From the absence of preserved arm crowns in both fossil Nautiloidea and Ammonoidea, we settle on rather delicate, thin arms with a low fossilization potential. The presence of a hood as in *Nautilus* is also speculative. In contrast, the dorsal shell and black layer, extending dorsally out of the body chamber is documented for *Manticoceras* (compare Keupp, 2000; Klug et al., 2004). This corroborates a mantle attachment comparable to Recent *Nautilus*. Color patterns are still unknown from Devonian ammonoids, and therefore, a spiral stripe pattern as known from several Mesozoic ammonoids, was chosen.

10 Conclusions

Representatives of the genus *Manticoceras* and its closest relatives (e.g., *Sphaeromanticoceras*) from various regions worldwide show a remarkable variability in many of their conch parameters throughout ontogeny. This has several far-reaching consequences, especially for taxonomy. Since other conch parameters do not differ much among the approximately 40 species, conch geometry was previously assumed as the most powerful tool to distinguish them. What might still be true for other ammonoid taxa, proved to be wrong for *Manticoceras*. Consequently, the currently valid species of *Manticoceras* have to be reviewed and their number probably has to be reduced significantly in the future. This again has consequences for the diversity changes of ammonoids at the Frasnian-Famennian boundary. Since the high ammonoid diversity of the Late Frasnian was previously largely based on the diversity of manticoceratids, this picture will significantly change after the reevaluation of this taxon. Considering ammonoids, the effect of the Frasnian-Famennian crisis on ammonoid diversity is less significant than previously assumed, because manticoceratid diversity was artificially inflated. Thorough testing of the taxonomic framework is mandatory prior to diversity examinations.

During late ontogeny, the body chamber length decreased. Simultaneously, the orientation of the aperture decreased from horizontal (approximately 90°) to upward oblique (approximately 60°). This implies a slight reduction of the swimming abilities as well as maneuverability among adult *Manticoceras*. The shortened but widened body chamber may be interpreted as a trade-off with the larger body chamber volume which could potentially have housed as many as 35,000 eggs. Such a large reproductive rate may have been the reason for the incredible success of *Manticoceras* at the end of the Frasnian and its mass occurrences in various regions worldwide. Its large conch size might be an indicator for longevity and this might also be the reason for their extinction. The smaller representatives of *Phoenixites* (which might have had a shorter life span and thus a shorter reproductive cycle) reached maturity probably earlier and might therefore have survived the Frasnian-Famennian boundary.

Acknowledgments

We are indebted to Raimund Feist (Montpellier) for access to the fossil collection in the Institut des Sciences de l'Evolution Montpellier (ISEM) and for the determination of conodonts. Zdzislaw Belka (Poznan) also contributed conodont determinations. Neil Landman (New York), Pascal Neige (Lyon), and Kazushige Tanabe (Tokyo) helped to improve the manuscript with numerous valuable comments in their review.

References

Aguirre-Urreta, M. B. 1998. The ammonites *Karakaschiceras* and *Neohoploceras* (Valanginian Neocomitidae) from the Neuquen basin, west-central Argentina. *Journal of Paleontology* **72**(1): 39–59.

Becker, R. T., and M. R. House. 1993. New early Upper Devonian (Frasnian) goniatite genera and the evolution of the "Gephurocerataceae." *Berliner geowissenschaftliche Abhandlungen E* **9**: 111–133.

Becker, R. T., and M. R. House. 2000. Devonian ammonoid zones and their correlation with established series and stage boundaries. *Courier Forschungsinstitut Senckenberg* **220**: 113–151.

Bhaumik, D., K. Datta, K. Jana-Sudipta, and S. Bardhan. 1993. Taxonomy and intraspecific variation of *Macrocephalites formosus* (Sowerby) from the Jurassic Chari Formation, Kutch, western India. *Journal of the Geological Society of India* **42**(2): 163–179.

Bogoslovsky, B. I. 1958. Devonskie ammonoidei Rudnogo Altaya. *Trudy Paleontologicheskogo Instituta Akademiya Nauk SSSR* **64**: 1–155.

Bogoslovsky, B. I. 1969. Devonskie ammonoidei. I. Agoniatity. *Trudy Paleontologicheskogo Instituta Akademiya Nauk SSSR* **124**: 1–341.

Böhm, R. 1935. *Études sur les faunes du Dévonien supérieur et de Carbonifère inférieur de la Montagne Noire*, pp. 1–203.

Buggisch, W. 1991. The global Frasnian-Famennian "Kellwasser Event." *Geologische Rundschau* **80**(1):49–72.

Clarke, J. M. 1899. The Naples Fauna (fauna with *Manticoceras intumescens*) in western New York. *New York State Museum, Annual Report of the Regents* **50** (for 1896): 31–161.

Clausen, C.-D. 1969. Oberdevonische Cephalopoden aus dem Rheinischen Schiefergebirge. II. Gephuroceratidae, Beloceratidae. *Palaeontographica A* **132**: 95–178.

Dagys, A. S., and W. Weitschat. 1993a. Intraspecific variation in boreal Triassic ammonoids. *In* S. Elmi, C. Mangold, and Y. Almeras (editors), *3ème symposium international sur les Céphalopodes actuels et fossils, Geobios, Mémoire Spécial* **15**: 107–109.

Dagys, A. S., and W. Weitschat. 1993b. Extensive intraspecific variation in a Triassic ammonoid from Siberia. *Lethaia* **26**(2): 113–121.

Doguzhaeva, L. A., and H. Mutvei. 1991. Organization of the softbody in *Aconeceras* (Ammonitina), interpreted on the basis of shell-morphology and muscle-scars. *Palaeontographica A* **218**: 17–33.

Doguzhaeva, L. A., and H. Mutvei. 1996. Attachment of the body to the shell in ammonoids. *In* N. H. Landman, K. Tanabe, and R. A. Davis (editors), *Ammonoid Paleobiology*, pp. 43–63. New York: Plenum Press.

Dzik, J. 1985. Typologic versus population concepts of chronospecies; implications for ammonite biostratigraphy. *Acta Palaeontologica Polonica* **30**(1–2): 71–92.

Glenister, B. F. 1958. Upper Devonian ammonoids from the *Manticoceras* zone, Fitzroy basin, western Australia. *Journal of Paleontology* **32**: 50–96.

House, M. R. 1993. Fluctuations in ammonoid evolution and possible environmental controls. *In* M. R. House, (editor), *The Ammonoidea: Environment, Ecology, and Evolutionary Change. The Systematics Association Special Volume* **47**: 13–34. Oxford: Clarendon Press.

House, M. R., W. T. Kirchgasser, J. D. Price, and G. Wade. 1985. Goniatites from Frasnian (Upper Devonian) and adjacent strata of the Montagne Noire. *Herzynica* **1**: 1–19.

Jacobs, D. K., and J. A. Chamberlain. 1996. Buoyancy and hydrodynamics in ammonoids. *In* N. H. Landman, K. Tanabe, and R. A. Davis (editors), *Ammonoid Paleobiology*, pp. 169–223. New York: Plenum Press.

Jacobs, D. K., and N. H. Landman. 1993. *Nautilus* – a poor model for the function and behavior of ammonoids? *Lethaia* **26**: 101–111.

Joachimsky, M. M., and W. Buggisch. 1996. The Upper Devonian reef crisis – insights from the carbon isotopic record. *In* J. Reitner, F. Neuweiler, and F. Gunkel (editors), *Global and Regional Controls on Biogenic Sedimentation. I. Reef Evolution. Research Reports, Göttinger Arbeiten zur Geologie und Paläontolgie, Sitzungsberichte* **2**: 365–370.

Keupp, H. 2000. *Ammoniten – Paläobiologische Erfolgsspiralen*. Stuttgart: Thorbecke.

Klapper, G. 1989. The Montagne Noire Frasnian (Upper Devonian) conodont succession. *In* N. J. McMillan, A. F. Embry, and D. J. Glass (editors), *Devonian of the World; Proceedings of the Second International Symposium on the Devonian System.* Volume III. *Paleontology, Paleoecology and Biostratigraphy. Memoir of the Canadian Society of Petroleum Geologists* **14**: 449–468.

Klapper, G., R. Feist, and M. R. House. 1987. Decision on the boundary stratotype for the Middle/Upper Devonian series boundary. *Episodes* **10**(2): 97–101.

Klapper, G., R. Feist, R. T. Becker, and M. R. House. 1993. Definition of the Frasnian/Famennian Stage boundary. *Episodes* **16**(6): 433–441.

Klug, C. 2001. Life-cycles of some Devonian ammonoids. *Lethaia* **34**: 215–233.

Klug, C., D. Korn, U. Richter, and M. Urlichs. 2004. The black layer in cephalopods from the German Muschelkalk (Middle Triassic). *Palaeontology* **47**: 1407–1425.

Korn, D. 2001. Morphometric evolution and phylogeny of Palaeozoic Ammonoids, Early and Middle Devonian. *Acta Geologica Polonica* **51**(3): 193–215.

Korn, D., and C. Klug. 2001. Biometric analyses of some Palaeozoic ammonoid conchs. *Berliner geowissenschaftliche Abhandlungen E* **36**: 173–187.

Korn, D., and C. Klug. 2002. *Ammoneae Devonicae. Fossilium Catalogus I: Animali* **138**. Leiden: Backhuys.

Landman, N. H. 1985. Early ontogeny of Mesozoic ammonites and nautiloids. *In* J. Wiedmann, and J. Kullmann (editors), *Cephalopods – Present and Past*, pp. 215–228. Stuttgart: E. Schweizerbart'sche Verlagsbuchhandlung.

Landman, N. H., K. Tanabe, and Y. Shigeta. 1996. Ammonoid embryonic development. *In* N. H. Landman, K. Tanabe, and R. A. Davis (editors), *Ammonoid Paleobiology*, pp. 343–405. New York: Plenum Press.

Lehmann, U. 1971. Jaws, radula, and crop of *Arnioceras* (Ammonoidea). *Palaeontology*, **14**: 338–241.

Mangold, K. 1987. Reproduction. In P. R. Boyle (editor), *Cephalopod Life Cycles*. Volume II. *Comparative Reviews*, pp. 157–200. London: Academic Press.

Miller, A. K. 1938. Devonian ammonoids of America. *Geological Society of America Special Papers* **14**: 1–262.

Mitta, V. V. 1990. Intraspecific variability in the Volgian ammonites. *Paleontological Journal* **24**(2): 10–15.

Mutvei, H. 1964. Remarks on the anatomy of recent and fossil cephalopoda. *Stockholm Contributions in Geology* **11**(4): 79–112.

Okamoto, T. 1996. Theoretical modelling of ammonoid morphology. *In* N. H. Landman, K. Tanabe, and R. A. Davis (editors), *Ammonoid Paleobiology*, pp. 225–251. New York: Plenum Press.

Paproth, E., R. Feist, and G. Flajs. 1991. Decision on the Devonian-Carboniferous boundary stratotype. *Episodes* **14**(4): 331–336.

Petter, G. 1959. Goniatites Dévoniennes du Sahara. *Publications du Service de la Carte Géologique de l'Algérie, Nouvelle Série, Paléontologie* **2**: 1–313.

Raup, D. M. 1967. Geometric analysis of shell coiling: coiling in ammonoids. *Journal of Paleontology* **41**(1): 43–65.

Richter, U. 2002. Gewebeansatz-Strukturen auf pyritisierten Steinkernen von Ammonoideen. *Geologische Beiträge Hannover* **4**: 1–113.

Reitner, J., and M. Urlichs. 1983. Echte Weichteilbelemniten aus dem Untertoarcium (Posidonienschiefer) Südwestdeutschlands. *Neues Jahrbuch für Geologie und Palaeontologie, Abhandlungen* **165**: 450–465.

Sakai, M., N. E. Brunetti, B. Elena, and Y. Sakurai. 1998. Embryonic development and hatchlings of *Illex argentinus* derived from artificial fertilization. *In* A. I. L. Payne, M. R. Lipinski, M. R. Clarke, and M. A. C. Roeleveld (editors), *Cephalopod Biodiversity, Ecology and Evolution. South African Journal of Marine Science* **20**: 255–265.

Saunders, W. B., and E. A. Shapiro. 1986. Calculation and simulation of ammonoid hydrostatics. *Paleobiology* **12**: 64–79.

Sprey, A. 2002. Morphometrie und Paläoökologie von Ammonoideen vor, während und nach globalen Faunenkrisen. *Münstersche Forschungen zur Geologie und Paläontologie* **95**: 1–218.

Tanabe, K. 1993. Variability and mode of evolution of the Middle Cretaceous ammonite *Subprionocyclus* (Ammonitina; Collignoniceratidae) from Japan. *In* S. Elmi, C. Mangold, and Y. Almeras (editors), *3ème symposium international sur les Céphalopodes actuels et fossils. Geobios, Mémoire Special* **15**: 347–357.

Trauth, F. 1935. Die Aptychen des Paläozoikums. *Jahrbuch der Preußischen Geologischen Landesanstalt* **55** (for 1934): 44–83.

Trueman, A. E. 1941. The ammonite body chamber, with special reference to the buoyancy and mode of life of the living ammonite. *Quarterly Journal of the Geological Society of London* **96**: 339–383.

Wedekind, R. 1913. Die Goniatitenkalke des unteren Oberdevon von Martenberg bei Adorf. *Sitzungsberichte der Gesellschaft Naturforschender Freunde zu Berlin* **1913**: 23–77.

Wedekind, R. 1918. Die Genera der Palaeoammonoidea (Goniatiten). Mit Ausschluβ der Mimoceratidae, Glyphioceratidae und Prolecanitidae. *Paläontographica* **62**: 85–184.

Weitschat, W. 1986. Phosphatisierte Ammonoideen aus der Mittleren Trias von Central-Spitzbergen. *Mitteilungen aus dem Geologisch-Paläontologischen Institut der Universität Hamburg* **61**: 249–279.

Westermann, G. E. G. 1996. Ammonoid life and habitat. *In* N. H. Landman, K. Tanabe, and R. A. Davis (editors), *Ammonoid Paleobiology*, pp. 607–707. New York: Plenum Press.

Chapter 4
GONIAT – The Current State of the Paleontological Database System on Paleozoic Ammonoids

Jürgen Kullmann

Institut für Geowissenschaften, Universität Tübingen, Sigwartstrasse 10, D-72076 Tübingen, Germany, juergen.kullmann@uni-tuebingen.de

Keywords: GONIAT, database, Devonian, Carboniferous, Permian, Paleozoic ammonoids, mediator architecture

1 Introduction

In 1988, a research project with the title: *The Development of Diversity and Provincialism of Paleozoic Ammonoid Faunas* was started at the University of Tübingen (Germany). The German Science Foundation (Deutsche Forschungsgemeinschaft) had to be convinced that a powerful database was one of the prerequisites for such a research program. It was finally accepted that finding out the requirements of paleontological databases for research on systematics and evolution was pioneering work.

N. H. Landman et al. (eds.), *Cephalopods Present and Past: New Insights and Fresh Perspectives*, 86–96.

2 Scope of the Database System GONIAT

The scope of the database system GONIAT was focused from the beginning on the complete recording of the ammonoid faunas of the world, from the Devonian until the Triassic, comprising about 150 million years, almost the first half of the history of ammonoids. Intended audiences include all scientists interested in ammonoid paleontology and distribution. GONIAT is free, easily accessible by downloading from GONIAT home pages in German, English, and Spanish at the University of Tübingen; it is entirely in English (Kullmann et al., 1993; Korn et al., 1994; Kullman and Korn, 2000). GONIAT treats all ammonoid orders and suborders, families, genera, and species known from the Devonian, Carboniferous, and Permian systems. After 15 years of uninterrupted work, 4,000 valid species and more than 700 genera and subgenera of about 120 families have been included. Approximately 7,500 localities are described, based on 1,760 publications; many taxa and localities are accompanied by pictures showing stratigraphic sections, goniatite suture lines, and cross sections. Taxa are currently covered at present to at least 95% of those known. Some recent publications are not yet fully included.

GONIAT is designed as a tool for investigations on the classification of the Paleozoic ammonoids, their paleogeographic distribution, and their biostratigraphic range. This database management system provides not only determinations based on morphological characteristics but also pertinent information regarding occurrence, duration, literature, and phylogenetic relationships of every taxon at the family, genus, and species levels.

Platform/Software: GONIAT was established in dBASE IV; it was transferred to Microsoft ACCESS in the year 2000 using a database front-end developed by the ROG enterprise. This enterprise and ELEMENTEC, both in Germany, are sponsoring GONIAT. The database was programmed by Peter S. Kullmann.

The Paleozoic Ammonoidea (in short, goniatites) represents a large and well-studied fossil group (Becker and Kullmann, 1996). The goniatites experienced several periods with high turnover rates and intervals with short-lived species and genera. During the Devonian, Carboniferous, and Permian times, a number of geologic events occurred affecting the environment, including the movements of the Variscan and Appalachian Orogenies, shifting of continents in many parts of the world, ice ages, coal producing periods, and varied drastic facies changes due to eustatic sea level changes. The prime aim was the creation of a database system providing the highest time resolution possible to reveal the time-dependent influences on evolution, like mass extinctions or recovery intervals. Therefore, every record of the fossils, as well as of the localities, includes data on the beginning or first appearance and the end or last occurrence. The estimated radiometric ages of the time-planes of the boundaries are also given. So every record of a taxon or locality is accompanied by two radiometric dates stored in the BOUNDARIES section of the database indicating the longevity of a taxon (family, genus, or species) or the supposed range of the beds of an outcrop.

3 Data Model

GONIAT contains six interconnected relational databases providing the data for the four main data complexes: (1) Taxonomy including Morphology, (2) Boundaries, (3) Literature (References), and (4) Geographic data (see Fig. 4.1 left column).

3.1 Taxonomy Including Morphology

The taxonomy section of the database covers data on the actual state of classification and morphological characteristics. They are stored in special records of the families, genera, species, and specimens in open nomenclature. The relational character of the databases allows the user to switch easily from the taxa database to the databases with the Geographic data or the Literature. The records of most families and genera contain extended descriptions with the current state of their diagnosis and definition. Synonymy lists of all taxa are stored in the note attached with each record.

But what do we really observe? Certainly not genera: nobody knows the extension of a genus; there is no generally accepted definition. The species concept provides the basic information about the diagnosis in which at least one character or one combination of characters is described as unique in this unit. But we do not know very much about the variability of the species under consideration unless we have a sufficient number of individuals. As can be seen in GONIAT, a large number

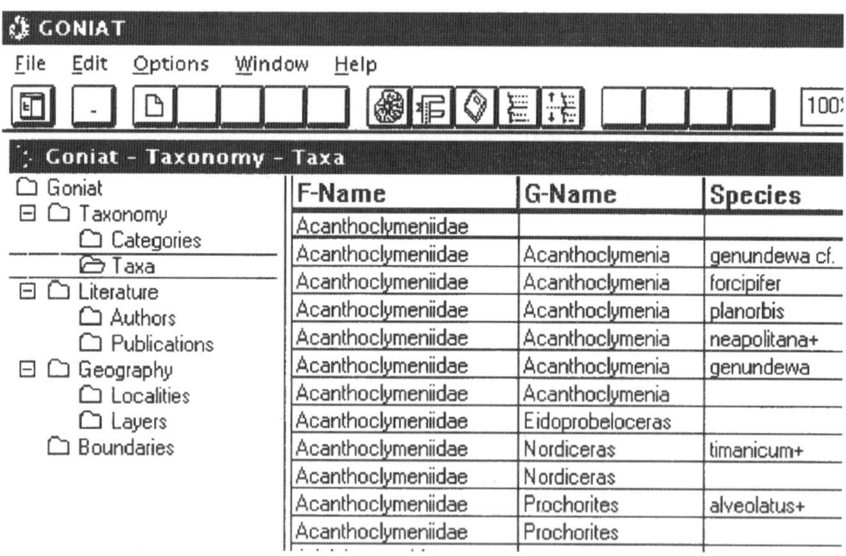

Fig. 4.1 *Main window and part of the table view of the Taxa database.*

of species are based just on a single specimen, the holotype, without any paratype. Because we have normally only conchs of the animal, we do not know if there are several species using the same conch shape. Ontogenetic investigations are rather rare. What we only have in reality are the characteristics of the shell. With respect to mass extinction, we must keep in mind that we are speaking only of characters that disappear or come up later in times of recovery. GONIAT is especially designed for investigation of the qualitative or morphologic changeover of the ammonoids. It concentrates on the duration of morphologic characteristics as expressed in the data on first appearance, last appearance, general frequency, and extant taxa. Two different databases, one for the early and middle stages of ontogeny (if known), the other for the adult conch shape, provide information on the ontogeny of the species.

3.2 Boundaries

Age determinations are an essential part of GONIAT. The time-stratigraphic limits of each taxon are documented by information on the first appearance and the last occurrence. More than 250 time-planes within the Devonian, Carboniferous, and Permian can be distinguished by means of biostratigraphic information. Because of the scarcity of radiometric ages, most age assignments are estimated according to present knowledge and must be modified as better information becomes available. The absolute dates presently used in GONIAT are based on timescales published recently (Tucker et al., 1998; Wardlaw, 1999; Menning et al., 2000; Gradstein et al., 2004).

3.3 References

Each record of the Taxonomy and Geography part of GONIAT allows an immediate connection with the Literature database. Recent publications not only list title and bibliographic citations but also the English abstract. Special databases of Authors and Publications facilitate the retrieval of remote publications. Some records in the Authors database contain bibliographic information.

3.4 Geographic Data

Many taxa are based on old material of the 19th and early 20th centuries without any information on the environment; many taxa completely lack information about the origin of the fossil. What can be done with the notation: Kazakhstan–Karaganda? A number of genera and species are based on such short locality descriptions.

The age, however, is often precisely known because of accompanying time-indicative species. Investigation on provincialism is easy in GONIAT, but is expressed by region. There are no data on latitude/longitude, because the paleolatitudes and -longitudes in the Paleozoic are still insufficiently known. Special remarks on occurrence, type of outcrop, and information on lithology and taphonomy are stored as far as described in the note attached to the respective record. Age assignments of localities are estimated in cases of insufficient knowledge.

4 Applications

GONIAT has already been used widely for the creation of comprehensive catalogues. The main data source of the work on Devonian ammonoids (Korn and Klug, 2002) was based on GONIAT. The Carboniferous/Permian volume of the Treatise on Invertebrate Paleontology, the revision of part L, will be partly connected with GONIAT.

Critical intervals in the history of ammonoids can be easily described by using GONIAT. The quantitative aspect of ammonoid evolution during the Pennsylvanian (Upper Carboniferous) may serve as an example (Kullmann, 2002). On the family level, high turnover rates (Fig. 4.2) are evident at three turning points (Table 4.1): (1) at the Yeadonian/Langsettian boundary [YLB, between Namurian/Westphalian (Eurasia) or Halian-Bloydian (America)], four families became extinct and five appeared; (2) around the Moscovian/Kasimovian boundary [MKB, Eurasia or Desmoinesian/ Missourian (America)], two or three families became extinct, three families appeared, and two or three families started to proliferate; and (3) at the Carboniferous–Permian boundary (CPB), five families became extinct and two appeared. The unusually high changeover rate of genera corroborates this impression: in the Yeadonian, 57% of the genera became extinct (16 out of 28), in the Kasimovian (Missourian), 42.3% (11 out of 26), and in the upper Gzhelian (late Virgilian), 46.5% (13 out of 28); in cases (1) and (3), the rate of surviving genera was rather low.

The qualitative analysis of the critical intervals shows a considerable change in the main characters of ornamentation and suture line, whereas the conch form does not vary fundamentally. Below the YLB, half of the species are strongly ornamented. In the early Moscovian, provincialism expanded; the neritic, strongly ornamented gastrioceratids were replaced by the smooth-shelled pelagic glaphyritids.

Approximately 90% of the existing species became extinct at the MKB and were replaced by goniatitids and prolecanitids showing a considerable increase in suture elements, frequently with subdivided and partly digitate sutures. At the CPB, the number of simple goniatites with a low number of lobes was strongly reduced and replaced by multilobate goniatitids with enormously complicated sutures. In general, the fluctuations of the ammonoid faunas during the critical intervals showed comparable patterns: a stepwise decline in diversity, a low point, an initiation point for new characters, and later stepwise increased evolutionary diversification.

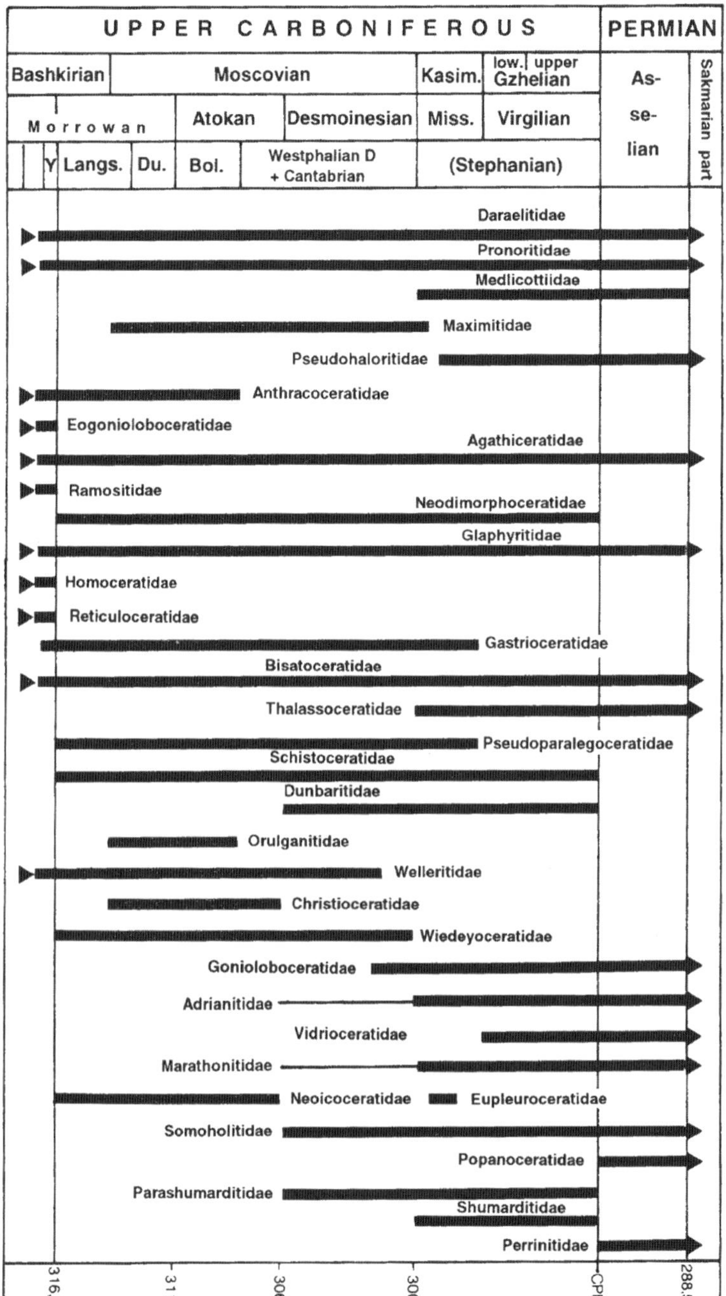

Fig. 4.2 *Chart showing the range of Late Carboniferous and Early Permian ammonoid families. Abbreviations: Y, Yeadonian substage; Langs., Langstettian substage (=West-phalian B); Du., Duckmantian substage (=Westphalian B); Bol., Bolsovian substage (=Westphalian C); Kasim., Kasimovian stage; Miss., Missourian stage.*

Table 4.1. *Global time planes as used in GONIAT version 3.0 (left of dash: below boundary; right of dash: above boundary).*

288	Asselian–Sakmarian
292	CPB,Carboniferous–Permian Boundary,base of Permian
295	Lower Gzhelian–upper Gzhelian
297	Kasimovian–Gzhelian, Missourian–Virgilian
301	Moscovian–Kasimovian, Desmoinesian–Missourian
306	Atokan–Demoinesian (Kashirian–Podolskian)
308	Bolsovian–Westphalian D
311	Morrowan–Atokan, Duckmantian–Bolsovian
313,5	Langsettian–Duckmantian
314,5	Bashkirian–Moscovian
316,5	Namurian–Westphalian, Yeadonian–Langsettian, Halian–Bloydian
317,5	Marsdenian–Yeadonian

5 Problems and Limitations

A number of limitations have to be taken into account in using GONIAT. The biostratigraphic records in many cases allow only vague estimates of the duration of taxa and provide at best a rough approximation of the time spans involved. The work with GONIAT revealed many inconsistencies and partly the incompleteness of the ammonoid record (Kullmann and Nikolaeva, 1999). The biostratigraphic ranges of the species are in many cases not well-documented or imprecisely described and figured. The total longevity of species often tends to be overestimated (see Fig. 4.3).

Another major point concerns the quality of the systematic treatment. Different approaches to taxonomy, which are sometimes uncritical with excessive splitting of taxa have resulted in fictitious taxa richness. This is especially true for the counts of species and genera. A number of species and genera are apparently synonyms due to excessive splitting or lack of data on the variability of the group under consideration. Many species are based on very few specimens and do not allow observation of variability, some are based only on a fragmentary holotype. Larger sample sizes would certainly reduce the number of recognised species. Because of these shortcomings, the use of raw data in these studies seems to be indispensable. The documentation of quantitative (numeric) changes may be a good starting point for the investigation of the qualitative (morphologic) changeover of the ammonoids. The change in morphology is, however, what we really observe in evolution. Taxonomy is only a form of grouping to interpret the relationships of organisms.

6 Future Aspects

The German Science Foundation was worried initially about the increasing number of independent small databases without interconnection that appeared to be "island solutions." Each database is focused on the specific interests of the responsible

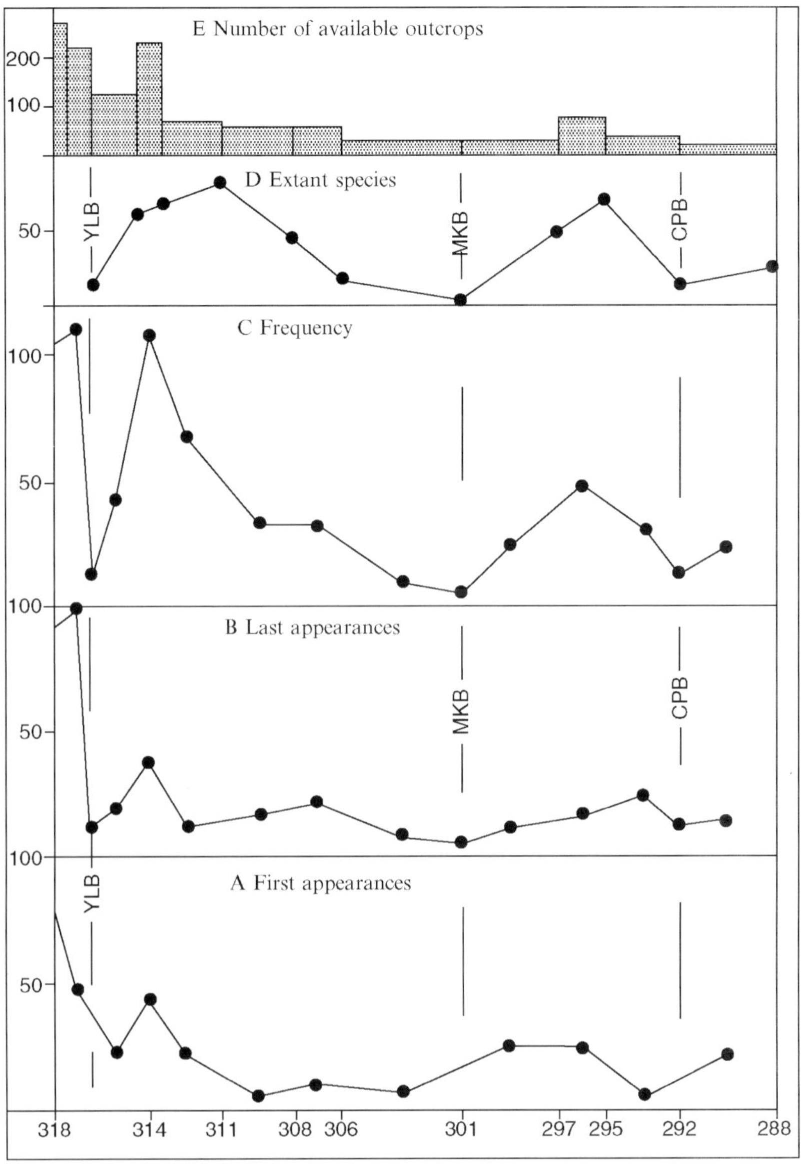

Fig. 4.3 *Frequency distribution of ammonoid species per million years (Ma.) in the late Carboniferous (between 318 and 288 Ma). Abbreviations: C = Frequency of species, D = Frequency of species crossing top of interval, E = number of available outcrops, for comparison of local availability. YLB: Yeadonian–Langsettian boundary; MKB: Moscovian–Kasimovian boundary; CPB: Carboniferous–Permian boundary. Radiometric timescale as used in GONIAT version 3.0 (From Kullmann, 2002).*

research group. Such databases are autonomous in the sense that they are completely under the control of the responsible group. Different opinions tend to lead to different concepts of the underlying knowledge.

For some time it was hoped to encourage other scientific groups to consider the possibility of interfacing GONIAT with other databases. The time-planes and the localities would be often identical in such databases. An attempt was made to relate the American conodont database set up by Sandberg and Ziegler, with GONIAT, but this database was not advanced enough.

In the long run, a uniform paleontological database could be accomplished along the following lines. The problem of integrating heterogeneous information is presently addressed in computer science research. Computer scientists have assured us that it is possible to work out an integrated access to distributed heterogeneous paleontological databases (P. S. Kullmann and J. Kullmann, 2002). The objective of "Intelligence Information Integration" is to find solutions for supplying uniform access to various heterogeneous and possibly broadly distributed information systems (see Fig. 4.4).

A widely accepted approach is the so-called mediator architecture (Gruber, 1995), which uses a software component that is dedicated to the processing of user queries with respect to a set of information sources. Without doubt, it would be desirable to have a uniform and consistent conceptual basis available that facilitates uniform access to all information systems. The consequences of integrated access for paleontological research could be a great improvement in the precision of biostratigraphic statements. The combination of the currently scattered information

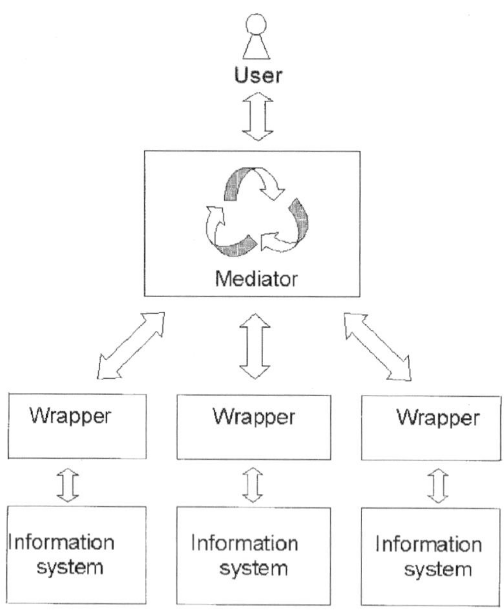

Fig. 4.4 *Scheme of the mediator architecture (From Kullmann and Kullmann, 2002).*

would yield new insights into the major open questions of palaeogeographic relations, such as configuration of oceans, climatic distribution, and variation.

7 Summary

The Data model of GONIAT contains six interconnected relational databases providing detailed data and figures for four essential data complexes: (1) taxonomy including morphology, (2) boundaries, (3) literature (references), and (4) geography (localities).

GONIAT contains with few exceptions an almost complete set of records of all Devonian, Carboniferous, and Permian taxa, with species being covered, at present, to at least 95%. Updates will be provided yearly by downloading from the GONIAT web site at the University of Tübingen (Germany):
http://www.uni-tuebingen.de/geo/gpi/mitarbeiter/kullmann/seiten/goniaten.html.

Acknowledgments

Thanks are due to H. Vollmer for assistance in designing the drawings. The reviewers, N. H. Landman and L. K. Meeks are thanked for many valuable suggestions.

References

Becker, R.T., and J. Kullmann. 1996. Paleozoic ammonoids in space and time. *In* N. H. Landman, R. A. Davis, and K. Tanabe (editors), *Ammonoid Paleobiology, Topics in Geobiology* **13**: 711–753. New York: Plenum Press.

Gruber, T. R. 1995. Toward principles for the design of ontologies used for knowledge sharing. *International Journal of Human-Computer Studies* **43**: 907–928.

Gradstein, F. M., J. G. Ogg, and A. G. Smith. 2004. *A Geologic Time Scale 2004*. Cambridge: Cambridge University Press.

Korn, D., and C. Klug, C. 2002. Ammoneae Devonicae. *In* W. Riegraf (editor), *Fossilium Catalogus, I: Animalia, pars* **138**. Leiden: Backhuys.

Korn, D., J. Kullmann, P. S. Kullmann, and M. S. Petersen. 1994. GONIAT, a computer-retrieval system for Paleozoic ammonoids. *Journal of Paleontology* **68**: 1257–1263.

Kullmann, J. 2002. Critical intervals in the evolution of Late Carboniferous ammonoids. *Coral Research Bulletin* **7**: 87–93.

Kullmann, J., and D. Korn. 2000. GONIAT Database System, version 3.0, Tübingen, www.uni-tuebingen.de/geo/gpi/mitarbeiter//kullmann/seiten/goniaten.html.

Kullmann, J., D. Korn, P. S. Kullmann, and M. S. Petersen. 1993. The Database System GONIAT – a tool for research on systematics and evolution of Paleozoic ammonoids. *In* S. Elmi, C. Mangold, and Y. Alméras (editors), *3ème Symposium international sur les Céphalopodes actuels et fossiles. Géobios, Mémoire spéciale* **15**: 239–245.

Kullmann, P. S., and J. Kullmann. 2002. PaleoWeb – Integrierte Nutzung verteilter heterogener paläontologischer Datenbanken. *Mathematische Geologie* **6**: 57–62.

Kullmann, J., and S. V. Nikolaeva. 1999. Ammonoid turnover at the Mid-Carboniferous boundary and the biostratigraphy of the early Upper Carboniferous. *In* A. Yu. Rozanov, and A. A. Shevyrev (editors), *Iskopaemye tsefalopody: noveishie dostizheniia v ikh izuchenii*, pp. 169–194.

Menning, M., D. Weyer, G. Drozdzewski, H. W. J. van Amerom, and I. Wendt. 2000. A Carboniferous time scale 2000: discussion and use of geological parameters as time indicators from Central and Western Europe. *Geologisches Jahrbuch* **A156**: 3–44.

Tucker, R. D., D. C. Bradley, A. G. Ver Straeten, J. R. Ebert, and S. R. McCutcheon. 1998. New U-Pb zircon ages and the duration and division of Devonian time. *Earth and Planetary Science. Letters* **158**: 175–186.

Wardlaw, B. R. 1999. Notes from the SPS Chair. *Permophiles* **35**: 1–3, Boise: Idaho.

Chapter 5
Ornamental Polymorphism in *Placenticeras kaffrarium* (Ammonoidea; Upper Cretaceous of India): Evolutionary Implications

Tapas K. Gangopadhyay[1] **and Subhendu Bardhan**[2]

[1] Department of Geology, Bengal Engineering and Science University Shibpur, Howrah 711103, India, tapasgeolbe@yahoo.com;

[2] Department of Geological Sciences, Jadavpur University, Kolkata 700032, India, s_bardhan01@yahoo.co.uk

1 Introduction

Genetically controlled discontinuous variation within a single population is termed polymorphism (Ford, 1940). According to the McGraw Hill Encyclopedia (1984: 1364), genetic polymorphism is "A form of genetic variation, specifically a discontinuous variation, occurring within plant and animal species in which distinct forms exist together in the same population. ... Distinct forms must be controlled by some switch which can produce one form or the other without intermediates such as those arising from environmental differences. This clear-cut control is provided by the recombination of genes." Genetic polymorphism may produce two or more discontinuous forms of a phenotypic feature due to functional or structural variation in a population. In the present-day organic world this term is used in reference to colonial organisms such as bryozoans and cnidarians and castes in bees, ants, and wasps. Recognition of polymorphism requires study of large populations of the organism in question. That is why it is frequently observed in the living organic world but is rarely recorded in fossils.

N. H. Landman et al. (eds.), *Cephalopods Present and Past: New Insights and Fresh Perspectives*, 97–120.

The phenomenon of polymorphism is commonly displayed by discontinuous differences in ornament together with continuously expressed variations in size and shape (Reyment, 1988). Polymorphism, as a modality of genetic intraspecific discontinuous variability (besides sexual dimorphism), has been the subject of much attention by many workers in paleontological studies, e.g., in ostracods by Kamiya (1992) and Reyment (1988), and in graptolites by Janusson (1973). Ornamental polymorphism is also reported from fossil cephalopods. For example, Halder et al. (1998) recorded this phenomenon in the Middle Jurassic nautiloid *Paracenoceras*. Within ammonites, intraspecific ornamental polymorphism has been described by Tintant (1963) in the family Kosmoceratidae, and by Tintant (1980) in Dactylioceratidae, and by Melendez and Fontana (1993) in Perisphinctidae.

In the present endeavour, we discuss ornamental polymorphism within a placenticeratid ammonite species *P. kaffrarium* from the Upper Cretaceous (Coniacian) Bagh Beds, Central India. The nature of polymorphism varies from a completely smooth shell to a morph having three rows of tubercles. Ontogeny of each morph is described in terms of heterochrony. Polymorphism has evolutionary implications. The role of ornamental polymorphism in *P. kaffrarium* in the evolution of the subsequent placenticeratid lineage is also discussed.

Placenticeras Meek, 1876, belongs to the family Placenticeratidae Hyatt and it ranges from the Upper Albian to Upper Campanian (Wright et al., 1996; Kennedy et al., 1996). It consists of numerous dimorphic species (Klinger and Kennedy, 1989; Ganguly and Bardhan, 1993) showing latitudinally clustered distribution patterns, but the genus itself has a pandemic distribution (Bardhan et al., 2002). The species *P. kaffrarium* has also been reported from the Turonian–Coniacian of Madagascar (Collignon, 1965a, b) and from the precisely dated Middle Coniacian of Zululand, South Africa (Klinger and Kennedy, 1989: 242). The *P. kaffrarium* population discussed here comes from the Upper Cretaceous Bagh Group of shallow marine carbonates of Central India (Fig. 5.1).

The carbonate sequence consists of two mappable formal units – the lower one, the Nodular Limestone and the upper one, the Bryozoan Limestone (Bardhan et al., 2002). Taylor and Badve (1992) informally subdivided the Bryozoan Limestone into two units – Deola Chirakhan Marl and Coralline Limestone and showed categorically that the Coralline Limestone always overlies the Nodular Limestone. Recently, Kennedy et al. (2003) showed alternations of the Nodular Limestone and Coralline Limestone in a particular section, but this cannot be regionally applicable for the entire Bagh Sequence. Specimens of *Placenticeras kaffrarium* are abundant in the middle subunit of the Nodular Limestone (Fig. 5.2). The Bagh sediments are exposed as discrete outcrops essentially on the northern part of the Narmada River, the course of which follows a mid-continental rift that was associated with the Karoo Rift System (Boselini, 1989). The carbonates were deposited in a narrow, intracratonic Narmada trough as the eastern arm of the Tethys transgressed (Chiplankar and Badve, 1973; Jafar, 1982). The Nodular Limestone horizons yield numerous ammonite specimens, which were previously grouped into at least 23 species belonging to four different families of varying age within the Late

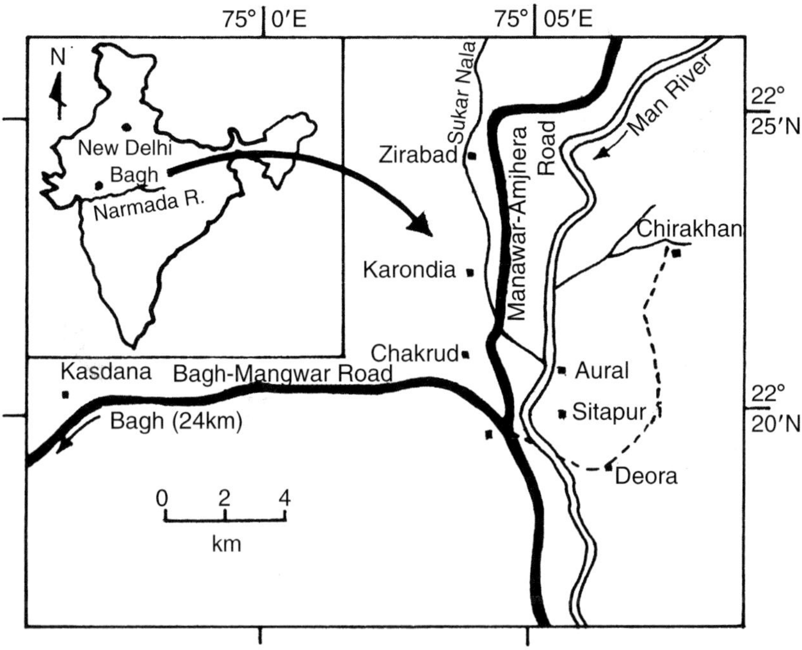

Fig. 5.1 *Outcrops of Bagh Group of rocks lying north of the Narmada River (inset) and geographic localities around the Man River valley showing important fossil-bearing stratigraphic sections (after Bardhan et al., 2002).*

Cretaceous (Chiplankar and Ghare, 1977a, b). Recent studies (Klinger and Kennedy, 1989; Ganguly and Bardhan, 1993), however, have shown that the majority of the ammonite population actually belongs to the highly variable species *Placenticeras kaffrarium*. The previously assigned multitude of species names perhaps resulted from the failure to recognize sexual dimorphism and wide intraspecific variability especially in ornamental patterns, which led to polymorphism in the population.

2 Ornamental Polymorphism in *Placenticeras kaffrarium*

Klinger and Kennedy (1989) provided the first detailed taxonomic account of the species and showed that the variation within *P. kaffrarium* is constrained by both genetic and non-genetic factors. The ontogeny of the species is marked by strong allometry: the early whorls are smooth, the intermediate stage is marked by the appearance of umbilical tubercles and ventral clavi, and the late phragmocone – early body whorl shows the presence of both umbilical and lateral tubercles connected by

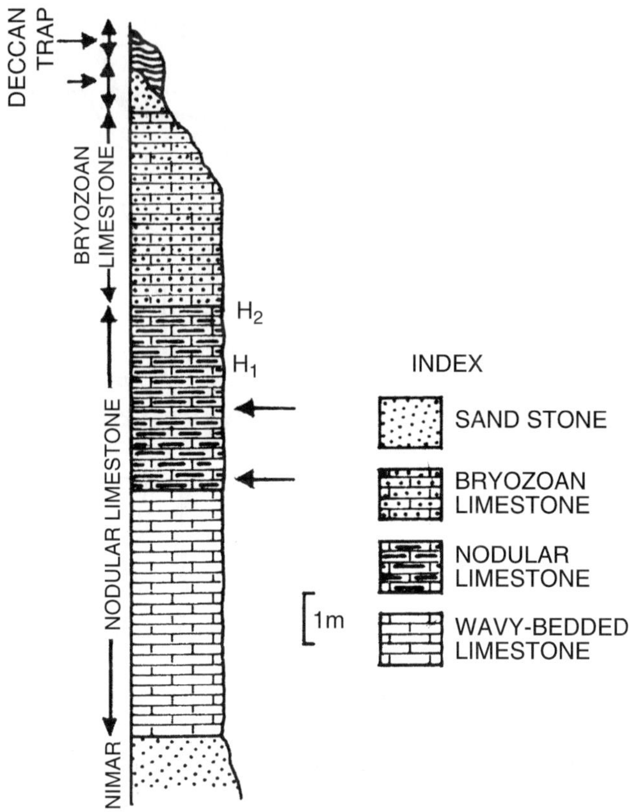

Fig. 5.2 *Generalized stratigraphy of the Bagh Group along the Man River valley. H1 and H2 are intraformational and interformational hardgrounds respectively (modified after Gangopadhyay and Bardhan, 2000). Horizontal long arrows indicate levels of maximum occurrence of* P. kaffrarium.

bifurcating ribs and strong alternating ventral clavi. Ornament weakens or disappears on the later part of the body chamber, which becomes rounded on the venter. This ontogenetic scaling of ornament is, however, not observed in all adult specimens of the population. Instead, each major ontogenetic stage may be retained up to the adult stage. Thus, some adult variants are smooth and devoid of any ornament. Some specimens retain only umbilical tubercles and ventral clavi in their adult outer whorl. In the majority of specimens in the Bagh population, the outer whorl is characterized by strong umbilical and lateral tubercles along with ventral clavi. All adult variants are of similar size, and the later part of the body chamber is marked by smoothening of the shell and rounding of the venter, especially in macroconchs. These three morphs, therefore, exhibit discontinuous variation with respect to ornament. In addition, there are many morphotypes, which represent intermediate stages, but do not show continuous variation. For example, some adult specimens retain the smooth stage for a longer period and develop umbilical tubercles and ventral clavi only during later ontogeny. Other variants may have umbilical tubercles in early ontogeny and subsequently develop weak lateral tubercles. Klinger and Kennedy (1989), in fact, recognized seven distinct morphotypes. The above mentioned three morphs show qualitatively different aspects of ornament and represent non-sexual intraspecific polymorphism. They correspond to typical *umkwelanense*, *subkaffrarium,* and *kaffrarium* variants described by Klinger and Kennedy (1989). *Placenticeras kaffrarium*, moreover, is strongly sexually dimorphic. It shows both size and ornamental dimorphism in Bagh. In addition, Bagh records many specimens with the peristome preserved – a feature which is rarely found in the family Placenticeratidae throughout the world (see Gangopadhyay and Bardhan, 1998). Adult microconchs show a peristomal modification, with a ventral rostrum, whereas adult macroconchs have a simple aperture (for details on adult modifications in ammonites, see Davis et al., 1996). Each morph is dimorphically paired, and both antidimorphs exhibit polymorphism. On the one hand, the *umkwelanense* polymorph shows only size dimorphism; on the other hand, the *kaffrarium* polymorph has well-marked size and ornamental dimorphism. The microconch is relatively evolute and coarsely ornate, and the ornament continues till to end.

There is strong covariation between the strength of ornament and degree of inflation and involution among polymorphs, and this provides a distinctive look for each variant. *Umkwelanense* is oxyconic involute, *subkaffrarium* is relatively evolute and inflated, with a subtrigonal whorl section, and *kaffrarium* is more evolute and highly inflated with a polygonal whorl section. It appears, therefore, that *P. kaffrarium* shows both sexual and non-sexual polymorphism. The variants intergrade morphometrically; this suggests that they belong to the same biological species (Fig. 5.3). Type material of *P. kaffrarium* from Zululand is also included in the bivariate plots and forms a homogeneous distribution with the present population in Bagh. This indicates that both the Bagh and Zululand populations belong to the same species (for a detailed morphometric analysis, see Bardhan et al., 2002). The other contemporaneous species, *P. mintoi*, in Bagh is closely related, but morphologically quite distinct (see Ganguly and Bardhan, 1993). Here, we provide a brief taxonomic description of each polymorphic variant.

2.1 *Placenticeras kaffrarium* Morph *umkwelanense* Etheridge, 1904

Microconch. The shell is essentially compressed (WB/H = c. 0.54), oxyconic up to the adult stage. The venter is narrow and tabulate, with the aperture trigonal in cross section. Laterals are smooth, flat, and strongly converging. Adult shell diameter varies from 72–132 mm (Av. = 94.76; S.D. = 10.03). The shell is involute (U/D = 0.26–0.35; Av. = 0.31, S.D. = 0.031). The umbilicus has a sharp margin and a steep wall. The maximum width of the shell lies at or near the umbilical margin. The adult body chamber occupies about three-fourths of the last whorl, and the beginning of the body chamber is marked by egression of the umbilical seam.

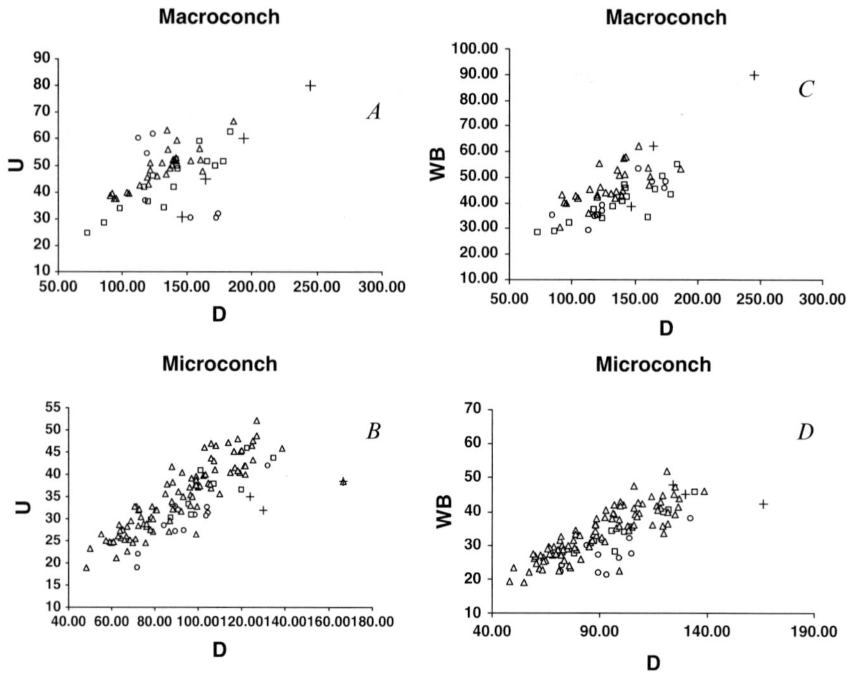

Fig. 5.3 *Bivariate analysis involving umbilical diameter and shell diameter (A, B), and whorl breadth and shell diameter (C, D) in different polymorphic variants of* P. kaffrarium. *Type specimens of all polymorphic variants from Zululand, South Africa, have also been plotted.*
D = Diameter of shell; U = Umbilical Diameter; WB = Whorl Breadth. All are measured in millimeters. ● umkwelanense, ▪ subkaffrarium, Δ kaffrarium, +*Types.*

Macroconch. The shell is morphologically similar to the microconch in all aspects except shell diameter, the largest shell being about 173 mm in diameter (Range = 84–173; Av. = 134.82; S.D. = 27.53). Adult body chamber becomes slightly inflated, with a weakly convex flank and a rounded venter.

2.2 *Placenticeras kaffrarium* Morph *subkaffrarium* Spath, 1921

Microconch. The shell is compressed to slightly inflated (WB/H = c. 0.71) with subtrigonal whorl section; it is relatively evolute (U/D = 0.3–0.4; Av. = 0.34; S.D. = 0.031). The venter is flat, with clavi. The adult body chamber occupies more than two-thirds of the last whorl. This variant is characterized by shells, each with one row of umbilical tubercles. Umbilical tubercles are pyramidal and clavi appear almost simultaneously in early whorls. Clavi on opposite sides are always alternate

Fig. 5.4 Umkwelanense *type of ornament. 1, lateral view, microconch with adult phragmocone and partially preserved body chamber; note uncoiling of body chamber. 2, lateral view, microconch, adult phragmocone; note trace of body chamber showing scaphitoid uncoiling. 3, apertural view. 4, lateral view, macroconch, almost complete. 5, ventral view, note tabulate venter. Scale bar = 1 cm.*

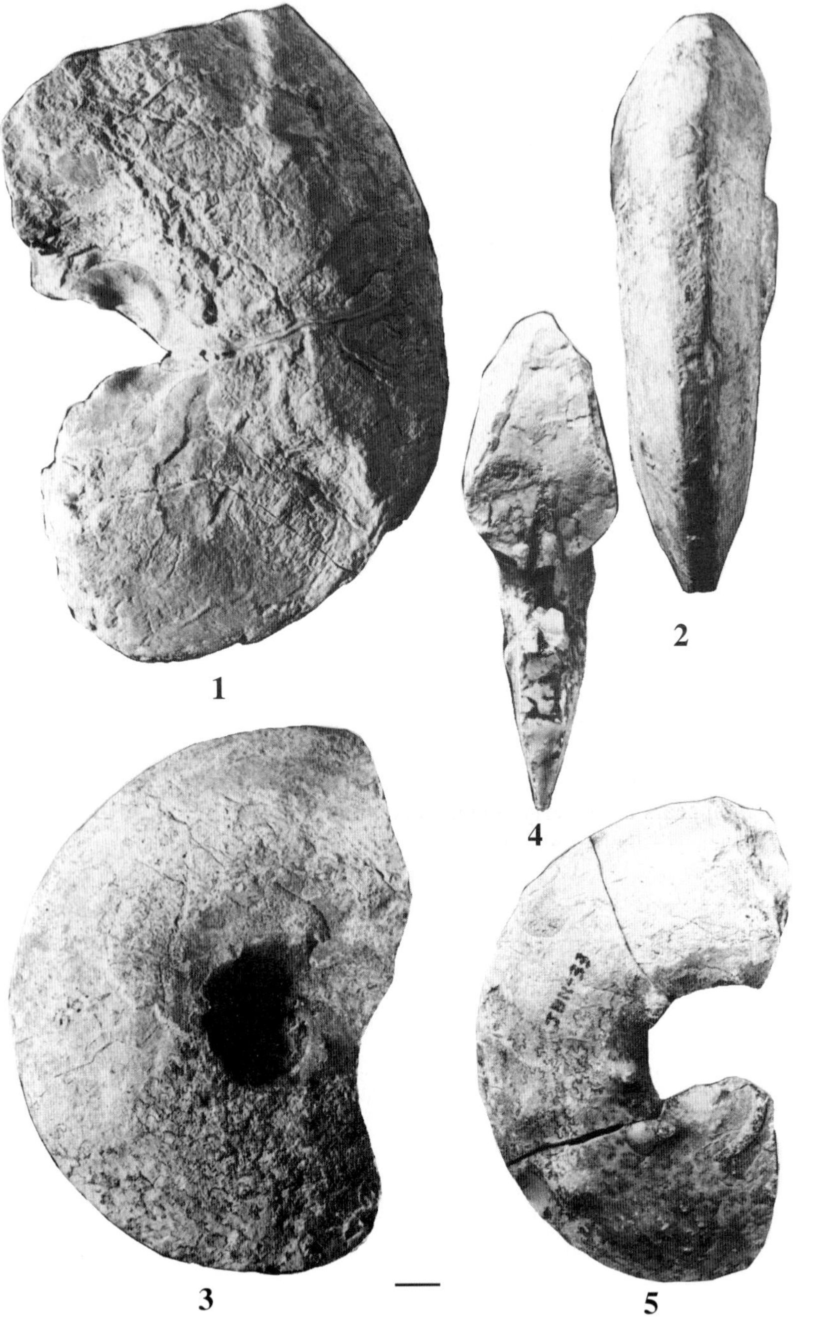

Fig. 5.5 Umkwelanense *(1) and* subkaffrarium *(2–5) types of ornament. 1, lateral view, microconch, body chamber; note feeble umbilical tubercles on inner whorl. 2, ventral view, microconch. 3, lateral view, microconch with incompletely preserved adult body chamber. 4, apertural view. 5, lateral view, microconch, adult with partially preserved body chamber; note strong umbilical tubercles. Scale bar = 1 cm.*

in position and persist to the end of the adult body chamber. Maximum diameter is about 134 mm (Range = 87–134; Av. = 104.97; S.D. = 17.90).

Macroconch. The inner whorls of the macroconch strongly resemble those of the microconch. Shell diameter exceeds 160 mm (Range = 85–160; Av. = 136.28; S.D. = 23.61). The body chamber occupies more than two-thirds of the last whorl. The venter is clavate up to the early mature body chamber; later it becomes broadly rounded. Umbilical tubercles appear early to mid-phragmocone stage and, at the late phragmocone stage, they become subdued and finally disappear on the early part of the body chamber. The whorl section is subtrigonal. Ventral clavi may persist up to the early part of the mature body chamber; after that, the venter becomes smooth and rounded.

2.3 *Placenticeras kaffrarium* Morph *kaffrarium* Etheridge, 1904

Microconch. The shell is tumid (WB/H = c. 0.95) with a broad clavate venter. The shell is strongly evolute (U/D = 0.32–0.47; Av. = 0.38; S.D. = 0.05), the flanks are flattened to weakly curved in the early stage, and may become strongly convex on the adult body chamber. Some compressed specimens exhibit a trigonal whorl section in the early stage, then become subtrigonal, and, finally, polygonal in the adult body chamber. However, in some inflated forms, the whorl section ranges from subtrigonal to polygonal. In highly ornate forms, robust ornament appears at an early stage. Ribs are prominent; primaries are short and bifurcate near the umbilical tubercles; secondaries are straight to slightly flexuous and terminate at the ventrolateral tubercles, which are nodose to bullate. The peristome has a ventral rostrum. The diameter may be up to 138 mm (Range = 72–138; Av. = 89; S.D. = 21.44).

Macroconch. The shell is similar to that of the microconch in the inner whorls, but shows variation in the adult whorls with respect to diameter, degree of involution, and strength of ornament. Maximum diameter may be up to 186 mm (Range = 90–186; Av. = 130.29; S.D. = 23.27). The venter is strongly clavate; this persists up to a variable length on the adult body chamber, which finally becomes smooth and rounded. Shell tumidity (WB/H = c. 1.02) and degree of involution (U/D = 0.29–0.47; Av. = 0.38; S.D. = 0.027) are less than those in the microconch. Again, there is covariation between inflated, evolute shells and the strength of the orn amentation. The strength of the lateral tubercles increases up to the first quarter of the adult body chamber; afterwards, they persist with attenuated strength. In the early to middle phragmocone stage, the umbilical tubercles are spinose, and more robust than the lateral tubercles, but, in the late phragmocone stage to the early part of the mature body chamber, the lateral tubercles appear to be stronger than the umbilical tubercles. The shell becomes smooth with a rounded venter near the aperture. The peristome is simple.

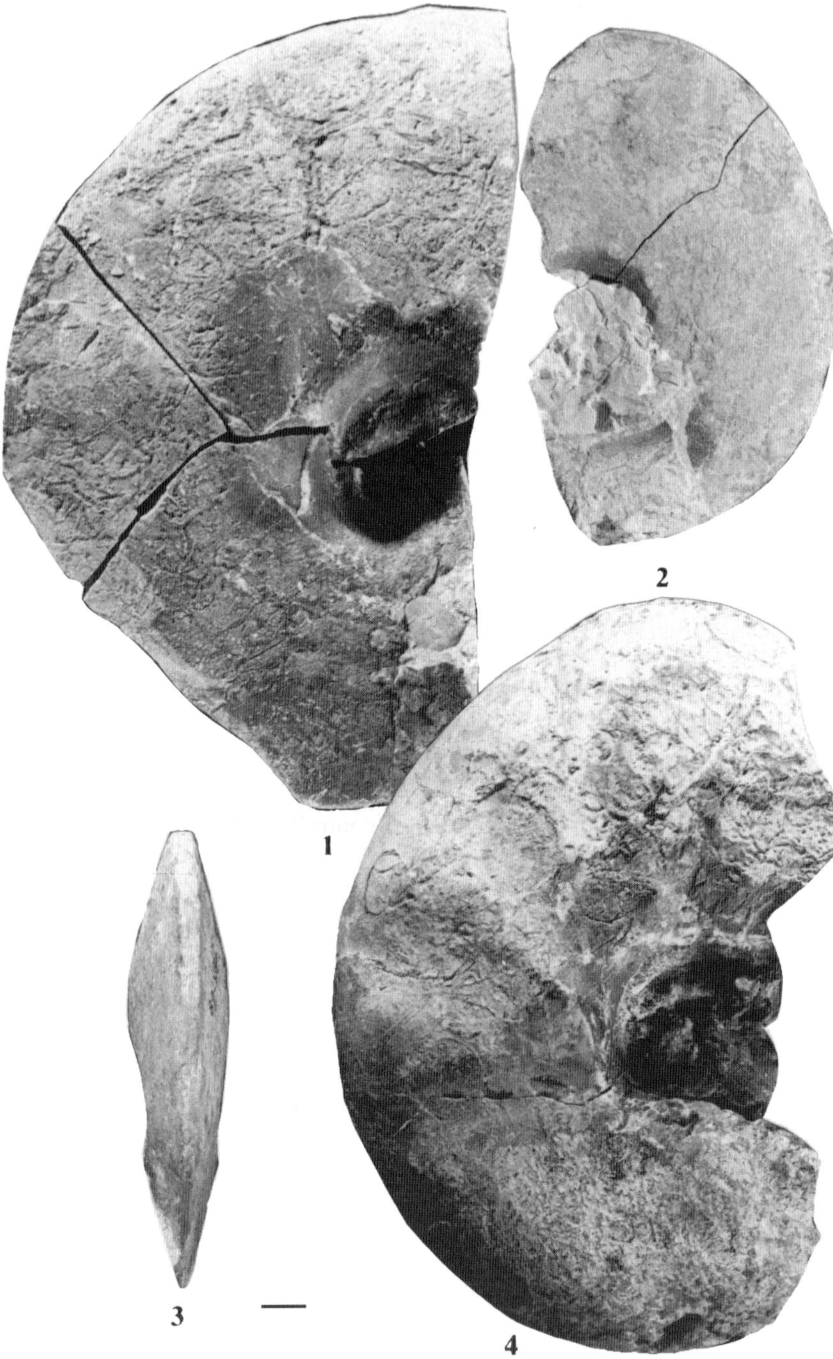

Fig. 5.6 Subkaffrarium *type of ornament. 1, lateral view, macroconch, incomplete body chamber. 2, lateral view, macroconch, adult phragmocone; note part of the body whorl preserved* 3, *ventral view, macroconch; note weakly developed clavi. 4, lateral view, macroconch, adult with major part of the body chamber preserved; note strength of umbilical tubercles decreases towards the aperture. Scale bar = 1 cm.*

3 Evolutionary Mechanisms of Polymorphism in *Placenticeras kaffrarium*

Detailed ontogenetic study of the *kaffrarium* morph of *Placenticeras kaffrarium* reveals a strong allometric change of development of ornament with age. There is a sequential development of ornament. The inner whorls are smooth and oxyconic with flattened flanks. Umbilical tubercles and ventral clavi appear simultaneously in the intermediate stage, which is followed by the development of lateral tubercles. The appearance and rate of development of tubercles and ventral clavi, however, vary greatly. The sequential and coordinated development of different aspects of ornament perhaps indicates that these features are constrained by a pleiotropic gene (see also Stanley, 1979), which, by switching off and on, can produce smooth or ornate adults. All polymorphic variants have more or less similar adult shell diameters and number of whorls, especially, of the macroconchs. This implies that they have a similar rate of shell growth, same biological age, and that the onset of sexual maturity takes place at the same time. In this perspective, the appearance of the smooth adult variant (*umkwelanense*) means that the activities of the gene or genes controlling ornament development, were completely stopped (recessive), and that the adult individuals neotenously retained the smooth, oxyconic early stage of the full ontogenetic spectrum. On the other hand, in the *subkaffrarium* type, the gene or genes were activated only in a later stage, and resulted in the development of only umbilical tubercles and ventral clavi. In the *kaffrarium* type of polymorphism, the early onset of the activities of controlling genes has predisplaced (*sensu* McNamara, 1990) the appearance of tubercles in early ontogeny and adult shells, therefore, bear strong ornament. Hence, it appears that the realization of different non-sexual adult variants is the product of allometry-induced heterochrony (cf. Gould, 1966, 1977).

4 Paleobiogeography and Paleoecology of *Placenticeras kaffrarium*

The genus *Placenticeras* has a wide geographic distribution, but species in any particular time are endemic and show clustered distribution patterns (Bardhan et al., 2002). They appear to have been constrained by latitudinal factors and, at times, were highly provincialized. Throughout their geological distribution, placenticeratids flourished best in the subtropical region on either side of the equator. For example, the earliest groups during the Albian-Cenomanian were mostly clustered in Central Asia or in Europe and America in areas having more–or–less similar latitudes (see Bardhan et al., 2002: Fig. 5.3). In Turonian-Coniacian times, the Central Asian stock migrated towards the southern margin of Tethys and occupied shallow epicontinental seas in India, Madagascar, and

Fig. 5.7 Kaffrarium *type of ornament, microconch. 1, lateral view, complete adult; note peristome showing ventral rostrum. 2, ventral view; note well-developed ventral clavi. 3, lateral view, small adult; note stronger lateral tubercles. 4, apertural view; note typical polygonal whorl section. 5, lateral view, adult; note increase in strength of ornamentation during ontogeny. Scale bar = 1 cm.*

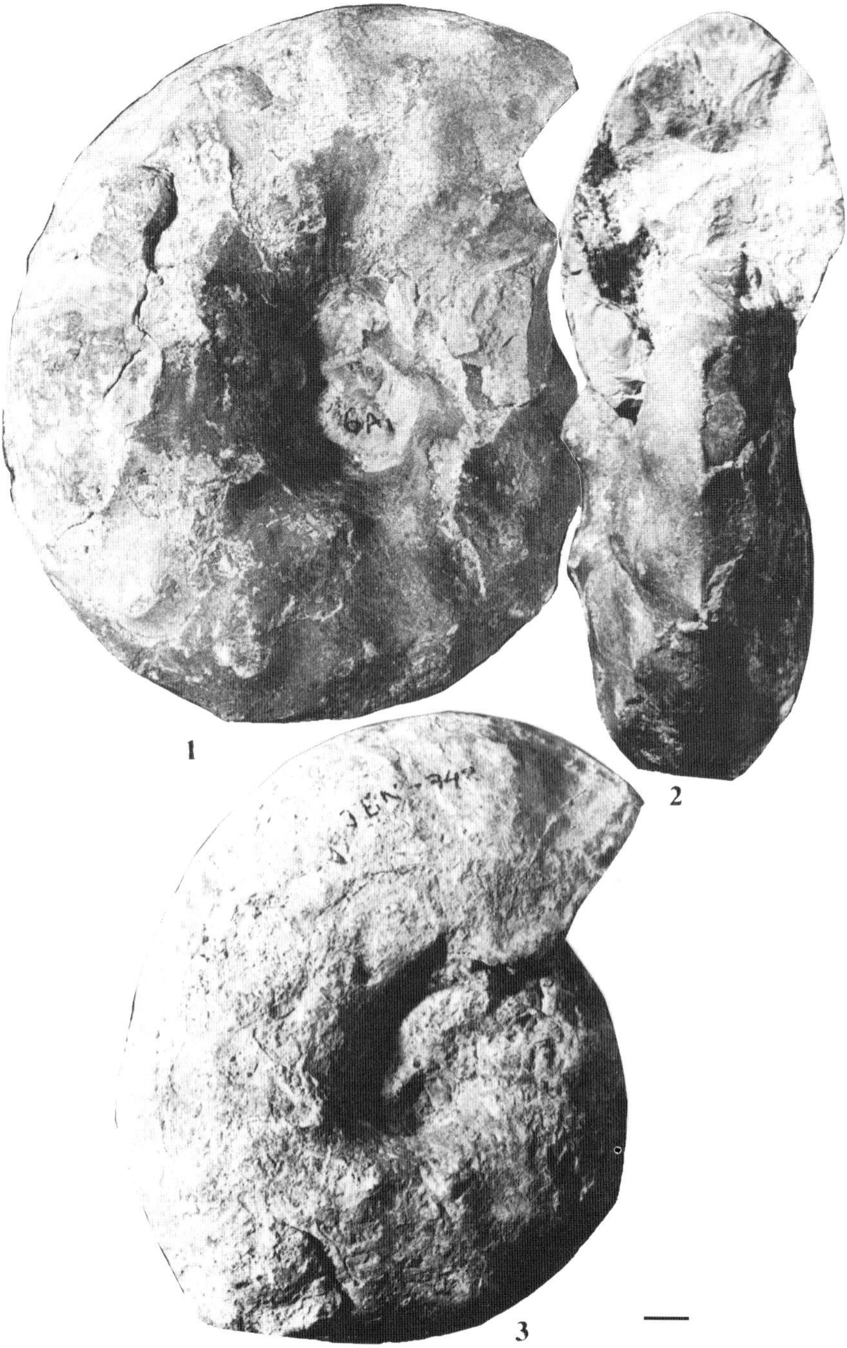

Fig. 5.8 Kaffrarium *type of ornament, macroconch. 1, lateral view, nearly complete adult showing strong umbilical and lateral tubercles, which are connected by coarse blunt ribs. 2, apertural view. 3, lateral view, adult; note attenuation of tubercles and scaphitoid uncoiling of the body chamber. Scale bar = 1 cm.*

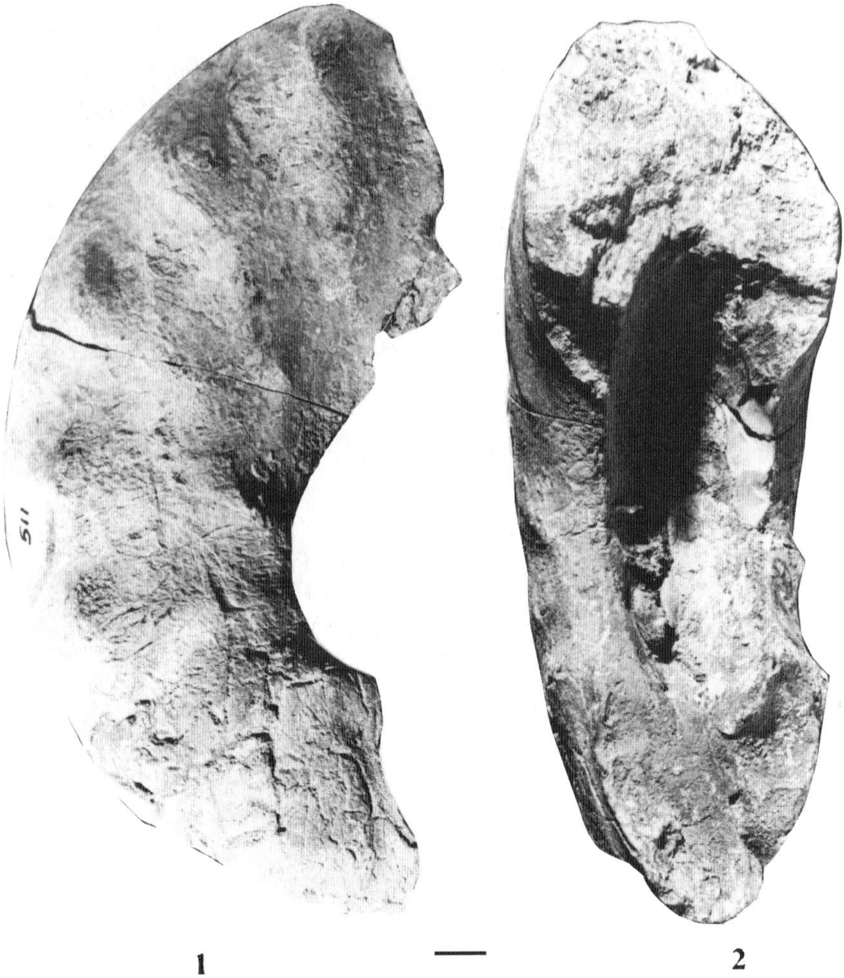

Fig. 5.9 Kaffrarium *type of ornament, macroconch. 1, lateral view, large body whorl fragment; note presence of faint rib. 2, apertural view. Scale bar = 1 cm.*

South Africa. It speciated quickly to produce *P. kaffrarium*, the most dominant species, which virtually monopolized the Indo–Madagascan Faunal Province (Bardhan et al., 2002: Fig. 5.4).

P. kaffrarium does not show any sedimentary facies control. In Zululand, it occurs mostly in siltstone and sandstone, whereas, in Bagh, the population is found only in carbonate facies. Klinger and Kennedy (1989: 383) interpreted the paleoenvironment of the Coniacian of Zululand as "indicative of a near shore protected environment with absence of strong currents and wave action but, nevertheless, normal salinity." In Bagh, the carbonates were deposited in a narrow intracratonic trough during the Coniacian transgression. The sedimentary facies analysis, as well as stable isotope (carbon) study, reveals that carbonate horizons bearing *P. kaffrarium* were deposited during the maximum flooding zone and the sedimentary facies and faunas indicate a normal marine environment. The Bagh basin was basically an inland seaway far from the open ocean. The associated faunas also include normal marine groups, such as echinoids, articulate brachiopods, and cheilostome bryozoans. However, all are characterized by exceptionally low diversity, almost monospecific; individuals nevertheless are highly abundant. The faunal association and low diversity pattern suggest that the environment was perhaps stressful, particularly with respect to seasonal salinity fluctuations. Tsujita and Westermann (1998) believed that *Placenticeras* could tolerate even a brachyhaline environment. The environment was placid from the biological and taphonomic point of view. Echinoids, brachiopods, and bivalves (mainly inoceramids) are mostly found articulated and in an exceptionally high state of preservation, without showing any injury or repair marks. In the history of the placenticeratids, the peristome was rarely found elsewhere in the fossil record and Westermann (1990) believed that they were highly predated. But, in the *P. kaffrarium* population of Bagh, many specimens have their apertural margins intact (Gangopadhyay and Bardhan, 1998: 195, 199). This indicates that the Bagh sea was environmentally, both physically and biologically, less stressful than other placenticeratid populations inhabiting the open shelf.

Ecologically isolated regions promote great diversity. Islands and lakes of the present day are areas of adaptive radiation (Stanley, 1979: 169–174). These regions experience much less biotic pressure, and organisms living there show a great genetic variability due to eco-insular conditions (Stanley, 1979). We believe that the release of stabilizing selection pressure due to the absence of competitors and predators, triggered the great intraspecific variability within the *P. kaffrarium* population, which ultimately led to polymorphism. For example, Janusson (1973) has also found the widespread presence of polymorphism in graptolite communities, which facilitated great evolutionary changes. Another example is pharyngeal teeth of cichlid fishes (Sage and Selander, 1975). The polymorphic differentiation of teeth enables the species as a whole, to get access to various foods. We shall see later that the different morphs of *P. kaffrarium* indicate niche partitioning and not a "biological extravaganza" as Arnould-Saget (1956, in Klinger and Kennedy, 1989: 383) believed for a similar variability in a population of *Knemiceras* Böhm, 1898. Because other polymorphic variants (*umkwelanense* and *subkaffrarium* morphs) are the scaled-up versions of different ontogenetic stages of the *kaffrarium* morph, they were "pre-adapted" and had no difficulties in finding different niches.

Ammonite shell forms and sculptures are functionally significant (Kennedy and Cobban, 1976; Ward, 1981; Westermann, 1990, and references therein). *Placenticeras* shows wide interspecific diversity with respect to size and ornament, but the shape (broadly involute and oxyconic) and sutural patterns (complex) show phylogenetic conservatism. Patterns of sedimentary facies and paleobiogeographic distribution indicate a shallow marine habitat for *Placenticeras*. Functional morphological analysis suggests that placenticeratids have the weakest shells known and were one of the shallowest ammonites having lived in the proximal sublittoral zone less than

50 m depth (see Westermann, 1990; Tsujita and Westermann, 1998). Analysis of the functional morphology of *P. kaffrarium* (Bardhan et al., 2002) reveals that individuals each have a very thin shell, thin septa with a relatively large siphuncle, and complex septal sutures with relatively low height of the sutural elements. This implies an exceptionally low SRI (Sutural Reinforcement Index) and a shallow water mode of life (Batt, 1986, 1993). The functional significance of ammonite ornament has been variously interpreted: camouflage (Kennedy and Cobban, 1976), an effective buoyancy device (Kennedy, 1986), predator avoidance (Tsujita and Westermann, 1998), and hydrodynamic device (Chamberlain and Westermann, 1976). The presence of great intraspecific variability with respect to ornament and form suggests that the population was not under strong selection pressure (Seilacher, 1972), and the absence of septate spines indicates that buoyancy regulation was not a constraint for the *P. kaffrarium* population (Kennedy, 1986). It is generally believed that individuals of ammonite species showing marked ontogenetic changes are vertical migrants during their lives (Westermann, 1990; Batt, 1993). Batt (1993) has shown that ammonite morphotypes are bathymetrically controlled. The smooth, oxyconic form (*umkwelanense* type) inhabited the shallowest part of the basin margin (see also Tsujita and Westermann, 1998), whereas strongly sculpted, nodose shells with polygonal whorl section (*kaffrarium* type) lived in the relatively deeper zone. Chamberlain and Westermann (1976) suggested that ammonite ornament also has hydrodynamic significance. Strong sculpture, especially located on the outer flanks (like the *kaffrarium* morph) can produce drag, thereby making the animal a slow swimmer. On the other hand, smooth forms, like the *umkwelanense* type with compressed shells, would have experienced minimum drag and, therefore, an enhanced swimming ability.

5 Remarks

Polymorphism may prompt evolutionary diversity resulting in adaptive radiation within a lineage (Stanley, 1979: 171). We have already mentioned widespread presence of polymorphism in graptolite communities, which facilitated great evolutionary changes (Janusson, 1973). Recently, Dommergues (1990) has shown the role of polymorphism in shaping the evolutionary trends of the Middle Liassic ammonite family Polymorphidae. Kennedy and Cobban (1976) have given an example of such a trend where polymorphic variants of the Late Albian *Neogastroplites haasi* Reeside and Cobban, 1960, show shifting proportions of various ornamental types in the evolution of successive species. We, here, have tried to outline the broad and major trends in placenticeratid evolution subsequent to the rise of extraordinary polymorphism in *P. kaffrarium* (Fig. 5.10).

Placenticeras has a geological range from the Upper Albian to Upper Campanian. Although the present species exhibits high intraspecific variability, the genus and even the entire family show evolutionary stasis of shell shape and coiling – involute and oxyconic (see also Kennedy and Wright, 1985). During phylogenetic

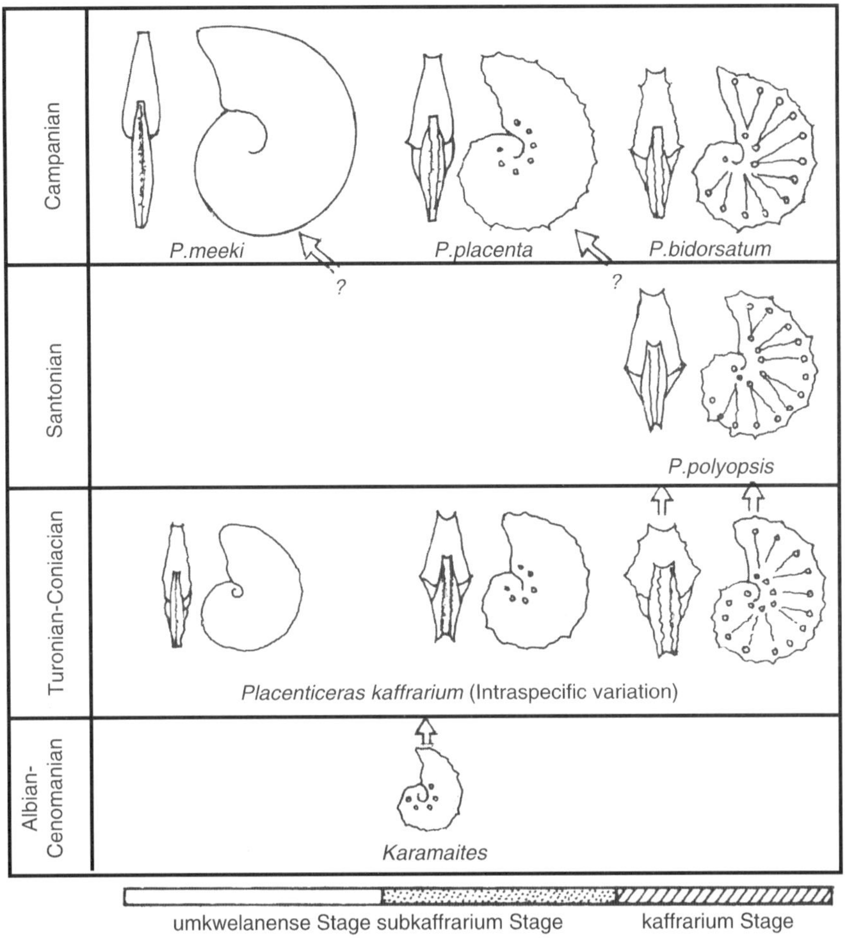

Fig. 5.10 *Some of the major evolutionary trends within the genus* Placenticeras *are shown. Note how intraspecific variability of* P. kaffrarium *influenced subsequent evolution. Different ontogenetic stages of* P. kaffrarium, *which have been variously recapitulated by the younger species, are shown at the base.*

history, species show only "fine tuning" of this basic design. What varies significantly in evolution is the body size and strength and nature of ornament. A general, sometimes weak, covariation exists among the degree of involution, inflation, and strength of ornament. Involute shells are oxyconic, compressed, and weakly ornamented, whereas coarsely ornate shells are relatively evolute with polygonal whorl sections. However, no such covariation is seen with increase in body size. Different species have different ranges of variability of the full spectrum of this covariation and their intraspecific variability sometimes intergrades; this created taxonomic confusion for many early workers. Now it is understood that species discrimination lies also in stratigraphic and paleobiogeographic distribution (Callomon, 1985;

Jana et al., 2005). Bardhan et al. (2002) demonstrated that *Placenticeras* is a wandering genus, and species have different spatiotemporal distribution patterns.

Bardhan et al. (2002) made a paleobiogeographical area cladogram based on phylogenetic systematics to illustrate the relationships among the major provincial groups of placenticeratid species. The earliest species are found mainly around Central Asia during the Late Albian–Cenomanian. They are characteristically involute and oxyconic, with weakly developed umbilical tubercles and ventral clavi on the phragmocone, which fade or disappear completely on the adult body chamber (see for example Marcinowski et al., 1996: pls. 12, Figs. 5 a, b). Falcoid ribs and feeble lateral bullae may appear in individuals of some species during ontogeny, but none is completely smooth. Size shows much variation; individuals of some, e.g., "*Karamaites*" *grossouvrei* (Semenov, 1899) may attain 50cm diameter (Klinger and Kennedy, 1989) and others, e.g., *K. mediasiaticum* (Luppov, 1963), 12cm diameter (Marcinowski et al., 1996). The Late Cenomanian–Turonian development of swelling or true lateral tubercles may appear in some forms, leading to the characteristic "*Proplacenticeras*" type of ornament (Klinger and Kennedy, 1989). In *P. cumminsi* Cragin, 1893, from the Upper Cenomanian of Texas (Kennedy 1988: 29–39), variation with respect to ornament is present within the population, but no distinct lateral tubercles are developed. Shell shape and coiling remain essentially oxyconic and involute. Size, however, may exceed 29cm (range = 19.5–29.5; Av. = 25.37; S.D = 43.97).

During the Turonian-Coniacian, *P. kaffrarium* evolved from Central Asian stock. It exhibited a stunning range of intraspecific variability as if the "whole genetic tool box" of the genus *Placenticeras* opened up within the range of variability of *P. kaffrarium*. Not only previous phylogenetic trends had been recapitulated; the completely smooth shell or shell with stronger and dominant lateral tubercles, coupled with coarser ribbing, developed as novelties, which contributed greatly to the subsequent evolution. Ancestral morphological characters not only were recapitulated during ontogeny, but they also were realized as different phenotypic adults – some are compressed and devoid of any ornament and have trigonal apertures, other may acquire strong ornament, and tumid shells with polygonal to rounded whorl sections. More importantly, dimorphism is manifested in body size and ornamental disparity. In addition, microconchs have peristomal modifications.

Why was there such a stunning display of intraspecific variability in *P. kaffrarium*? It is not an evolutionary corollary of a phylogenetic trend that started from feebly tuberculate "*Karamaites*" of the Albian–Cenomanian by way of Turonian ribbed and outwardly tuberculate forms (for example *P. cumminsi* Cragin, 1893). During the Coniacian, placenticeratid species show two dominant trends; one is Indo-Madagascan *P. kaffrarium*, which may be strongly sculpted, and another is European *P. fritschi* de Grossouvre, 1894, which exhibit a narrow range of variation, dominated by weakly ornamented forms with small umbilical tubercles, but no ventral clavi. Similar weakly ornamented forms are also found in the coeval rocks of the Western Interior, which Kennedy and Cobban (1991) described as *P. kaffrarium*; we, however, believe that it is a different species and closely related to, if not conspecific with *P. fritschi* (see also Bardhan et al., 2002: 203).

Klinger and Kennedy (1989) called upon the effect of geographic isolation and environmental control to explain the development of phenotypic variation within the *P. kaffrarium* population of Zululand, South Africa. We add that lack of predation and strong dimorphism are also equally responsible for producing this kind of great intraspecific variability (see Bardhan et al., 2002). *Placenticeras kaffrarium* even extends the range of variability inherited from its ancestor. It has a completely smooth adult shell, namely, *umkwelanense*. This smooth, non-tuberculate morph actually represents the plesiomorphic inner whorls that characterize the entire family and never appeared before in adults of any species in the phylogeny. The *kaffrarium* morph has lateral tubercles, which are stronger than umbilical tubercles – another novelty. Thus, it appears that the whole genetic plasticity of the genus *Placenticeras* has been released in this *P. kaffrarium* population. Klinger and Kennedy (1989: 259) also observed that "some…of the *Placenticeras* populations we have studied show as much morphological variation at a single stratigraphic level as the whole genus throughout its entire history." However, *P. kaffrarium* is smaller in diameter in Bagh, with a maximum diameter of about 19 cm, whereas in Zululand it is slightly larger, with a maximum diameter of 23 cm.

Subsequent evolutionary history of *Placenticeras* is simple. Individuals of younger species retain features of one or the other variant of *P. kaffrarium*, with minor modifications. Body size shows great fluctuation. Individuals of some species, for example, *P. semiornatum* (d'Orbigny, 1850) of the Late Coniacian (Kennedy, 1984: 44–45) appear even smaller than at the starting point, but up the sequence, maximum body size moves to larger values and in the Campanian, *Placenticeras meeki* Böhm, 1898, attains a gigantic size of about 1 m (Tsujita and Westermann, 1998). This kind of trend according to Gould (1988) is due to an increase in variance which results from an increase in speciation rate. This is called Cope's Rule in broad sense (cf. Jablonski, 1996). However, so far as ornament is concerned, the evolutionary trend followed no particular pattern. Individuals of the species may be completely smooth or strongly tuberculate. For example, *P. semiornatum* strongly resembles the smooth *umkwelanense* morph. The *kaffrarium* type of ornament is best observed in *P. polyopsis* Dujardin, 1837, of France in the Santonian (Kennedy and Wright, 1983). The Campanian of Europe is typified by the presence of large *P. bidorsatum* (Roemer, 1841). This species has a non-tuberculate, entire venter developed to a great extent. According to some workers (Kennedy, 1986; Klinger and Kennedy, 1989) *P. bidorsatum* evolved from the *P. polyopsis–P. paraplanum* Wiedmann, 1978, group of species and they have similar ornament on the inner whorls. It was believed that the evolution was induced through hypermorphosis. Alternatively, Ulbrich (1971) envisioned evolution of *P.bidorsatum* from *P. guadulupae* (Roemer, 1852) via *P. radiatum* Riedel, 1937. The evolutionary changes include progressive disappearance of ribs and tubercles. Pending the controversy of the evolutionary legacy of *P. bidorsatum*, it can be said that both ancestral *P. polyopsis* and *P. guadulupae* (this species is now designated as *P. syrtale* Morton, 1834, see Klinger and Kennedy, 1989: 385) had *P. kaffrarium*-like ornament and shell shape and *P. bidorsatum* represented the *umkwelanense* type having smooth shell or feeble ornament and tabulate venter up to a great extent. *Placenticeras polyopsis* has been found below the *P. syrtale* zone in Alabama (Kennedy, 1987; Kennedy and Cobban, 1993). The occurrence of *P. polyopsis* in both Europe and America speaks for migration. *Placenticeras syrtale* strongly resembles *P. polyopsis* and they may form an evolutionary lineage. Both species are simply dimorphic and exhibit a wide range of intraspecific variability like the present *P. kaffrarium* population.

Above, we have discussed the macroevolutionary changes in *Placenticeras* which shows the sudden appearance of many species having morphological novelty. This evolutionary pattern seems to correspond well with the punctuational model of evolution (Eldredge and Gould, 1972). However, within species, changes are gradual. Klinger and Kennedy (1989) observed gradual shifts of the population from the *kaffrarium* type of variant in lower stratigraphic levels to the *umkwelanense* type in younger stratigraphic horizons. They even were inclined to subdivide the species into evolutionary subspecies. Similar, gradual evolutionary changes have been documented by Wolleben (1967) in the Senonian *P. syrtale* of TransPecos. He documented many species and found numerous transitional forms between them. The intermediates exist both spatially and temporally. It has recently been found that these are nothing but polymorphic variants of the same species as *P. syrtale* (Klinger and Kennedy, 1989: 385–388).

Later evolutionary changes in the American placenticeratid lineage are exemplified by large and smooth or weakly ornamented species, e.g., *P. placenta* Dekay, 1828, of the Upper Santonian to Lower Campanian (Kennedy and Cobban, 1993). It strongly resembles the *umkwelanense* morph. During the Late Campanian several species have been described from the Bearpaw Formation of the Western Interior. They are *P. meeki* Böhm, 1898, *P. intercalare* (Meek and Hayden, 1860), and *P. costatum* Hyatt, 1903. All of these are giants approaching nearly 1 m diameter (see Tsujita and Westernman, 1998). *Placenticeras meeki* has the *umkwelanense* type of involute coiling and smooth shell, possibly retained through neoteny, whereas *P. intercalare* and *P. costatum* have robust ornament. We believe that they may belong to the same species and represent polymorphic variants (also Cobban, personal communication, 2005). While explaining the evolutionary mechanism of large sized *P. bidorsatum*, Klinger and Kennedy (1989) inferred its origin from *P. paraplanum* through hypermophosis. *P. bidorsatum* retains the entire venter up to a great extent, but is nothing more than the retention of a juvenile ancestral feature into the adult stage. We also believe that evolution of giant placenticeratid species of the Bearpaw Formation involve hypermorphosis.

Thus, it appears that the great intraspecific variability of *P. kaffrarium*, has subsequently contributed to the lineage mostly through neotenous retention or by hypermorphosis.

Acknowledgments

We are grateful to two reviewers, W. A. Cobban and another anonymous reviewer, and the editor R. A. Davis for their valuable suggestions. We thank S. Shome, P. Roy, S. Das, and S. Paul for help in various stages. Mrs. N. Sengupta helped in German translation. T. K. G. acknowledges the organizing committee of the Sixth International Symposium on Cephalopods – Present and Past for local hospitality and the authorities of the Bengal Engineering and Science University Shibpur, Howrah, for providing travel support, S. B. acknowledges the partial financial support provided by the CAS, Jadavpur University.

References

Bardhan, S., T. K. Gangopadhyay, and U. Mondal. 2002. How far did India drift during the late Cretaceous? *Placenticeras kaffrarium* Etheridge, 1904 (Ammonoidea) used as a measuring tape. *Sedimentary Geology* **147**: 193–217.

Batt, R. 1986. A test of the effects of paleoecological factors on the distribution of ammonite shell morphotypes. Greenhorn Cyclothem, Cretaceous Western Interior Seaway. *In* E. G. Kauffman (editor), *Cretaceous Biofacies of the Central Part of Western Interior Seaway: A Field Guidebook*, pp. 16–52. Colorado: Fourth North American Paleontological Convention. University of Colorado.

Batt, R. 1993. Ammonite morphotypes as indicators of oxygenation in a Cretaceous epicontinental sea. *Lethaia* **26**: 49–63.

Böhm, J. 1898. Über *Ammonites pedernalis* V. Buch. *Zeitschrift der Deutschen Geologischen Gesellschaft* **50**: 183–201.

Boselini, A. 1989. The continental margins of Somalia: their structural evolution and sequence stratigraphy. *Memorie di Scienze Geologiche* **41**: 373–458.

Callomon, J. H. 1985. The evolution of the Cardioceratidae. *Special Papers in Palaeontology* **33**: 49–90.

Chamberlain, J. A., and G. E. G. Westermann. 1976. Hydrodynamic properties of cephalopod shell ornament. *Paleobiology* **2**: 316–331.

Chiplankar, G. W., and R. M. Badve. 1973. Age and affinities of the Bagh fauna – a reassessment. *Bulletin of the Indian Natural Science Academy* **45**: 19–29.

Chiplankar, G. W., and M. A. Ghare. 1977a. Palaeontology of the Bagh Beds. Part X. Scaphitidae. *The Proceedings of the Indian Academy of Sciences* **85B**(2): 67–76.

Chiplankar, G. W., and M. A. Ghare. 1977b. Comments on *Proplacenticeras stantoni* (Hyatt) and its variety *bolli* (Hyatt). *Journal of the University of Poona, Science and Technology* **50**: 221–226.

Collignon, M. 1965a. Atlas des fossiles caractéristiques de Madagascar (Ammonites). **XII** (Turonien). Tananarive: Service Géologique. iv + 82pp., pls. 376–413.

Collignon, M. 1965b. Atlas des fossiles caractéristiques de Madagascar (Ammonites). **XIII** (Coniacien). Tananarive: Service Géologique. vii + 88pp., pls. 414–454.

Davis, R. A., N. H. Landman, J. L. Dommergues, D. Marchand, and H. Bucher. 1996. Mature modifications and dimorphism in ammonoid cephalopods. *In* N. H. Landman, K. Tanabe, and R. A. Davis (editors), *Ammonoid Paleobiology*, pp. 464–539. New York: Plenum Press.

DeKay, J. E. 1828. Report on several fossil multilocular shells from the state of Delaware etc. *Annals of the Lyceum of Natural History of New York* **2**: 273–279, pl. 5 (Figs. 2–5).

Dommergues, J. L. 1990. Ammonoids. *In* K.J. McNamara (editor), *Evolutionary Trends*, pp. 162–187. London: Belhaven Press.

Dujardin, F. 1837. Mémoire sur les couches du sol en Touraine et description des coquilles de la craie et des Faluns. *Mémoires de la Société géologique de France* **2**: 211–311, pls. 15–20.

Etheridge, R. 1904. Cretaceous fossils of Natal, 1, The Umkwelane Hill Deposit. *Report of the Geological Survey of Natal and Zululand* **1**: 71–93, pls. 1–3.

Ford, E.B. 1940. Polymorphism and Taxonomy. *In* J. S. Huxley (editor), *The New Systematics*, pp. 493–513. London: Oxford University Press.

Gangopadhyay, T., and S. Bardhan. 1998. Apertural modifications and jaw structures of placenticeratid ammonites from the Upper Cretaceous Bagh Group, Central India. *Neus Jahrbuch Geologie und Palaeontologie Monatschafte* **4**: 193–202.

Ganguly, T., and S. Bardhan. 1993. Dimorphism in *Placenticeras mintoi* from the Upper Cretaceous Bagh Beds, Central India. *Cretaceous Research* **14**: 747–756.

Gould, S. J. 1966. Allometry and size in ontogeny and phylogeny. *Biological Review* **41**: 587–640.

Gould, S. J. 1977. *Ontogeny* and *Phylogeny*. Cambridge: Harvard University Press.

Gould, S. J. 1988. Trends as changes in variance: a new start on progress and directionality in evolution. *Journal of Paleontology* **62**: 319–29.

Grossouvre, A. de. 1894. Recherches sur la craie supérieure, 2: paléontologie. Les ammonites de la craie supérieure. *Mémoires pour servir à l'explication de la carte géologique détaillée de la France*, 264pp., 39 pls.

Halder, K., S. K. Jana, and S. Bardhan. 1998. Ornamental polymorphism in *Paracenoceras prohexagonum* (Nautiloidea) from the Jurassic Chari sequence, Kutch, Gujarat. *Indian Minerals* **52**: 89–94.

Hyatt, A. 1903. Pseudoceratites of the Cretaceous. *Monograph. United States Geological Survey* **44**: 1–351.

Jablonski, D. 1996. Body size and macroevolution. *In* D. Jablonski, D. H. Erwin, and J. Lipps (editors), *Evolutionary Paleobiology*, pp. 256–289. Chicago: University of Chicago Press.

Jafar, S.A. 1982. Nannoplankton evidence of Turonian transgression along Narmada valley, India and Turonian–Coniacian boundary problem. *Journal of Paleontolgical Society of India* **27**: 17–30.

Jana, S.K., S. Bardhan, and K. Halder. 2005. Eucycloceratin ammonites from the Callovian Chari Formation, Kutch, India. *Palaeontology* **48**(4): 883–924.

Janusson, V. 1973. Morphological discontinuities in the evolution of graptolite colonies. *In* R. S. Boardman, A. H. Cheetham, and W. A. Oliver (editors), *Animal Colonies: Development* and *Function through time*, pp. 515–521. Stroudsburgh, Pennsylvania: Dowden, Hutchinson and Ross.

Kamiya, T. 1992. Heterochronic dimorphism in *Loxoconcha vranouchiensis* (ostracoda) and its implication for speciation. *Paleobiology* **18**: 221–236.

Kennedy, W. J. 1984. Systematic Palaeontology and stratigraphic distribution of the ammonite faunas of the French Coniacian. *Special Papers in Palaeontology* **31**: 1–160.

Kennedy, W. J. 1986. Campanian and Maastrichtian ammonites from Northern Aquitaine, France. *Special Papers in Palaeontology* **36**: 1–145.

Kennedy, W. J. 1987. Ammonites from the type Santonian and adjacent parts of Northern Aquitaine, Western France. *Palaeontology* **30**: 765–782.

Kennedy, W. J. 1988. Late Cenomanian and Turonian ammonites from north-east and central Texas. *Special Papers in Palaeontology* **39**: 1–131.

Kennedy, W. J., and W. A. Cobban 1991. Coniacian ammonite faunas from the United States Western Interior. *Special Papers in Palaeontology* **45**: 1–96.

Kennedy, W. J., and W. A. Cobban. 1993. Lower Campanian (Upper Cretaceous) ammonites from the Merchantville Formation of New Jersey, Maryland, and Delaware. *Journal of Palaeontology* **67**: 828–849.

Kennedy, W. J., W. A. Cobban, and N. H. Landman. 1996. Two species of *Placenticeras* (Ammonitina) from the Upper Cretaceous (Campanian) of the Western Interior of the United States. *American Museum Novitates* **3173**: 1–13.

Kennedy, W. J., and C. W. Wright. 1983. *Ammonites polypsis* Dujardin, 1837 and the Cretaceous ammonite family Placenticeratidae Hyatt, 1900. *Palaeontology* **26**: 855–873.

Kennedy, W. J., and C. W. Wright. 1985. Evolutionary patterns in Late Cretaceous ammonites. *Special Papers in Palaeontology* **33**: 131–294.

Kennedy, W. J., V. G. Phansalkar, and I. Walaszczyk. 2003. *Prionocyclus germari* (Reuss, 1845), a Late Turonian monster fossil from the Bagh Beds of Central India. *Cretaceous Research* **24**: 433–438.

Klinger, H. C., and W. J. Kennedy. 1989. Cretaceous faunas from Zululand and Natal, South Africa, The ammonite subfamily Placenticeratidae Hyatt, 1900: with comments on the systematic position of the Genus *Hypengonoceras* Spath, 1924. *Annals of the South African Museum* **98**(9): 241–408.

Luppov, N. P. 1963. New Cenomanian and Turonian ammonites of the genus *Placenticeras* from Middle Asia. *Trudy VSEGEL, N.S.* **109**: 142–158.

Marcinowski, R., I. Walaszczyk, and D. Olszewska–Nejbert. 1996. Stratigraphy and development of the mid Cretaceous (Upper Albian through Coniacian) of the Mangyshlak Mountains, Western Kazakhstan. *Acta Geologica Polonica* **46**(1–2): 1–60.

McGraw–Hill Concise Encyclopaedia of Science & Technology. 1984. S.P. Parker (editor). New York: McGraw-Hill.

McNamara, K. J. 1990. The role of heterochrony in evolutionary trends. *In* K. J. McNamara (editor), *Evolutionary Trends*, pp. 59–74. London: Belhaven Press.

Meek, F. B. 1876. A report on the invertebrate Cretaceous and Tertiary fossils of the Upper Missouri country. *In* F. V. Hayden, *Report of the United States Geological Survey of the Territories* **9**: lxiv + 629pp., 45 pls.

Meek, F. B., and F. V. Hayden. 1860. Descriptions of new Lower Silurian (Primordial), Jurassic, Cretaceous and Tertiary fossils, collected in Nebraska Territory with some remarks on the rocks from which they were obtained. *Proceedings of the Philadelphia Academy of Sciences* **13**: 415–447.

Melendez, G., and Fontana, B. 1993. Intraspecific variability, sexual dimorphism and non-sexual polymorphism in the ammonite genus *Larcheria* Tintant (Perisphinctidae) from the Middle Oxfordian of Western Europe. *In* M. R. House (editor), *The Ammonoidea: Environment, Ecology* and *Evolutionary Change*, pp. 165–186. Oxford: Clarendon Press.

Morton, S.G. 1834. *Synopsis of the organic remains of the Cretaceous groups of the United States.* Illustrated by nineteen plates to which is added an appendix containing a tabular view of the Tertiary fossils discovered in America. Philadelphia: Key and Biddle.

Reeside, J. B, and W. A. Cobban. 1960. Studies of the Mowry Shale (Cretaceous) and contemporary formations in the United States and Canada. *U.S. Geological Survey Professional Paper* **355**: 1–126.

Reyment, R. A. 1988. Does sexual dimorphism occur in Upper Cretaceous ammonites? *Senckenbergiana Lethaea* **69**(1, 2): 109–119.

Riedel, I. 1937. Die Salzbergmergel und ihre äquivalente in Westfalen. *Jahrbuch der preussischen geologischen Landesanstalt Bergakademie* **58**: 207–229.

Roemer, F. A. 1841. Die Versteinerungen des morddeutschen Kreidegebirges. *Hahn'schen Hofbuchandlung*: 49–145.

Roemer, F. A. 1852. Die Kreidebildungen von Texas und ihre organischen Einschilusse. Bonn.

Sage, R. D., and R. K. Selander. 1975. Trophic radiation through polymorphism in cichlid fishes. *Proceedings of the National Academy of Sciences U.S.A.* **72**: 4669–4673.

Seilacher, A. 1972. Divaricate patterns in pelecypod shells. *Lethaia* **5**: 325–343.

Semenov, W. P. 1899. The fauna of the Cretaceous deposits of Mangyschlak and some other localities in the Transcaspian Province. *Travaux de la Société Impérial de St. Petersbourg* (*Section Géologie et Minéralogie*) **28**: 1–178. [In Russian]

Spath, L. F. 1921. On Cretaceous Cephalopoda from Zululand. *Annals of the South African Museum* **12**(7): 217–321.

Stanley, S. M. 1979. *Macroevoluton–Patterns and Processes.* San Francisco: W.H. Freeman.

Taylor, P.D., and R.M. Badve. 1992. The mid Cretaceous bryozoan fauna from the Bagh Beds of Central India: composition and evolutionary significance. *In* P. J. Hayword, J. S. Ryland, and P. D. Taylor (editors), *Biology and Palaeobiology of Bryozoans*, pp. 1–35. Fredersberg: Olsen & Olsen.

Tintant, H. 1963. Les Kosmocératidès du Callovien inférieur et moyen d'Europe occidentale. *Publications de l'Université de Dijon* **29**: 500pp.

Tintant, H. 1980. Problématique de l'espèce en Paléozoologie. *In* Les problèmes de l'espèce dans le monde animal. *Mémoire Société Zoologique de France* **40**: 321–372.

Tsujita, T., and G. E. G. Westermann. 1998. Ammonoid habitats and habits in the Western Interior Seaway: a case study from the Upper Cretaceous Bearpaw Formation of Southern Alberta, Canada. *Palaeogeography, Palaeoclimatology, Palaeoecology* **144**(1–2): 135–160.

Ulbrich, V. H. 1971. Zur Paläozoikums und Mesozoikums Europas. *Freiberger Forschungshefte*: 47–71.

Ward, P. D. 1981. Shell sculpture as a defensive adaptation in ammonites. *Paleobiology* **7**: 96–100.

Westermann, G. E. G. 1990. New developments in ecology of Jurassic–Cretaceous ammonoids. *In* G. Pallini, F. Cecca, S. Cresta, and M. Santantonio (editors), *Atti del Secondo Convegno Internazionale Fossili, Evoluzione, Ambiente*, pp. 459–478. Pergola.

Wiedmann, S. 1978. Eine paläogeographisch interessante Ammonitenfaunula aus der alpinen Gosau. *Eclogae Geologicae Helvetiae* **71**: 663–675.

Wolleben, J. A. 1967. Senonian (Cretaceous) mollusca from Trans–Pecos Texas and northeastern Chihuahua, Mexico. *Journal of Paleontology* **41**(5): 1150–1165.

Wright, C. W., J. H. Calloman, and M. K. Howarth. 1996. Cretaceous Ammonoidea. *In* R. L. Kaesler (editor), *Treatise on Invertebrate Paleontology*, Part L, Mollusca 4, Revised, pp. 1–362. Lawrence, Kansas, and Boulder, Colorado: University of Kansas Press and Geological Society of America.

Chapter 6
A Late Carboniferous Coleoid Cephalopod from the Mazon Creek *Lagerstätte* (USA), with a Radula, Arm Hooks, Mantle Tissues, and Ink

Larisa A. Doguzhaeva,[1] Royal H. Mapes,[2] and Harry Mutvei[3]

[1]Paleontological Institute of the Russian Academy of Sciences, St. Profsoyuznaya, 123, Moscow 117997, Russia, ldoguzhaeva@rambler.ru;

[2]Department of Geological Sciences, Ohio University, Athens, OH 45701, USA, mapes@ohio.edu;

[3]Department of Palaeozoology, Swedish Museum of Natural History S-10405 Stockholm, Sweden, harry.mutvei@nrm.se

Keywords: coleoids, Carboniferous, radula, arm hooks, ink

1 Introduction

In the light of the current scanty knowledge of their origin and early evolutionary history, and our inadequate understanding of the higher-level phylogenetic relationships of the Coleoidea, any Carboniferous shell, which might have belonged to a coleoid cephalopod, is of exceptional interest. This report describes two relatively small (total shell and cephalic area is about 64 mm long in one specimen) cephalopod fossils from the Mazon Creek *Lagerstätte* (Middle Pennsylvanian–Desmoinesian) in Illinois, USA: specimen FMNH PE 32521 (Saunders and Richardson, 1979:

N. H. Landman et al. (eds.), *Cephalopods Present and Past: New Insights and Fresh Perspectives*, 121–143.

Fig. 9a, b, d) and specimen FMNH PE 20808; the latter as a gift to the Field Museum of Natural History by the collector David Young. Specimen PE 32521 originally was referred by Saunders and Richardson in 1979 to the only known Carboniferous non-rostrum-bearing coleoid *Jeletzkya douglassae* Johnson and Richardson, 1968. The diagnosis of the holotype of *J. douglassae* was based mainly on a moderately well-preserved crown of ten arms and a small, poorly preserved phragmocone. Johnson and Richardson (1968) interpreted this specimen as the earliest known teuthid. This assignment was questioned by Gordon (1971), who was convinced that the disposition of the arm hooks in pairs pointed to a belemnitid-phragmoteuthid affinity. In 1994, Doyle et al. in their "Phylogeny and Systematics of the Coleoidea," considered *Jeletzkya* as a taxon of undetermined status.

In contrast to the holotype of *J. douglassae*, the two specimens [PE 32521 and PE 20808 described by Saunders and Richardson (1979) from the Mazon Creek reexamined herein with SEM] each has a moderately well-preserved shell, but lacks clearly outlined tentacles. The lack of tentacles has been an insurmountable obstacle to the reconstruction of the arm crown and has hindered comparison of all other Mazon Creek and other coleoids recovered in the Upper Paleozoic with *J. douglassae*. Reexamination of these two specimens reveals new crucial data on the morphology of the coleoid shell and radula, presence of arm hooks, muscular mantle tissue, and traces of ink. These data, as well as those in the presently known diverse Carboniferous phragmocone-bearing coleoids (see Doguzhaeva, 2002c; Doguzhaeva et al., 2003 for summary) favor a reevaluation of the systematic placement of the specimens under discussion. These specimens are redescribed as *Saundersites illinoisiensis,* g. and sp. n., and the new genus is placed in the family Donovaniconidae (Doguzhaeva et al., 2003). This is the only coleoid family known to combine a moderately breviconic shell, a long body chamber, a short phragmocone, a short proostracum-like structure, and a substantial ink sac. This family is placed in the new order Donovaniconida, ord. n.

With the erection of *Saundersites illinoisiensis,* g. and sp. n., additional information is made available to evaluate the morphological plasticity and evolutionary trends of Carboniferous coleoids.

2 Studied Material, State of Preservation, and Methods

We have been able to borrow three specimens (FMNH PE 32521, PE 20808, and PE 28955) of the six specimens assigned to *Jeletzkya* by Saunders and Richardson (1979). One of the specimens (?*Jeletzkya douglassae* Saunders and Richardson, 1979: Fig. 10, PE 28955) appears to have been assigned to the Cephalopoda erroneously. Specimen PE 32521 (Saunders and Richardson, 1979: Fig. 9 a, b, d) preserves the arm hooks and an unusual radula structure. In addition, this specimen shows remnants of muscular mantle and probable ink that were not observed previously. The specimen PE 20808 shows a well-preserved adoral projection. In both specimens the shell has a similar gross morphology and size. However, in neither fossil is the shell material preserved.

Specimen PE 32521 (Fig. 6.1A, B) shows its ventral side on split halves of a medium-sized, flat, brownish ironstone concretion. The shell is weakly compressed because of compaction. It has a triangular shape, with a pointed adapical end and almost straight sides. It consists of a proostracum-like structure, a body chamber, phragmocone, and an irregularly calcified rostrum. A radula is situated approximately in the middle of the length of the proostracum-like structure (Figs. 6.2–4). More than 20 arm hooks are preserved in front of the shell aperture, and most are more-or-less randomly dispersed (Figs. 6.5, 6), except in two cases where they occur in pairs. It is not possible to reconstruct the arm crown from the arm-hook distribution. Most of the hooks are exposed in cross, or longitudinal sections, but some show their sides and provide details of their gross morphology. The outer surface of the body chamber bears a dispersed black substance that SEM study suggests is probably ink (Fig. 6.7C, E). SEM analysis of the material bordering the body chamber and excavated from the middle of its length revealed numerous soft-tissue debris (Fig. 6.7A), some of which presumably belonged to the muscular mantle (Figs. 6.8–10). Specimen PE 20808 is also exposed on its ventral side and shows a nearly complete proostracum-like structure.

Initially, both specimens were examined under a dissecting microscope (x7 to x35 magnification) under alcohol. In specimen PE 32521, the structure of the radula, arm hooks, presumed ink, and muscular-mantle tissues were examined with SEM. To prevent destruction while making the SEM preparation, the specimen was trimmed from the concretion as part of a slab about 10 mm thick. The length of the slab (about 80 mm) was much longer than typical SEM preparations (usually 10–20 mm). To diminish charging that could occur under high voltage in the SEM, the opposite side of the slab was coated with colloidal silver.

The occurrence of fossilized ink in specimen PE 32521 was revealed by SEM examination of the black material on the surface of the specimen and by comparison with the previously studied ink and other kinds of organic substance in present-day and extinct cephalopods (Doguzhaeva et al., 2002b, c, 2003, 2004a, b, c; Doguzhaeva and Mutvei, 2003, 2005). The characteristic structure of the ink was compared with the structure of the sediment in the concretion around the specimen. In order to detect fossilized ink substance in specimen PE32521, pieces of dark material were sampled from the shell surface. In order to display presumed mantle remnants, tiny pieces (less than 5 mm in size) were removed from the groove between the shell and the sediment. For their SEM analyses, the selected fragments of the specimen were coated with gold. The same pieces also were used for EDAX analysis to determine the chemical composition and diagenesis of the presumed muscular mantle debris and ink during fossilization. For EDAX analysis seven spectra were taken, four from the remnants of soft tissue and ink and three from the sediment.

SEM and EDAX analyses were carried out with Hitachi S-4300 at the Swedish Museum of Natural History, Stockholm.

The studied material is reposited at the Field Museum of Natural History in Chicago, USA (FMNH PE 32521, PE 20808, and PE 28955).

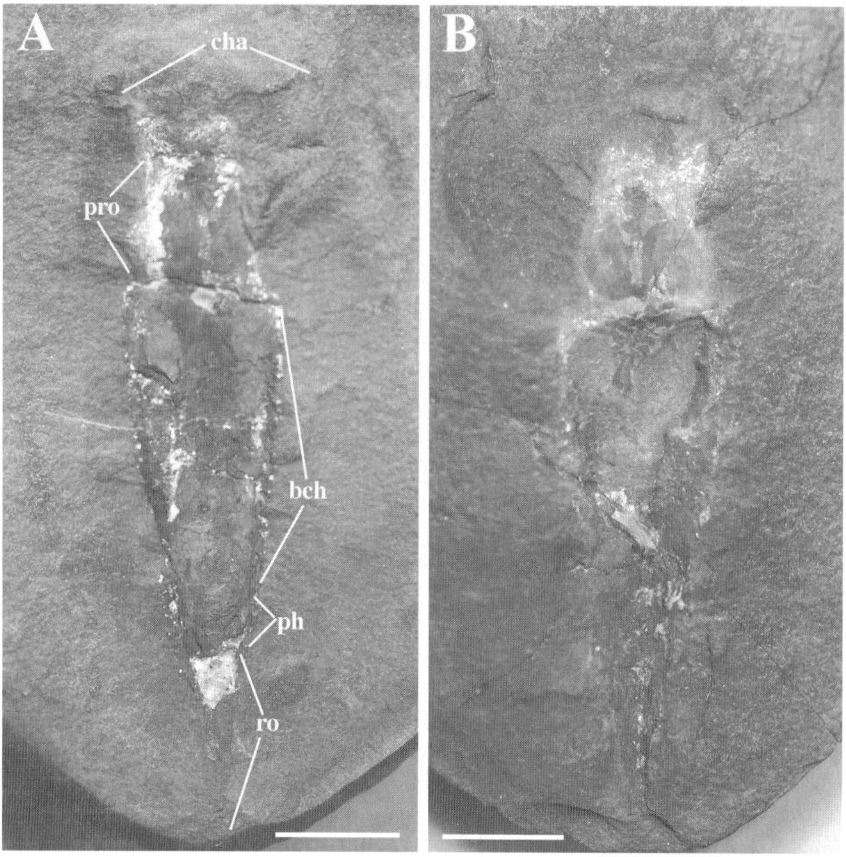

Fig. 6.1A, B. Saundersites illinoisiensis *g. and sp. n., PE 32521. General view (A – part and B – counterpart) shows the cephalic area (CHA), proostracum like structure (PRO), body chamber (BCH), probable phragmocone (PH), and probable rostrum (RO), Upper Carboniferous, Middle Pennsylvanian, Desmoinesian (Westfalian D), Francis Creek Shale, Mazon Creek, northwestern Illinois, USA. Scale bar is 1 cm.*

3 Comparative Morphology

3.1 Shell

3.1.1 Morphology

The shell length in specimens PE 32521 (Fig. 6.1A, B) and PE 20808 is 56 mm and 46 mm, respectively, and the maximum diameter is 12 mm and 14 mm, respectively. In both specimens the shell has a narrow triangular shape with an apical angle of approximately 12°–15°. The shells appear to be exposed from the ventral side.

Fig. 6.2 Saundersites illinoisiensis *g. and sp. n., PE 32521. General view of a radula showing the approximate radula width (RW) and longitudinal rows of two marginal plates on each side (MP1, MP2) and marginal teeth (M). Scale bar is 1.25 mm.*

Four portions can be distinguished in each shell. The most adoral portion forms a plate on the apparent dorsal side of the shell. The length of this plate is about one-fifth of the total shell length. In PE 32521, the anterior and lateral margins of this plate are fractured, and the surface of the plate is uneven due to compression. In PE 20808, the adoral plate is well preserved and shows a slightly forward curved, broadly rounded anterior margin. Its surface shows growth lines parallel to the anterior margin. Because it is developed only on one side of the shell the anterior plate is here interpreted as a proostracum-like structure. The middle portion of the shell is here interpreted as the body chamber. In both specimens, it is three times longer than the proostracum-like structure. It has a rounded cross section that is slightly compressed in PE 32521. As in all other cephalopod shells from the Mazon Creek area (Richardson and Johnson, 1971; Saunders and Richardson, 1979), the shell material is not preserved. The third portion, immediately apical of the body chamber, can be clearly distinguished from the rest of the shell, because, in PE 32521, it is completely filled and, in PE 20808, partially filled with white calcite and has a sharp boundary toward the middle portion (the body chamber) of the shell. The length of this portion is about one-sixth of the total shell length. It is here interpreted as a phragmocone, although septa cannot be clearly distinguished. Because the phragmocone is short (one-sixth of the shell length only), it could have been truncated during the lifetime of the animal. The most apical portion of the shell is a narrow, short structure that in each specimen shows irregularly shaped longitudinal ridges. This portion is here interpreted as a partly calcified, partly organic rostrum. It is interesting to note that, in

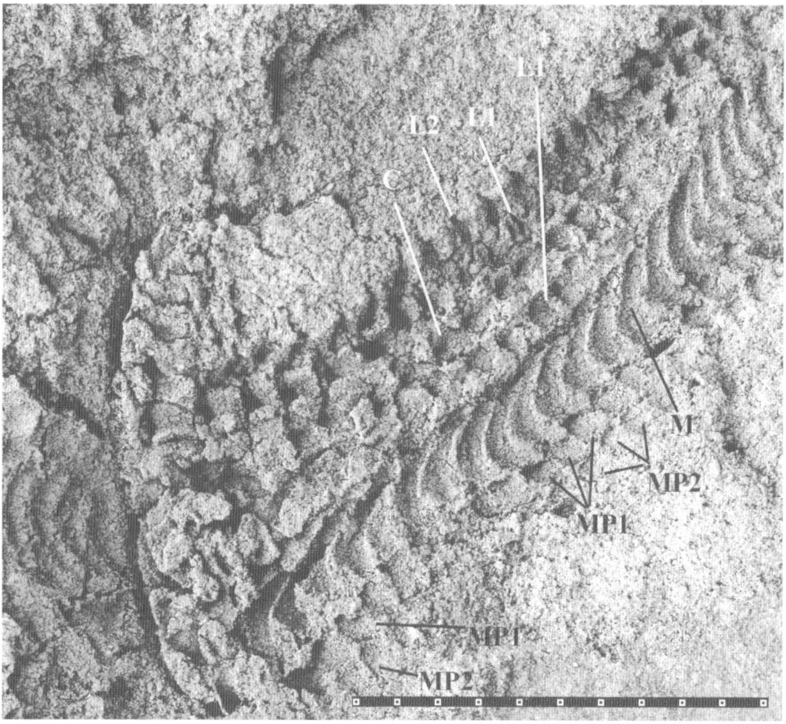

Fig. 6.3 Saundersites illinoisiensis *g. and sp. n., PE 32521. Enlarged detail of Fig. 2. The anterior portion of the radula shows two rows of marginal plates (MP1, MP2), a row of well exposed marginal teeth (M) that have a broad basal part, two rows of lateral teeth (L1, L2), which are partly preserved, and a row of central teeth (C) of which only the tips are exposed. Scale bar is 1.25 mm.*

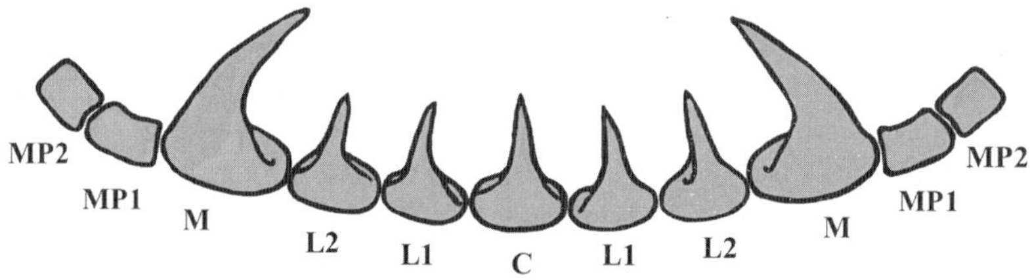

Fig. 6.4 Saundersites illinoisiensis *g. and sp. n., PE 32521. Model of a radula with 11 elements in each transverse row: on each side of the central tooth (C) there are two lateral teeth (L1, L2), a marginal tooth (M), and two marginal plates (MP1, MP2).*

these two Mazon Creek specimens, the apical portion of the phragmocone and rostrum shows certain similarities to those in the Early Carboniferous coleoid *Hematites* Flower and Gordon, 1959 (Doguzhaeva et al., 2002a: Fig. 6.4; pl. 7, Figs. 1, 2). In specimens of this genus, the apical portion of the phragmocone that is totally

Fig. 6.5 Saundersites illinoisiensis *g. and sp. n., PE 32521. Arm hooks (AH) preserved in front of the radula (R, left bottom corner). Two hooks with tips turned to the left show their general shape (right bottom corner). Other hooks, exposed in cross section, demonstrate that they are hollow and have a thin, originally organic wall. The hooks are arranged in pairs and in longitudinal rows. Each division of the scale bar is 0.5 mm.*

surrounded by the rostrum is almost always truncated, and the internal ultrastructure of the rostrum suggests that it was constructed of both organic and calcified material. In addition, the surface of the rostrum of specimens of this genus has longitudinal ridges that vary in size and spacing.

3.1.2 Comparison

From the Desmoinesian deposits of the Mazon Creek *Lagerstätte*, Saunders and Richardson (1979: Fig. 1d) illustrated an orthoconic shell and identified it as *Bactrites* sp. (specimen PE 25584). This shell is 52 mm long and ca 20 mm in diameter at the aperture. It has a long body chamber, seven short camerae in the phragmocone, a ventral marginal siphuncle, straight sutures, and an apical angle of about 20°. With the exception of this specimen, bactritoids have not been found in the Mazon Creek fauna. Assignment of this shell to *Bactrites* sp. is questioned herein, and its relationship to the studied specimen PE 32521 is left open. However, the overall lack of distinct morphological characteristics and the poor preservation of PE 32521 makes its taxonomic assignment to either the coleoids or bactritoids difficult, if not impossible.

Another comparable specimen of Desmoinesian age is the shell of the coleoid *Donovaniconus oklahomensis* Doguzhaeva et al., which also has a proportionally long, moderately breviconic body chamber, comparatively short phragmocone, and a short proostracum-like structure (Doguzhaeva et al., 2002b, 2003).

Fig. 6.6 Saundersites illinoisiensis *g. and sp. n., PE 32521. Enlarged detail of Fig. 5 (right bottom corner). Five arm-hooks (AH) are arranged in rows. Scale bar is 0.2 mm.*

3.2 Cephalic Remains

3.2.1 Radula

Morphology

In specimen PE 32521, a long portion of the radula (Figs. 6.2–4) and several dispersed arm-hooks (Figs. 6.5, 6.6) are preserved. The radula is located on the proostracum-like structure close to its margin. A few irregular ridges can be distinguished in front of the proostracum-like structure (Fig. 6.1A, B). Several arm hooks occur close to these ridges (Figs. 6.5, 6.6). It is probable that the ridges are impressions of the arms. To judge from the distribution of the arm hooks in front of the proostracum-like structure, the arms probably were short. Short arms also occur in the holotype of *Jeletzkya douglassae* (Johnson and Richardson, 1968: Figs. 1, 2). The radula is about 3.5 mm in length and 1.5 mm in width (Fig. 6.2). In front of the radula, there are two poorly preserved carbonaceous, triangular elements oriented transverse to the radula axis (Fig. 6.2). They may be poorly preserved remains of the jaws. About 30 transverse rows of the radula are visible; one side of the radula is better preserved than the other side. SEM examination shows that the radula has two marginal plates on each side; both are situated outside the marginal teeth, in contact with each other (right side of Fig. 6.3). The two marginal plates and the

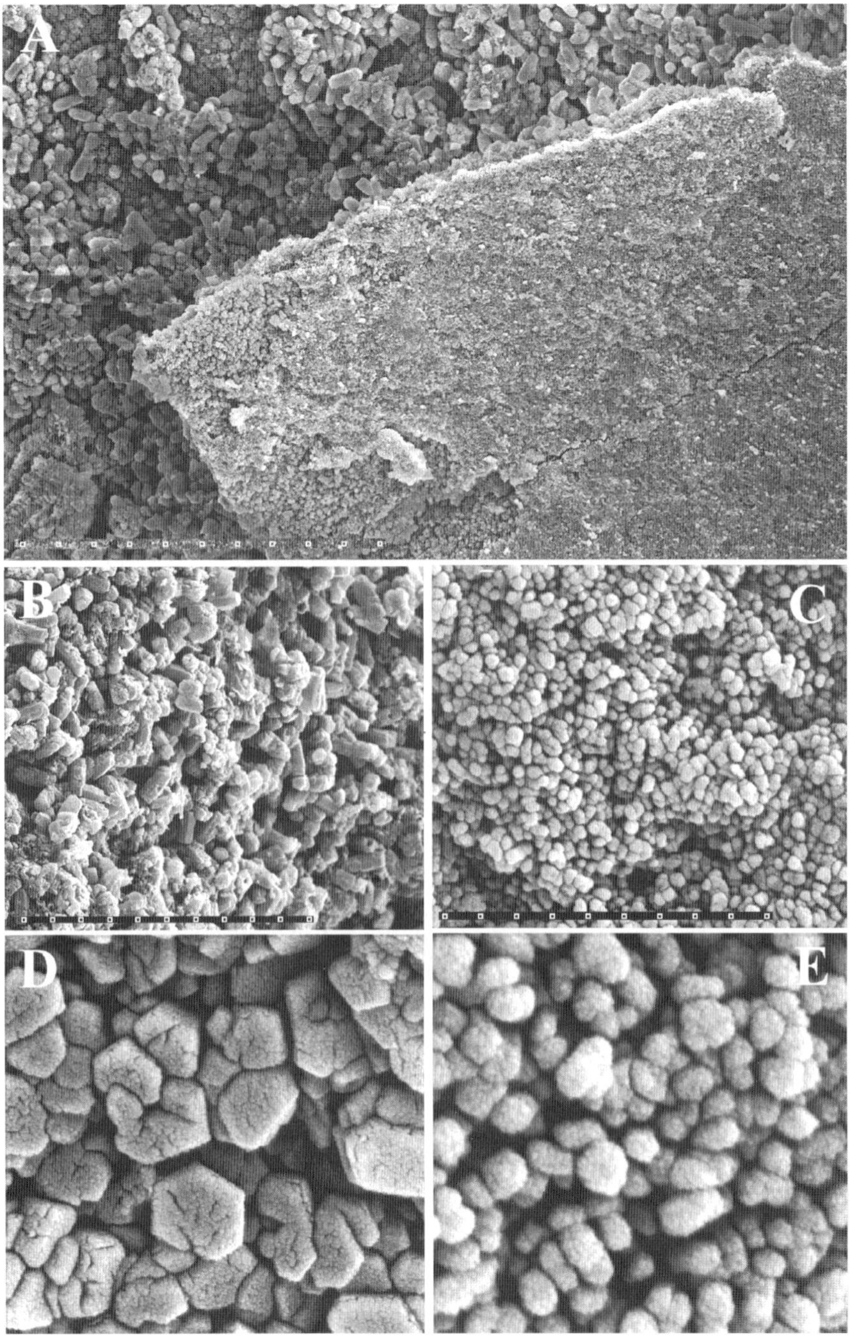

Fig. 6.7 Saundersites illinoisiensis *g. and sp. n., PE 32521. A. A fragment of the soft tissue remnant from the external surface of the shell showing traces of ink substance and from the shaly sediment around it. Each division of scale bar is 3 μm. B, D. Two types of crystals in the shaly sediment surrounding the soft tissue remnants, enlarged details of A. Each division of scale bar is 1.2 μm and 0.12 μm, respectively. C, E. Globular ultrastructure of the ink, enlarged detail of A. Each division of the scale bar is 0.3 μm and 60 nm, respectively.*

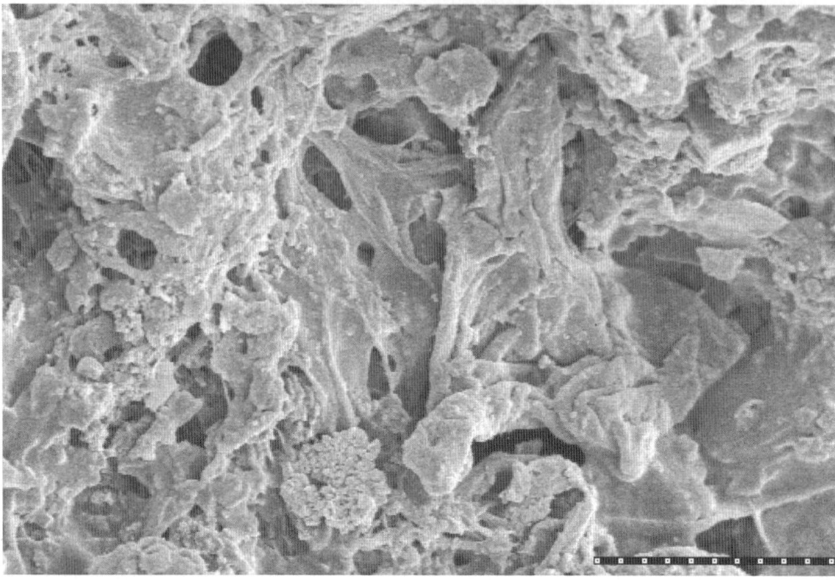

Fig. 6.8 Saundersites illinoisiensis *g. and sp. n., PE32521. General view of the fossilized soft tissue debris preserved on the shell surface. Each division of scale bar is 1–2 μm.*

large marginal tooth are distinct. The marginal tooth has a broad, semilunar shape in cross section; its basal portion, about one-third of the total height of the tooth, is massive; the rest of the tooth is pointed and curved backward; its sides form an angle of about 20°. The ratio between the height and maximum diameter of the tooth is ca 1:1. The central (or rachidian) and lateral teeth are poorly preserved and cannot be described in detail. This is particularly the case with the inner lateral tooth of which only indistinct remains can be seen. The radula formula is C L1 L2 M MP1 MP2 (11 elements in total) (Fig. 6.4).

Comparison

In present-day molluscs a radula of 11 elements in each transverse row is known in the class Monoplacophora (Starobogatov, 1990a). This class is considered a possible ancestor of the Cephalopoda (Yochelson et al., 1973). Two or more marginal plates along the periphery of the radula are known in the class Polyplacophora, and there are 17 elements in each transverse row (Ivanov and Sirenko, 1990; Starobogatov, 1990a, b).

In addition to the present specimen, several radulae have been recovered from Paleozoic cephalopods: (1) in an unidentified Late Ordovician orthocone the radula has five teeth in each transverse row; the radula configuration is more similar to that of ammonoids and coleoids than to that of nautiloids (Gabbott, 1999); (2) in the

Silurian orthocerid *Michelinoceras* the radula is reported to have seven elements in each transverse row (Mehl, 1984), but this report has been questioned (Nixon, 1988: 113); (3) two separately preserved radulae with 13 elements in each transverse row have been found from the Late Carboniferous (Desmoinesian) Mazon Creek deposits; they are described as *Paleocadmus herdinae* (Solem and Richardson, 1975) and *Paleocadmus pohli* (Saunders and Richardson, 1979) and have the radula formula: C L1 L2 M1 MP1 M2 MP2; both *Paleocadmus* radulae have two marginal plates on each side, but they are separated by a second marginal tooth; the similarity of these two Carboniferous radulae with the 13-element radula in present-day *Nautilus* suggests that there is a strong evolutionary stability of the radula with 13 elements in each transverse row; (4) radulae with nine elements in each transverse row are reported in the following Carboniferous goniatites: *Glaphyrites* (Closs and Gordon, 1966), *Cravenoceras* (Tanabe and Mapes, 1995) and *Girtyoceras* (Doguzhaeva et al., 1997).

Present-day coleoids and fossil ammonoids have either seven or nine elements in each transverse radula row (see Nixon, 1988). Individuals of these taxa each have a single marginal plate on each side. The occurrence of radulae with two marginal plates that are located outside the marginal tooth makes the radula structure in the coleoid specimen under discussion unique in the class Cephalopoda.

Evolutionary Development of the Radula in Molluscs

In present-day molluscs, the number of elements in each transverse row in the radula is different in different classes: Polyplacophora – 17, Monoplacophora – 11, Scaphopoda – 5, Cephalopoda – 13, 9, 7, or none, Gastropoda – variable (Ivanov and Sirenko, 1990; Ivanov and Starobogatov, 1990; Rossolimo, 1990). Starobogatov (1990a, b) concluded that radulae in monoplacophorans evolved through oligomerization and that the number of teeth was about 80 in Paleozoic genera instead of 11 in present-day monoplacophorans. The evolutionary reduction of the number of elements in the radula of monoplacophorans was believed by Starobogatov to have been caused by migration of these molluscs from the littoral zone, characterized by coarse substrate, to abyssal zones with soft substrate.

At present, it is uncertain whether the evolutionary transformation of radula elements in cephalopods occurred by oligomerization. The possible scenario is as follows: at early stages of coleoid evolution, the radula with 11 elements (seven teeth and four plates) was derived from the nautiloid radula with 13 elements (Nixon, 1988: Fig. 6.2e) by the elimination of the second marginal tooth. That resulted in a changed position of the two marginal plates to side by side, which is the case with the present specimen (PE 32521). At the next evolutionary stage, the two marginal plates fused into one plate, and as a consequence, the radula had nine elements per transverse row. Finally, the marginal plates became entirely eliminated, and a radula with seven elements in each transverse row was formed.

One can also assume that the radula possessing 11 elements in each transverse row (like that in the studied specimen) was inherited from the monoplacophoran radula, and that the radula of the studied specimen is similar to the archaic radula of early cephalopods. In this case, the nautiloid type of radula, with 13 elements in each row, might have arisen from a radula with 11 elements by adding an additional marginal tooth between the two marginal plates on each side. The radula in present-day coleoids could have been formed by oligomerization of the two marginal plates.

3.2.2 Arm Hooks

Morphology

The hooks (Figs. 6.5, 6.6) are small (about 0.2–0.3 mm in length), shiny, hollow structures composed of black horny material. The shaft is short and thick. The distal part, extending from the maximum curvature to the tip, is well developed and long. The angle between the shaft and the distal part is about 90°. The cross section through the basal part is triangular.

Comparison

The arm hooks of the present specimen and the separately preserved hooks from the Mazon Creek area, illustrated by Saunders and Richardson (1979: Fig. 9c), show significant morphological differences. The separate hooks appear to be similar to the hooks seen in the holotype of *Jeletzkya douglassae*. They are bigger than those seen in the specimen studied here (ca 1.0 mm versus 0.2–0.3 mm, respectively). In addition, the shaft in each of the hooks of the holotype of *J. douglassae* and the hooks illustrated by Saunders and Richardson (1979: Fig. 9c) are thicker and swollen on the internal side, and the distal part is relatively shorter than those in the studied specimen.

The hooks of the present specimen show some similarity to the hooks of the Lower Jurassic "fossil teuthid" *Loligosepia aalensis* Zieten (Doguzhaeva and Mutvei, 2003: Fig. 1c, d, e). In both, the shafts are short and thick, and the distal parts are long. The approximately 90° angle between the shaft and distal part is also similar. Both the Carboniferous and the Jurassic hook-specimens have cross sections that are triangular, and the ratio of wall thickness to hook diameter is 1:3.

3.3 Fossilized Soft-Tissue Remnants

3.3.1 Chemical Composition

In spite of significant ultrastructural differences between the fossilized soft tissues and the sediment around the shell (Fig. 6.7A–E), the EDAX analysis did not reveal noticeable differences in their chemical composition with the exception of a slightly higher content of Si in the former. Both contain O, Si, Fe, C, Mg, and Ca, but have no P.

3.3.2 Muscular Mantle

Studies with SEM show that the material collected from the shell surface (Figs. 6.6, 6.7A) is porous and contains diverse microorganisms. It includes debris of plastically deformed (folded or rolled) material (Fig. 6.9). At higher magnifications (×10,000 to ×20,000), the material is seen to consist of numerous superimposed, sheets of irregular thickness. On the surface, these sheets are disrupted in numerous places and give the impression that they were exposed to strong tension during fossilization. The sheets exhibit distinct, parallel fibers about 0.1μm in diameter (Fig. 6.9B). The fibers have a fine granular ultrastructure, and they are arranged into bundles (Fig. 6.10A). The latter indicates that the sheets under examination represent a fossilized muscular mantle.

In some aspects, the mantle tissue of the Mazon Creek specimen is similar to the fossilized mantle in the Upper Triassic ceratitid *Austrotrachyceras* (Doguzhaeva et al., 2004c), but in other aspects it is also similar to the mantle in the Late Jurassic *Belemnoteuthis* (Owen, 1844) and the Early Jurassic squid-like coleoid *Loligosepia* (Kear et al., 1995; Doguzhaeva and Mutvei, 2003). The mantle in the ceratitid *Austrotrachyceras* (Fig. 6.11A) also has a sheet-like pattern (Doguzhaeva et al., 2004c: Fig. 6.4A) with a fine lamination and a fibrous ultrastructure (Doguzhaeva et al., 2004c: Fig. 6.2A, B), but it does not have the fibers arranged into bundles as in the squid-like *Loligosepia* (Doguzhaeva et al., 2004c: Fig. 6.6A, B). However the mantle in the Mazon Creek specimen has no distinct transverse striations formed by alternation of circular, longitudinal, and transverse muscle fibers. Therefore, the level of development of the mantle structure in the Mazon Creek specimen seems to have been somewhere between that in the less muscular mantle in ceratitid *Austrotrachyceras* and those seen with a more advanced mantle structure such as those in the squid-like coleoid *Loligosepia*.

In spite of a different chemical substitution during fossilization (phosphatized in soft-bodied squids from the Jurassic Oxford Clay, England (Allison, 1988), and *Loligosepia* from the Posidonian Schiefer, Holzmaden, Germany, bituminous in *Austrotrachyceras* from the Reihgraben Shales, Schindelberg locality, Lower Austria, and rich in Si in the specimen from the Mazon Creek area, Illinois, USA), the mantle in specimens of all taxa shows a granular ultrastructure. This was probably caused by different bacteria that caused the precipitation of different elements (in the Posidonian Schiefer by P-accumulating bacteria, in the Reihgraben Shales by C-accumulating bacteria, and in the Mazon Creek area by Si-accumulating bacteria).

3.3.3 Ink Substance

The substance that is interpreted to be dispersed ink (Fig. 6.7C) consists of a mass of tiny, globular granules, 0.1–0.4μm in diameter. Each granule is an agglomerate of

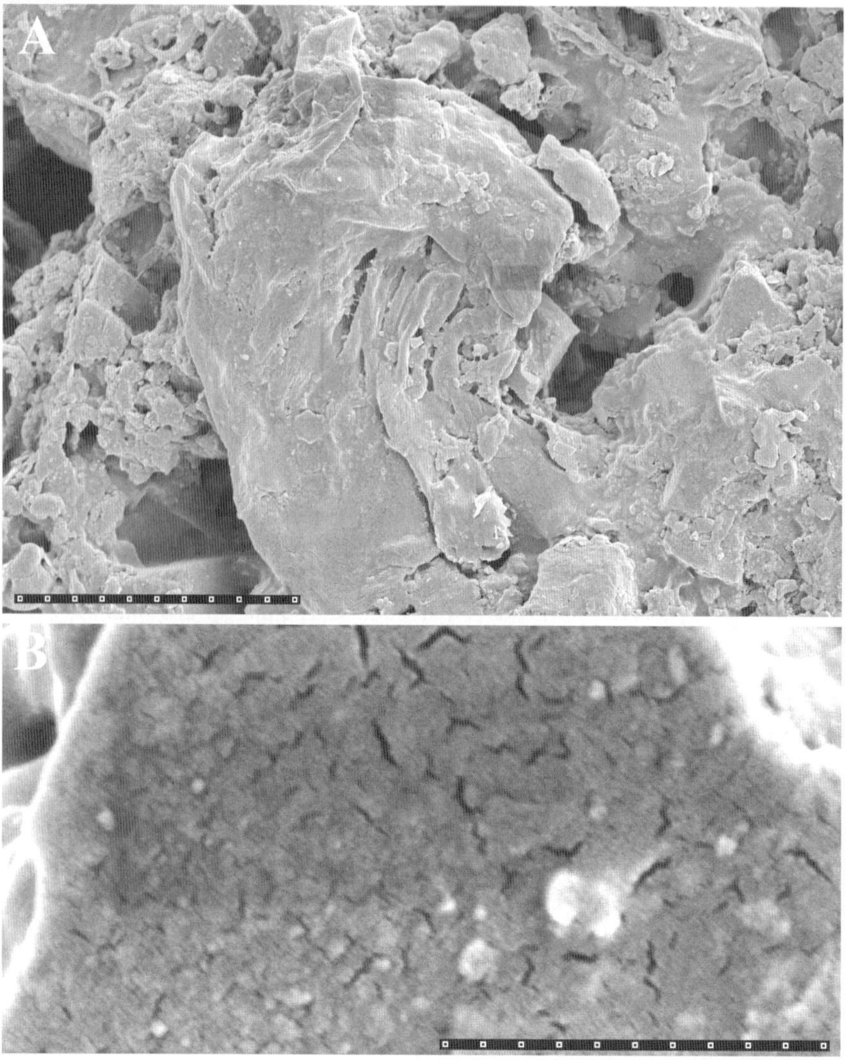

Fig. 6.9A, B. Saundersites illinoisiensis *g. and sp. n., PE32521. A. Fragment of fossilized muscular mantle among debris of soft tissues. B. Fibrous ultrastructure of the muscular mantle; the fibers form a criss-cross pattern, enlarged view of Fig. A. Each division of the scale bar in A is 1.2 μm and in B is 0.12 μm.*

smaller particles (Fig. 6.7E). These granules do not form layers or fibers as do the granules in fossilized soft-tissues, and they are ultrastructurally identical to the ink in present-day and fossil coleoids (Fig. 6.11B–D). The matrix of the concretion around the specimen has crystal-shaped grains (Fig. 6.7B, D) indicating its abiotic origin.

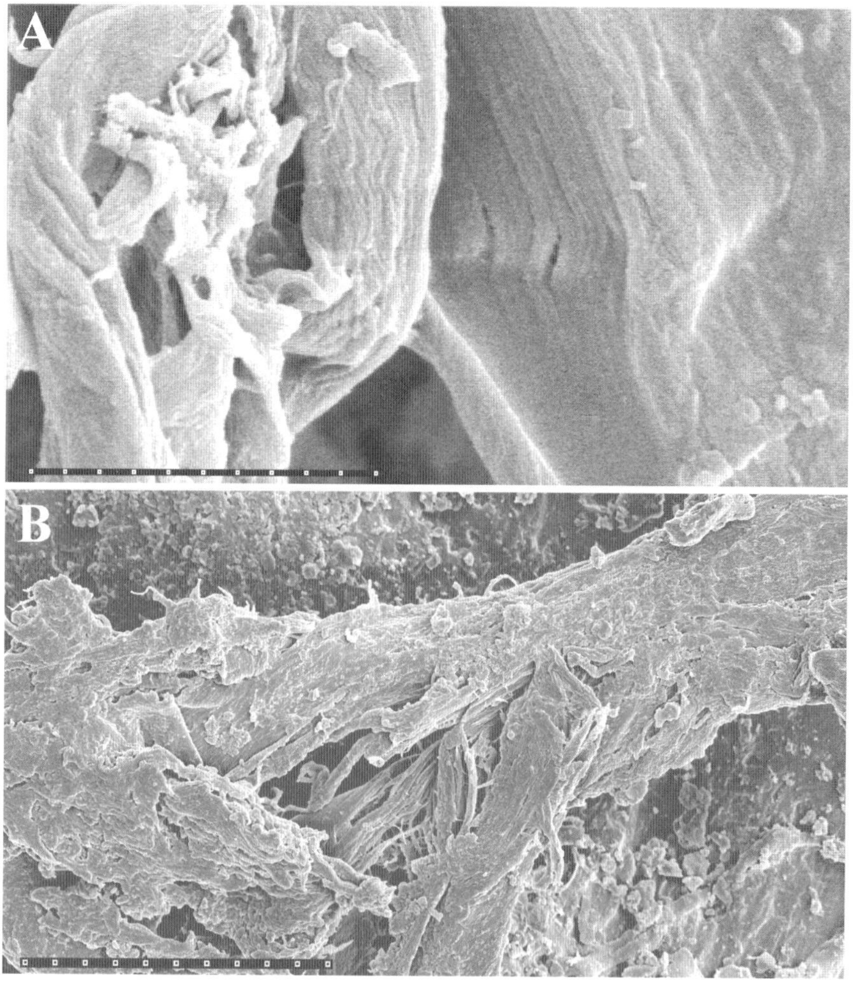

Fig. 6.10A Saundersites illinoisiensis *g. and sp. n., PE 32521. Fragment of fossilized muscular mantle. Each division of scale bar is 0.6 μm.* ***B.*** Austrotrachyceras sp., *NHMW 2005z/0006/0001, Upper Triassic, Lower Carnian, Lower Austria, locality Schindelberg near Lunz. Fragment of fossilized muscular mantle. Each division of the scale bar is 3 μm.*

4 Systematic Paleontology

The following features of the morphology and preservation in the specimens under discussion were analyzed to elucidate their systematic assignment: (1) division of the shell into a proostracum-like structure, body chamber, phragmocone, and rostrum; (2) presence of fossilized soft-tissues covering the shell; (3) presence of the arm hooks, (4) presence of ink, and (5) a unique radula structure.

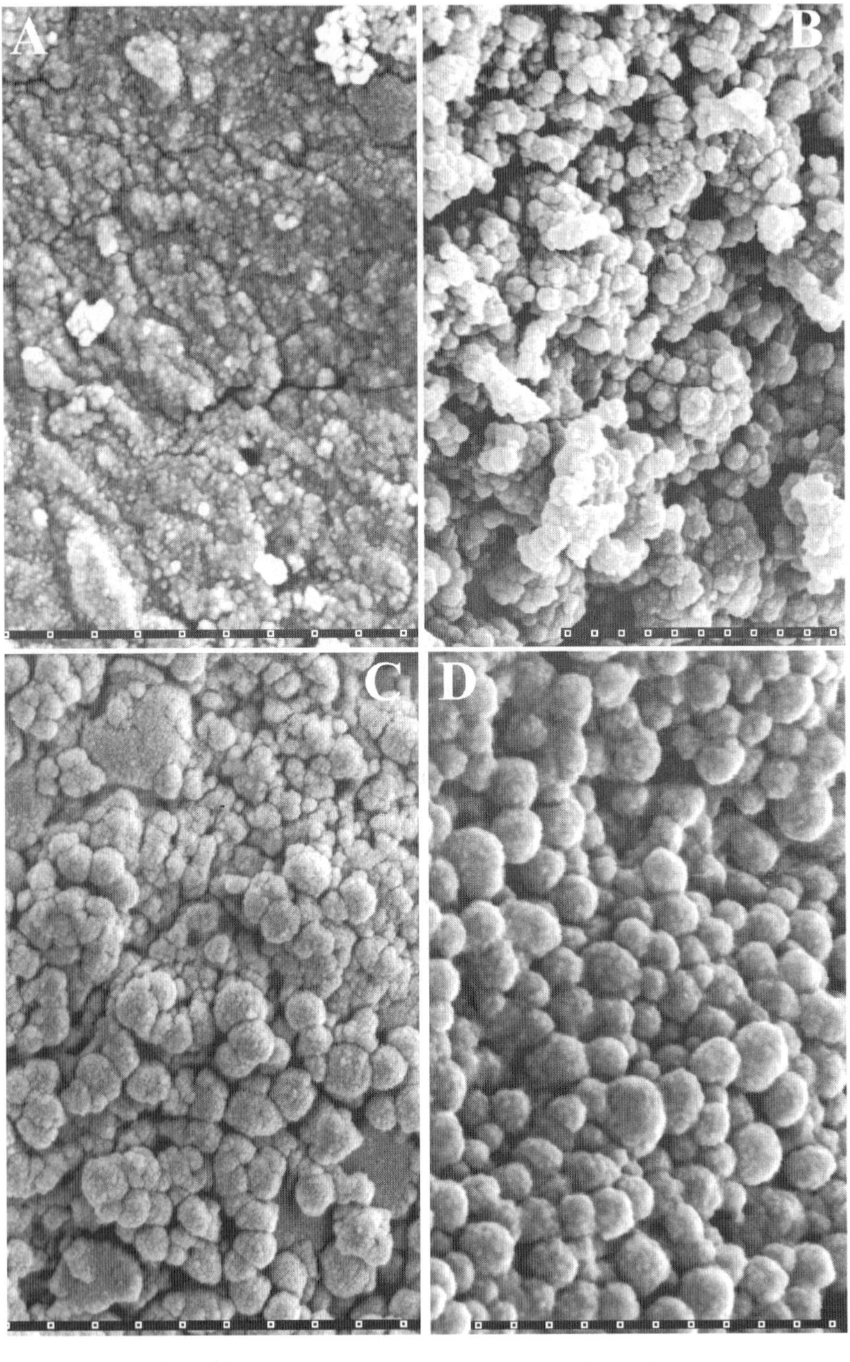

Fig. 6.11A, B Austrotrachyceras sp., *NHMW 2005z/0006/0001, Upper Triassic, Lower Carnian, Lower Austria, locality Schindelberg near Lunz.* **A.** *Fossilized mantle tissues preserved*

Subclass COLEOIDEA Bather, 1888
Superorder is unknown
Order DONOVANICONIDA, ord. n.

Diagnosis. Small to medium-sized, about 30–60 mm in length, phragmocone-bearing coleoids each with long body chamber, relatively short and broad proostracum-like structure at the aperture, ink sac, arm hooks, and small rostrum or sheath; siphuncle narrow, ventral; connecting rings thin, organic.

Comparison. The erection of a new order is required because no existing coleoid order can accommodate all of the characteristics of a new genus being described from the Lower Carboniferous Bear Gulch Formation in Montana (Mapes, personal communication, 2007, and submitted) and two described genera, one from the Upper Carboniferous Mazon Creek locality (Westphalian D), Francis Creek Shale, in Illinois (described herein), and the other from the Wewoka Formation in Oklahoma (Doguzhaeva et al., 2002b). This order better accommodates the family Donovaniconidae that was provisionally placed in the order Phragmoteuthida (Doguzhaeva et al., 2003). Members of the new order Donovaniconida have a relatively long body chamber, a relatively short rostrum, and a short broad proostracum-like structure.

Comparison of all the presently recognized Upper Paleozoic coleoid orders are in general as follows: Members belonging to the order Hematitida are readily distinguished from the Donovaniconida by having a very short body chamber and breviconic phragmocone, a massive, short rostrum that covers the entire phragmocone, and being without a proostracum-like structure. The order Aulacocerida has a longiconic phragmocone, a proostracum-like structure, and a relatively long, massive rostrum on more mature specimens and in the new order the phragmocone is very short and the rostrum is short and weakly developed. The order Spirulida in the Carboniferous can be separated from the new order by the presence of a longiconic phragmocone, and the lack of a proostracum-like structure. The order Phragmoteuthida differs from the new order by having an exceptionally long, broad tripartite proostracum, no enclosed body chamber, and a short phragmocone with closely spaced chambers. The new order can be separated from the order Octopoda because the latter lacks an internal shell.

Discussion. The geologic range of the new order is presently confined to the Carboniferous.

Fig. 6.11A, B (continued) *in the body chamber.* ***B.*** *Globular ultrastructure of possible ink dispersed in soft tissues from the body chamber. Each division of scale bar is: A – 0.12 μm; B – 0.15 μm.* ***C.*** *Unnamed phragmocone-bearing coleoid, Missourian, Middle Pennsylvanian, Nebraska, USA, University of Iowa Paleontology collections, SUI 62497. Globular ultrastructure of ink preserved within the ink sac. Each division of scale bar is 0.3 μm.* ***D.*** Loligo forbesi *Steenstrup, 1856, globular ultrastructure of dried Recent ink. Each division of the scale bar is 0.12 μm.*

Family DONOVANICONIDAE Doguzhaeva et al., 2003
Genus *SAUNDERSITES*, gen. n.

Derivation of name. In honor of W. Bruce Saunders for his valuable contributions to the study of fossil cephalopods and present-day *Nautilus*.
Type species. Saundersites illinoisiensis gen. and sp. n.

Diagnosis. Orthocone about 60 mm in length and 12 mm in diameter, apical angle of about 12°–15°; length ratio of proostracum-like structure/body chamber/phragmocone together with rostrum is ca. 3:9:5; rostrum irregularly calcified; arms apparently short, with arm-hooks arranged in double rows; radula formula C:L1: L2:M:MP1:MP2.

Comparision. The closest coleoid taxon in the Upper Paleozoic to *Saundersites* is *Donovaniconus* from time equivalent beds located in the southern part of the American mid-continent in Oklahoma. These two taxa differ in the apical angle of the phragmocone; *Donovaniconus* has an apical angle of about 20° and *Saundersites* has an apical angle of about 15° degrees. Another difference is that the phragmocone of *Donovaniconus* makes up about one-third (33%) of the conch length, whereas *Saundersites* has a phragmocone that makes up about 40% of the conch length.

Discussion. Preservational differences of *Saundersites* and *Donovaniconus* prevent additional comparisons. In *Saundersites* the radula and arm hooks are preserved; ink is present as a dispersed material. With the exception of the dispersed ink, these structures are not preserved in *Donovaniconus*. However, *Donovaniconus* does preserve the shell ultrastructure, shell ornament, cameral spacing, ink in an ink sac, and siphuncular details. These features are not well preserved in either *Donovaniconus* or *Saundersites*. In both genera the proostracum-like structure is short and broad, and it can be determined that this feature is longer in *Saundersites* than in *Donovaniconus*. An ultrastructural comparison of this structure cannot be made because in *Saundersites* this structure is preserved as an impression, while on *Donovaniconus* the presence of this structure was identified by the shape of the growth lines on shell fragments and by the shape of a partly complete aperture.

Saundersites illinoisiensis, sp. n.

Holotype. Specimen PE 32521.

Paratype. Specimen PE 20808.

Type locality. Mazon Creek, northwestern Illinois, USA.

Type Horizon. Upper Carboniferous, Francis Creek Shale (Westphalian D), Middle Pennsylvanian, Desmoinesian.

Derivation of name. Named for the state of Illinois.

Description. The holotype is an orthocone 56 mm long and a short cephalic area 8 mm long. Apical angle of the shell is about 15°. Body chamber is long (ca. 30 mm in length and 12 mm in diameter). Proostracum-like structure is short (ca. 1/3 body chamber length) and broad (ca. 1/2 circumference length). Short phragmocone (apparently truncated) and short rostrum are about 1/2 body chamber length together. Rostrum is irregularly mineralized. The radula has 11 elements and two marginal plates, on each side, located outside the marginal tooth (Figs. 6.2, 6.3). The arm hooks are arranged in double rows. The hooks are about 0.2–0.3 mm in length. They show a short and thick shaft. A distal part, extending from the maximum curvature to the tip, is long. The angle between the shaft and the distal part is about 90° (Fig. 6.4). The paratype shows a well-preserved proostracum-like structure at the aperture. The shell has a similar size and shape as that in the holotype.

5 Morphological Plasticity and Evolutionary Trends in Carboniferous Coleoids

The early appearance of coleoids in cephalopod evolution is no longer doubted as it was prior to the report of Flower and Gordon (1959), who discovered an extensive Early Carboniferous collection of rostrum-bearing coleoids they named *Hematites*. Recently, additional Carboniferous coleoids have been discovered from the mid-continent of North America (Johnson and Richardson, 1966, 1968; Gordon, 1964, 1971; Saunders and Richardson, 1979; Allison, 1987; Kluessendorf and Doyle, 2000; Doguzhaeva et al., 1996, 2002a, c, 2003), southern Urals in Russia (Doguzhaeva, 2002c), and northern Moravia, Czech Republic (Kostak et al., 2002). The currently known diversity of Carboniferous coleoid cephalopods is limited to fewer than ten described genera; almost all of these genera are monospecific.

Doguzhaeva (2002a) recognized the remarkable morphological plasticity in Carboniferous coleoids. Her conclusions were based on recently obtained morphological and ultrastructural data, and she demonstrated that different coleoids had the capability to combine different morphological elements into the same shell. This kind of recombination has not been observed in Mesozoic coleoids. The morphological elements are present either in the external shell, such as a long body chamber, or in the internal shell, such as the rostrum, proostracum-like structure, and lamello-fibrillar nacre in the septa. From the classical point of view (Naef, 1922; Jeletzky, 1966), the rearrangement of the morphological elements within the early coleoid shells was not considered to be possible. However, the recombination of these elements indicates a great morphological plasticity within Carboniferous coleoids, and it had a great impact on the early diversification within this cephalopod lineage.

The evolutionary plasticity of Carboniferous coleoids confirms the empirical rule in evolutionary theory according to which morphological plasticity is higher at early evolutionary stages, but it tends to become less variable at later evolutionary stages, when stabilization control created more stable morphotypes (Mamkaev, 1968).

The following early evolutionary trends can be distinguished in the subclass Coleoidea:

The order Hematitida (for example, *Hematites*, Flower and Gordon, 1959), in which the longiconic phragmocone was not completely covered by the mantle (and rostrum) during early ontogeny, but became covered in later stages of growth; the body chamber was already eliminated, and the terminal chamber was about 1. 5 camerae long; the rostrum is prominent with numerous longitudinal ridges separated by deep grooves, its anterior portion has a short dorsal protrusion, the conotheca is multilayered, and it lacks a nacreous layer (Doguzhaeva et al., 2002a);

The order Spirulida (for example, *Shimanskya*, Doguzhaeva et al., 1999), in which the body chamber and phragmocone are long and tubular, the shell wall lacks a nacreous layer and is covered by the outer plate instead of a rostrum; a proostracum-like structure is not developed (Doguzhaeva et al., 1996, 1999);

The order Aulacocerida (*Mutveiconites*, Doguzhaeva, 2002c), in which the rostrum is a small, cone-like structure surrounding the protoconch and the initial portion of the longiconic phragmocone; the conotheca has a nacreous layer; the body chamber is long and tubular, at least in early ontogenetic stages (Doguzhaeva, 2002a, c; Doguzhaeva et al., 2006).

The order Donovaniconida, ord. n. (*Rhiphaeoteuthis*, Doguzhaeva, 2002c; *Donovaniconus*, Doguzhaeva et al., 2003, *Saundersites*, gen. n.) in which the moderately breviconic body chamber is present either through the whole ontogeny (*Donovaniconus, Saundersites*), or during early ontogeny only (*Rhiphaeoteuthis*); the proostracum-like structure is comparatively short with a broadly rounded anterior edge with growth lines *(Donovaniconus, Saundersites)*; a substantial ink-sac is present *(Donovaniconus)*; the breviconic phragmocone is about the same length as the body chamber or shorter; the rostrum is weakly calcified (*Saundersites*), or sheet-like, or missing (*Rhiphaeoteuthis*); the conotheca has a principal nacreous layer; mural parts of septa are long, up to 1/2 camera length, they are embedded into a shell wall and seem to be organic-rich; sutures are nearly straight or weakly undulated; septa consist of lamello-fibrillar nacre *(Donovaniconus)* (Doguzhaeva, 2002a; Doguzhaeva et al., 2002c, 2003); arm hooks are in double rows (*Saundersites*).

Acknowledgments

We thank the Royal Swedish Academy of Sciences for providing grants supporting L. Doguzhaeva's yearly visits since 1980 to the Department of Palaeozoology, Swedish Museum of Natural History, Stockholm, Sweden. This support made it possible to carry out our joint, long-term study of the evolution of coleoid cephalopods. We are greatly indebted to staff of the Field Museum of Natural History in Chicago who provided the material, and to Thomas Denk (Swedish Museum of Natural History) for scanning the images with the digital camera used in this study. This research was partly supported by NSF grant EAR-0125479 to R.H.M. We cordially thank R. Davis, N. Landman, W. B. Saunders, and D. Fuchs for their helpful comments on earlier drafts of this manuscript.

References

Allison, P. A. 1987. A new cephalopod with soft parts from the Upper Carboniferous Francis Creek Shale of Illinois, USA. *Lethaia* **20**: 117–121.

Allison, P. A. 1988. Phosphatized soft-bodied squids from the Jurassic Oxford Clay. *Lethaia* **21**: 403–410.

Bather, F. A. 1888. Professor Blake and shell-growth in Cephalopoda. *Annals and Magazine of Natural History* **1**: 421–427.

Closs, D., and M. Gordon. 1966. An Upper Paleozoic goniatite radula. *Rio Grande do Sul, Universidade, Escola de Geologia, Notas e Estudos* **1, 2**: 79–80.

Doguzhaeva, L. A. 2002a. Evolutionary trends of Carboniferous coleoids: the ultrastructural view. *In* K. Warnke (editor), *International Symposium, Coleoid Cephalopods Through Time, Program and Abstracts, Berliner Paläobiologische Abhandlungen* **1**: 29–33.

Doguzhaeva, L. A. 2002b. Adolescent bactritoid, orthoceroid, ammonoid and coleoid shells from the Upper Carboniferous and Lower Permian of south Urals. *In* H. Summesberger, K. Histon, and A. Daurer (editors), *Cephalopods – Present and Past, Abhandlungen Geologische Bundesanstalt* **57**: 9–55.

Doguzhaeva, L. A., R. H. Mapes, and E. Dunca. 2006. A Late Carboniferous adolescent cephalopod from Texas (USA), with a short rostrum and a long body chamber. *Acta Universitatis Carolinae, Geologica, Special Paper*. **49**: 59–67.

Doguzhaeva, L. A., R. H. Mapes, and H. Mutvei. 1996. Ultrastructural comparison of the shell in Carboniferous *Bactrites* sp. (Russia) and *Bactrites postremus* (USA). *In* F. Olóriz and F. J. Rodríguez-Tovar (editors), *IV International Symposium, Cephalopods – Present and Past, Granada 1996*. Abstracts Volume. 51–52.

Doguzhaeva, L. A., R. H. Mapes, and H. Mutvei. 1997. Beaks and radulae of Early Carboniferous goniatites. *Lethaia* **30**: 305–313.

Doguzhaeva, L. A., R. H. Mapes, and H. Mutvei. 1999. A Late Carboniferous spirulid coleoid from the southern mid-continent (USA). *In* F. Olóriz and F. J. Rodríguez-Tovar (editors), *Advancing Research on Living and Fossil Cephalopods*, pp. 47–57. New York: Kluwer Academic/Plenum Publishers.

Doguzhaeva, L. A., R. H. Mapes, and H. Mutvei. 2002a. Early Carboniferous coleoid *Hematites*, Flower and Gordon, 1959 (Hematitida ord. nov.) from midcontinent (USA). *In* H. Summesberger, K. Histon, and A. Daurer (editors), *Cephalopods – Present and Past, Abhandlungen Geologische Bundesanstalt* **57**: 299–320.

Doguzhaeva, L. A., R. H. Mapes, and H. Mutvei. 2002b. The coleoid with an ink sac and a body chamber from the Upper Pennsylvanian of Oklahoma, USA. *In* K. Warnke (editor), *International Symposium, Coleoid Cephalopods Through Time, Program and Abstracts, Berliner Paläobiologische Abhandlungen* **1**: 34–38.

Doguzhaeva, L. A., R. H. Mapes, and H. Mutvei. 2003. The shell and ink sac morphology and ultrastructure of the Late Pennsylvanian cephalopod *Donovaniconus* and its phylogenetic significance. *Berliner Paläobiologische Abhandlungen* **3**: 61–78.

Doguzhaeva, L. A., R. H. Mapes, and H. Mutvei. 2004a. Occurrence of ink in Paleozoic and Mesozoic coleoids (Cephalopoda). *Mitteilungen aus dem Geologisch-Paläontologischen Institut der Universität Hamburg* **88**: 145–156.

Doguzhaeva, L. A., R. H. Mapes, and H. Mutvei. 2004b. Reevaluation of a coleoid from the Mazon Creek (Middle Pennsylvanian) *Lagerstätte. VI International Symposium, Cephalopods – Present and Past, 16–19 September, Arkansas,* Abstracts Volume: 35–37.

Doguzhaeva, L. A., R. H. Mapes, H. Mutvei, and R. K. Pabian. 2002c. The Late Carboniferous phragmocone-bearing orthoconic coleoids with ink sacs: their environment and mode of life. *First International Palaeontological Congress, Geological Society of Australia, Abstracts* **68**: 200.

Doguzhaeva, L. A., and H. Mutvei. 2003. Gladius composition and ultrastructure in extinct squid-like coleoids: *Loligosepia, Trachyteuthis* and *Teudopsis. Revue de Paléobiologie* **22**: 877–894.

Doguzhaeva, L. A., and H. Mutvei. 2005. The original composition of the gladius in the Aptian plesioteuthid *Nesisoteuthis*, based on its ultrastructure. *In* M. Kostak and J. Marek (editors), *2nd International Symposium, Coleoid Cephalopods Through Time, Prague 2005, Short Papers*/Abstract Volume: 50–53.

Doguzhaeva, L. A., H. Mutvei, H. Summesberger, and E. Dunca. 2004c. Bituminous soft body tissues in Late Triassic ceratitid *Austrotrachyceras. Mitteilungen aus dem Geologisch-Paläontologischen Institut der Universität Hamburg* **88**: 37–50.

Doyle, P., D. T. Donovan, and M. Nixon. 1994. Phylogeny and systematics of the Coleoidea. *University of Kansas Paleontological Contributions, New Series* **5**: 1–15.

Flower, R. H., and M. Gordon, Jr. 1959. More Mississippian belemnites. *Journal of Paleontology* **33**: 809–842.

Gabbott, S. E. 1999. Orthoconic cephalopods and associated fauna from the Late Ordovician Soom Shale *Lagerstätte*, South Africa. *Palaeontology* **42**: 123–148.

Gordon, M., Jr. 1964. Carboniferous Cephalopods of Arkansas. *United States Geological Survey Professional Paper* **460**: 320pp.

Gordon, M., Jr. 1971. Primitive squid gladii from the Permian of Utah. *United States Geological Survey Professional Paper* **750C**: 34–38.

Ivanov, D. L., and B. I. Sirenko. 1990. Radula in polyplacophoran molluscs (Class Polyplacophora). *Evolutionary Morphology of Mollusks. Archives of Zoological Museum of Moscow State University* **28**: 149–158. [In Russian]

Ivanov, D. L., and Y. I. Starobogatov. 1990. Radula in cephalopods. *Evolutionary Morphology of Mollusks. Archives of Zoological Museum of Moscow State University* **28**: 142–149. [In Russian]

Jeletzky, J. A. 1966. Comparative morphology, phylogeny, and classification of fossil Coleoidea. *University of Kansas Paleontological Contributions, Mollusca* **7**: 1–162.

Johnson, R. G., and E. S. Richardson. 1966. A remarkable Pennsylvanian fauna from the Mazon Creek area, Illinois. *Journal of Geology* **74**: 626–631.

Johnson, R.G., and E. S. Richardson. 1968. Ten-armed fossil cephalopod from the Pennsylvanian in Illinois. *Science* **159**: 526–528.

Kear, A. J., D. E. G. Briggs, and D. T. Donovan. 1995. Decay and fossilization of non-mineralized tissue in coleoid cephalopods. *Palaeontology* **38**: 105–131.

Kostak, M., J. Marek, P. Neumann, and M. Pavela. 2002. An early Carboniferous? Coleoid (Cephalopoda Dibranchiata) fossil from the Kulm of Northern Moravia (Czech Republic). *In* K. Warnke (editor), *Program and Abstracts, Coleoid Cephalopods Through Time, Berliner Paläobiologische Abhandlungen* **1**: 58–60.

Kluessendorf, J., and P. Doyle. 2000. *Pohlsepia Mazonensis*, an early "octopus" from the Carboniferous of Illinois, USA. *Palaeontology* **43**: 919–926.

Mamkaev, J. V. 1968. Comparison of the morphological differences in lower and higher groups of one and the same stem. *Zhurnal obschey biologii* **29**: 48–55. [In Russian]

Mehl, J. 1984. Radula und Fangarme bei *Michelinoceras* sp. aus dem Silur von Bolivien. *Paläontologische Zeitschrift* **58**: 211–229.

Naef, A. 1922. Die Fossilen Tintenfische. Jena: Gustav Fischer.

Nixon, M. 1988. The buccal mass of fossil and recent cephalopods. *In* M. R. Nixon and E. R. Trueman (editors), *The Mollusca. Palaeontology and Neontology*. **12**: 103–122. San Diego, CA: Academic Press.

Owen, R. 1844. A description of certain belemnites, preserved with a great proportion of their soft parts, in the Oxford Clay, at Christian-Malford, Wilts. *Philosophical Transactions of the Royal Society* **134**: 65–85.

Richardson, E. S., and R. G. Johnson. 1971. The Mazon Creek faunas. *Proceedings of North American Paleontological Convention* **2**: 1223–1235. Lawrence, KS: Allen Press.

Rossolimo, O. L. 1990. Evolutionary morphology of mollusks (regularities of morpho-functional transformation of radula apparatus). *Archives of Zoological Museum of Moscow State University* **28**: 223 pp. [In Russian]

Saunders, W. B., and E. S. Richardson. 1979. Middle Pennsylvanian (Desmoinesean) (sic) Cephalopoda of the Mazon Creek fauna, northeastern Illinois. *In* M. H. Nitecki (editor), *Mazon Creek Fossils*, pp. 333–359. New York: Academic Press.

Solem, A., and E. S. Richardson. 1975. *Paleocadmus*, a nautiloid cephalopod radula from the Pennsylvanian Francis Creek Shale of Illinois. *The Veliger* **17**: 233–242.

Starobogatov, Y. I. 1990a. Evolutionary trends of radula in modern mollusks. Evolutionary morphology of mollusks. *Archives of Zoological Museum of Moscow State University* **28**: 37–47. [In Russian]

Starobogatov, Y. I. 1990b. Evolutionary development of radulae. Evolutionary morphology of molluscs. *Archives of Zoological Museum of Moscow State University* **28**: 48–90. [In Russian]

Tanabe, K., and R. H. Mapes. 1995. Jaws and radula of the Carboniferous ammonoid *Cravenoceras*. *Journal of Paleontology* **69**: 703–707.

Yochelson, E. L., R. H. Flower, and G. F. Webbers. 1973. The bearing of the new Late Cambrian monoplacophoran genus *Knightoconus* upon the origin of the Cephalopoda. *Lethaia* **6**: 275–310.

Chapter 7
On the Species Status of *Spirula spirula* (Linné, 1758) (Cephalopoda): A New Approach Based on Divergence of Amino Acid Sequences Between the Canaries and New Caledonia

Kerstin Warnke

Freie Universität Berlin, FR Paläontologie, Malteserstr. 74-100, Haus D, 12249 Berlin, Germany, spirula@web.de

Keywords: *Spirula spirula*, 16s rDNA, cytochrome oxidase subunit III (COIII)

1 Introduction

Spirula is one of the most unusual Recent cephalopods, with a unique chambered shell related to an osmotically regulated buoyancy control, a unique photophore at the tip of the mantle, an oegopsid eye, and a greatly reduced radula (Nixon and Young, 2003). The external structure of the early *Spirula* shell starting from a spherical initial chamber resembles very much that of Ammonoidea, indeed more than that of Belemnoidea (Bandel and Boletzky, 1979; Bandel, 1982). Thus, *Spirula* is likely to provide interesting insights concerning both biology and paleontology.

Well preserved material of *Spirula* is scarce, however, despite the countless shells found on oceanic beaches (Dauphin, 1979a, b). As a consequence, the question whether there is more than one species of *Spirula* has not yet been answered definitively. The chief aim of the present study is to discuss the reliability of molecular data for a distinction of inter- versus intraspecific divergence of *Spirula*.

N. H. Landman et al. (eds.), *Cephalopods Present and Past: New Insights and Fresh Perspectives*, 144–155.

2 Taxonomy

In early papers on *Spirula*, up to five species were described (Owen, 1879; Huxley and Pelseneer, 1885; Lönnberg, 1896). Only one species was accepted by Chun (1915), Naef (1923), and Bruun (1943). Nesis (1987) and Norman (2000) also described one living species of *Spirula*. A major problem in the description of more than one species of *Spirula* is that different "species" are represented by only a few, often incomplete specimens. Moreover, most type repositories are unknown (Young and Sweeney, 2002). At present, the taxonomic status of *Spirula* is still called into question. Indeed, Young and Sweeney (2002) listed most species of *Spirula* as either undetermined or *nomina nuda*. Young and Sweeney (2002) considered only *Spirula spirula* as a valid species (Table 7.1). Nonetheless, the type repository of this holotype remains unknown and the type locality is extremely vague.

No doubt, *Spirula* is widely distributed. Live specimens occur in the waters of Indonesia, Melanesia, Australia, southeastern Africa, between the Canary Islands and northwestern Africa (Fig. 7.1), in the Caribbean Sea, and in the Gulf of Mexico (Bruun, 1943; Clarke, 1966; Dauphin, 1979a, b; Nesis, 1987).

Apparently the geographic range of *Spirula* is disjunct like that of many other epi- and mesopelagic, circumtropical species that are widely distributed in the Atlantic and Indo-West Pacific. Probably there is no gene flow between the Atlantic and Indo-Pacific populations living around South Africa (Nesis, 1998). Thus, one can infer the occurrence of geographic subspecies, one in the Atlantic and the other in the Indo-West Pacific (Nesis, 1998). Subspecies are rarely used in cephalopod systematics (Voss, 1977). However, Bruun (1943) already considered the

Table 7.1 *Status of* Spirula *species after Young and Sweeney (2002).*

Species name	Determination	Type locality	Comments
Spirula peronii atlantica	[*forma*] Girard, 1890: 250.	Algarve (Portugal); Azores; etc.	Undetermined
Spirula australis	Lamarck, 1816: pl 465, Figs. 5a, b.	Not designated	Undetermined
Spirula blakei	Lönnberg, 1896: 100.	West Indies	Undetermined
Spirula fragilis	Lamarck, 1801: 102.		Nomen nudum
Spirula peronii indopacifica	[*forma*] Girard, 1890: 250.	New Caledonia; Indian Ocean; etc.	Undetermined
Spirula peronii	Lamarck, 1822 in 1815–1822: 601.	"l'Océan austral et celui des Moluques"	Undetermined
Spirula prototypus	Lesueur and Petit, 1807: pl 30, Fig. 4.	Not designated	Undetermined
Spirula reticulata	Owen, 1848: 14, pl. 4, Figs. 3, 9.	Off Timor [*fide* Lönnberg (1896: 99)]	Undetermined
Loligo spiralis	d'Orbigny, 1826: 153	–	Nomen nudum
Nautilus spirula (Spirula spirula)	Linné, 1758: 710 [*fide* Bruun (1943: 3)]	"America"	Valid species

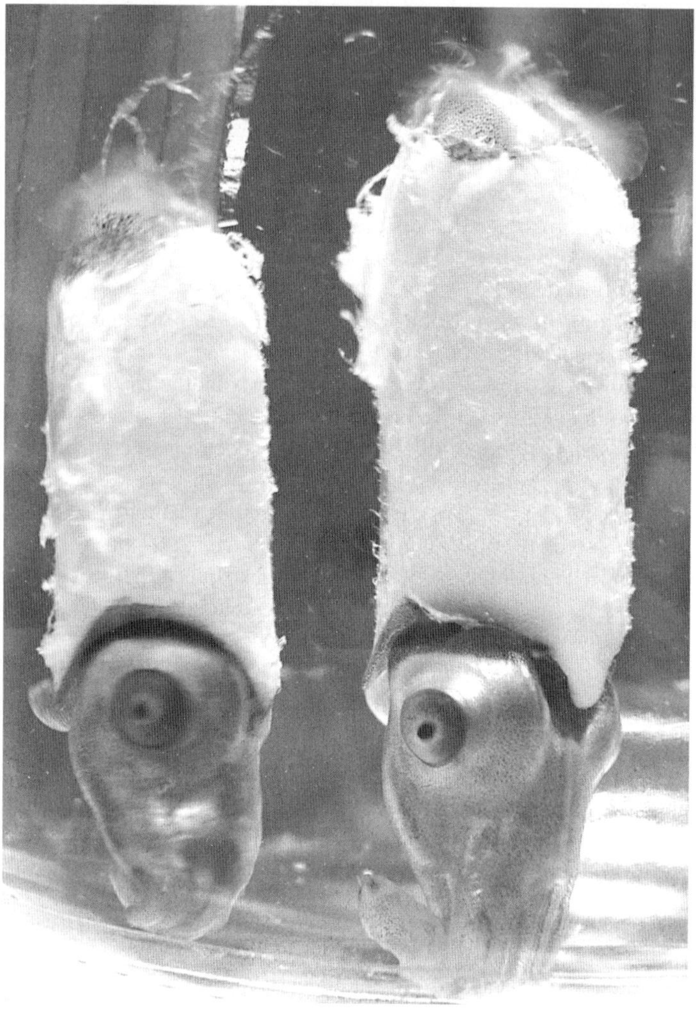

Fig. 7.1 *Living* Spirula spirula *(left, female [mantle length = 4 cm], right, male [mantle length = 5 cm) caught between the Canary Islands and the coast of Morocco. The animals were caught with an Isaacs Kidd mid-water trawl and placed in an aquarium to take photos.*

possibility of subspecies or races of *Spirula* based on morphological characters. He compared preserved soft parts from the Atlantic and the Indo-Pacific and analyzed them with regard to pigmentation, size, sexual arm differentiation, similarity of external characters, and some further morphometric measurements, but he did not find any definite morphological difference between them.

Nesis (1998) again emphasized the potential importance of biochemical or molecular data to distinguish intra- and interspecific differences, especially regarding species with a distribution pattern like that of *Spirula*.

3 DNA Sequence Data

In most cases, molecular data provide informative phylogenetic indications for determining evolutionary relationships between sibling species or at least morphologically nearly identical species. A famous example is the study of teleost fish species by Sturmbauer and Meyer (1992). Numerous, nearly morphological identical cichlid species recognized in Lake Tanganyika are genetically divergent. The genetic divergence within these species displays twice as much as the genetic divergence within the morphologically different cichlid species from Lake Malawi. Many further examples can be found in the literature of the last 20 years (e.g., Palumbi and Benzie, 1991; Knowlton et al., 1993).

Using allozyme markers, several studies have been carried out to detect the genetic divergence of especially closely related cephalopod species, cryptic speciation, and population structure (Augustyn and Grant, 1988; Levy et al., 1988; Carvalho and Pitcher, 1989; Brierley et al., 1993; Katugin, 1993; Yeatman and Benzie, 1993; Allcock et al., 1997; Pérez-Lozada et al., 1999; Maltagliati et al., 2002). Using randomly amplified polymorphic DNA markers (RAPD) or restriction fragment length polymorphisms (RFLPs), the genetic distinction between cephalopod species or between subunits inside the species themselves can also be established (Warnke et al., 2000; Herke and Folz, 2002; Chapela et al., 2003; Sands et al., 2003). Shaw (2002) reviewed all these papers as well as the studies based on microsatellite markers, which are used to analyze the population structures of cephalopods.

The DNA sequence data used for direct analyses were mostly obtained from mitochondrial genes. The advantage of mitochondrial DNA (mtDNA) over nuclear DNA is that this small extra-nuclear part of the genome can be found in multiple copies. Consequently, a small amount of tissue (e.g., 10–15 mg from the mantle) is sufficient to isolate the required initial amount of DNA for PCR-based methods. In the present case, this constitutes a major advantage, since only few freshly caught animals were available and they would not have to be destroyed so that they could be used for further morphological studies. MtDNA is known to evolve relatively fast; it is still used widely today to assess taxonomic relationships and differences between populations within species. MtDNA sequences provide the most direct means for measuring genetic diversity to determine the sequences of bases in the DNA (Beebee and Rowe, 2004; Frankham et al., 2004). MtDNA also encodes for rRNA. The 16s rRNA gene has one further advantage in that universal primers are available (Kocher, 1992).

Based on mtDNA sequences, several studies have been devoted to the inter- and intraspecific variations within Recent Cephalopoda. MtDNA sequence data were used by Wray et al. (1995) among others. These authors traced the geographically related diversifications of some genetically distinct lineages of *Nautilus* using a portion of the mt16S rRNA gene. Molecular data were also used to check genetic relationships within the genus *Octopus* (Barriga Sosa et al., 1995; Hudelot, 2002). Other studies examined in particular the distribution of *O. vulgaris*. Using mtCOIII (cytochrome oxidase subunit III) and mt16S rDNA sequences (Warnke, 1999; Söller et al., 2000; Warnke et al., 2002, 2004), the distribution of *O. vulgaris* was confirmed for the Mediterranean as well as for the eastern and western Atlantic. *Octopus vulgaris* could not be found in the eastern Pacific.

Studies concerning interspecific variations in the genus *Pareledone* were performed by Allcock and Piertney (2002) again, using mt16S rDNA sequences. During study of higher-level relationships of coleoids, intra- or interspecific sequence variations of Decabrachia and Vampyropoda were also analyzed as well (Bonnaud et al., 1994; Boucher-Rodoni and Bonnaud, 1996; Bonnaud et al., 1996, 1997; Carlini and Graves, 1999; Carlini et al., 2001). For example, the sequence divergences between two different species of the genus *Loligo* were examined focusing on partial sequences of the mitochondrial 16S rDNA and COIII (Bonnaud et al., 1996). Analysis of the phylogeny and biogeography of loliginid squids was made by Anderson (2000) based on mtDNA (COI, 16S) sequences. Nishiguchi et al. (2004) analyzed the phylogentic relationships between Sepiolidae species using mtDNA (COI, 12S, 16S) and a partial sequence of a nuclear gene (28s rRNA).

The phylogenetic position of *Spirula* within the Decabrachia is still called into question, but a position next to *Loligo* or next to *Sepia* is discussed (Bonnaud et al., 1996; Warnke et al., 2003; Strugnell et al., 2005). Genetic studies within the genus *Sepia* are less frequent. For example, Zheng et al. (2001) described the genetic variation within the common Chinese cuttlefish *Sepiella maindroni* using the cytochrome oxidase subunit I (COI).

To date, no molecular investigation of intra- or interspecific variations of *Spirula* has been made. Therefore, a preliminary molecular analysis of *Spirula* is presented in order to obtain some clues about whether there is more than one *Spirula* species. MtCOIII and mt16S rDNA were used in this analysis because corresponding sequences of *Spirula* from a different part of their distribution area already exist in the EMBL data bank, and because these sequence data proved useful earlier on in determining evolutionary relationships between nearly indistinguishable sibling species.

4 Material and Methods

Live animals of *Spirula* were caught near Fuerteventura (Canary Islands, Spain, Fig. 7.2). They were preserved in 95–100% ethanol. A small tissue sample was taken from the arm tip of each animal. DNA was isolated from these samples following the Chelex method (Walsh et al., 1991) modified by Söller et al. (2000). Chelex supernatant was purified with the DNeasy Kit (Quiagen, Hilden).

A fragment of mitochondrial ribosomal 16S RNA gene (16S) and the cytochrome oxidase III gene (COIII) were used as target sequences. The DNA of mt16S rDNA was amplified by PCR using universal primers 16Sar and 16Sbr (Simon et al., 1991). For the COIII fragment, primers from Barriga Sosa et al. (1995) and Warnke

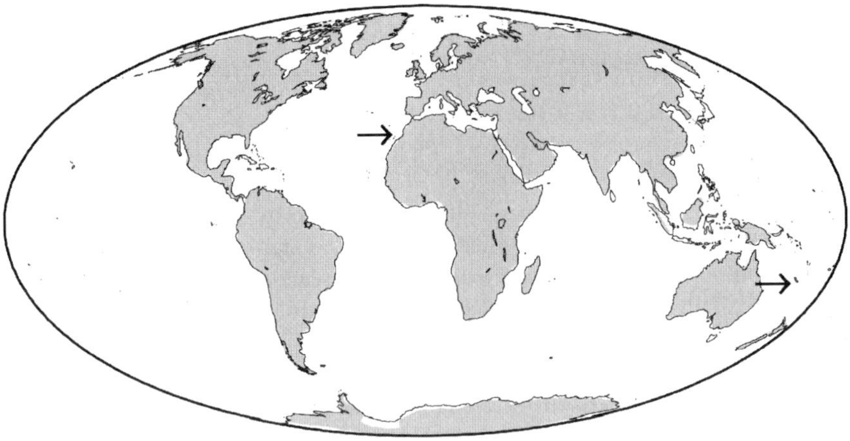

Fig. 7.2 *Map of localities of* Spirula *sample (arrows show sample location, map modified after Scotese, 2001).*

(1999) were used. These primers were designed for obtaining the DNA of COIII from various *Octopus* species. A polymerase chain reaction (PCR) was performed in 50 µl reaction volumes containing 10 mM Tris-HCL, pH 8.8; 25 mM KCL; 2 mM $MgSO_4$; 0.2 mM each of dATP, dCTP, dGTP, dTTP; 0.2 µM of both forward and reverse primers; 0.5 U *Taq* polymerase (Pharmacia Biotech, Freiburg i. Br.), 1–10 µl DNA solution (purified Chelex supernatant).

The fragments were sequenced and arranged together with three further *Spirula* sequences from the EMBL database [EMBL Acc-Nr.: X79574, Bonnaud et al. (1994), fragment of mt16S DNA; EMBL Acc-Nr.: X97957, Bonnaud et al. (1996), fragment of mtCOIII; EMBL Acc-Nr.: AY293659, Nishiguchi et al. (2004), fragment of mt16S DNA)] in a multiple sequences alignment. The *Spirula* specimens used by Bonnaud et al. (1994, 1996) were recovered from New Caledonia. For the animal mentioned by Nishiguchi et al. (2004), the Atlantic Ocean was given as the collecting site. The sequences of this study have been deposited in the EMBL database (accession numbers follow: AJ966784 *Spirula spirula* isolate 1, fragment of mt16S DNA; AJ966785, *Spirula spirula* isolate 2, fragment of mt16S DNA; AJ966786, *Spirula spirula* isolate 2, fragment of mtCOIII; AJ966787, *Spirula spirula* isolate 1, fragment of mtCOIII). The sequences of the animal 1, 3, and 4 were identical. Because of this, just the sequence of animals 1 and 2 were deposited in the database.

Sequence divergences were examined using the PAUP* program (Swofford, Smithonian Institution, Washington, DC). Additionally, the nucleotide sequences of COIII were translated into the amino acid sequence using the invertebrate mitochondrial code. The putative phylogenetic relationships were calculated applying both distance- and character-based analyses of the data. The trees were rooted with *Octopus vulgaris* (EMBL Acc-Nr.: AJ012121, Söller et al., 2000) as an outgroup representative.

5 Results

From each sample, fragments of up to 500 base pairs of mt16S rDNA and additionally 500 base pairs of COIII were amplified by PCR. The sequence divergence of 16S was zero or nearly so (0–0.2%) for all animals. The nucleotide divergence detected for COIII was also relatively low between populations of the Canary Islands (0–0.4%). The divergences between the nucleotide sequences of *Spirula* from Fuerteventura and New Caledonia were high (0.7–1.3%). However, all these substitutions were transversions. Transversions are mutations in which a purine nucleotide (Adenin, Guanin) is replaced by a pyrimidine nucleotide (Thymin, Cytosin). Compared to transitions, transversions do not occur often. Transitions are mutations in which a purine is substituted by the other purine nucleotide or a pyrimidine nucleotide by another pyrimidine (Beebee and Rowe, 2004). Moreover, when comparing the amino acid sequences of this gene (as obtained by the invertebrate mitochondrial code), the divergence between *Spirula* populations from the Atlantic and the Pacific increased considerably to reach a value of 3.9–4.6% while the divergence within the population of Fuerteventura remained low (0–0.66%). Calculation of an MP tree was not possible. PAUP found only two informative characters. These are not sufficient for a reliable parsimonious analysis (Hillis and Huelsenbeck, 1992).

6 Discussion

The relatively high divergence between the *Spirula* amino acid sequences of animals from New Caledonia and Fuerteventura suggests that the genus *Spirula* consists of more than one species. Using the gene COIII, the mentioned divergence is normally found among different species of cephalopods (Barriga Sosa, 1995; Bonnaud et al., 1996; Warnke, 1999; Söller et al., 2000). This observation notwithstanding, it is unusual to find such an enormous difference between the amino acid sequences of COIII. The limited data of this study are not unambiguous, however, in particular because only one sequence is available for *Spirula* from New Caledonia. Furthermore, it seems desirable to get more *Spirula* from distribution areas to conduct more detailed morphological investigations. In addition to the scarcity of morphological and molecular data, a discussion of the distribution of *Spirula* is difficult because data about spawning and the mode of life of *Spirula* hatchlings are not available (Bruun, 1943; Clarke, 1966). In particular, we know nothing about how far the paralarvae may drift in the open sea. But in general terms, it seems unwise to assume a recent gene flow between *Spirula* populations of the Canary Islands and of New Caledonia because *Spirula* has just a short life span of about 20 months (Clarke, 1970). Knowlton (2000) emphasized that the tools of molecular genetics have a great potential for clarifying the species boundaries in marine organisms. As compared with other molluscs or different invertebrates, cephalopod species boundaries have been examined rather little, so much remains to be done.

Acknowledgments

I would like to thank Dr. Sigurd von Boletzky (Laboratoire Arago, Banyuls-sur-Mer, France) for his support, encouragement, and discussions relating to the manuscript. Dr. Roger A. Hewitt (Essex, UK), Prof. Royal Mapes (Ohio University, Athens, Ohio), Dr. Neil Landman (American Museum of Natural History, New York) and one anonymous reviewer were greatly appreciated for making helpful comments on the manuscript. The assistance of Robert Schreiber (Humboldt University, Berlin) with DNA sequencing is especially acknowledged. I am grateful to the Canarian Institute of Marine Sciences (Instituto Canario de Ciencias Marinas, Spain). Special thanks to José Ignacio Santana, Maria José Rueda, and Dr. Octavio Llinás for making arrangements for my participation in several field operations of the research vessel "P/O Taliarte" and for assisting me on board. Moreover I am very grateful to the ship's crew for their kind support during the operations when catching *Spirula*. This work was supported by grants WA 1454/1-1 and WA 1454/1-2 from the Deutsche Forschungsgemeinschaft (DFG) and by subsidies related to the fishing operations of the Canarian Institute of Marine Sciences (Instituto Canario de Ciencias Marinas).

References

Allcock, A. L., A. S. Bierley, J. P. Thorpe, and P. G. Rodhouse. 1997. Restricted gene flow and evolutionary divergence between geographically separated populations of the Antarctic octopus *Pareledone turqueti*. *Marine Biology* **129**: 97–102.

Allcock, A. L., and S. B. Piertney. 2002. Evolutionary relationships of Southern Ocean Octopodidae (Cephalopoda: Octopoda) and a new diagnosis of *Pareledone*. *Marine Biology* **140**: 129–135.

Anderson, F. E. 2000. Phylogeny and historical biogeography of the loliginid squids (Mollusca: Cephalopoda) based on mitochondrial DNA sequence data. *Molecular Phylogenetics and Evolution* **15**: 191–214.

Augustyn, C. J., and W. S. Grant. 1988. Biochemical and morphological systematics *of Loligo vulgaris vulgaris* Lamarck and *Loligo vulgaris reynaudii* D'Orbigny *Nov. Comb*. (Cephalopoda: Myopsida). *Malacologia* **29**: 214–233.

Bandel, K. 1982. Morphologie und Bildung der frühontogenetischen Gehäuse bei conchiferen Mollusken. *Facies* **7**: 1–198.

Bandel, K., and S. v. Boletzky. 1979. A comparative study of structure, development and morphological relationships of chambered cephalopod shells. *Veliger* **21**: 313–354.

Barriga Sosa, I. D. L. A., K. Beckenbach, B. Hartwick, and M. J. Smith. 1995. The molecular phylogeny of five eastern north Pacific *Octopus* species. *Molecular Phylogenetics and Evolution* **4**: 163–174.

Beebee, T., and G. Rowe. 2004. *An Introduction to Molecular Ecology*. Oxford: Oxford University Press.

Bonnaud, L., R. Boucher-Rodoni, and M. Monnerot. 1994. Phylogeny of decapod cephalopods based on the partial 16S rDNA nucleotide sequence. *Comptes rendus de l'Académie des sciences* III **317**: 581–588.

Bonnaud, L., R. Boucher-Rodoni, and M. Monnerot. 1996. Relationship of some coleoid cephalopods established by 3′end of the 16S rDNA and cytochrome oxidase III gene sequence comparison. *American Malacological Bulletin* **12**: 87–90.

Bonnaud, L., R. Boucher-Rodoni, and M. Monnerot. 1997. Phylogeny of cephalopods inferred from mitochondrial DNA sequences. *Molecular Phylogenetics and Evolution* **7**: 44–54.

Boucher-Rodoni, R., and L. Bonnaud. 1996. Biochemical and molecular approach to cephalopod phylogeny. *American Malacological Bulletin* **12**: 79–85.

Brierley, A. S., P. G. Rodhouse, J. P. Thorpe, and M. R. Clarke. 1993. Genetic evidence of population heterogeneity and cryptic speciation in the ommastrephid squid *Martialia hyadesi* from the patagonian shelf and antarctic polar frontal zone. *Marine Biology* **116**: 593–602.

Bruun, A. F. 1943. The biology of *Spirula spirula* (L.). *Dana Report*. **24**: 1–44, 2 pls.

Carlini D. B., and J. E. Graves. 1999. Phylogenetic analysis of the cytochrome oxidase I sequences to determine higher level relationships within the coleoid cephalopods. *Bulletin of Marine Science* **64**: 57–76.

Carlini, D. B., R. E. Young, and M. Vecchione. 2001. A molecular phylogeny of the Octopoda (Mollusca: Cephalopoda) evaluated in light of morphological evidence. *Molecular Phylogenetics and Evolution* **21**: 388–397.

Carvalho, G. R., and T. J. Pitcher. 1989. Biochemical genetic studies on the Patagonian squid *Loligo gahi* d'Orbigny. II. Population structure in Falkland waters using isozymes, morphometrics and life history data. *Journal of Experimental Marine Biology and Ecology* **126**: 243–258.

Chapela, M. J., C. G. Sotelo, and R. I. Pérez-Martín. 2003. Molecular identification of cephalopod species by FINS and PCR-RFLP of a cytochrome b fragment. *European Food Research and Technology* **217**: 524–529.

Chun, C. 1915. Die Cephalopoden. 2. Teil; Myopsida, Octopoda. *Wissenschaftliche Ergebnisse der Deutschen Tiefsee-Expedition "Valdivia" 1898–1899* **18**: 311–552.

Clarke, M. R. 1966. A review of the systematics and ecology of oceanic squids. *Advances in Marine Biology* **4**: 91–300.

Clarke, M. R. 1970. Growth and development of *Spirula spirula*. *Journal of the Marine Biological Association of the United Kingdom* **50**: 53–64.

Dauphin, Y. 1979a. Contribution a l'étude de la formation des gisements de Cephalopodes. I- Les coquilles de spirules (Dibranchiata, Cephalopoda) de Nouvelle-Calédonie. *Cahiers de l'Indo-Pacifique* **1, 2**: 165–194.

Dauphin, Y. 1979b. Contribution a l'étude de la formation des gisements de Cephalopodes. II- Les coquilles de spirules (Dibranchiata, Cephalopoda) de l'Ile Maurice (Océan indien). *Cahiers de l'Indo-Pacifique* **1, 3**: 335–351.

Frankham, R., J. D. Ballou, and D. A. Briscoe. 2004. *A Primer of Conservation Genetics*. Cambridge: Cambridge University Press.

Girard, A. A. 1890. Revision des Mollusques du Museum de Lisbonne. 1. Céphalopodes. *Journal de Sciencias Mathematicas, Physicas e Naturaes (Series 2)* **4**: 233–268, 1 pl.

Herke, S.W., and D. W. Foltz. 2002. Phylogeography of two squid (*Loligo pealei* and *L. plei*) in the Gulf of Mexico and northwestern Atlantic Ocean. *Marine Biology* **140**: 103–115.

Hillis, D. M., and J. P. Huelsenbeck. 1992. Signal, noise, and reliability in molecular phylogenetic analyses. *Journal of Heredity* **83**: 189–195.

Hudelot, C. 2002. Molecular techniques and phylogeny of octopods. *In* P. R. Boyle, M. A. Collins, J. G. Pierce, and J. J. Watson (editors), *Cephalopod Biomass and Production. Bulletin of Marine Science* **71**: 1128.

Huxley, T, H., and P. Pelseneer. 1885. Report on the specimen of the genus *Spirula* Collected by H.M.S. "Challenger." *Report of the Scientific Results of the Voyage of the H.M.S. Challenger. Zoology* ***83***.

Katugin, O. N. 1993. Genetic variation in the squid *Berryteuthis magiister* (Berry 1913) (Oegopsida: Gonatidae). *In* D. Okutani, R. K. O'Dor, and T. Kubodera (editors), *Recent Advances in Cephalopod Fisheries Biology*, pp. 201–213. Tokyo: Tokai University Press.
Knowlton, N., L. A. Weigt, L. A. Solórzano, D. K. Millis, and E. Bermingham. 1993. Divergence in proteins, mitochondrial DNA, and reproductive compatibility across the Isthmus of Panama. *Science* **260**: 1629–1632.
Knowlton, N. 2000. Molecular genetic analyses of species boundaries in the sea. *Hydrobiologia* **420**: 73–90.
Kocher, T. D. 1992. PCR, direct sequencing, and the comparative approach. *PCR Methods Applications* **1**: 217–221.
Lamarck, J. B. 1801. *Système des animaux sans vertèbres, ou tableau général des classes, des ordres et des genres de ces animaux*. Paris.
Lamarck, J. B. 1815–1822. *Histoire naturelle des animaux sans vertèbres, presentant les caractères géneraux et particuliers de ces animaux, leur distribution, leurs classes, leurs familles, leurs genres, et la citation des principales espèces qui s'y raportent*. 7 volumes [Volume 7, 1822]. Paris.
Lamarck, J. B. 1816. Tableau encyclopédique et méthodique des trois règnes de la nature. [Volume 4, pls. 391–488]. Paris: Mme. veuve Agasse.
Lesueur, C. A., and N. M. Petit. 1807. Atlas Historique, 41 pls. *In* F. Peron (editor), *Voyage de découvertes aux terres Australes sur les corvettes le Géographe, le Naturaliste, et la Goëlette le Casuarina, pendant 1800–1804*. 2 volumes, atlas and 14 maps, 1807–1816.
Linné, C. 1758. *Systema naturae per regna tria naturae, secundum classes, ordines, genera, species cum characteribus, differentiis, synonymis, locis*. Holmiae.
Levy, J. A., M. Haimovici, and M. Conceicao. 1988. Genetic evidence for two species to the genus *Eledone* (Cephalopoda: Octopidae) in South Brazil. *Comparative Biochemistry and Physiology B. Biochemistry and Molecular Biology* **90**: 275–277.
Lönnberg, E. 1896. Notes on *Spirula reticulata* Owen and its phylogeny. *In* J. M. Hulth (editor), *Zoologiska Studier. Festskrift W. Lilljeborg tillegnad pa hans attionde fodelsedag af Svenska Zoologer*, pp. 99–119. Uppsala.
Maltagliati, F., P. Belcari, D. Casu, M. Casu, P. Sartor, G. Vargiu, and A. Castelli. 2002. Allozyme genetic variability and gene flow in *Octopus vulgaris* (Cephalopoda, Octopodidae) from the Mediterranean Sea. *In* P. R. Boyle, M. A. Collins, J. G. Pierce, and J. J. Watson (editors), *Cephalopod biomass and production. Bulletin of Marine Science* **71**: 473–486.
Naef, A. 1923. *Die Cephalopoden. Fauna und Flora des Golfes von Neapel und der angrenzenden Meeresabschnitte*. Monographie **35**(A): 149–863.
Nesis, K. N. 1987. *Cephalopods of the World. Squids, Cuttlefishes, Octopuses and Allies*. Neptune City, NJ: Tropical Fish Hobbyist Publications.
Nesis, K. N. 1998 Biodiversity and systematics in cephalopods: unresolved problems require an integrated approach. *South African Journal of Marine Science* **20**: 165–173.
Nishiguchi, M. K., J. E. Lopez, and S. V. Boletzky. 2004. Enlightenment of old ideas from new investigations: more questions regarding the evolution of bacteriogenic light organs in squids. *Evolution & Development* **6**: 41–49.
Nixon, M., and J. Z. Young. 2003. *The Brain and Lives of Cephalopods*. Oxford: Oxford University Press.
Norman, M. 2000. *Cephalopods – A World Guide*. Hackenheim: ConchBooks.
d'Orbigny. A. 1826. Tableau méthodique de la Classe de Céphalopodes [Introduction, pp. 96–120 by A. E. Ferussac]. *Annales des Sciences Naturelles, Paris* (series 1) **7**: 95–169, 245–314, 8 pls.
Owen, R. 1848. Description of two mutilated specimen of *Spirula peronii*, with some observations on *S. australis* and *reticulate. In* A. Adams, and L. Reeve, 1848–1850, *The Zoology of the Voyage of HMS "Samarang" under the command of Captain Sir Edward Belcher*, pp. 6–17. Paris.
Owen, R. 1879. Supplementary observations on the anatomy of *Spirula australis* Lam. *Annals and Magazine of Natural History* **3**(5).
Palumbi, S. R., and J. Benzie. 1991. Large mitochondrial DNA differences between morphologically similar Penaeid shrimp. *Molecular Marine Biology and Biotechnology* **1**: 27–34.

Pérez-Lozada, M., A. Guerra, and A. Sanjuan. 1999. Allozyme differentiation in the cuttlefish *Sepia officinalis* (Mollusca: Cephalopoda) from the NE Atlantic and Mediterranean. *Heredity* **83**: 280–289.

Sands, C. J., S. N. Jarman, and G. D. Jackson. 2003. Genetic differentiation in the squid *Moroteuthis ingens* inferred from RAPD analysis. *Polar Biology* **26**: 166–170.

Scotese, C. R. 2001. *Atlas of Earth History, Volume 1, Paleogeography*. Arlington, TX: PALEOMAP Project.

Shaw, P. W. 2002. Past, present and future applications of DNA-based markers in cephalopod biology. *In* P. R. Boyle, M. A. Collins, J. G. Pierce, and J. J. Watson (editors), *Cephalopod Biomass and Production*. *Bulletin of Marine Science* **71**: 67–78.

Simon, C., A. Franke, and A. Martin. 1991. The polymerase chain reaction: DNA extraction and amplification. *In* G. M. Hewitt, A. W. B. Johnston, and J. P. W. Young (editors), *Molecular Techniques in Taxonomy. NATO ASI series H* **57**: 329–355.

Söller, R., K. Warnke, U. Saint-Paul, and D. Blohm. 2000. Sequence divergence of mitochondrial DNA indicates cryptic biodiversity in *Octopus vulgaris* and supports the taxonomic distinctiveness of *Octopus mimus* (Cephalopoda: Octopodidae). *Marine Biology* **136**: 29–35.

Strugnell, J., M. Norman, J. Jackson, A. J. Drummond, and A. Cooper. 2005. Molecular phylogeny of coleoid cephalopods (Mollusca: Cephalopoda) using a multigene approach; the effect of data partitioning on resolving phylogenies in a Bayesian framework. *Molecular Phylogenetics and Evolution* **37**: 426–441.

Sturmbauer, C., and A. Meyer. 1992. Genetic divergence, speciation and morphological stasis in a lineage of African cichlid fishes. *Nature* **359**: 578–581.

Swofford, D. L. 2001. *PAUP*. Phylogenetic Analysis Using Parsimony (*and Other Methods)*. Version 4. Sunderland, MA: Sinauer Associates.

Voss, G. L. 1977. Present status and new trends in cephalopod systematics. *Symposia of the Zoological Society of London* **38**: 49–60.

Walsh, P. S., D. A. Metzger, and R. Higushi. 1991. Chelex 100 as a medium for simple extraction of DNA for PCR-based typing from forensic material. *Biotechniques* **10**: 506–513.

Warnke, K. 1999. Diversität des Artenkomplexes *Octopus* cf. *vulgaris* Cuvier, 1797 in Beziehung zu seiner Verbreitung an der Ost- und Westküste Lateinamerikas. Doctoral dissertation, Universität Bremen, Aachen: Shaker Verlag..

Warnke, K., J. Plötner, J. I. Santana, M. J. Rueda, and O. Llinas. 2003. Reflections on the phylogenetic position of *Spirula* (Cephalopoda): preliminary evidence from the 18S ribosomal RNA gene. *Berliner Paläobiologische Abhandlungen* **3**: 253–260.

Warnke, K., R. Söller, D. Blohm, and U. Saint-Paul. 2000. Rapid differentiation between *Octopus vulgaris* Cuvier, 1797 and *Octopus mimus* Gould 1852, using randomly amplified polymorphic DNA. *Journal of Zoological Systematics and Evolutionary Research* **38**: 119–122.

Warnke, K., R. Söller, D. Blohm, and U. Saint-Paul. 2002. Assessment of the phylogenetic relationship between *Octopus vulgaris* Cuvier, 1797 and *O. mimus* Gould 1852, using mitochondrial 16S rRNA. *In* H. Summesberger, K. Histon, and A. Daurer (editors), *International Symposium "Cephalopods Present and Past." Abhandlungen der Geologischen Bundesanstalt Wien* **57**: 401–405.

Warnke, K., R. Söller, D. Blohm, and U. Saint-Paul. 2004. A new look at geographic and phylogenetic relationships within the species group surrounding *Octopus vulgaris* (Mollusca: Cephalopoda): indications from mitochondrial DNA nucleotide sequences. *Journal of Zoological Systematics and Evolutionary Research* **42**: 306–312.

Wray, C. G., N. H. Landman, W. B. Saunders, and J. Bonacum. 1995. Genetic divergence and geographic diversification in *Nautilus*. *Paleobiology* **21**: 220–228.

Yeatman, J., and J. A. H. Benzie. 1993. Cryptic speciation in *Loligo* from northern Australia. *In* D. Okutani, R. K. O'Dor, and T. Kubodera, (editors), *Recent Advances in Cephalopod Fisheries Biology*, pp. 641–652. Tokyo: Tokai University Press.

Young, R. E., and M. J. Sweeney. 2002. Taxa associated with the family Spirulidae Owen, 1836. Tree of Life, web projekt, http://tolweb.org/accessory/Spriulidae_Taxa?acc_id=2337#about.

Zheng, X., R. Wang, X. Wang, S. Xiao, and B. Chen. 2001. Genetic variation in populations of the common Chinese cuttlefish *Sepiella maindroni* (Mollusca: Cephalopoda) using allozymes and mitochondrial DNA sequence analysis. *Journal of Shellfish Research* **20**: 1159–1165.

Part II
Morphology of Soft and Hard Tissues

Chapter 8
Understanding Ammonoid Sutures

New Insight into the Dynamic Evolution of Paleozoic Suture Morphology

Emily G. Allen

Department of Geology, Bryn Mawr College, Bryn Mawr, PA 19010, USA, egallen@brynmawr.edu

Keywords: ammonoid, suture, Paleozoic, extinction, morphospace

1 Introduction

Paleozoic ammonoids showed remarkable diversity in the morphology of the internal septa that subdivide the shell into a series of chambers, despite a notable lack of innovation in external shell geometry over the same time period (Saunders et al., 2004). It has long been recognized that the complexity of these septa generally increases through time, and that trend has been well documented using the suture formed by the intersection of the internal and external shell walls as a proxy for septal morphology. Recent work has indicated that the trend for increasing suture complexity resulted from a pervasive bias, which favored the origination of more complex forms and simultaneously eliminated taxa possessing the simplest suture morphologies (Saunders et al., 1999; see also Allen, 2005). This trend persisted despite significant losses in complexity across two of the three mass extinctions that interrupted Paleozoic ammonoid evolution.

N. H. Landman et al. (eds.), *Cephalopods Present and Past: New Insights and Fresh Perspectives*, 159–180.

Additionally, origination of more complexly sutured morphotypes coincided more closely with periods of net origination than recovery from this taxonomic turnover. Postcrisis rebound was instead characterized by a recovery of preextinction complexity levels and apparent refilling of previously occupied morphospace (Saunders et al., 1999).

However, measures of complexity are single-valued proxies for suture line morphology. As such, they frequently confound disparate line shapes (Lutz and Boyajian, 1995; Allen, 2006) and are not universally applicable to all forms, e.g., fractal dimension and nonammonitic sutures (Pérez-Claros et al., 2002; for more discussion, see Allen, 2005). Thus, a recovery of preextinction complexity levels does not necessarily document morphological equivalency, as very different sutures may be equally complex. Likewise, there is more than one way to increase complexity (e.g., add lobes, subdivide main lobes, subdivide subdivisions, vary lobe amplitudes, or suture curvature), and while similar trends toward increasingly complex sutures have been documented within most Paleozoic ammonoid subclades (Saunders et al., 1999), a focus on complexity fails to emphasize their morphological diversity.

2 Assessing Suture Morphology

2.1 Background

The intricacy and irregularity of many septal folds has made rigorous quantification of suture line morphologies difficult. Thus, discussions of suture evolution have been largely limited to qualitative and quantitative assessments of line complexities. As a result, suture evolution is frequently considered independently of, or only qualitatively in, comprehensive large-scale studies of ammonoid evolution, despite their recognized phylogenetic and evolutionary significance.

Suture shape is difficult to both qualify and quantify for three reasons. First, sutures are *morphologically complex*. But, as McShea (1991) argued, while most of us will agree that complexity increases in evolution, we are not quite sure what constitutes complexity or how to quantify it. The same holds true for the specific case of ammonoid sutures: it is apparent that their complexity increases through time, ontogeny and phylogeny, but whether that complexity is a property of lobe elaboration, number of lobes, variety of lobe morphologies, arrangement of lobes, or some combination thereof is debatable. This fundamental quandary is apparent even in the general descriptive terminology used to discuss suture form; for example, what is the difference between suture "morphology," "complexity," "intricacy," or even "geometry," all of which are used in this chapter to describe the character of suture lines?

The term *morphology* is a general term referring to the form and structure of a suture line, while the *geometry* of that same line is a precise mathematical description of its morphology, technically one based upon geometric or trigonometric principles.

On the other hand, the *intricacy* of a suture refers specifically to the quality or extent of its complex character (i.e., its parts, details, and their arrangement). Suture *complexity* is technically equivalent to suture intricacy. However, in the literature it has taken on a more quantitative connotation and is usually only meaningful in relative terms, i.e., one suture is more or less complex than another.

Second, ammonoid sutures may be mathematically as well as morphologically complex. The *mathematical complexity* of a suture refers to its morphological characteristics that hinder mathematical representation, such as nonperiodicity or variation in lobe morphology along suture length. Due to this type of complexity, many sutures fail to satisfy the mathematical requirements necessary to employ standard morphometric techniques for assessing variation in forms defined by a curve (see Allen [2006] for detailed discussion). Finally, homology is difficult to assess except in a limited phylogenetic context. In particular, mature sutures may range in shape from a single period of a low frequency sinusoid, through a series of moderate-frequency folds, to fractal-like geometries (see Figs. 8.1–3). This makes direct point-to-point (landmark) comparisons along the suture line length inappropriate, since a single lobe in one suture may not be functionally or developmentally homologous to multiple lobes occupying the same relative position in a suture with higher-frequency folds. It may be possible to assess regional homology; ventral, lateral, and umbilical segments could theoretically be compared. However, the relative extent of those regions is both subjective and difficult to establish using published suture traces.

Thus, the study of ammonoid suture morphology is a nontrivial problem, and it is not surprising that significant work has been dedicated to both qualifying and quantifying variation among forms. Methods for describing suture morphology can be divided into four main categories: (1) Descriptive Methods, (2) Complexity Indices, (3) Pattern-Matching, and (4) Morphometric Methods, all of which are discussed in more detail in the following paragraphs.

2.2 Qualitative and Quantitative Descriptions: A Review

Descriptive methods are those that qualitatively generalize either suture complexity or morphology. They are primarily useful for systematic description, as they are both intuitive and generally applicable to all morphological variations. These methods include both suture formulae and relative classification schemes. The suture formula is a schema based upon lobe position and ontogeny and is perhaps the most widely used method of suture classification (see Wiedmann and Kullmann [1981] for historical review). These formulae are helpful in systematic study as they provide a standard for representing all suture variations, while describing both the form and development of the suture. However, they are not useable in more quantitative analyses. Relative classification schemes are less specific than suture formulae and work by assigning related morphotypes to categories (e.g., nautilitic, goniatitic, ceratititc, ammonitic). While also purely qualitative, this sort of classification can be useful for incorporating suture morphology or complexity in comprehensive analyses of ammonoid evolution (e.g., Dommergues et al., 1996, 2001).

Complexity indices are quantitative measures of morphological complexity. They are useful for studies requiring a numerical, but approximate, description of suture morphology. Such indices are univariate, which makes them simple to calculate and easy to incorporate into comprehensive analyses. However, as discussed earlier, they are subjective and disparate suture morphologies may be equally complex. There are two indices traditionally used to quantify suture complexity: fractal dimension (FD) and suture complexity index (SCI). FD quantifies the topological dimension of the suture, measuring both its self-similarity and sinuosity, i.e., FD considers suture complexity a property of the arrangement, not the number of lobes (Pérez-Claros et al., 2002; see also Boyajian and Lutz, 1992; Lutz and Boyajian, 1995). While useful for discriminating among ammonitic and other sutures with fractal-like morphologies, as mentioned earlier, FD is not applicable to all suture morphotypes. In particular, fractal measures do not adequately characterize sinusoidal nautilitic and goniatitic sutures, which are lacking in fractal character (i.e., a topological dimension greater than their true dimension).

In contrast to FD, SCI measures complexity primarily in terms of the number, and not the arrangement of lobes. SCI, as introduced by Saunders (1995), is calculated as the product of two measures: a sinuosity index (SI) and complexity factor (CF). SI is the ratio between the suture length to whorl circumference, measured as the distance from venter to umbilical seam (Westermann, 1971, 1975; Ward, 1980; Hewitt, 1985). Because this measure confuses relative number and amplitude of lobes, Saunders (1995) introduced CF to emphasize the importance of lobe number in Paleozoic suture complexity. CF is the sum of the number of primary elements, such as lobes, and secondary subdivisions, such as bifurcations, serrations, and digitizations. Like SI, CF is easy to calculate; however, it is very subjective and all subdivisions, regardless of size, are equivalent. Although SCI does compound the problems with SI and CF (i.e., confuses sutures with many low-amplitude lobes and those with few high-amplitude lobes, and gives equal weight to bifurcations and serrations), it still greatly outperforms the fractal dimension in discriminating among Paleozoic ammonoid suture morphologies (Allen, 2005). At the same time, it is difficult to apply the SCI to the more complex Mesozoic suture morphologies, because the recognition of levels of subdivision is subjective.

Pattern-matching is the newest type of method employed to investigate variation in suture morphology and refers to a set of techniques that evaluates the degree of similarity between two or more sutures. Such methods are useful primarily for classification; however, they also show some promise for evaluating the extent of ontogenetic and other within-taxon variation (e.g., Waggoner and Manship, 2004). However, these methods do require an assumption of positional homology to align samples with a reference template, which as discussed earlier, is functionally and developmentally invalid for ammonoid sutures, except in a limited phylogenetic context. Researchers interested in pattern-matching problems have introduced a Geographical Information System (GIS)-based method that uses GIS software to provide a graphical interface and evaluates similarity in terms of overlap between scaled, rotated traces, and a reference template (Manship, 2004). One drawback to this method is the reliance on GIS, but other software solutions are freely available for pattern-matching analysis. However, all require significant numbers of specimens to establish reference templates and the sensitivity of this method to sampling error is still unknown.

Morphometric methods differ from the others as they provide unique, nonrelative, mathematical representations of suture morphologies. They are useful for any analysis requiring a detailed, specimen-specific study of morphological variation. In particular, several researchers have attempted to use Fourier analysis to quantify morphological variation among ammonoid sutures (e.g., Canfield and Anstey, 1981; Gildner and Ackerly, 1985; Gildner, 2000; Allen and Gildner, 2002; Gildner, 2003). Fourier analysis works by calculating the series of sine and cosine functions that sum to generate a suture line. This method yields a unique trigonometric (and thus geometric) mathematical representation for each suture morphology, and the original trace can be reconstructed from the Fourier coefficients that define the sine-cosine series. However, because of the mathematical complexity of ammonoid sutures, traditional Fourier analysis has failed to yield robust results in repeated trials (Allen, 2006), severely limiting its application in large-scale studies of suture evolution.

2.3 The Windowed, Short-Time Fourier Transform: A New Perspective

Here I use a modified Fourier method, the windowed short-time Fourier transform (STFT), to directly and uniquely quantify the morphology of a set of Paleozoic and Lower Triassic ammonoid sutures. The STFT is a specialized Fourier method that accounts for starting point and other processing errors introduced by the mathematical complexity of the suture line (e.g., nonperiodicity and changing frequency of folds along the suture length) by subdividing the sample into overlapping segments or windows. Within each window, the data are smoothed using a tapering function to reduce edge effects. A localized Fourier transform is then performed on the smoothed data to calculate the complex Fourier coefficients describing the unique sine-cosine series defining the segment. The frequency of each constituent sine or cosine wave corresponds to a harmonic, or multiple of the fundamental (most dominant) frequency of the sampled curve. As harmonic number increases, so does the relative frequency of the constituent curve: the second harmonic has twice the number of lobes in one repetition of the signal as the fundamental; the third has three times; and so on.

The relative contribution, or power, of each harmonic to the total shape information is equal to the square of the coefficients. Power values are useful for distinguishing among morphologies as they are equivalent to a measure of within- and among-feature variance (Allen, 2006; see also Rohlf and Archie, 1984; Schweitzer et al.,

1986). With the STFT approach, local power spectrum densities (PSD) are estimated for each window (Press et al., 1992) and then averaged to generate a smoothed spectral estimate for the entire sampled suture. This quantifies the relative contribution of specific frequencies to total shape, without making any assumptions of positional homology. The details of this method are discussed elsewhere (Allen, 2005, 2006; see also Welch, 1967).

Here, I use results from a rigorous morphometric analysis of ammonoid suture geometry to examine the pattern of suture morphospace occupation through the 168 myr duration of the Paleozoic and Lower Triassic ammonoids. The goal of this analysis is threefold: (1) to present a first look at the evolution of Paleozoic ammonoid suture morphology; (2) to examine effects of mass extinction events (Frasnian-Famennian [F-F], end-Devonian [Hangenberg], and Permo-Triassic [P-T]) on that evolution; and (3) to compare the evolution of suture morphology with those reported for shell geometry.

3 Material and Methods

Data analyzed in this study were derived from mature sutures traced from specimens of 641 Paleozoic and Lower Triassic ammonoid genera. Included taxa comprise four Paleozoic orders: Anarcestida (82 genera), Clymeniida (63 genera), Goniatitida (363 genera), and Prolecanitida (40 genera; one short-lived Triassic survivor); and one order that spanned the Permo-Triassic (P-T) boundary: Ceratitida (93 genera). For the purpose of this study, complexity is measured by SCI (Saunders, 1995), or the product of the normalized suture length and a weighted count of lobes and subdivisions. For the Paleozoic taxa, complexity values were obtained from Saunders et al. (1999). Shell coiling morphology is defined in terms of the Raup parameters: shape of generating curve (S), distance from coiling axis (D), and whorl expansion rate (W) (Raup, 1967). Parameter values were obtained from Saunders et al., 2004.

Occurrence data for all Paleozoic genera were obtained from the GONIAT 2004 database (Kullmann, 2004). The timescale used in this study is derived from that established by the authors of the database; Devonian and Carboniferous ammonoid zones correspond to recent revisions by Becker and House (2000) and Amler and Gereke (2002), respectively. Dates in million years, are taken from the "Stratigraphic Table of Germany 2002" from the German Stratigraphic Commission (DSK; J. Kullmann, personal communication, 2004). Systematic and range data for all sampled taxa are available upon request.

Suture traces were obtained from the literature and from preparations made by David Work for the Saunders et al. (1999) study. Following the method outlined by Allen (2006), traces were scanned and then sampled from venter to umbilical seam by digitizing inflection points; 512 evenly spaced points were interpolated along the length of each suture by fitting a parametric cubic spline to the sampled landmarks. These pseudolandmarks were then parameterized by calculating the tangent angle function of each line, a mapping of position along the suture to fold amplitude. Finally, the STFT method was applied to the tangent angle parameterization of all sampled sutures.

Using a Blackman tapering function, a moving window of 128 points, a frame-skip of a quarter the window width, and a 256-point zero-padded Fourier transform, the powers of 128 harmonics describing the morphology of each parameterized suture were calculated. Calculated spectra were normalized by the power of each suture's fundamental frequency. Because of the high dimensionality of these results, a principal components analysis (PCA) was performed to reduce the number of variables and facilitate comparison of line geometries.

PCA was performed on the spectral data for all genera. Spectral power is proportional to the square of the amplitude of the constituent waveform. Consequently, for most sampled sutures the power of the low-frequency folds are several orders of magnitude higher than that of the high-frequency folds (i.e., lobe subdivisions and serrations have smaller amplitudes and contribute less to total shape information than the containing lobes). To account for these scaling differences and to ensure that all frequencies were given equal weight, PCA was performed using the correlation matrix of the log-transformed spectral data as the input. All data transformations and statistical analyses were done using MatLab ver. 6.5r13 and the JMP-IN ver. 5.1 (SAS Institute, Inc.) statistical software package. Code for the STFT is available upon request.

4 Results

4.1 Paleozoic Ammonoid Suture Morphology

PCA identified six factors with eigenvalues >1.0. Ordination and original spectral data for each sampled suture are listed in Allen (2005) and are available upon request. The first three components account for nearly 92% of the total variation in the data set. PC1 itself explains ~83% of the total variation and was loaded primarily by the first 20 harmonics. This is not surprising as the majority of Paleozoic suture morphologies are low-frequency sinusoids; relatively few cases exhibit significant contributions from mid-range (bifurcations-trifurcations) and higher-order folding (ammonitic detail and serrations). Because of this simplicity, the general pattern and frequency of most Paleozoic ammonoid sutures can be described solely by the lower portion of the spectrum and by the first principal component. However, the variation represented by the second and third principal components (~7% and 2%, respectively) is also informative, since these components represent variation in the higher-order harmonics (lobe elaboration, subdivision, and serration).

Two sets of bivariate plots, generated for each period, nicely summarize the tempo of morphospace exploration by the Paleozoic and basal-Triassic ammonoids (PC1–2, Fig. 8.1, and PC2–3, Fig. 8.2). As power spectra are nonlinear summaries

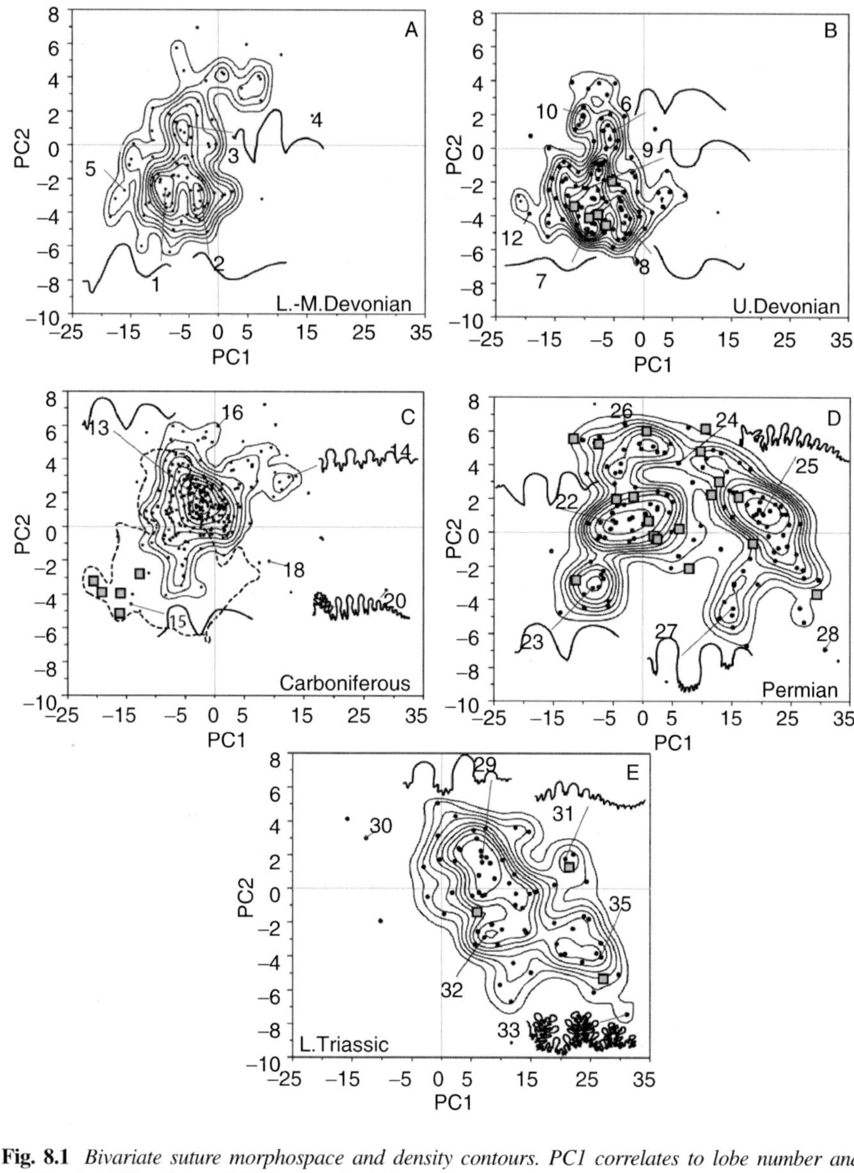

Fig. 8.1 *Bivariate suture morphospace and density contours. PC1 correlates to lobe number and amplitude; PC2 to lobe subdivision, with elaboration inversely correlated to score. Numbers refer to sutures illustrated in either Figs. 8.1 or 8.2 and correspond to taxa listed in Table 8.1. Sutures were reconstructed from sampled landmarks. Extinction survivors are indicated by gray squares. L.-M. Devonian morphospace (A; n = 90) is characterized by sutures possessing few broad, shallow, sinusoidal lobes, with the exception of a few unique, regular-sinusoidal multilobate forms, e.g.,* Beloceras *[4]. F-F extinction survivors fell within the mode of morphospace occupation during the L.-M. and U. Devonian (B; n = 110). End-Devonian extinction and Carboniferous morphospace (C; n = 245) reveals a significant shift away from typical Devonian forms; extinction survivors fell at the extremes of the U. Devonian morphospace, indicated by the dashed gray line. No ammonoid extinction is associated with the C-P boundary; holdover Pennsylvanian genera are numerous (n = 17) and cover the full range of Carboniferous morphologies. Novel taxa appearing within the Permian (D; n = 151) radiated near these points, establishing a bimodal distribution. Taxa surviving the P-T extinction fell at the extremes of the Permian occupied space. Postextinction recovery (E; n = 73) failed to recover the Permian mode corresponding to PC1 < 0, forms that had persisted since the L. Carboniferous.*

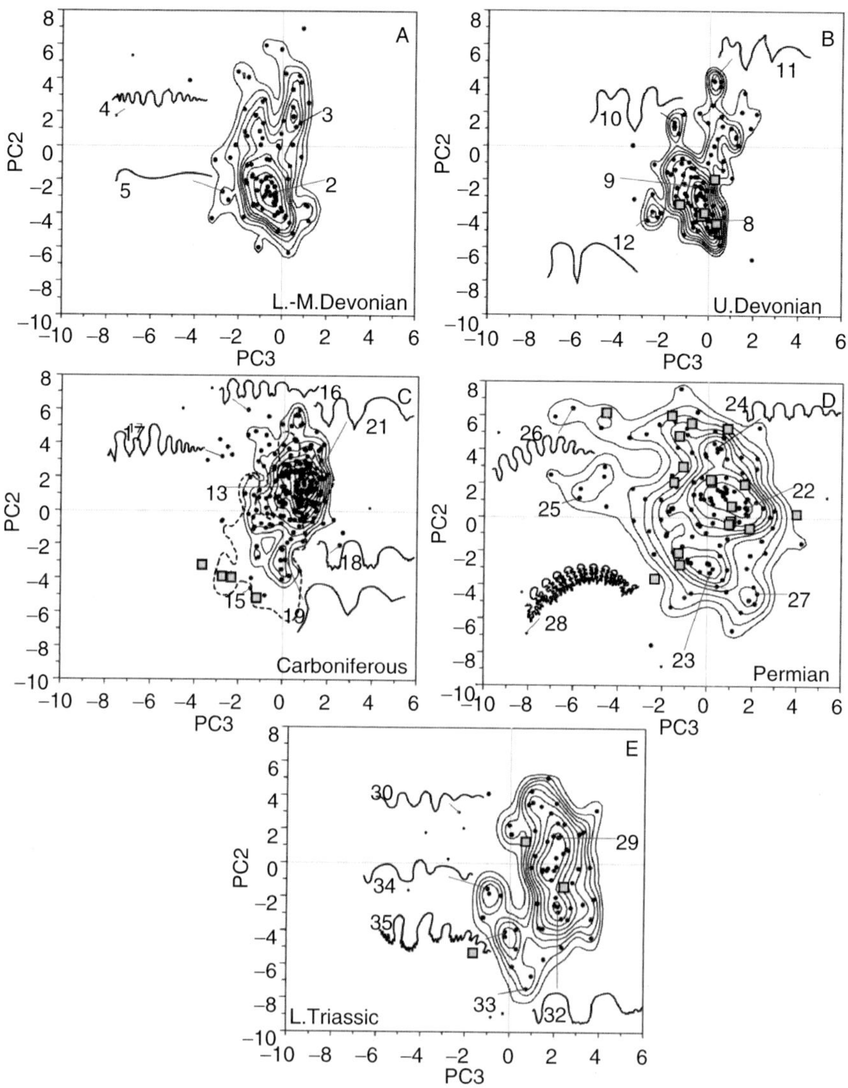

Fig. 8.2 *Bivariate plots and density contours summarizing PC2–3 suture morphospace. Conventions as in Fig. 8.1. Although PC3 accounts for ~2% of the variation in the data set, it is still informative, differentiating self-similar (−PC3) from more irregular morphologies (+PC3). As in the PC1–2 space, F-F survivors (A, B) possessed modal L.-M. Devonian suture morphologies and new forms were generated within the limits established before the extinction. Also, recovery from the end-Devonian event (C) resulted in a near complete shift away from preexisting morphologies similar to that seen in PC1–2 space. During the Permian (D), the most extreme suture morphologies seen in the Paleozoic appeared, including ammonitic, yet sinusoidal forms, such as* Timorites *[28], which occupied a region of morphospace noticeably avoided by all other examined taxa (PC2 < 0 and PC3 < 0). Survivors of the P-T extinction were also distantly placed in PC2–3 morphospace, and recovery from the extinction (E) was typified by irregular suture forms and the failure to regenerate the multilobate, yet regularly sinusoidal morphologies that appeared in the late Paleozoic (PC3 < 0).*

of suture shape, it is difficult to assign morphological states to the principal axes. However, the first principal component seems to correlate with the relative frequency and amplitude of lobes along the suture length, with number of lobes increasing, and amplitude decreasing, with larger scores. The second principal component correlates most strongly with the shape of the ventral portion of the suture; positive values correspond mostly to sutures possessing pronged ventral lobes, (e.g., *Pseudopronorites* [suture 17]; Fig. 8.2C) or ventral saddles (e.g., *Schartymites* [suture 21]; Fig. 8.2C), and the negative values to sutures with ventral lobes (e.g., *Costimitoceras* [suture 15]; Fig. 8.1C). Presence or absence of very-high-frequency subdivisions also load on PC2, with negative values coinciding with more highly serrated forms. The third principal component discriminates sutures resembling regular, periodic sinusoids from irregular forms, with irregularity increasing with increasing PC3 score.

Density contours, calculated using JMP-IN's default algorithm, were added to the PC plots to highlight patterns of space occupation through time. Also illustrated are sutures representative of the extremes and modes of the sampled geometries, with their taxonomic affiliations listed in Table 8.1. These data reveal a strikingly dynamic pattern of suture evolution: the occupied region of suture morphospace shifts significantly across period and extinction boundaries. These changes are described in more detail below.

Frasnian-Famennian extinction. Five genera survived the Frasnian-Famennian extinction (376 Ma), all of which possessed the modal suture morphotypes of the lower-middle Devonian (Figs. 8.1A, B, 8.2A, B; survivors indicated as gray squares on upper Devonian plots, Figs. 8.1B, 8.2B). After the extinction, the majority of the lower-middle Devonian suture morphospace was reoccupied, with the modal morphotypes being maintained. Noticeable differences between the two distributions included the failure of highly multilobate forms (e.g., *Beloceras* [suture 4]; Fig. 8.2A) to reevolve, and the appearance of irregular clymeniid morphotypes (e.g., *Falciclymenia* [suture 8], *Otoclymenia* [suture 10], *Discoclymenia* [suture 11]; Figs. 8.1B and 8.2B).

End-Devonian extinction. In contrast, the rebound and subsequent filling of suture morphospace in the Carboniferous after the end-Devonian extinction saw a near-complete shift away from pre-existing geometries. Extinction survivors fell at the extreme of the Devonian suture morphospace (survivors indicated as squares on Carboniferous plots; Figs. 8.1C, 8.2C). These taxa also fell well outside the main Carboniferous-occupied space. When compared to suture complexity, as measured by SCI, this morphological changeover is somewhat surprising. Both intervals were dominated by ammonoids with very simple sutures; ~80% of Devonian and ~60% of Carboniferous genera had SCI values less than 5 (i.e., ~3 lobes in the sampled region), and average Mississipian suture complexities closely mirrored those of middle and late Devonian taxa (Saunders et al., 1999; see also Allen, 2005). Despite this, overlap between the two faunas in all three dimensions of the PC space is minimal. Devonian taxa had broader, shallower lobes than their Carboniferous counterparts. Additionally, the lobes of Devonian sutures were more rounded (resembling a true sine wave; on average, values less than zero on PC3) than those of the Carboniferous forms, which tapered more strongly to form a point, i.e., typical "goniatitic" suture geometries (Figs. 8.1C, 8.2C).

Table 8.1 *Genera associated with sutures referred to in text and illustrated in Figs. 8 1, 2. L.– M. Devonian: 1–5; U. Devonian: 6–12; Carboniferous: 13–21; Permian: 22–28. L. Triassic: 29–36. ID corresponds to specimen number in Saunders* et al., *1999, 2004; Allen, 2005.*

No.	Genus	Order	ID
1.	*Delephinites*	Anarcestida	57
2.	*Anetoceras*	Anarcestida	1
3.	*Sphaeromanticoceras*	Anarcestida	52
4.	*Beloceras*	Anarcestida	88
5.	*Mimagoniatites*	Anarcestida	14
6.	*Prionoceras*	Goniatitida	201
7.	*Uraloclymenia*	Clymeniida	96
8.	*Falciclymenia*	Clymeniida	133
9.	*Oxytornoceras*	Goniatitida	178
10.	*Otoclymenia*	Clymeniida	112
11.	*Discoclymenia*	Clymeniida	181
12.	*Mimitoceras*	Goniatitida	209
13.	*Nuculoceras*	Goniatitida	333
14.	*Marathonites*	Goniatitida	479
15.	*Costimitoceras*	Goniatitida	205
16.	*Vidrioceras*	Goniatitida	486
17.	*Pseudopronorites*	Prolecanitida	542
18.	*Eothalassoceras*	Goniatitida	403
19.	*Kazakhastania*	Goniatitida	216
20.	*Artinskia*	Prolecanitida	554
21.	*Schartymites*	Goniatitida	351
22.	*Atsabites*	Goniatitida	456
23.	*Shangracoceras*	Goniatitida	232
24.	*Almites*	Goniatitida	480
25.	*Artioceratoides*	Prolecanitida	563
26.	*Sosiocrimites*	Goniatitida	523
27.	*Tapashanites*	Ceratitida	573
28.	*Timorites*	Goniatitida	496
29.	*Subinyoites*	Ceratitida	1201
30.	*Beatities*	Ceratitida	1235
31.	*Pseudoaspenites*	Ceratitida	1236
32.	*Tirolites*	Ceratitida	1128
33.	*Rhacophyllites*	Ceratitida	1293
34.	*Proavites*	Ceratitida	1149
35.	*Clypopoceras*	Ceratitida	1168

Carboniferous–Permian. By the onset of the Permian, the relative proportion of simple sutured taxa was greatly reduced (~12% with SCI < 5; Saunders et al., 1999). In the PC-suture morphospace, this deficit is reflected by thinning in the region of modal Carboniferous morphospace occupation across the Carboniferous-Permian (C-P) boundary (Figs. 8.1C, D, 8.2C, D). The 17 ammonoid genera that crossed the C-P boundary were scattered evenly throughout the Carboniferous suture morphospace, falling both within the mode and at the extremes (holdovers indicated as gray squares on Permian plots; Figs. 8.1D, 8.2D).

During the Permian, a wide range of novel suture morphologies originated, most of which fell within the boundaries established by the Carboniferous ammonoids. This morphological radiation led to a bimodal distribution in PC1-PC2 morphospace with each mode centered near the holdover morphotypes. This differentiation highlights the separation between groups characterized by serrated and fractal-like ammonitic lines (more positive PC1 values; Carboniferous extremes) and more typical "Paleozoic" sutures (Carboniferous mode; Figs. 8.1D, 8.2D).

Permian–Triassic. The early Triassic saw a continuation of the evolutionary shift initiated in the Permian. The three genera that survived the end-Permian extinction event (*Xenodiscus*, *Otoceras*, and *Episageceras*) fell at the extremes of the Permian morphospace (survivors indicated as gray squares on Triassic plots; Figs. 8.1E, 8.2E). *Xenodiscus* and its descendants formed the rootstock of the Triassic Ammonoidea, while *Otoceras* and its descendants had short-lived success and the prolecanitid *Episageceras* survived only briefly (McGowan, 2004).

The early Triassic was typified by increasing lobe serration, but also saw the reemergence of both "simple" forms (albeit with irregular folding; e.g., *Beatities* [suture 30]) and subammonitic ones (e.g., *Rhacophyllites* [suture 33]). There are also notable gaps in the distribution. In particular, sutures with PC1 scores less than zero rarely reappeared, indicating a failure to regenerate Devonian and most Carboniferous morphotypes (Fig. 8.1E).

4.2 Timing of Morphological Change

The change in suture shape over time is illustrated in more detail in Fig. 8.3. PC1 describes ~83% of the total variance in the data set and therefore is a good proxy for the general pattern of suture evolution during the Paleozoic. PC1 scores for all sampled genera (light gray dots) are plotted for each one myr time bin. Data points are connected vertically (light gray lines) to emphasize the range of morphospace occupied during each time interval. The mean trend in morphospace occupation through time is emphasized by the solid black line.

Average PC1 scores remained fairly conserved throughout the early Devonian, but increased significantly by the middle of the period. The morphologies causing that shift were lost at the Frasnian-Famennian extinction, returning the morphospace to a more typical early Devonian state (Fig. 8.3). This changed again during the initial rebound after the end-Devonian extinction and by the late Tournaisian (353 Ma), the mean PC1 value had shifted significantly when compared to the Devonian values (Wilcoxon rank sum: $Z = -7.55$, $p < 0.0001$), and remained fairly conserved until the Moscovian-Kasimovian boundary (305 Ma). The interval following that boundary marked the onset of a morphologic diversification, later supplemented by taxonomic radiation, reflected in the steady increase in mean PC1 values through the end of the Permian.

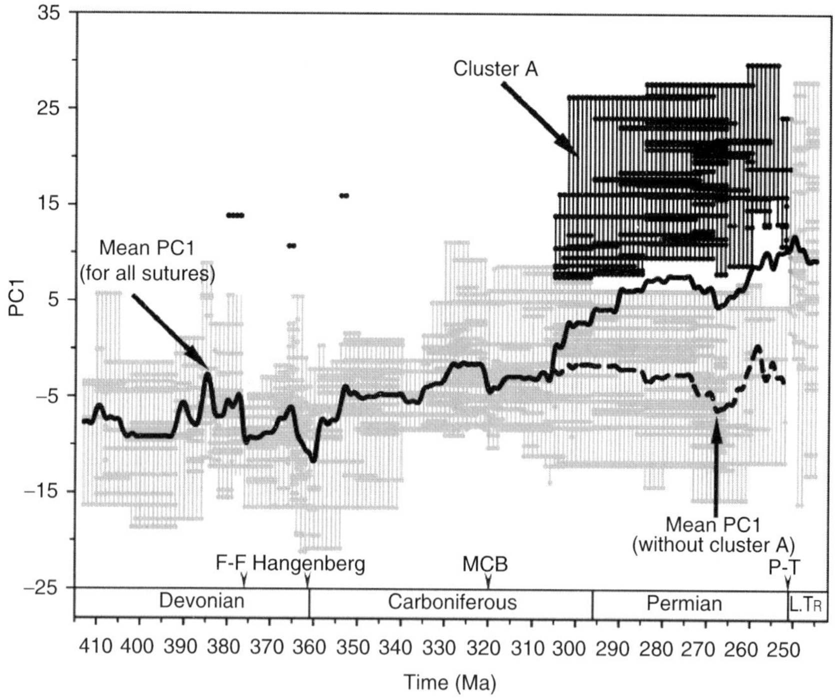

Fig. 8.3 *Detailed view of the change in suture morphology through time, using PC1 scores as a proxy. Data are binned into 168 1-myr time intervals. Points correspond to the score for each genus and are connected vertically to emphasize total range of values per interval. Black points highlight a statistically distinct cluster of morphologies (A) generated mainly during the Upper Carboniferous–Permian radiation. Comprising taxa belong almost exclusively to three subclades (Goniatitina, Prolecanitida, Ceratitida), with the middle Devonian anarcestid* Beloceras *[4] being the only notable exception. The solid black line tracks the mean evolutionary trajectory for all sampled taxa. Note how the average runs through unoccupied space during the Permian due to the effect of the outlying Cluster A taxa. The dotted black line tracks the trend after removal of those taxa, emphasizing the morphological stasis that persisted until the P-T extinction.*

The late Pennsylvanian to early Permian morphological radiation led to a bifurcation in the distribution of taxa within morphospace (Figs. 8.1D, 8.3). To assess the significance of this division a *k*-means cluster analysis was performed on the original spectral data for all taxa. This revealed two statistically distinct morphological groups within the ammonoids, interestingly segregated by apparent complexity: highly multilobate forms, or those with serrated or ammonitic lobes ("complex" morphologies; Cluster A) and those with sutures containing ~7 or fewer undifferentiated lobes (true "goniatitic" morphologies; Cluster B). The two clusters are highlighted in the plot of PC1 through time (Fig. 8.3), where Cluster A is emphasized in black. A recalculated mean trend in suture morphology that disregards the statistically distinct morphologies (A) is indicated by the dashed line. This reanalysis reveals a striking pattern in the trajectory of mean suture morphology over time.

When calculated over all taxa, the average fails to accurately trace the evolution of Permian suture morphology; it falls within a region of morphospace that was unoccupied until the Triassic! After excluding the extreme morphologies, the average PC1 value drops back down to Carboniferous levels, emphasizing the conservation of Carboniferous suture geometries until the end-Permian extinction.

5 Discussion

5.1 Overview of Paleozoic Suture Morphology

The occupied region of suture morphospace shifted dramatically as ammonoids progressed through the Paleozoic. Most striking of these shifts was the almost complete loss of the Devonian morphotypes after the end-Devonian extinction event (Figs. 8.1C, 8.2C). A similar loss of morphotypes was observed across the P-T boundary where the modal morphologies that appeared in the Carboniferous and persisted through the Permian were lost and never reinvented (PC1 < 0; Figs. 8.1E, 8.2E). Moreover, despite considerable overlap in the region of occupied morphospace from the Carboniferous to lower Triassic, multivariate analysis of variance suggests that the distributions of taxa within morphospace were statistically different from one another (Wilk's $\lambda = 0.114$; $p < 0.001$). Only one region of morphospace was largely avoided during the Paleozoic: the combination of negative PC2 (possessing a ventral lobe and/or finely subdivided lobes) and PC3 (sinusoidal) scores (Figs. 8.1, 8.2). A noticeable exception was the Permian cyclolobids (e.g., *Timorites* [suture 28]; Fig. 8.2D), the only Paleozoic ammonoids to possess sutures resembling true ammonitic morphotypes.

5.2 Shell Coiling Versus Suture Morphology

The succession of multiple suture morphologies contrasts strikingly with the reported iterative evolution of coiling geometries. Despite the "boom and bust" nature of their taxonomic evolution, dramatic postextinction shifts in the overall pattern of shell coiling were rare; exceptions include the middle Emsian replacement of agoniatitid morphs with anarcestid morphs (Korn and Klug, 2003) and end-Devonian replacement of clymeniid morphs with true goniatitids (Saunders et al., 2004). Instead, the majority of newly derived taxa reoccupied previously inhabited regions of coiling morphospace. It has been argued that this continued reemergence of modal external shell morphotypes reflects a persistent functional constraint on ammonoid shell geometry (for a review, see Saunders et al., 2004). The maintenance of modal Carboniferous suture morphologies through the end of the Paleozoic, coupled with the rare exploration of the "ammonitic" region of the morphospace argues for a similar functional restriction on suture form, i.e., ammonitic morphologies were rare, but developmentally possible. At the same time, the complete reorganization of the morphospace following the end-Devonian extinction and the failure to regenerate the Carboniferous geometries after the P-T extinction (Figs. 8.1C, 8.2C) implies some phylogenetic control on potential morphological variation.

The function of septal folding is the subject of much debate and many hypotheses have been proposed, e.g., buttressing (Hewitt and Westermann, 1986, 1987, 1997); cameral liquid transfer (Daniel et al., 1997); developmental epiphenomena (García-Ruiz et al., 1990; Checa and Garcia-Ruiz, 1996; Hammer, 1999; Hammer and Bucher, 1999); and locomotion (Seilacher and LaBarbera, 1995). Several of these, based upon observations of Mesozoic ammonoids, depend on functional or developmental coupling between external shell morphology and septal shape. Among the Paleozoic clades, significant associations between suture complexity and shell morphology have only been reported for the Prolecanitida, while the Goniatitida showed no significant relationship between the two (Saunders and Work, 1997). The data presented here corroborate this observed disjunction between shell and suture geometries for most Paleozoic ammonoids.

Overall, significant correlations between individual shell and suture geometries only existed within a subset of Paleozoic ammonoids ($n = 73$ genera, ~11% of those sampled; Allen, 2004). These genera all possessed suture morphotypes falling within Cluster A (group segregated from the typical Carboniferous morphologies), including the rare outlying Devonian multilobate forms ($r^2 = 0.39$; $p < 0.001$; Fig. 8.3). Interestingly, these taxa also possessed a restricted set of coiling geometries when compared to those of the remaining genera (Table 8.2). In particular, the forms were generally more tightly coiled and compressed ($D \leq 0.3$, t-value: 3.142, p-value: 0.002; $S \leq 1.15$, t-value: 2.488, p-value: 0.0132).

These observations imply that ammonoid suture and shell evolution were more or less decoupled until the Late Carboniferous (~305 Ma) when a novel region of suture morphospace was explored (serrated, bi- and tri-furcating sutures). Taxa possessing suture morphologies falling within this statistically distinct cluster (A; Fig. 8.3) belonged to three ammonoid clades: the Prolecanitida, Ceratitida, and Goniatitina (Goniatitida). The Ceratitida, which formed the rootstock of all Mesozoic ammonoids, were derived from within the Prolecanitida during the Permian, well after the association between internal and external morphology was established. However, the Prolecanitida and Goniatitina first appeared after the end-Devonian extinction, ~55 Ma before the morphological radiation began. Thus, the relationship between internal and external morphology was likely partially due to functional limitations on form, i.e., once a "complexity" threshold was reached, changes in external shell morphology required associated changes in septal form to maintain functional integrity, regardless of phylogenetic affiliation. It may be this relationship that allowed exploration of regions of morphospace (second PC1 mode, rare ammonitic morphologies; Fig. 8.1D) previously avoided.

Table 8.2 *Distribution of the Raup coiling parameters – whorl expansion rate (W), distance from coiling axis (D), and aperture shape (S) – for ammonoids exhibiting two statistically segregated clusters of suture morphologies. Minimum values for both groups are equitable. Note how taxa belonging to Cluster A (almost exclusively Permian, with ceratitic, subammonitic and other complex suture morphologies) have a more restricted range of values for D and S than those in Cluster B (with more typical, simple, goniatitic suture morphologies).*

	Mean ± 1 STD			Max		
Group	W	D	S	W	D	S
Cluster A	2.1 ± 0.42	0.17 ± 0.13	0.90 ± 0.48	2.89	0.55	2.15
Cluster B	2.0 ± 0.58	0.24 ± 0.17	1.08 ± 0.52	5.01	0.71	2.86

5.3 Response to Mass Extinction Events: Selectivity and Rebound

Across the Frasnian-Famennian extinction boundary, the dominant gephuroceratid (Anarcestida) clade went extinct, while five of the seven genera belonging to a single, marginal goniatitid family (Tornoceratidae) survived. These holdovers possessed similar suture geometries, all falling within the modal form of the preextinction fauna. Likewise, these taxa were on average involute (low D), but also within the mode of preextinction shell-coiling geometries (Saunders et al., 2004). Together, these data suggest that the extinction was highly selective phylogenetically, but not necessarily morphologically.

While not considered one of the big five extinctions, the end-Devonian Hangenberg event devastated taxonomic diversity for several clades, including the ammonoids (Korn et al., 2004). Nine ammonoid genera, belonging to both the tornoceratid (n = 4; gray squares, Figs. 8.1C, 2C) and clymeniid (n = 5) clades, survived the initial extinction. As with the F-F event, the end-Devonian extinction was apparently phylogenetically selective: four of the five surviving clymeniids belonged to a single family (Clymeniidae) and the few tornoceratid survivors all belonged to the family Prionoceratidae. Of this set, only the tornoceratids persisted into the early Mississippian (Saunders et al., 2004). However, the significance of this selectivity is difficult to assess without more rigorous testing because clymeniid generic diversity had begun to decline before the Hangenberg event.

In contrast to the F-F extinction, all end-Devonian survivors possessed a localized subset of relatively extreme suture morphologies in Devonian morphospace. Tornoceratid boundary-crossers also shared similar shell coiling geometries (D ~ 0.05, S ~ 0.5; Saunders et al., 2004), while the clymeniids that persisted past the extinction boundary were widely distributed within the mode of the clade WDS morphospace. This suggests that certain morphologies increased the likelihood of survival during the Hangenberg event.

Such selectivity is an important issue when considering the macroevolutionary effect of mass extinctions on character and taxonomic evolution, as clustering has the potential to affect the degree to which the evolutionary clock is reset by extinction events (Jablonski, 2001). The differing response of suture morphology to the three mass extinctions (F-F, end-Devonian, and P-T) that interrupted the evolution of Paleozoic ammonoids, and the potential influence of selectivity on those responses is of special interest because of the trend toward increasing suture complexity. For example, the persistence of the trend despite a complete taxonomic turnover across the F-F boundary and morphological turnover during the end-Devonian extinction, suggests that the factors driving the trend were independent of starting-point morphology or phylogeny.

In comparison, potential survivorship across the Permo-Triassic extinction was apparently more random with respect to ammonoid taxonomy and morphology. Two of three ammonoid genera that survived the Permo-Triassic extinction were ceratitid ammonoids, the most diverse clade in the preextinction interval. Additionally, holdovers fell at disparate extremes of both suture and shell-coiling morphospace, with no simple-sutured survivors (Saunders et al., 2004). The low number of survivors makes it difficult to assess the significance of the survivorship of extreme over modal morphotypes, although McGowan (2004) and Villier and Korn (2004) suggested that the taxonomic and morphologic (external shell morphology) loss was much greater than expected due to chance. This observation implies that there may have been extinction selectivity on a character or quality not yet considered.

To fully understand the effect of mass extinctions on morphological evolution, it is also necessary to consider the postextinction recovery and resulting pattern of morphological diversification. For Paleozoic ammonoid sutures, the pattern of rebound from the extinctions is highlighted in the behavior of average suture morphology (using mean PC1 as a proxy) through time. Morphological recovery from the Frasnian-Famennian extinction resulted in radiation around the holdover taxa, and much of the original morphospace was reoccupied. At the same time, there was a failure to redevelop the extreme suture morphologies (e.g., *Beloceras* [suture 4]; Fig. 8.2A) that cause the spike in average PC1 values during the middle Devonian. Together, the maintenance of the modal forms and the loss of the extremes reduced average PC1 back to early Devonian levels (Fig. 8.3).

The similarity between the pre- and postextinction morphospaces is surprising, as many taxa with irregular suture geometries appeared during the late Devonian clymeniid radiation (e.g., *Otoclymenia* [suture 10]; Fig. 8.2B). While some of these strongly resemble forms generated during the early and middle Devonian (e.g., *Falciclymenia* [suture 8]; Fig. 8.1B), the coincidence between the two morphospaces may in some part be a consequence of method error. In particular, the STFT method of power spectrum estimation quantifies relative contribution of a specific frequency to the total shape, regardless of its actual position along the suture length. This creates some ambiguity in comparisons; differently shaped sutures can have similar spectra, if the composite wave forms exist in similar proportions. These forms can easily be distinguished by maximizing the positional resolution of the analysis by varying window and overlap lengths, or by using a more specialized version of the STFT, such as the wavelet transform (Daubechies, 1992). However, doing so imposes an assumption of positional homology, which is only valid in a restricted phylogenetic framework.

In contrast, the morphological recovery from the end-Devonian extinction led to a complete shift away from preextinction forms, significantly changing the pattern of morphospace occupation (Figs. 8.1C, 2C, 3). This may have been due to the complete loss of incumbent Devonian morphologies across the extinction boundary, allowing for the exploration of previously uninhabited regions of morphospace. The initial extinction rebound led to an increase in average suture morphology (PC1) by the late Tournaisian, which was maintained until the radiation following the Moscovian-Kasimovian boundary (305 Ma; Fig. 8.3).

Upon segregating the cluster of novel morphologies that appeared during the Late Carboniferous radiation (Cluster A) from the more typical goniatitid morphologies (Cluster B), the general pattern of morphospace occupation (i.e., the Carboniferous mode) and the average PC1 score remain unchanged from the D-C extinction to the P-T boundary (Fig. 8.3). Similar morphological stasis has been reported for shell-coiling geometries (Saunders et al., 2004). These modal Carboniferous suture geometries were eliminated during the Permo-Triassic extinction and did not reappear during the initial Triassic rebound. Instead, recovery from the extinction involved rapid refilling of much of the Permian morphospace, as well as the introduction of new suture morphotypes not seen in the Paleozoic (those falling between the two Permian clusters; Fig. 8.3). This shifted the overall pattern of early Triassic suture morphospace toward that exhibited by the extreme Permian morphotypes (Cluster A; Fig. 8.3). It is important to note that many of the suture geometries appearing during the P-T rebound are unique when compared to Paleozoic forms (e.g., *Rhacophyllites* [suture 33], *Pseudoaspenites* [suture 31]; Fig. 8.1E). However, these forms are characterized by high-frequency detail not well differentiated by the principal axes considered in this analysis.

These data reinforce claims made by Saunders et al. (1999, 2004) that without the disruptions caused by the mass extinctions, Paleozoic ammonoid shell morphology (internal, external, and suture complexity) would probably have maintained the primitive Devonian morphotypes throughout the Paleozoic. The pattern of loss of well-established clades followed by diversification of formerly marginal taxa is a characteristic pattern of mass extinctions (Jablonski, 2001). In the case of the ammonoids, the extinctions reset the stage and, while lost shell morphologies could be reevolved, suture morphologies usually could not.

Additionally, these results suggest that the evolution of Paleozoic ammonoid sutures is an interesting case for comparing morphological and phylogenetic selectivity in background versus mass extinction and rebound intervals. Jablonski (1986, 2001) argued that many traits correlated with survivorship during background extinction tend to be of little importance during the unusual environmental and ecological conditions imposed by mass extinction events. The fine detail of the Paleozoic ammonoid fossil record, rapid taxonomic turnover at all levels, new availability of comprehensive morphological data for shell and suture shape, and known associations between shell-coiling geometries and ecological-functional constraints together make the ammonoids a prime model system for assessing the relative indifference of survivorship to factors promoting success in background intervals. These and related questions are the subject of a follow-up study.

6 Summary

During their ~170 myr history, the Paleozoic and lower Triassic ammonoids showed remarkable variation in suture morphology, from the subsinusoidal nautilitic forms to true ammonitic morphotypes. As had been argued for external shell geometries, the pattern of evolution and origin of these various morphotypes resulted from a combination of phylogenetic, functional, and contingent factors. At fine timescales, selective replacement of dominant groups by formerly marginal subclades, random morphologic extinctions, and extinction-independent diversification into outlying or previously unoccupied regions of morphospace all shifted the pattern of occupied suture morphospace significantly during the evolutionary history of Paleozoic ammonoids. For the most part these changes occurred independently of corresponding evolution in external shell morphology, although increased functional coupling between external shell and septal folding during the Permian may have facilitated development of more complex morphotypes.

Finally, mass extinctions consistently redirected the trajectory of suture evolution, e.g., loss of multilobate forms and complete taxonomic turnover during the F-F extinction, replacement of clymeniid and ancestral goniatitid morphologies with standard tapered goniatitid forms during the end-Devonian extinction, and loss of those same morphologies and goniatitids during the P-T extinction. Judging by the survival of the Carboniferous morphotypes through the end-Permian despite extensive taxonomic turnover, the lower and middle Devonian suture geometries would have likely persisted through the end of the Paleozoic in the absence of major extinction events. Together, these data show that ammonoid evolution was shaped by a complex interplay of background and mass extinction which gave rise to a dynamic in sutures that contrasted dramatically with external shell form.

Acknowledgments

The author wishes to thank W. B. Saunders for manuscript review and for preparing and providing along with D. Work, the suture traces used in this study. Additional thanks to D. Jablonski, M. Foote, M. LaBarbera, P. Wagner, and M. Webster for their advice and for manuscript review. Special thanks as well to the attendees of the VI International Cephalopod Symposium for discussion and constructive criticism concerning the ideas and methods presented in this paper. This research was supported with funds from the Beinecke Brothers Memorial Scholarship.

References

Allen, E. G. 2004. Preliminary comparison of Paleozoic ammonoid suture morphology quantified using a windowed, short-time Fourier transform to external shell coiling. *Geological Society of America. Abstracts with Programs* **36**: 315.

Allen, E. G. 2005. New perspectives on a classic macroevolutionary trend: complexity, extinction selectivity, and the evolution of Paleozoic ammonoid suture morphology. Unpublished Ph.D. thesis, University of Chicago, Chicago, IL: 156pp.

Allen, E. G. 2006. New approaches to Fourier analysis of ammonoid sutures and other complex, open curves. *Paleobiology* **32**: 299–315.

Allen, E. G., and R. F. Gildner. 2002. A new Fourier approach to interpreting ammonoid suture morphology. *Geological Society of America. Abstracts with Programs* **24**: 354.

Amler, M. R. W., and M. Gereke. 2002. Karbon-Korrelationstabelle (KKT)/Carboniferous Correlation Table (CCT). *In* K. Weddige, and W. Ziegler (editors), *Advances on Conodonts, Devonian and Carboniferous Research Special Issue of Senckenbergiana lethaea* **82**(2): 691.

Becker, R. T., and M. R. House. 2000. Devonian ammonoid zones and their correlation with established series and stage boundaries. *Courier Forschungs-Institut Senckenberg* **220**: 113–151.

Boyajian, G., and T. Lutz. 1992. Evolution of biological complexity and its relation to taxonomic longevity in the Ammonoidea. *Geology* **20**: 983–986.

Canfield, D. J., and R. L. Anstey. 1981. Harmonic analysis of cephalopod suture patterns. *Mathematical Geology* **13**: 23–35.

Checa, A. G., and J. M. Garcia-Ruiz. 1996. Morphogenesis of the septum in ammonoids. *In* N. H. Landman, K. Tanabe, and R. A. Davis (editors), *Ammonoid Paleobiology*, pp. 253–296. New York: Plenum Press.

Daniel, T. L., B. S. Helmuth, W. B. Saunders, and P. D. Ward. 1997. Septal complexity in ammonoid cephalopods increased mechanical risk and limited depth. *Paleobiology* **23**: 470–481.

Daubechies, I. 1992. Ten lectures on wavelets. *CBMS-NSF Regional Conference Series in Applied Mathematics* **61**. Philadelphia, PA: Society for Industrial and Applied Mathematics (SIAM).

Dommergues, J.-L., B. Laurin, and C. Meister. 1996. Evolution of ammonoid morphospace during the early Jurassic Radiation. *Paleobiology* **22**: 219–240.

Dommergues, J.-L., B. Laurin, and C. Meister. 2001. The recovery and radiation of Early Jurassic ammonoids; morphologic versus palaeobiogeographical patterns. *Palaeogeography, Palaeoclimatology, Palaeoecology* **165**: 195–213.

García-Ruiz, J. M., A. Checa, and P. Rivas. 1990. On the origin of ammonite sutures. *Paleobiology* **16**: 349–354.

Gildner, R. F. 2000. A method to quantify and analyze suture patterns. *Geological Society of America. Abstracts with Programs* **32**: 371.

Gildner, R. F. 2003. A Fourier method to describe and compare suture patterns. *Palaeontologia Electronica* **6**: 12pp.

Gildner, R. F., and S. Ackerly. 1985. A Fourier technique for studying ammonoid sutures. *Geological Society of America. Abstracts with Programs* **17**: 592.

Hammer, Ø. 1999. The development of ammonoid septa: an epithelial invagination process controlled by morphogens? *Historical Biology* **13**: 153–171.

Hammer, Ø., and H. Bucher. 1999. Reaction-diffusion processes: application to the morphogenesis of ammonoid ornamentation. *Geobios* **32**: 841–852.

Hewitt, R. A. 1985. Numerical aspects of sutural ontogeny in the Ammonitina and Lytoceratina. *Neues Jahrbuch für Geologie und Paläontologie, Abhandlungen* **170**: 273–290.

Hewitt, R. A., and G. E. G. Westermann. 1986. Function of complexly fluted septa in ammonoid shells. I. Mechanical principles and functional models. *Neues Jahrbuch für Geologie und Paläontologie, Abhandlungen* **172**: 47–69.

Hewitt, R. A., and G. E. G. Westermann. 1987. Function of complexly fluted septa in ammonoid shells. II. Septal evolution and conclusions. *Neues Jahrbuch für Geologie und Paläontologie, Abhandlungen* **174**: 135–169.

Hewitt, R. A., and G. E. G. Westermann. 1997. Mechanical significance of ammonoid septa with complex sutures. *Lethaia* **30**: 205–212.

Jablonski, D. 1986. Background and mass extinctions: the alternation of macroevolutionary regimes. *Science* **231**: 129–133.

Jablonski, D. 2001. Lessons from the past: evolutionary impacts of mass extinctions. *Proceedings of the National Academy of Sciences USA* **98**: 5393–5398.

Korn, D., and C. Klug. 2003. Morphological pathways in the evolution of Early and Middle Devonian ammonoids. *Paleobiology* **29**: 329–348.

Korn, D., B. Zdzislaw, S. Fröhlich, M. Rücklin, and J. Wendt. 2004. The youngest African clymeniids (Ammonoidea, Late Devonian) – failed survivors of the Hangenberg Event. *Lethaia* **37**: 307–315.

Kullmann, J. 2004. GONIAT Version 3.2, Paleozoic ammonoid database system (University of Tübingen).

Lutz, T. M., and G. E. Boyajian. 1995. Fractal geometry of ammonoid sutures. *Paleobiology* **21**: 329–342.

Manship, L. L. 2004. Pattern matching: classification of ammonoid sutures using GIS. *Palaeontologia Electronica* **7**: Art.6A, 15p, 736KB; http://palaeo-electronica.org/paleo/2004_2/suture/issue2_04.htm.

McGowan, A. J. 2004. Ammonoid taxonomic and morphological recovery patterns after the Permian-Triassic. *Geology* **32**: 665–668.

McShea, D. W. 1991. Complexity and evolution: what everybody knows. *Biology and Philosophy* **6**: 303–324.

Pérez-Claros, J. A., P. Palmqvist, and F. Olóriz. 2002. First and second orders of suture complexity in ammonites: a new methodological approach using fractal analysis. *Mathematical Geology* **34**: 323–343.

Press, W. H., S. A. Teukolsky, W. T. Vetterling, and B. P. Flannery. 1992. *Numerical Recipes in C: The Art of Scientific Computing*. New York: Cambridge University Press.

Raup, D. M. 1967. Geometric analysis of shell coiling: coiling in ammonoids. *Journal of Paleontology* **41**: 43–65.

Rohlf, F. J., and J. W. Archie. 1984. A comparison of Fourier methods for the description of wing shape in mosquitoes (Diptera: Culcidae). *Systematic Zoology* **33**: 302–317.

Saunders, W. B. 1995. The ammonoid suture problem: relationships between shell- and septal-thickness and suture complexity in Paleozoic ammonoids. *Paleobiology* **21**: 343–355.

Saunders, W. B., and D. M. Work. 1997. Evolution of shell morphology and suture complexity in Paleozoic prolecanitids, the rootstock of Mesozoic ammonoids. *Paleobiology* **23**: 301–325.

Saunders, W. B., D. M. Work, and S. V. Nikolaeva. 1999. Evolution of complexity in Paleozoic ammonoid sutures. *Science* **286**: 760–763.

Saunders, W. B., D. M. Work, and S. V. Nikolaeva. 2004. The evolutionary history of shell geometry in Paleozoic ammonoids. *Paleobiology* **30**: 19–43.

Schweitzer, P. N., R. L. Kaesler, and G. P. Lohmann. 1986. Ontogeny and heterochrony in the ostracode *Cavellina coryell* from Lower Permian rocks in Kansas. *Paleobiology* **12**: 290–301.

Seilacher, A., and M. LaBarbera. 1995. Ammonites as cartesian divers. *Palaios* **10**: 493–506.

Villier, L., and D. Korn. 2004. Morphological disparity of ammonoids and the mark of Permian mass extinctions. *Science* **306**: 264–266.

Waggoner, K. J., and L. L. Manship. 2004. Sutural variation in ammonite ontogeny: applying GIS for paleontologic analyses. *Geological Society of America. Abstracts with Programs* **36**: 422.

Ward, P. D. 1980. Comparative shell shape distribution in Jurassic-Cretaceous ammonites and Jurassic-Tertiary nautilids. *Paleobiology* **6**: 32–43.

Welch, P. D. 1967. The use of the fast Fourier transform for the estimation of power spectra: a method based on time averaging over short, modified periodograms. *IEEE Transactions on Audio and Electroacoustics* **15**: 70–73.

Westermann, G. E. G. 1971. Form, structure and function of shell and siphuncle in coiled Mesozoic ammonoids. *Life Science Contributions of the Royal Ontario Museum* **78**: 1–39.

Westermann, G. E. G. 1975. Model for origin, function, and fabrication of fluted cephalopod septa. *Paläontologische Zeitschrift* **49**: 235–253.

Wiedmann, J., and J. Kullmann. 1981. Ammonoid sutures in ontogeny and phylogeny. *In* M. R. House, and J. R. Senior (editors), *The Ammonoidea: The Evolution, Classification, Mode of Life and Geological Usefulness of a Major Fossil Group*, pp. 215–255. New York: Academic Press.

Chapter 9
Cameral Membranes in Carboniferous and Permian Goniatites: Description and Relationship to Pseudosutures

Kristin Polizzotto,[1] Neil H. Landman,[2] and Royal H. Mapes[3]

[1]Division of Paleontology (Invertebrates), American Museum of Natural History, Central Park West at 79th Street, New York, NY 10024–5192, USA and Department of Biological Sciences, Kingsborough Community College, 2001 Oriental Boulevard, Brooklyn, NY 11235, USA, kpolizzotto@kingsborough.edu;

[2]Division of Paleontology (Invertebrates), American Museum of Natural History, Central Park West at 79th Street, New York, NY 10024–5192, USA, landman@amnh.org;

[3]Department of Geological Sciences, 316 Clippinger Laboratories, Ohio University, Athens, OH 45701–2979, USA, mapes@ohio.edu

Keywords: goniatites, Carboniferous, Permian, cameral membranes, siphuncular membranes, translocation, pseudosutures, pseudosepta

1 Introduction

During the last century, cameral membranes have been reported in many different ammonoids (Grandjean, 1910; Schoulga-Nesterenko, 1926; Hölder, 1952, 1954; Schindewolf, 1968; Erben and Reid, 1971; Westermann, 1971; Bayer, 1977;

N. H. Landman et al. (eds.), *Cephalopods Present and Past: New Insights and Fresh Perspectives*, 181–204.

Kulicki, 1979; Weitschat and Bandel, 1991; Tanabe and Landman, 1996). Cameral membranes include two main varieties: chamber linings coating the inside surfaces of each chamber, and three-dimensional structures suspended within the chambers (Landman et al., 2006). In the Mesozoic, cameral membranes have been observed in phylloceratids, lytoceratids, and ammonitids. More recently, similar membranes have been observed in Paleozoic prolecanitids (Mapes et al., 2002; Tanabe et al., 2005). Until recently, however, it was unclear whether such membranes occur in goniatites (Landman et al., 2006). Schoulga-Nesterenko (1926) reported membranes associated with the siphuncle in the Artinskian goniatite *Agathiceras uralicum* (Karpinsky, 1889); however, the illustration of the specimen, which shows a central siphuncle, puts its identification as a goniatite in doubt.

We describe suspended cameral membranes in the phragmocones of two species of goniatites: (1) *Crimites elkoensis* Miller et al., 1957, from the Permian Arcturus Formation near Buck Mountain, Nevada, and (2) *Cravenoceras fayettevillae* Gordon, 1965, from the Carboniferous (Mississippian) Fayetteville Shale in northwestern Arkansas. The membranes are similar in some ways to those observed in other ammonoids, but are much less complex and are present only in the immediate vicinity of the siphuncle. We refer to them as siphuncular membranes, which are defined as cameral membranes attached to and surrounding the siphuncle (Landman et al., 2006).

The presence of pseudosutures in several specimens of *Cravenoceras fayettevillae* further allowed us to investigate the relationship between siphuncular membranes and pseudosutures. Pseudosutures are markings between septal sutures that form incomplete replications of the suture. Some workers have hypothesized that pseudosutures are the preserved vestiges of pseudosepta, that is, thin, organic structures, which in most cases decomposed early in diagenesis (Hewitt et al., 1991; Westermann, 1992). Pseudosepta have been interpreted as membranes that resulted from the desiccation of a cameral gel and replicated the shape of the septum (Hewitt and Westermann, 1987; Hewitt et al., 1991), while pseudosutures are the preserved margins of the pseudosepta where they contacted the interior of the shell wall (Hewitt et al., 1991). Some authors have suggested that siphuncular membranes such as the ones described herein are in fact the remains of pseudosepta (Hewitt and Westermann, 1987; Hewitt et al., 1991; Landman et al., 1993). Although siphuncular membranes have been observed in a wide variety of ammonoids, the identification of these membranes as pseudosepta is unclear. Because both siphuncular membranes and pseudosutures are present in some of the goniatites we examined, we investigated the relationship between them.

1.1 Background

Three main types of suspended cameral membranes have been described: transverse, horizontal, and siphuncular (Weischat and Bandel, 1991). Of these, siphuncular membranes are the most common, and they are the only type documented in the goniatites described in this study. Weitschat and Bandel

(1991) described siphuncular membranes as sheets extending from the outermost layer of the siphuncle to the inside surface of the ventral wall, and commonly attached to the septal surface as well. They further noted that the edges of the sheets are projected adorally. This description is similar to what Landman et al. (2006) noted in prolecanitids from Nevada. Although the siphuncular membranes in goniatites are much less complex than those in other ammonoids, these membranes also extend between the siphuncle and ventral floor of the chamber, and are commonly projected adorally.

Pseudosutures were first described by John (1909) as markings ("*Pseudolobenlinie*") between sutures on the steinkerns of Triassic ceratites. Pseudosutures have since been described in a variety of Paleozoic and Mesozoic ammonoids, and they most often occur on the venter or flanks (Hölder, 1954; Vogel, 1959; Schindewolf, 1968; Bayer, 1977; Zaborski, 1986; Hewitt et al., 1991; Landman et al., 1993; Lominadze et al., 1993).

The significance of pseudosutures for understanding the mode of chamber formation also has been examined (Zaborski, 1986; Seilacher, 1988; Hewitt et al., 1991; Landman et al., 1993; Lominadze et al., 1993). The most detailed model proposed that the animal secreted a cameral gel, which supported the back of the body during translocation and formed cameral membranes (=pseudosepta) due to differences in the viscosity of the gel, and subsequent desiccation (Hewitt et al., 1991). The pseudosutures are interpreted as the margins of these membranes on the inner surface of the shell wall. Of the two goniatite species we examined, pseudosutures are present only in *Cravenoceras fayettevillae*.

2 Material

2.1 *Crimites elkoensis* Miller et al., 1957

Crimites elkoensis Miller et al., 1957, is an adrianitid goniatite from the Lower Permian (Wolfcampian = Sakmarian) of eastern Nevada. The genus *Crimites* has a cosmopolitan distribution in the Permian (Sakmarian to Kazanian). Several species of this genus occur in Lower Permian deposits in Nevada (Miller et al., 1957). Although other ammonoids are more common at this site, *Crimites elkoensis* is the most abundant goniatite.

The shell of *Crimites elkoensis* is relatively small, globular, and involute, with reticulate ornament and four or five constrictions per whorl (Fig. 9.1A). The whorl width is commonly greater than the whorl height. The suture is relatively simple, with about 14 lobes (Fig. 9.1B).

Specimens were collected near Buck Mountain in east-central Nevada (Figs. 9.2, 9.3). The strata containing ammonoids are part of the Lower Permian Arcturus Formation and are located on the south side of the mountain in three

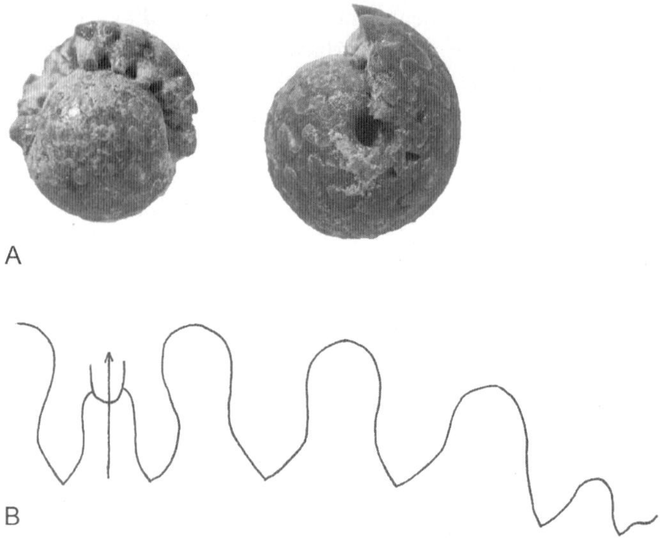

Fig. 9.1 Crimites elkoensis *Miller et al., 1957, SUI 39000, Rib Hill-Arcturus Formation, Lower Permian, Buck Mountain, Nevada.* ***A****. Apertural view and right side. Maximum diameter = 8.8 mm.* ***B****. Partial suture of the same specimen at shell diameter = 8.8 mm. From Lee (1975: 105).*

Fig. 9.2 *Localities of* Crimites elkoensis *Miller et al., 1957, and* Cravenoceras fayettevillae *Gordon, 1965. Map of the USA with asterisks indicating the sites at Buck Mountain, Nevada, and Fayetteville, Arkansas.*

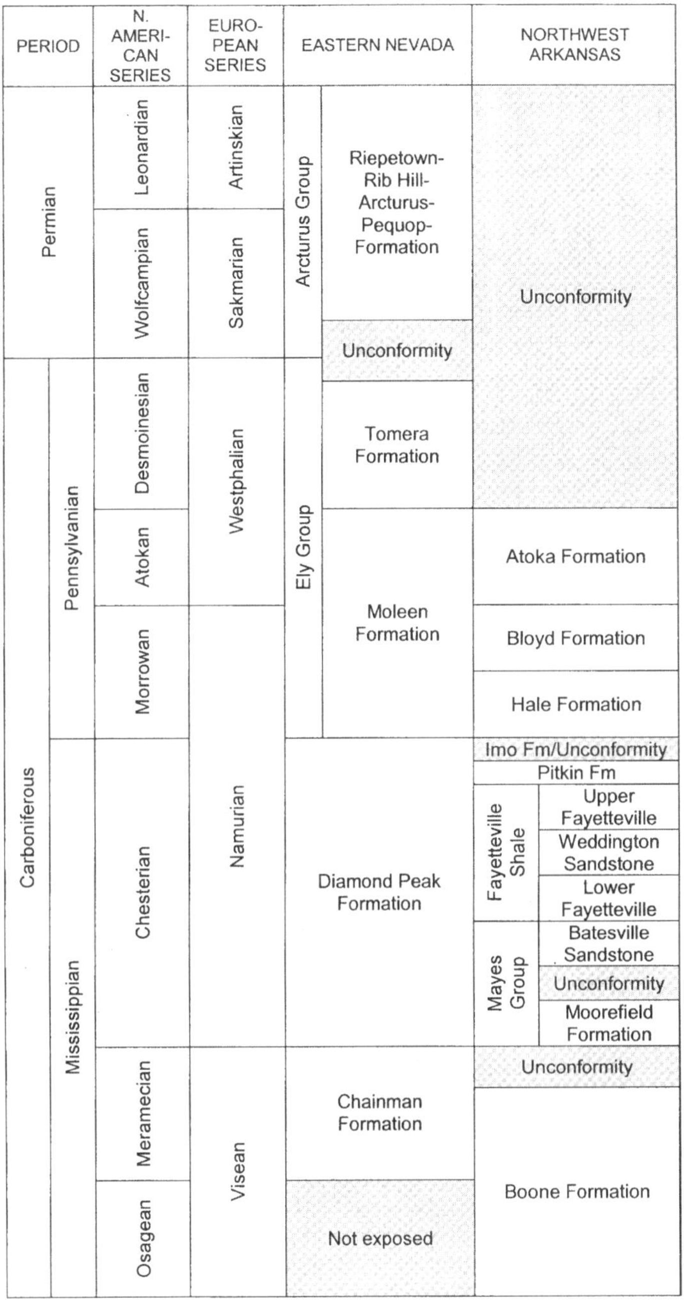

Fig. 9.3 *Stratigraphic column for Buck Mountain, Nevada, and Fayettevillae, Arkansas. The specimens of* Crimites elkoensis *Miller et al., 1957, are from the Arcturus Group (Permian). The Arcturus Group has been referred to by various names, as shown. The specimens of* Cravenoceras fayettevillae *Gordon, 1965, are from the lower Fayetteville Shale (Mississippian). Modified From Tomastik (1981: 18) and Manger (2004: 5, 14).*

ridges just east of Beck Springs. The fossils, which include marine invertebrates and vertebrates, occur in carbonate and phosphate concretions. Prolecanitid ammonoids are the most abundant fossils, followed by goniatites and nautiloids. During deposition of these strata, the lower part of the water column may have been anoxic and slightly acidic (Landman et al., 2006). These conditions would have allowed for rapid phosphatization of cameral membranes in ammonoids very early in diagenesis (Briggs, 2003; Landman et al., 2006). The originally aragonitic shell material of the goniatites was transformed into calcite, while the originally organic membranes were replaced by phosphate.

2.2 *Cravenoceras fayettevillae* Gordon, 1965

Cravenoceras fayettevillae Gordon, 1965, is a goniatite from the Upper Mississippian (Chesterian = early Namurian) of northwest Arkansas. The shell is relatively small, evolute, and globular, with a small umbilicus, thin, closely spaced ribs, and three or four constrictions per whorl (Fig. 9.4A). The body chamber is approximately 1.5 whorls long, and the suture is simple and very similar to that of *Cravenoceras articum* (Fig. 9.4B). Growth occurs in three stages (Fig. 9.5). This species is present only in the Fayetteville Shale (Fig. 9.3), although the genus has a cosmopolitan distribution (Eurasia, North Africa, and North America) during the Late Mississippian and Early Pennsylvanian.

The specimens we studied came from the lower part of the Fayetteville Shale along the White River near Durham, Arkansas (Fig. 9.2). Many of the specimens were obtained from carbonate concretion "halos" that occasionally surround the body chambers of large (up to 3 m in length) actinoceratids. The concretions were slabbed with a diamond saw to expose the goniatites (Mapes and Dalton, 2002). Most of the ammonoids appear to be *Cravenoceras fayettevillae*, although specimens of other taxa, such as *Tumulites* and *Paracravenoceras*, have been observed.

Fig. 9.4 A. Cravenoceras fayettevillae *Gordon, 1965, UA 77–205–1, lower Fayetteville Formation, Upper Mississippian (Chesterian), Fayetteville, Arkansas. Apertural view and left side, maximum diameter = 24 mm. From Saunders et al. (1977: 129).* ***B****. Partial suture of a closely related species,* Cravenoceras articum *Librovich, 1938, Middle Carboniferous (Namurian), Novaya Zemlya, Russia. From Ruzhentsev (1962: 589).*

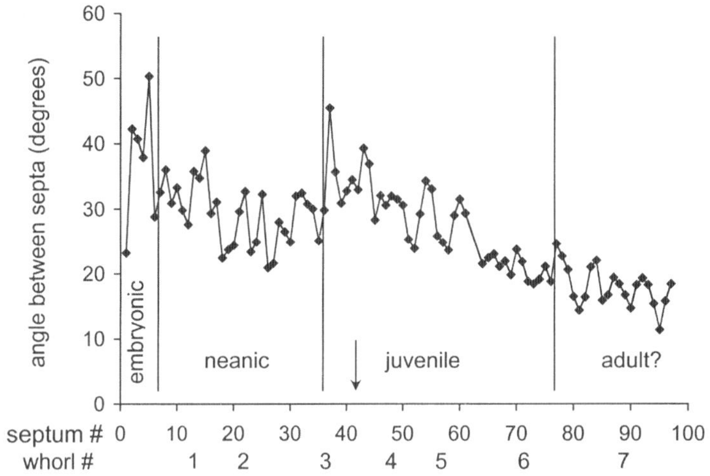

Fig. 9.5 *Plot of septum number versus septal angle (angle between septa) in* Cravenoceras fayettevillae *Gordon, 1965, AMNH 51239, lower Fayetteville Formation, Durham, Arkansas. Vertical lines demarcate three growth stages as described by Bucher et al. (1996). The arrow indicates the point at which siphuncular membranes appear at about 3.5 whorls. Siphuncular membranes may have appeared even earlier, but this is impossible to determine as the siphuncle is missing between septa 30 and 40.*

The Fayetteville Shale is a black, concretionary marine shale with a primarily molluscan fauna dominated by cephalopods (Gordon, 1965; Saunders et al., 1977). Nautiloids and ammonoids occur as pyritized casts, or alternatively, in carbonate concretions. The environment most likely was a deep, muddy shelf with anoxic bottom conditions. As with the specimens of *Crimites* from Nevada, such conditions were conducive to the preservation of soft tissue. The shells of *Cravenoceras fayettevillae* are calcitic. Rapid burial and an abundance of phosphorus resulted in the preservation of the originally organic membranes, through replacement by phosphate or through the development of phosphatic coatings on the surfaces.

2.3 Phosphatization of Organic Material

Preservation of soft tissues and delicate organic structures, such as cameral membranes, is relatively rare in the fossil record. Briggs (2003) discussed the circumstances under which such preservation is likely to occur: warm, shallow, marine environments with high organic productivity, episodic sedimentation, fluctuating salinity, and low oxygen. The presence of active bacterial decomposers or microbial mats appears to be a key factor in mediating the rapid mineralization of soft tissues (Briggs and Kear, 1993; Briggs, 2003; Martin et al., 2004). Phosphatization of organic material such as siphuncular membranes most likely occurs where bottom conditions are poor (probably anoxic and slightly acidic). This would allow postmortem precipitation of phosphate in the ammonoids as well as precipitation of phosphate concretions around suitable nuclei. The following scenario represents one possible pathway for the fossilization of the organic membranes in the goniatites we studied. After death, some of the ammonoids sank to the bottom and were rapidly buried. This rapid burial prevented destruction of soft tissue by scavengers. Dissolved phosphate in the sediment diffused into the phragmocone via the siphuncle (Weitschat and Bandel, 1991), and additional phosphate was liberated by bacteria from the soft tissues of the ammonoids. Decay resulted in the production of bicarbonate, which combined with phosphate to form apatite. Briggs (2003) has shown that when the concentration of phosphate is high, calcite or aragonite precipitation is inhibited, and apatite replaces soft tissue. Briggs (2003) further has shown that bacterial mediation and a significant external source of phosphate (e.g., from decomposing phytoplankton) are probably necessary for soft-tissue fossilization.

3 Methods

To study siphuncular membranes, we prepared 26 specimens in three different orientations. Sixteen specimens (10 *Crimites elkoensis* and 6 *Cravenoceras fayettevillae*) were ground in a median section coincident with the plane of the siphuncle; this provided a lateral cross section of the membranes. One specimen of *C. elkoensis* and one of *C. fayettevillae* were ground in a dorsoventral cross section perpendicular to the plane of symmetry to reveal the siphuncular membranes in transverse cross section. The eight remaining specimens (all *C. fayettevillae*) were left intact. We etched the specimens with dilute acetic acid (5%) to expose the membranes. Because the membranes are phosphatic, etching revealed the preserved membranes in three dimensions. The etching process was monitored and stopped as soon as the membranes were adequately exposed. Etching time varied from 2 min up to 3 h. Further etching resulted in the destruction of the membranes.

The exposed membranes included the organic lining of the chamber, the wall of the siphuncle, and the siphuncular membranes. In whole (unground) specimens, the only structure visible after dissolution is commonly the smooth outer surface of the chamber lining. This lining must be partly removed to observe the siphuncle and its associated membranes. Removal of the delicate lining usually resulted in destruction of the underlying siphuncle and associated membranes, but sometimes the lining was serendipitously removed, exposing the siphuncular membranes. The specimens with siphuncular membranes were examined by scanning electron microscopy (SEM).

To examine pseudosutures, four specimens of *Cravenoceras fayettevillae*, in which only the phragmocones had been preserved, were prepared by carefully separating the phragmocone from the surrounding matrix. In general, the shell wall adhered to the matrix, resulting in an external and internal mold (steinkern). Both the external and internal molds of two specimens were etched, while the molds of the two other specimens were not etched. By etching, we hoped to determine the original nature of the pseudosutures (mineralized or organic). The molds were examined by SEM. All illustrated specimens are reposited in the American Museum of Natural History (AMNH), the Department of Geology of the University of Arkansas (UA), and the Department of Geoscience of the University of Iowa (SUI).

4 Observations

4.1 *Crimites elkoensis* Miller et al., 1957

The membranes in this species consist of a series of small, adorally concave sheets that extend from the outer layer of the siphuncle to the inner surface of the ventral part of the chamber (Figs. 9.6, 9.7).

They are usually evenly spaced along the siphuncle in each chamber, although in some specimens they are more common in the adapical part of the chamber. The

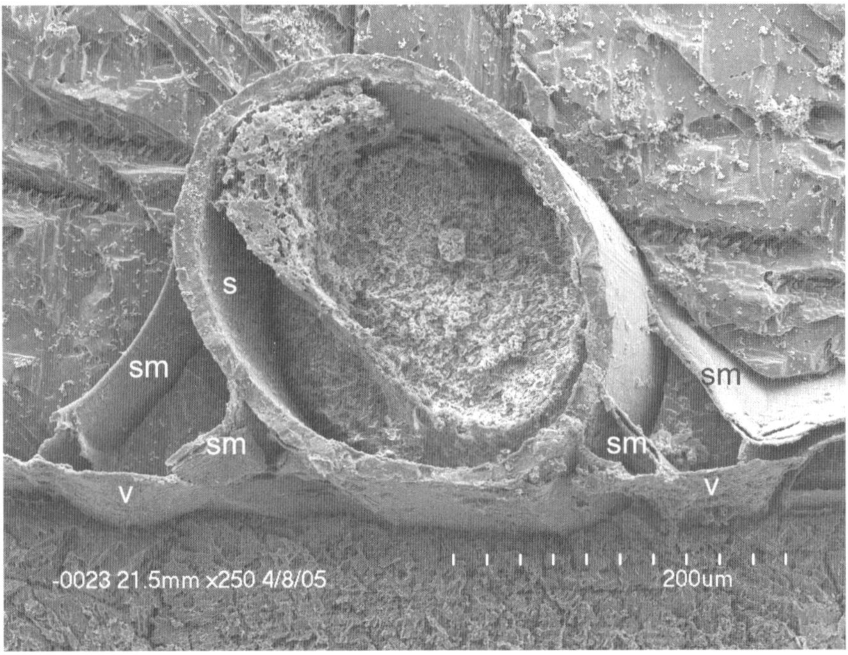

Fig. 9.6 Crimites elkoensis *Miller et al., 1957, AMNH 51223, Arcturus Formation, Buck Mountain, Nevada. Dorsoventral section showing the siphuncle (s) and siphuncular membranes (sm). The membranes form sheets between the siphuncle and the ventral floor of the chamber (v).*

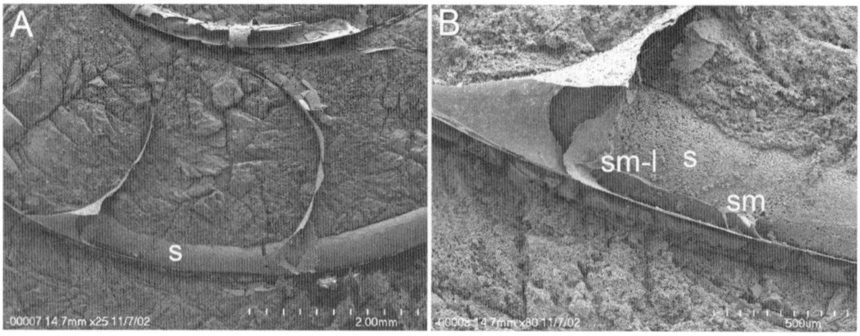

Fig. 9.7 Crimites elkoensis *Miller et al., 1957, AMNH 51240, Arcturus Formation, Buck Mountain, Nevada. Median views of the siphuncle and siphuncular membranes. A. The siphuncle (s) is visible in two chambers. Adoral direction is to the right. B. Close-up of the siphuncular membranes (sm) in A. A fragment of a siphuncular membrane (sm-l) that connects multiple siphuncular membranes is visible.*

membranes sometimes appear to have two layers (Fig. 9.6), although in most instances, they appear as one solid sheet. The bilayered structure may indicate that the original membrane was coated with phosphate on both sides during early diagenesis. When the original membrane decomposed, the two phosphatic layers remained with a small space between them.

Figure 9.7 indicates that the membranes may merge dorsally into a thin sheet parallel to the floor of the chamber. However, the fragmentation of the specimen makes it difficult to determine this for certain. Such sheets, formed by the merging of several elements, are common in ceratites and prolecanitids (Weitschat and Bandel, 1991; Landman et al., 2006).

Perhaps because goniatites are relatively rare in the Buck Mountain concretions, we were unable to find a specimen with preservation adequate to determine the morphology of the siphuncular membranes throughout ontogeny. The goniatite *Crimites elkoensis* is present in the same deposits as the prolecanitid *Akmilleria electraensis* Plummer and Scott, 1937, whose membranes do not appear until the end of the neanic stage (at or near the beginning of the third whorl, corresponding to a shell diameter of approximately 3 mm). The membranes in *A. electraensis* first appear as simple sheets, similar to those of the goniatites, and then become more complex during ontogeny (Landman et al., 2006). Until better-preserved specimens are examined, it is impossible to state whether there is any consistent ontogenetic pattern in *C. elkoensis*.

4.2 *Cravenoceras fayettevillae* Gordon, 1965

The siphuncular membranes in *Cravenoceras fayettevillae* also appear as a series of simple, evenly spaced, adorally concave sheets between the siphuncle and ventral

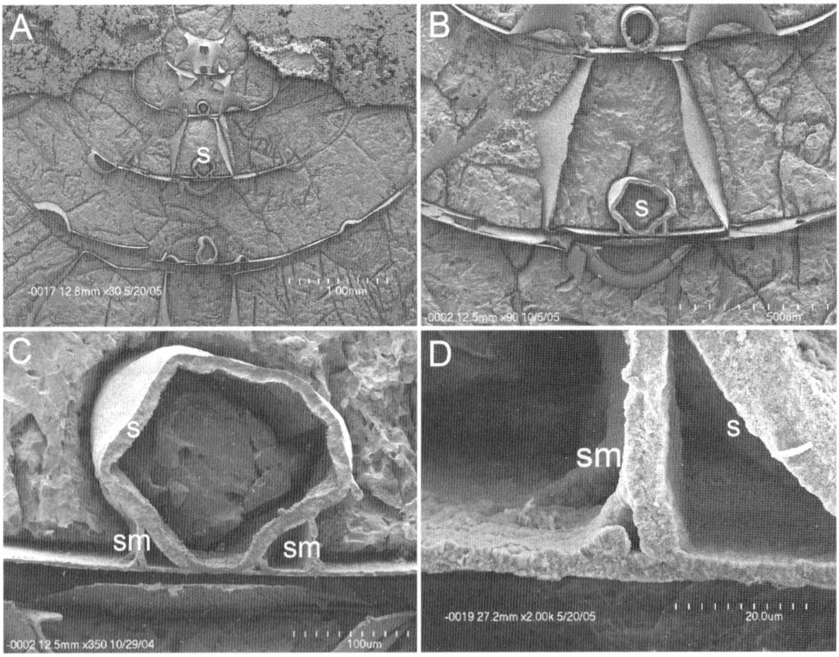

Fig. 9.8 Cravenoceras fayettevillae *Gordon, 1965, AMNH 51233, lower Fayetteville Shale, Fayetteville, Arkansas. Dorsoventral views of the siphuncle and siphuncular membranes. A. Overview of several whorls. B. Close-up view of a portion of the siphuncle (s), showing siphuncular membranes. C. Close-up view of the central portion of B, showing the siphuncular membranes (sm) on either side of the siphuncle (s). D. Close-up view of the siphuncular membranes in C, showing tiny podial attachments along the length of the siphuncular sheet.*

floor of the chamber (Figs. 9.8–9.10). In the dorsoventral view in Fig. 9.8, tiny "podia" are visible attaching the membrane to the inner lining of the shell. Such podia are very common on the siphuncular membranes in prolecanitids (Landman et al., 2006). In the specimen of *C. fayettevillae* in Fig. 9.8, the membranes are oriented perpendicular to the chamber floor, with no adoral projection. The median sections of *C. fayettevillae* confirm, however, that the siphuncular membranes are adorally concave and projected forward, as in specimens of *Crimites elkoensis* (Fig. 9.9).

In Fig. 9.9C, D, a thin, fragmented sheet is visible along the ventrolateral portion of the siphuncle. This is probably similar to the sheet described in *Crimites elkoensis*, in which the smaller, ventral membranes merge dorsally into a single sheet. Part of the siphuncle has been ground away, so the connection between the smaller sheets and the larger sheet is no longer visible. Figure 9F, H additionally shows that the larger sheet is not continuous, but shows numerous gaps (at least five in Fig. 9.9H).

Fig. 9.10 illustrates several examples of specimens in which the venter of the phragmocone was etched whole without first being ground, revealing the cameral membranes. In these specimens, the orientation of the siphuncular membranes in

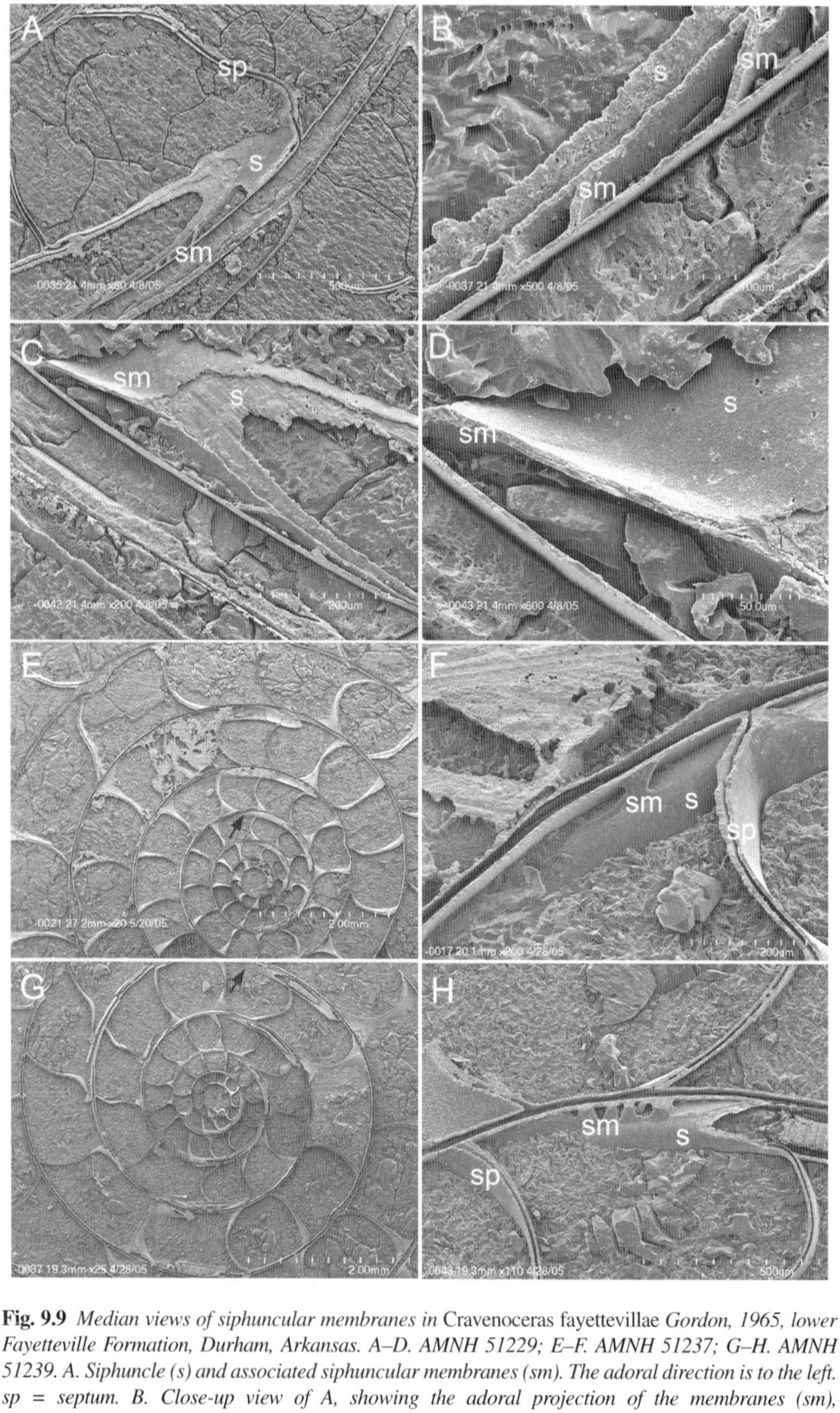

Fig. 9.9 *Median views of siphuncular membranes in* Cravenoceras fayettevillae *Gordon, 1965, lower Fayetteville Formation, Durham, Arkansas. A–D. AMNH 51229; E–F. AMNH 51237; G–H. AMNH 51239. A. Siphuncle (s) and associated siphuncular membranes (sm). The adoral direction is to the left. sp = septum. B. Close-up view of A, showing the adoral projection of the membranes (sm),*

relation to the siphuncle and chamber floor is visible. The siphuncular membranes appear as a series of adorally concave sheets between the siphuncle and the ventral lining of the chamber. The sheets are projected forward and commonly merge ventrolaterally.

We were able to examine changes in the morphology of the membranes in the ontogeny of two specimens (Fig. 9.9E, G). Although the siphuncle was missing in places, it is clear that membranes are not present in the earliest chambers. The earliest appearance of membranes is at approximately 3.5–3.75 whorls. This corresponds to the end of the neanic stage (Fig. 9.5). In the prolecanitid *Akmilleria electraensis* from Buck Mountain, Nevada, membranes also appear at the beginning of the third whorl, which marks the end of the neanic stage, that is, the end of the first postembryonic stage (Bucher et al., 1996).

In one specimen, two phosphatized, presumably originally organic, layers are present in the body chamber (Fig. 9.11). Because the specimen is fragmented, it is impossible to determine in which part of the body chamber the layers occur. One layer, located on the outside of the shell wall, is presumed to be the periostracum (p), while the other, on the inside of the shell wall, is most likely the chamber lining (cl).

We observed pseudosutures in many specimens of *Cravenoceras fayettevillae* (Figs. 9.12–9.14), though none are visible in *Crimites elkoensis*. The pseudosutures occur on the ventral and/or ventrolateral parts of the specimen, adoral or adapical of the first saddle and the first and second lobes. They are generally quite close to the suture. Pseudosutures do not occur in every specimen or in every chamber, although, if present at all, they generally occur in more than one chamber. No pattern in the number or location of pseudosutures within or among specimens is apparent.

The pseudosutures in *Cravenoceras fayettevillae* appear as ridges on the inside surface of the shell on the external mold (Fig. 9.13A, B), and as grooves on the corresponding steinkern (Fig. 9.13C, D). They are sometimes very well defined (Fig. 9.13B), but more commonly they are indistinct (Fig. 9.14). The ridges, especially when well developed, are asymmetrical in lateral cross section, with an abrupt adapical slope and a longer, more gradual adoral slope (Fig. 9.12C, D).

In etched specimens, we noted that the pseudosutural ridges on the inside surface of the external mold are sometimes dissolved and sometimes intact. Similarly, the

Fig. 9.9 (continued) *which is characteristic of ceratites, prolecanitids, and goniatites.* C. Siphuncular *membrane (sm) along the length of the siphuncle (s). The adoral direction is to the right. D. Close-up view of C, showing that the membranes sometimes consist of two layers and are continuous with the layer surrounding the siphuncle (s). E. Overview of a specimen with the initial chamber preserved. The arrow indicates the first appearance of siphuncular membranes. F. Close-up view of earliest siphuncular membranes (sm). The adoral direction is to the left and down. G. Overview of another specimen with the initial chamber preserved. The first appearance of siphuncular membranes is indicated by the arrow. H. Close-up view of the siphuncular membranes in G, showing how they appear when unbroken. Several siphuncular sheets may form parallel to the floor of the chamber, with the vertically directed sheets shown in A and B hidden underneath. The adoral direction is to the right.*

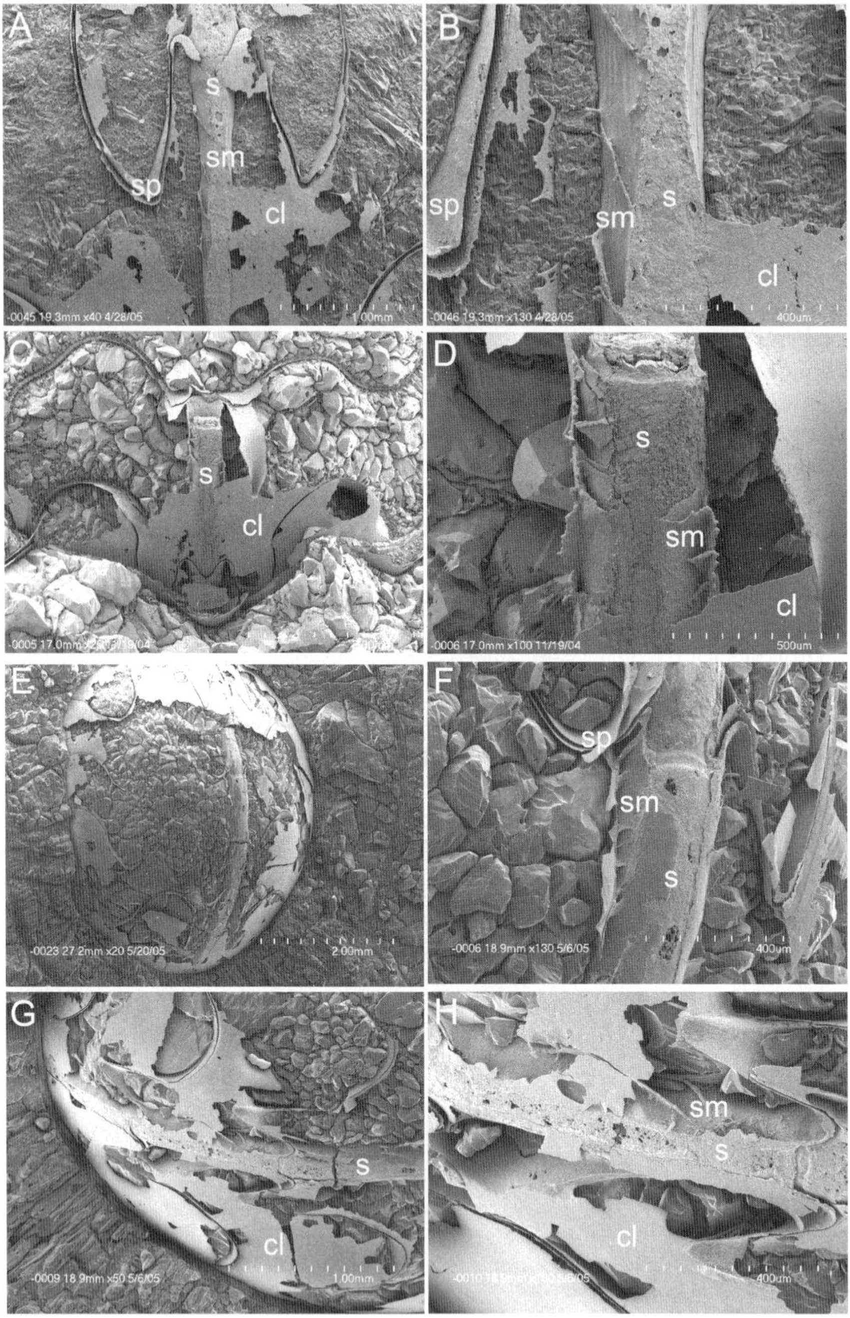

Fig. 9.10 Cravenoceras fayettevillae *Gordon, 1965, lower Fayetteville Formation, Durham, Arkansas. A–B. AMNH 51241; C–D. AMNH 51225; E–H. AMNH 51242. The venter has been etched to expose the siphuncle and siphuncular membranes. A. Siphuncle (s) with traces of siphuncular membranes (sm). Adoral direction is up. sp = septum, cl = chamber lining. B. Close-up*

pseudosutural grooves on the internal mold, which we expected to disappear as the calcite mold etched, are sometimes preserved. These observations suggest that either the mineral composition of the pseudosutures varies among specimens from the same location (and even from the same concretion), or, more likely, that a non-calcitic layer occurs between the shell wall on the inside surface of the external mold and the matrix of the internal mold (see section 5.2). The mineral composition of the pseudosutures is presumably calcitic, with regular crystals oriented perpendicular to the inner surface of the shell wall in well-developed pseudosutures (Fig. 9.12C), and with irregular crystals oriented at random in less well-developed pseudosutures (Fig. 9.12D). The pseudosutures may also be secondarily mineralized (see section 5.3).

5 Discussion

5.1 Origin of Membranes

The origin of siphuncular membranes has been debated (Kulicki, 1979; Weitschat and Bandel, 1991; Westermann, 1992; Checa, 1996; Landman et al., 2006).

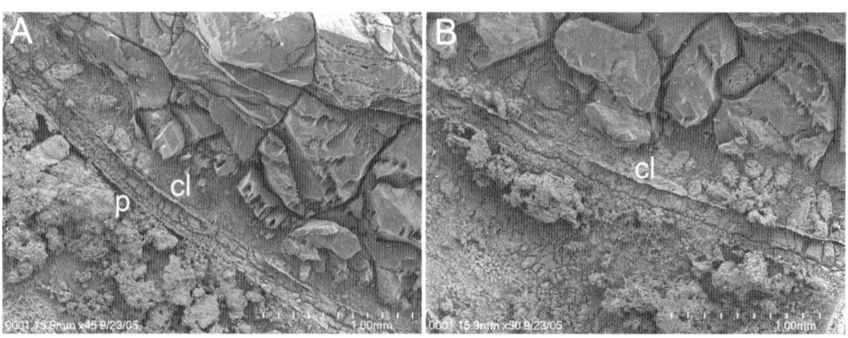

Fig. 9.11 Cravenoceras fayettevillae *Gordon, 1965, AMNH 51243, median cross section, lower Fayetteville Formation, Durham, Arkansas. Because the specimen is fragmented, it is unclear whether this is the adapical part of the body chamber or a more adoral portion. The shell wall exhibits two layers of phosphatized, presumably originally organic, layers. One of these layers (p) is assumed to be the periostracum, while the other (cl) is likely to be the chamber lining, similar to the chamber lining found in the phragmocone.*

Fig. 9.10 (continued) *view of siphuncular membranes (sm) in A. C. Fragmented siphuncular membranes along the siphuncle (s). Adoral direction is up. D. Close-up view of siphuncular membranes in C (sm). E. Etched phragmocone with the siphuncle exposed. Adoral direction is up. F. Close-up view of siphuncular membranes in E (sm). G. Same specimen as in E, F, rotated 90° clockwise to show the siphuncular membranes in another chamber. Adoral direction is to the right. H. Close-up view of siphuncular membranes in G (sm).*

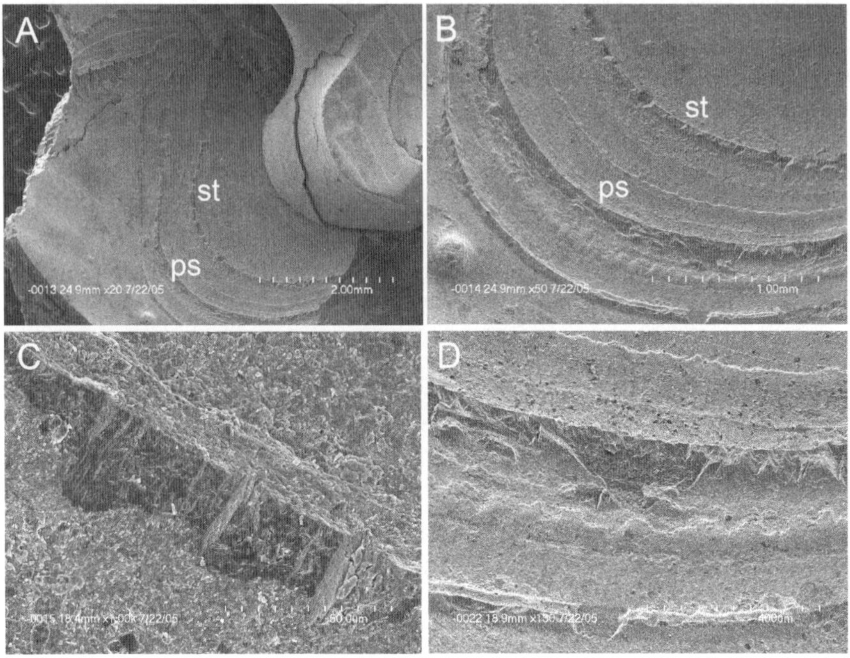

Fig. 9.12 *Pseudosutures in* Cravenoceras fayettevillae *Gordon, 1965, AMNH 51244, lower Fayetteville Formation, Durham, Arkansas. The adoral direction is down and to the left. A. The inner surface of the shell is exposed, showing the ridges or pseudosutures (ps). The specimen was not etched, st = suture. B. Close-up view of pseudosutures (ps), which appear on the adoral side of the suture (st). C. Close-up view of the most prominent pseudosuture, showing mineralization and an asymmetrical slope similar to the mural ridge in Nautilus. D. Close-up view of weaker pseudosutures, demonstrating the wide variability in the amount of material comprising each pseudosuture.*

Landman et al. (2006) argued that these membranes were secreted by the rear part of the mantle, and are not solely the result of desiccation of a cameral liquid or gel after formation of the septum. As evidence for this view, they noted: (1) the abrupt appearance of complex membranes at the end of the neanic stage; (2) the consistency of the ontogenetic pattern among individuals; (3) the surface morphology of the membranes, which lack features characteristic of desiccated gels; and (4) the presence of membranes of similar composition in the adapical portions of the body chambers of Cretaceous ammonoids (Tanabe et al. 2005: Fig. 9.1), which could not have formed by desiccation of cameral liquid after chamber formation.

The similarity in structure and composition of goniatite membranes to prolecanitid membranes strongly suggests that they were formed by similar processes. The four points of Landman et al. (2006) in support of the secretion hypothesis were also evaluated for the goniatites we studied.

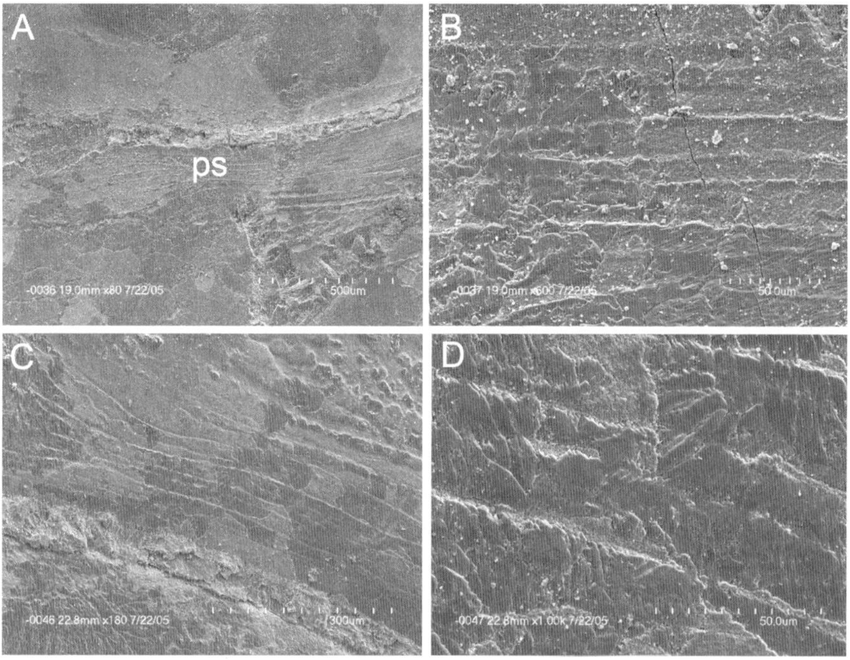

Fig. 9.13 *Pseudosutures on the external molds (A and B) and corresponding internal molds (C and D) of an unetched specimen of* Cravenoceras fayettevillae *Gordon, 1965, AMNH 51245, lower Fayetteville Formation, Durham, Arkansas. A, B. Overview and close-up of the pseudosutures (ps) on the inside surface of the shell (external mold). These pseudosutures are located on the adapical side of the saddle, and appear as ridges. The adoral direction is to the upper left. C, D. Pseudosutural grooves on the internal mold of the specimen. When specimens are broken, the pseudosutures appear as ridges on the external mold and as grooves on the internal mold.*

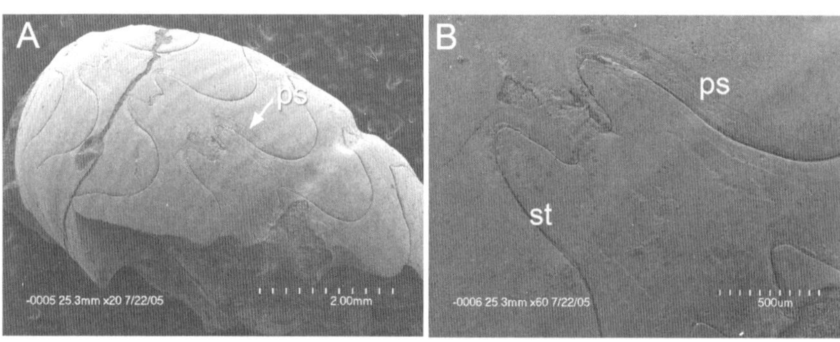

Fig. 9.14 Cravenoceras fayettevillae *Gordon, 1965, AMNH 51246, lower Fayetteville Formation, Durham, Arkansas. Internal mold with pseudosutures (ps) on the adapical side of the suture (st). The adoral direction is to the bottom right. A. Overview. B. Close-up of the pseudosutures (ps). The position and spacing of the pseudosutures do not correspond to the position and spacing of the siphuncular membranes.*

(1) If membranes were produced by desiccation after chamber formation, membranes would have formed in all chambers. However, the abrupt appearance of siphuncular membranes in *Cravenoceras fayettevillae* in the fourth whorl suggests that these features did not form by desiccation. Chamber linings are present starting from the initial chamber in both prolecanitids and in the goniatites we examined, which leaves open the question of whether the linings (as opposed to the siphuncular membranes) were formed by secretion or desiccation (but see point 4).
(2) In the goniatites we examined, the data are insufficient to determine whether a consistent ontogenetic pattern of membranes exists, although the uniform structure of the membranes in each chamber suggests secretion rather than the vagaries of desiccation.
(3) The surface of the membranes in goniatites is generally smooth, similar to that of membranes observed in other ammonoids. Features characteristic of a desiccated gel, such as wrinkles or tessellation, are absent.
(4) No body chambers were preserved in specimens of *Crimites elkoensis*, and so we cannot comment on the presence or absence of an organic lining in the body chambers of individuals of that species. Such linings, however, have been observed in body chambers of some individuals of *Cravenoceras fayettevillae* (Fig. 9.11). Thus, at least some chamber linings probably were secreted by the animal.

In summary, although additional data are needed, there is some evidence that the membranes formed by secretion rather than by desiccation.

5.2 Preservation of Pseudosutures

Whether or not pseudosutures are present in an etched specimen depends on the mineral composition of the pseudosutures and on how the fossil breaks. Pseudosutures that are preserved as calcite ought to dissolve, and in some cases this occurred. In a few instances, however, pseudosutures were not dissolved by acid etching. We attribute this to an overlying phosphatic coating sometimes left behind when the fossil broke out of the matrix (Fig. 9.15). When specimens are removed from the matrix, both the cameral lining and the shell wall may adhere to the external mold (a concave fragment; scenario 1, Fig. 9.15).

Alternatively, the cameral lining may adhere to the steinkern while the shell wall adheres to the surrounding matrix (scenario 2, Fig. 9.15). These different breakage patterns affect whether pseudosutures are visible after etching of the specimens. If the matrix breaks away from the internal mold with both the calcitic shell and the phosphatic chamber lining adhering to it (scenario 1), the pseudosutures will be visible as ridges on the inner surface of the external mold, and as grooves on the internal mold ("r" and "g," respectively, Fig. 9.15). If the pieces are then etched, the pseudosutures should still be visible on the external mold regardless of their

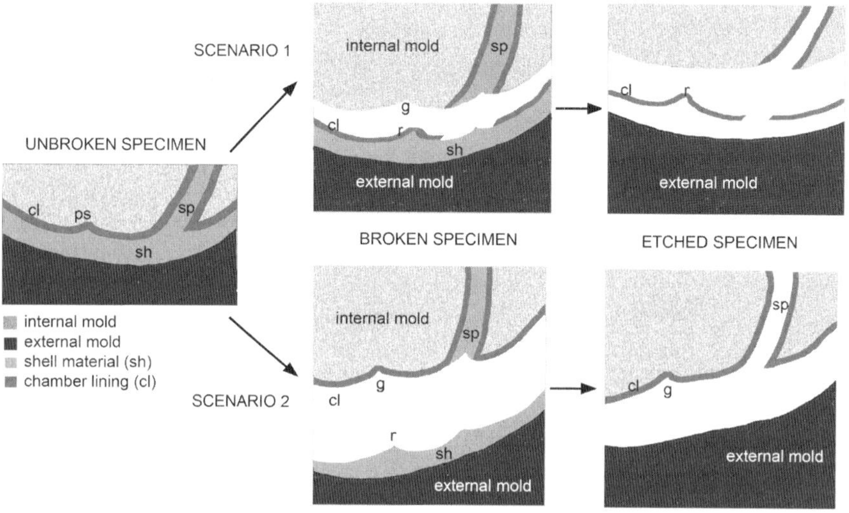

Fig. 9.15 *Illustration of two possibilities that may occur during preparation of a specimen, with the specimens shown in median cross section (parallel to the siphuncle). This diagram explains why pseudosutures sometimes remain visible after etching. In both scenarios 1 and 2, the unbroken specimen has a mineralized layer that includes the shell wall (sh) and the pseudosutures (ps) on the inside surface of the wall. There is a chamber lining (cl) overlying the mineralized layer. In scenario 1, both layers break away from the internal mold, and the pseudosutures are visible as grooves on the internal mold (g) and as ridges on the external mold (r). After etching, the grooves on the internal mold are no longer visible, but the impressions of the ridges on the external mold (r) are preserved by the overlying phosphatic layer. In scenario 2, the mineralized layer breaks away from the internal mold, but the phosphatic chamber lining (cl) adheres to the internal mold. As in scenario 1, the pseudosutural ridges (r) are visible on the external mold, and the impressions are visible on the internal mold as grooves (g). After etching, however, the mineralized ridges are dissolved away on the external mold, yet the impression of the ridges remains on the internal mold because the phosphatic lining (cl) resists etching.*

mineral composition, because the chamber lining protects and preserves the pseudosutures it overlies. On the internal mold, however, the grooves would disappear, as would all other calcitic material.

Alternatively (scenario 2, Fig. 9.15), if the specimen breaks between the calcitic layer and the phosphatic cameral lining, the calcitic layer and pseudosutural ridges (r) would adhere to the external mold, and the cameral lining would adhere to the internal mold. Before etching, the pseudosutures would be visible as ridges on the external mold and as grooves on the internal mold, as before. After etching, however, the pseudosutural ridges on the external mold, if they are calcitic, would disappear, leaving nothing on the internal surface of the shell wall fragment, while the pseudosutural grooves would still be preserved on the internal mold as impressions in the cameral lining (scenario 2, Fig. 9.15). This model thus explains our observations of the variation in position and presence of pseudosutures before and after etching.

5.3 Origin of Pseudosutures

Differences in position, extent, and mineral composition suggest that pseudosutures and siphuncular membranes may have had different origins. In goniatites, siphuncular membranes are present only in the immediate vicinity of the siphuncle, whereas pseudosutures commonly extend across the entire venter and onto lateral portions of the shell (Fig. 9.14). The spacing and extent of pseudosutures differ from that of the membranes in *Cravenoceras fayettevillae*. Pseudosutures are located near the suture itself, in closely spaced series, and extend around the margin from the venter to the flanks. Siphuncular membranes, in contrast, are spaced evenly throughout the chamber and are confined to the area immediately surrounding the siphuncle (compare the position of the membranes in Figs. 9.9, 9.10 and the position of the pseudosutures in Figs. 9.12–9.14).

This lack of correspondence suggests that siphuncular membranes and pseudosutures were secreted either at different times, or by two different portions of the rear mantle. Because pseudosutures retain the general shape of the suture, it can be supposed that they were secreted by the margin of the rear mantle, at its peripheral contact with the interior of the shell wall. Siphuncular membranes would have been secreted by the rear area of the body immediately surrounding the siphuncle, not necessarily around the entire margin of the mantle. If the siphuncular membranes were secreted separately from the pseudosutures, then the siphuncular membranes would not necessarily have had contact with the pseudosutures. However, none of the specimens we examined contains siphuncular membranes and pseudosutures in the same chamber. Therefore, it is impossible to determine their relationship in the present material. That determination must await the examination of median or paramedian sections of specimens in which both structures occur in the same chamber. If additional data confirm that pseudosutures and siphuncular membranes contact each other and correspond in position and spacing, then the siphuncular membranes would be equivalent to pseudosepta.

If the siphuncular membranes in *Crimites elkoensis* and *Cravenoceras fayettevillae* are equivalent to pseudosepta, then it is reasonable to assume that they were deposited by a single process and may therefore have an equal chance of preservation. However, although the preservation of siphuncular membranes and pseudosutures is relatively common, the preservation of pseudosepta is extremely rare (for the sole published observation of pseudosepta, see Checa, 1996). Some workers have proposed explanations for these apparent contradictions. Even if these structures were parts of the same original structure, pseudosepta may have been quite thin and fragile, with thicker concentrations of proteinaceous or chitinous material near the margins of the chambers and immediately surrounding the siphuncle (Hewitt and Westermann, 1987). Greater accumulations in the areas close to the shell wall, as opposed to lesser accumulations within the open space of the chambers, would explain why only marginal traces, such as siphuncular membranes or pseudosutures, are preserved. In addition, increased bacterial activity near the siphuncle would explain why organic deposits in this region would have a much greater probability of phosphatization in early diagenesis (Hewitt, 2005, personal communication). If siphuncular membranes and pseudosepta were parts of the same original structure, this would account for the preservation of siphuncular membranes even when pseudosepta are not preserved.

Nevertheless, the apparent differences in mineral composition of pseudosutures and siphuncular membranes are at odds with the hypothesis that these structures were formed by a single process and were originally continuous. It is, at first glance, unclear why the membranes are phosphatic whereas the pseudosutures are calcitic. It may be that the pseudosutures were originally secreted as a soft, organic substance that was secondarily mineralized (C. Kulicki, 2005, personal communication). Again, the proximity of the siphuncular membranes to bacteria that promoted phosphatization could explain why the membranes were phosphatized while the pseudosutures were not. Study of siphuncular membranes in specimens of the prolecanitid *Akmilleria electraensis* has confirmed that these membranes are phosphatic (Tanabe et al., 2000). An energy dispersive X-ray (EDX) analysis of the mineral composition of pseudosutures would shed light on their original composition. Analysis of pseudosutures in specimens with aragonitic preservation would show whether pseudosutures were originally mineralized (and therefore still aragonitic) or secondarily mineralized (and therefore calcitic). If pseudosutures were secondarily mineralized, a well-preserved aragonitic specimen ought to have pseudosutures composed solely of calcite. Future research should certainly include such specimens so that the question of mineralization of pseudosutures may be settled.

5.4 Implications for Chamber Formation

Some authors have suggested that, during the cycle of chamber formation, the soft body of the ammonoid crept forward incrementally (Seilacher, 1988). This creeping movement insured that the body was never entirely detached from the inner surface of the shell. Between movements, the animal would have paused within the chamber, which may have resulted in the accumulation of organic or mineral secretions from the mantle epithelium. It has been proposed that the siphuncular membranes formed as an accretion of organic material during such pauses (Landman et al., 2006). It is reasonable to assume that originally mineralized deposits or organic deposits that were later mineralized, could have accumulated at the peripheral contact of the rear body with the shell wall during such pauses as well. These ridges would be the pseudosutures, and could have served as sites for ephemeral attachment of the soft body.

Ward (1987: 85) proposed that the mural ridge in modern *Nautilus* serves as a site for temporary attachment of the periphery of the rear body during septal formation. The mural ridge is an aragonitic deposit formed as the first stage in septal morphogenesis (Blind, 1976, 1988; Ward, 1987: 85; Grégoire, 1987), and it is secreted by the mantle epithelium prior to the translocation of the body within the chamber. The mural ridge becomes the margin of the new septum. The cross section of the mural ridge mimics the shape of the back of the body, which moves forward and away from the mineralized deposit. The asymmetric shape of the pseudosutures in cross section in *Cravenoceras fayettevillae* (Fig. 9.12) recalls the morphology of the mural ridge in *Nautilus*, and the pseudosutures in goniatites may have served as sites of temporary attachment of the soft body during translocation.

Siphuncular membranes and pseudosutures both appear to have formed as accretions during the incremental forward motion of the animal in the course of translocation. If they were secreted by different parts of the rear mantle, the membranes would not necessarily be continuous with the pseudosutures, although they may merge along the venter. It is impossible to verify this, however, given the fact that pseudosutures have not yet been observed on the mid-venter in *Cravenoceras fayettevillae*.

Although reasonable explanations can account for the differences in the extent and mineral composition of siphuncular membranes and pseudosutures, it is difficult to explain why the spacing of structures with presumably the same origin is so different. Examination of specimens in which both structures are visible in the same chamber is the only way to solve this apparent contradiction.

We propose that, as the animal moved forward in the chamber, the rear mantle secreted both siphuncular membranes and pseudosutures. Both structures replicated the shape of the rear mantle at the time of deposition, and at least some pseudosutures and siphuncular membranes may have been continuous. Pseudosutures may have been originally mineralized or may have been secondarily mineralized, but siphuncular membranes were probably originally organic. Future studies of the mineral composition of pseudosutures, and of specimens in which both pseudosutures and siphuncular membranes occur in the same chamber, will clarify the relationship between these structures.

Acknowledgments

We are grateful to Walter Manger (University of Arkansas, Fayetteville) and Lisa Meeks (Exxon Mobil Development Company) for providing specimens of *Cravenoceras fayettevillae*, and to Roger Hewitt (Leigh-on-Sea, England), Cyprian Kulicki (Instytut Paleobiologii, Polska Akademia Nauk, Warsaw, Poland), and Kazushige Tanabe (University of Tokyo) for many helpful suggestions during the course of this project. We also thank Steve Thurston (AMNH) for technical advice on the figures. Jason Biederman took the photos in Fig. 9.1.

References

Bayer, U. 1977. Cephalopoden-Septen. I. Konstruktionsmorphologie des Ammoniten-Septums. *Neues Jahrbuch für Geologie und Paläontologie Abhandlungen* **154**: 290–366.

Blind, W. 1976. Die ontogenetische Entwicklung von *Nautilus pompilius* (Linné). *Palaeontographica A* **153**: 117–160.

Blind, W. 1988. Über die primäre Anlage des Siphos bei Ectocochleaten Cephalopoden. *Palaeotographica Abteiling A* **204**: 67–93.

Briggs, D.E.G. 2003. The role of decay and mineralization in the preservation of soft-bodied fossils. *Annual Review of Earth and Planetary Sciences* **31**: 275–301.

Briggs, D. E. G., and A. J. Kear. 1993. Fossilization of soft tissue in the laboratory. *Science* 259: 1439–1442.

Bucher, H., N. H. Landman, S. M. Klofak, and J. Guex. 1996. Mode and rate of growth in ammonoids. *In* N. H. Landman, K. Tanabe, and R. A. Davis (editors), *Ammonoid Paleobiology, pp.* 407–461. New York: Plenum Press.

Checa, A. 1996. Origin of intracameral sheets in ammonoids. *Lethaia* **29**: 61–75.

Erben, H. K., and R. E. H. Reid. 1971. Ultrastructure of shell, origin of conellae, and siphuncular membranes in an ammonite. *Biomineralization Research Reports* **3**: 22–31.

Gordon, M., Jr. 1965. Carboniferous cephalopods of Arkansas. *U.S. Geological Survey Professional Paper* **460**: 1–322.

Grandjean, F. 1910. Le siphon des ammonites et des belemnites. *Bulletin de la Societé Géologique de France* **10**: 496–519.

Grégoire, C. 1987. Ultrastructure of the *Nautilus* shell. *In* W. B. Saunders, and N. H. Landman (editors), *Nautilus – the Biology and Paleobiology of a Living Fossil, pp.* 463–486. New York: Plenum Press.

Hewitt, R. A., and G. E. G. Westermann. 1987. Function of complexly fluted septa in ammonoid shells II. Septal evolution and conclusions. *Neues Jahrbuch für Geologie und Paläontologie Abhandlungen* **174**: 135–169.

Hewitt, R. A., A. Checa, G. E. G. Westermann, and P. M. Zaborski. 1991. Chamber growth in ammonites inferred from colour markings and naturally etched surfaces of Cretaceous vasoceratids from Nigeria. *Lethaia* **24**: 271–287.

Hölder, H. 1952. Über Gehäusebau, insbesondere Hohkiel jurassicher Ammoniten. *Paläontographica Abteilung A* **102**: 18–48.

Hölder, H. 1954. Über die Sipho-Anheftung bei Ammoniten. *Neues Jahrbuch für Geologie und Paläontologie Monatshefte* **8**: 372–379.

John, R. 1909. Über die Lebenweise und Organisation des Ammoniten, Inaugural-Dissertation, Universität Tübingen, Stuttgart.

Karpinsky, A. 1889. Über die Ammoniten der Artinsk-Stufe und einige mit denselben verwandte caronische Formen. *Mémoires de l'Academie Impérial des Sciences de St. Pétersbourg, sér. 7* **37**(2): 1–104.

Kulicki, C. 1979. The ammonite shell: its structure, development, and biological significance. *Acta Palaeontologica Polonica* **39**: 97–142.

Landman, N. H., K. Polizzotto, R. H. Mapes, and K. Tanabe. 2006. Cameral membranes in prolecanitid and goniatitid ammonoids from the Permian Arcturus Formation, Nevada, USA. *Lethaia* **39**: 365–379.

Landman, N. H., K. Tanabe, R. H. Mapes, S. M. Klofak, and J. Whitehill. 1993. Pseudosutures in Paleozoic ammonoids. *Lethaia* **26**: 99–100.

Lee, C. 1975. Lower Permian ammonoid faunal provinciality. Unpublished M.Sc. thesis, University of Iowa, Iowa City, Iowa.

Librovich, L. S. 1938. Kamennougol'nye ammonei s yuzhnogo ostrova Novoi Zemli (Carboniferous ammonoids from the south island of Novaya Zemlya). *Paleontologiya Sovetskoi Arktiki 3, Trudy Arkticheskogo Instituta* **101**: 47–107.

Lominadze, T., M. Sharikadze, and I. Kvantaliani. 1993. On mechanism of soft body movement within body chamber in ammonites. *Geobios Mémoire spéciale* **15**: 267–273.

Manger, W. L. 2004. Middle Carboniferous lithostratigraphy and ammonoid successions, northwestern Arkansas. *Field Trip Guide for the Sixth International Symposium on Cephalopods – Present and Past*, University of Arkansas, Fayetteville, Arkansas.

Mapes, R. H. and R. B. Dalton. 2002. Scavenging or predation? Mississippian ammonoid accumulations in carbonate concretion halos around *Rayonnoceras* (Actinoceratoidea—Nautiloidea) body chambers from Arkansas. *In* H. Summesberger, and K. Histon (editors), *Cephalopods – Present and Past, Abhandlungen der Geologischen Bundesanstalt* **57**: 407–422.

Mapes, R. H., N. H. Landman, K. Tanabe, and H. Maeda. 2002. Intracameral membranes in Permian ammonoids from the Buck Mountain, Nevada, Lagerstätte. *Geological Society of America, Abstracts with Programs* **34**(6): 354.

Martin, D., D. E. G. Briggs, and R. J. Parkes. 2004. Experimental attachment of sediment particles to invertebrate eggs and the preservation of soft-bodied fossils. *Journal of the Geological Society, London* **161**: 735–738.

Miller, A. K., W. M. Furnish, Jr., and D. L. Clark. 1957. Permian ammonoids from the western United States. *Journal of Paleontology* **31**: 1057–1068.

Plummer, F. B., and G. Scott. 1937. Upper Paleozoic ammonites in Texas. The Geology of Texas. *The University of Texas Bulletin* **3701**: 1–516.

Ruzhentsev, V. E. (editor). 1962 (1974 translation). *Mollusca – Cephalopoda I. Nautiloidea, Endoceratoidea, Actinoceratoidea, Bactritoidea, Ammonoidea (Agoniatitida, Goniatitida, Clymeniida): Fundamentals of Paleontology*. Volume 5, Israel Program for Scientific Translation. Jerusalem: Keter Publishing House Jerusalem.

Saunders, W. B., W. L. Manger, and M. Gordon, Jr. 1977. Upper Mississippian and lower and middle Pennsylvanian ammonoid biostratigraphy of northern Arkansas. *Oklahoma Geological Survey Guidebook* **18**: 117–137.

Schindewolf, O. 1968. Analyse eines Ammoniten-Gehäuses. *Akademie der Wissenschaften und der Literatur, Abhandlungen der Mathematisch-Naturwissenschaftlichen Klasse in Mainz* **8**: 139–188.

Schoulga-Nesterenko, M. 1926. Nouvelles données sur l'organisation intérieure des conques des ammonites de l'étage d'Artinsk. *Bulletin de la Societé des Naturalistes de Moscou Section Géologique* **34**: 81–99.

Seilacher, A. 1988. Why are nautiloid and ammonoid sutures so different? *Neues Jahrbuch für Geologie und Paläontologie Abhandlungen* **177**: 41–69.

Tanabe, K., C. Kulicki, and N. H. Landman. 2005. Precursory siphuncular membranes in the body chamber of *Phyllopachyceras* and comparisons with other ammonoids. *Acta Palaeontologica Polonica* **50**(1): 9–18.

Tanabe, K. and N. H. Landman. 1996. Septal neck-siphuncular complex. *In* N. H. Landman, K Tanabe, and R. A. Davis (editors), *Ammonoid Paleobiology, pp.* 129–165. New York: Plenum Press.

Tomastik, T. 1981. The geology of the southern part of Buck Mountain, White Pine County, Nevada. Unpublished M.Sc. thesis, Ohio University, Athens, Ohio.

Vogel, K. P. 1959. Zwergwuchs bei Polyptychiten (Ammonoidea). *Geologisches Jahrbuch* **76**: 469–540.

Ward, P. D. 1987. *The Natural History of Nautilus*. Boston: Allen & Unwin.

Weitschat, W., and K. Bandel. 1991. Organic components in phragmocones of boreal Triassic ammonoids: implications for ammonoid biology. *Paläontologische Zeitschrift* **65**: 269–303.

Westermann, G. E. G. 1971. Form, structure, and function of shell and siphuncle in coiled Mesozoic ammonites. *Life Sciences Contributions, Royal Ontario Museum* **78**: 1–39.

Westermann, G. E. G. 1992. Formation and function of suspended organic cameral sheets in Triassic ammonoids – discussion. *Paläontologische Zeitschrift* **66**: 437–441.

Zaborski, P. M. P. 1986. Internal mould markings in a Cretaceous ammonite from Nigeria. *Palaeontology* **29**: 725–738.

Chapter 10
Soft-tissue Attachment of Middle Triassic Ceratitida from Germany

Christian Klug,[1] Michael Montenari,[2] Hartmut Schulz,[2] and Max Urlichs[3]

[1]Paläontologisches Institut und Museum, Universität Zürich, Karl Schmid-Strasse 4, CH-8006, Zürich, Switzerland, chklug@pim.uzh.ch;

[2]Institut für Geowissenschaften, Eberhard-Karls-Universität Tübingen, Sigwartstrasse 10, D-72076 Tübingen, Germany; Present address: Earth Sciences and Geography, Keele University, William Smith Building, Newcastle (Staffordshire) ST5 5BG, United Kingdom;

[3]Staatliches Museum für Naturkunde, Rosenstein 1, D-70191 Stuttgart, Germany, urlichs.smns@naturkundemuseum-bw.de

Keywords: Triassic, Germany, Ceratitida, muscle attachment, pseudosutures, black layer

1 Introduction

Soft-tissue attachment structures were found in various molluscs from the marls and limestones of the German Muschelkalk (Anisian, Ladinian). For instance, bivalves often display the sites where the adductors and the pallium were attached (e.g., Klug et al., 2005). For the nautiloid *Germanonautilus*, Klug and Lehmkuhl (2004) documented the attachment scars of the cephalic retractors which sometimes are preserved in great detail. Although representatives of the Ceratitida are very common (at least in some beds of some localities), soft-tissue attachment sites have never been described and figured previously. Muscle attachment scars (the dorsal and ventral scars) were listed as "unpublished observations" by Doguzhaeva and Mutvei (1996: 54), but, as far as we know, they never published these "observations" in greater detail.

N. H. Landman et al. (eds.), *Cephalopods Present and Past: New Insights and Fresh Perspectives*, 205–220.

Body parts that consisted of organic matter such as mandibles of *Ceratites* and *Germanonautilus* are moderately common in the marls and limestones of the German Muschelkalk. Although usually preserved as steinkerns, cephalopods from the German Muschelkalk sometimes display exceptionally well-preserved delicate morphological details such as the "black layer" (Klug et al., 2004), siphuncular membranes (Hagdorn and Mundlos, 1983; Wang and Westermann, 1993; Rein, 1995), three-dimensionally preserved nautiloid mandibles with their originally chitinous wings (Müller, 1963, 1969, 1974; Klug, 2001), three-dimensionally preserved *Ceratites* mandibles that lack extensive calcified components (Lehmann, 1985; Rein, 1993) and moulds of the buccal mass of *Germanonautilus* (Müller, 1969; Klug, 2001).

Soft-tissue attachment structures in recent nautiloids have been described by several authors (e.g., Ward, 1987; Isaji et al., 2002). In most cases, the functional interpretation of these structures was that they were for muscle attachment and fixation of the posterior mantle (e.g., Griffin, 1900; Doguzhaeva and Mutvei, 1991; Mutvei, 1957, 1964; Mutvei et al., 1993). Isaji et al. (2002) emphasized the essential constructional differences in the muscle–shell attachment in Recent *Nautilus* compared with benthic molluscs. According to these authors, the "myoadhesive epithelium-semi-transparent membrane junction in *Nautilus pompilius* seems to be physically weak against tensile stress caused by muscle movement." They interpreted this phenomenon as being an effect of adaptation to the nektonic mode of life in combination with the mode of shell growth.

Soft-tissue attachment structures in ammonoids have been described and analysed in numerous articles. These structures have been interpreted as attachment structures of muscles (e.g., Crick, 1898; Jones, 1961; Mutvei, 1964; Jordan, 1968; Rakús, 1978; Weitschat, 1986; Doguzhaeva and Kabanov, 1988; Doguzhaeva and Mutvei, 1991, 1996; Lominadze et al., 1993; Wang and Westermann, 1993; Tanabe et al., 1998; Richter, 2002), of preseptal cameral membranes (e.g., Jordan, 1968; Weitschat, 1986; Weitschat and Bandel, 1991, 1992; Checa and Garcia-Ruiz, 1996; Mapes et al., 1999; Richter, 2002), of the posterior mantle (e.g., Zaborski, 1986; Seilacher, 1988; Lominadze et al., 1993; Richter, 2002), and of the supracephalic mantle fold (Kulicki et al., 2001).

Newly discovered specimens now reveal some interesting details of the soft-tissue attachment structures in *Ceratites* from the German Muschelkalk. These will be described and discussed in this article.

2 Methods

The sediments of the Muschelkalk are of Anisian – Early Ladinian age (Aegean/Bithynian to Longobardian substages, 244–231 Ma according to Menning and Deutsche Stratigraphische Kommission, 2002). The sediments of the Lower Muschelkalk belong to the late Aegean/early Bithynian – early Illyrian substages, whereas the carbonates of the Upper Muschelkalk were correlated by Brack et al. (1999) with the late Illyrian – early Longobardian substages (see also Bachmann et al., 1999).

All samples subjected to investigation under a scanning electron microscope (SEM) have been mounted on conventional aluminum SEM-stubs and sputter-coated with carbon. Investigations have been carried out using a LEO VP 1450 SEM (Institut für Geowissenschaften, Tübingen University). Operating parameters were 15 kV EHT at 2.575. A beam current with a working distance of 15 mm. The elemental composition of the samples was measured using an energy-dispersive X-ray analyzer (EDAX) with a LINK Pentafet Si (Li)-drifted detector crystal (Oxford Instruments). Measure-time was 60 sec per element.

3 Material

This study is based on four representative specimens that all belong to the Ceratitidae (Ceratitoidea, Ceratitina, Ammonoidea):

The first specimen is a whorl fragment consisting of the majority of the body chamber and seven chambers (SMNS 65533). This specimen measures 63 mm in diameter and 20 mm in whorl width. It belongs to *Ceratites flexuosus* Philippi, 1901, and was collected by A. Ehmann in his own quarry at Neckarrems near Stuttgart, Germany. It was collected from marls of the *atavus* Zone, Lower Hauptmuschelkalk Formation, Upper Muschelkalk, late Illyrian, late Anisian, Middle Triassic. This specimen is stored in the Staatliches Museum für Naturkunde in Stuttgart (SMNS).

The second specimen is almost complete except for a little fragment of the phragmocone (MHI, 1856). The mature stage of this individual is manifested by the presence of narrowly spaced riblets near the aperture (see also Klug, 2004). It measures 83.5 mm in diameter and at the posterior end of the body chamber, it is 17.8 mm wide. This specimen of *Ceratites philippii philippii* Riedel, 1916, was collected by H. Hagdorn (Ingelfingen) in the quarry of Schön and Hippelein at Neidenfels near Crailsheim, Germany. It was extracted from approximately 4–8 m below the *Spiriferina*-bed. This interval is situated in the *pulcher* and *robustus* Zones, Lower Hauptmuschelkalk Formation, Upper Muschelkalk, late Illyrian, late Anisian, Middle Triassic (see section in Urlichs and Mundlos, 1988). It is housed in the Muschelkalkmuseum Hagdorn, Ingelfingen, MHI.

The third specimen is a representative of *Ceratites transgressor* Wenger, 1957, (SMNS 65537). Except for the internal whorls and the aperture, this specimen is complete. It measures 107 mm in diameter. Since it is incompletely filled, the precise thickness cannot be measured. This specimen was collected in the abandoned quarry of Wollmershausen near Crailsheim, Germany, from approximately 5–6 m below the *Spiriferina*-bed, *pulcher* to *robustus* Zone, Lower Hauptmuschelkalk Formation, Upper Muschelkalk, late Illyrian, late Anisian, Middle Triassic. This specimen is stored in the Staatliches Museum für Naturkunde in Stuttgart.

The fourth specimen (SMNS 18996–4) comprises two-thirds of the last whorl of the phragmocone and the posterior part of the body chamber of a *Ceratites dorsoplanus* Philippi, 1901. It is a well-preserved steinkern showing fine details on its surface. Its maximal diameter is 203 mm and at the anterior end, the phragmocone is 47 mm wide. The specimen was collected by G. Wagner (Künzelsau). The locality is unknown but speculatively, it comes from northern Baden-Württemberg, Germany. It is of early Longobardian age (early Ladinian) and most likely comes from the *dorsoplanus* Zone, Upper Hauptmuschelkalk Formation, Upper Muschelkalk. This specimen is stored in the Staatliches Museum für Naturkunde in Stuttgart (SMNS).

4 Soft-tissue Attachment Structures

4.1 Large Scar of the Cephalic Retractor

The cephalic retractor muscles represent the most massive muscles in *Nautilus* and probably also in most fossil ectocochleate cephalopods, i.e., nautiloids, bactritoids, and ammonoids. In *Nautilus*, they serve to retract the soft body and also play an important role in locomotion. These muscles insert in the capito-pedal cartilage and form a roof above the mantle cavity (see figures in Doguzhaeva and Mutvei, 1991). By contraction of its longitudinal muscle fibers, the head is pulled back into the shell, exerting pressure on the mantle cavity and thus expelling water out of it.

In contrast to the *Germanonautilus* figured by Klug and Lehmkuhl (2004), there is only one specimen displaying traces of this muscle scar and fewer details are preserved (MHI, 1856: Fig. 10.1). The specimen received attention because of a shining and reflective surface at the posterior end of the body chamber. This surface is subdivided in several irregular spots on both sides of the steinkern. This subdivision was caused by a burrowing organism which left its traces in a polygonal pattern in the posterior two-thirds of the body chamber. It is important to note here that exactly the same ichnotaxon co-occurs with phosphatized soft-tissues in schizodont bivalves of the same age (Klug et al., 2005). The presence of this trace fossil in combination with the shining surface (which is distributed over both sides of the body chamber including also dorsum and venter) led to the presumption that this structure is actually phosphatized. Some EDAX-analyses confirmed this presumption, showing the presence of a significant amount of phosphorus. The shining surface was analyzed at two spots (samples 1 and 2 in Fig. 10.1). These analyses yielded the following results for the elements oxygen, phosphorus, and calcium (in weight percent):

1. Sample 1: O - 48.66; P - 14.87; Ca - 36.46
2. Sample 2: O - 47.99; P - 13.37; Ca - 38.64

Consequently, it appears reasonable that this structure actually represents a formerly, at least partially, organic part of this organism. A map of all the phosphatic spots and surfaces (Fig. 10.2) roughly revealed an outline that very closely resembles the

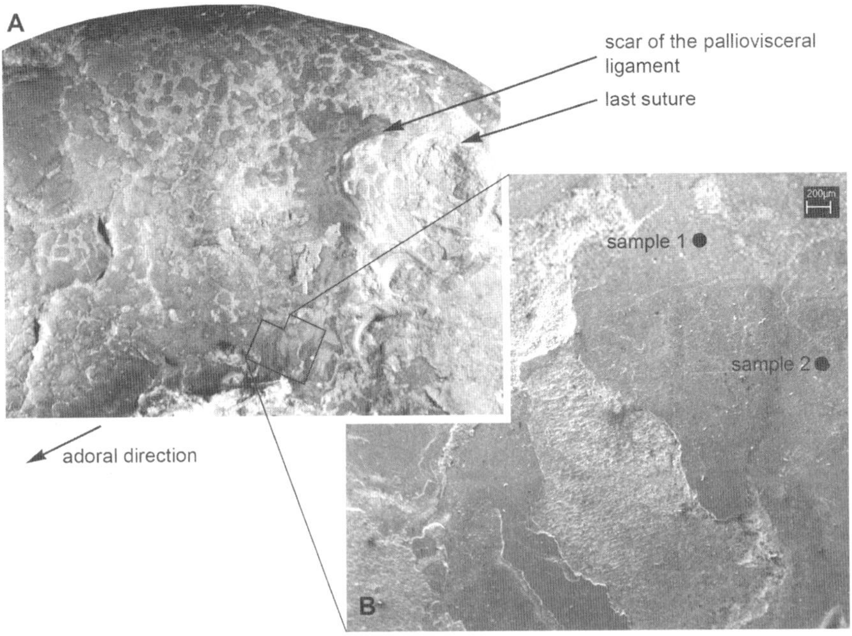

Fig. 10.1 Ceratites philippii philippii *Riedel, 1916, MHI 1856, col. H. Hagdorn, 4–8 m below the* Spiriferina-*bed,* pulcher *to* robustus *Zone, Upper Muschelkalk, late Illyrian, late Anisian, Middle Triassic, Schön and Hippelein quarry at Neidenfels near Crailsheim, Germany. (A) Lateral view of the posterior part of the body chamber showing the phosphatized remains of the annular elevation and the traces in the internal mould. The rectangle approximately demarcates the detail shown in B. x 2. (B) Detail of A to show the smooth surface of the phosphatized remains of the annular elevation. x 5.*

arrangement of muscle scars in other ammonoids such as *Aconeceras* (Doguzhaeva and Mutvei, 1991). According to this map, the shape and position of the "large scar" of the cephalic retractor could be identified in detail. It has a roughly oval outline with a semicircular indentation at its posterior end. The specimen investigated shows the scar of the right cephalic retractor of approximately 17 mm length from the posterior indentation to the most anterior spot and also from its dorsal and ventral limits demarcated by the scars of the myoadhesive band where it is running subparallel to the dorsum and venter, respectively.

In this context, a rare but repeatedly occurring structure found in ammonoids needs to be mentioned. As described and figured by Keupp (2000), internal moulds of ammonoids sometimes display remains of the – sometimes questionably aragonitic– nacreous layer in spots of varying size and shape. In some cases, they are associated with shell injuries. Keupp (2000) explains this fragmentary shell preservation by an increased amount of organic matter in the shell, somehow conserving the aragonite around the injury. Nevertheless, it is striking, that these spots with shell remains often are located in the posterior portion of the body chamber.

In the specimen figured by Keupp (2000: 142), the aragonitic spot is actually very clearly delimited at its posterior end. We suggest that these structures in some cases actually represent phosphatized parts of healed shell injuries and in some other cases they represent aragonitic or phosphatized shell remains of the annular elevation.

4.2 Mantle Myoadhesive Band

The phosphatic surface in the posterior end of the body chamber in specimen MHI 1856 shows the partially well-preserved attachment scar of the mantle myoadhesive band (corresponding to the "anterior band scar" in Isaji et al., 2002) at the anterior edge of the subcircular to bean-shaped attachment area of the cephalic retractor on the right side (Fig. 10.2). This structure was called "functional area of origin of retractor muscles (conchial zone I)" by Mutvei (1957: 225, Fig. 10.3; = "large scar" in Isaji et al., 2002). The scar of the mantle myoadhesive band is conspicuous ventral and dorsal of the large scar; anterior to the large scar, its course can be reconstructed only roughly.

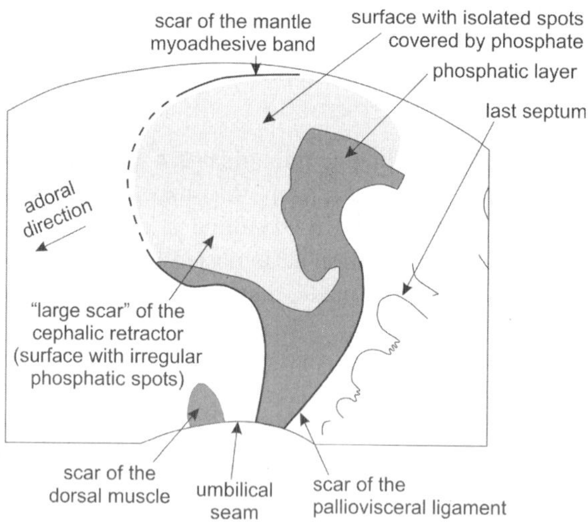

Fig. 10.2 *Line drawing of the posterior part of the body chamber of* Ceratites philippii philippii *Riedel, 1916, MHI 1856, col. H. Hagdorn, 4–8 m below the* Spiriferina*-bed,* pulcher *to* robustus *Zone, Upper Muschelkalk, late Illyrian, late Anisian, Middle Triassic, Schön and Hippelein quarry at Neidenfels near Crailsheim, Germany (see Fig. 1). x 3.*

4.3 Septal Myoadhesive Band

The muscles of the septal portion of the body wall originate at the posterior end of the body chamber ("posterior narrow scar" in Isaji et al., 2002). Between the scar of the mantle myoadhesive band and the last formed septum, an additional line can be seen on the steinkern of specimen MHI 1856. This line is visible on the right flank and the umbilical wall (Figs 10.1, 10.2; "scar of the palliovisceral ligament" in Fig. 10.2). Posterior to the large scar, this scar forms a semicircular bow and then runs more or less straight across the umbilical shoulder and umbilical wall to the umbilical seam. This line demarcates the posterior limit of the phosphatized surface. According to its position, dimension, course, and preservation, it is here supposed to be homologous with the scar of the septal myoadhesive band. Mutvei (1957: 225, Fig. 10.3) called this band "origin of subepithelial musculature of dorsal portion of body proper (conchial zone II)."

4.4 Paired Dorsal Muscle

Anterior to the scar of the mantle myoadhesive band, specimen MHI 1856 displays a parabolic spot which is covered by a phosphatic layer on the umbilical wall (Figs 10.1, 10.2). This spot is delimited by the umbilical seam. It measures 2.6 mm in the longitudinal direction and in the dorsoventral direction, it measures 4 mm. According to its position, shape, and preservation, it is interpreted as one of the scars of the paired dorsal muscle (Doguzhaeva and Mutvei, 1996; Richter, 2002).

4.5 Ventral Muscle

The scar of the unpaired ventral muscle was found in specimen SMNS 65533 (Figs 10.3–10.5). In this specimen, the surface of the internal mould is darkened anterior to the last formed septum ("Präseptalfeld" *sensu* Richter, 2002, compare pl. 7, Fig. 6 therein). This dark coating fades out anteriorly but posteriorly, it clearly delimits an oval spot which displays an area of lighter color. This spot is approximately 5 mm long and 3 mm wide and more or less symmetrical with the plane of symmetry being the same as that of the entire shell. According to its shape, position, and preservation, it is assumed that this spot represents the scar of the unpaired ventral muscle (Doguzhaeva and Mutvei, 1991, 1996; Richter, 2002). A similar structure consisting of a well-developed preseptal field (often dark-colored field in front of the last formed septum within the body chamber) with a lighter central oval spot is preserved in SMNS 65537. In this latter specimen, the preseptal field spans the external lobe and the adjacent ventrolateral lobe and fades out on the ventral flank of the subsequent saddle (Fig. 10.5).

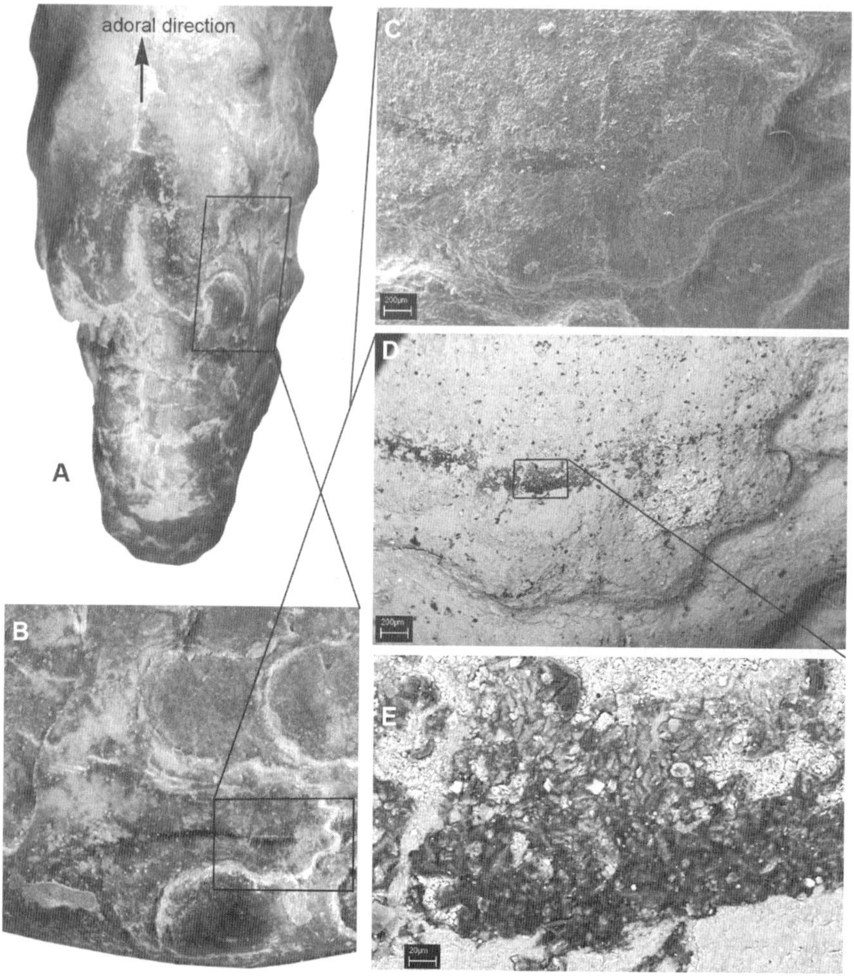

Fig. 10.3 Ceratites flexuosus *Philippi, 1901, SMNS 65533, col. A. Ehmann,* atavus *Zone, Upper Muschelkalk, late Illyrian, late Anisian, Middle Triassic, Neckarrems near Stuttgart, Germany. (A) Ventral view showing the dark preseptal field (adoral of the last suture) surrounding the scar of the unpaired ventral muscle. x 2. (B) Detail of the ventrolateral part of the flank showing the pairs of dark lines in the lobes. x 5. (C) Detail of B showing a dark line with the ventral half of a lobe in a normal SEM image. x 20. (D) Same detail, BSEI image showing the different mineralogical composition of the dark lines. x 20. (E) Detail of D. x 200.*

4.6 Pseudosutures, Drag Bands, and Tension Wrinkles

Pseudosutures (or phantom sutures; Seilacher, 1988) can frequently be seen on the internal moulds of ceratites from the Germanic Basin, but also in other ammonoids (Zaborski, 1986; Hewitt et al., 1991; Weitschat and Bandel, 1991, 1992; Westermann, 1992; Landman et al., 1993; Checa, 1996; Checa and Garcia-Ruiz,

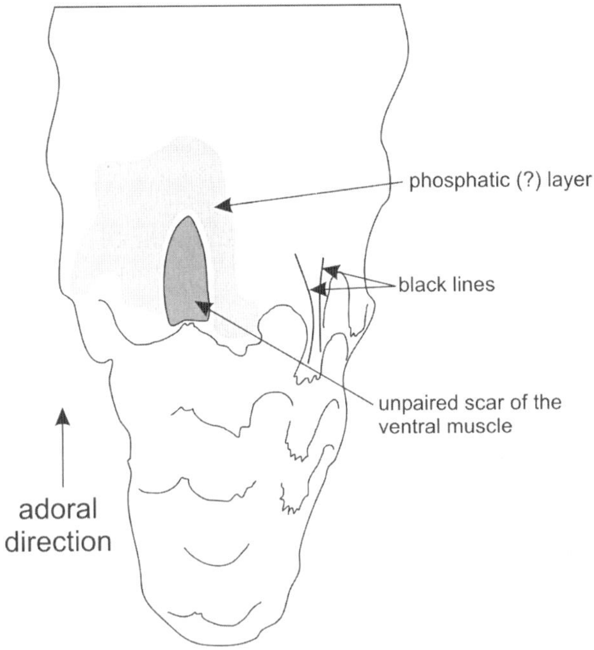

Fig. 10.4 *Line drawing of the posterior part of the body chamber of* Ceratites flexuosus *Philippi, 1901, ventral view, SMNS 65533, col. A. Ehmann, atavus Zone, Upper Muschelkalk, late Illyrian, late Anisian, Middle Triassic, Neckarrems near Stuttgart, Germany (see Fig. 10. 3).*

Fig. 10.5 *Oblique view of* Ceratites transgressor *Wenger, 1957, SMNS 65537, 5–6m below the* Spiriferina-*bed*, pulcher *to* robustus *Zone, Upper Muschelkalk, late Illyrian, late Anisian, Middle Triassic, Wollmershausen near Crailsheim, Germany. Note the dark preseptal field and the paired dark lines in some of the lobes of the last chambers. x 1.*

1996; Doguzhaeva and Mutvei, 1996; Tanabe et al., 1998; Richter, 2002). These are represented by fine lines running subparallel to sutures. Usually, a whole set of such lines is preserved between each pair of real sutures. Apparently, these structures were formed in the course of the translocation of the soft body by the insertion of organic pseudosepta before the insertion of a new septum. In some ammonoids, these lines do not only represent furrows in the steinkern surface but they additionally display a darker color than the surrounding sediment (Vogel, 1959; Hewitt et al., 1991; Richter, 2002).

In the ceratite specimen presented here (SMNS 18996–4), the pseudosutures are extraordinarily well developed (Fig. 10.6). Especially the saddles clearly display sets composed of several pseudosutures, whereas the lobes are associated with drag bands. The pseudosutures are more strongly developed and more closely spaced close to a real septum, possibly implying a faster rate of formation of pseudosutures and/or a deceleration and subsequent acceleration of soft body movement directly before and after the formation of a new septum. Each of the drag bands originates in one spike of the serration of a lobe.

The drag bands are strongest on the venter, ventrolaterally, and in the ventral half of the flanks of the last formed chambers. The fact that pseudosutures and drag bands are almost always combined with lobes and saddles in the same manner points at locally different processes involved in the translocation of the soft body and the formation of the sutural elements (for details see, e.g., Hewitt et al., 1991).

Additional to the pseudosutures and the drag bands, one saddle of the same specimen displays minute wrinkles (Fig. 10.6D). Approximately 10–12 such wrinkles were counted per millimeter. These wrinkles have an appearance similar to imprints of tension wrinkles (see Checa and Garcia-Ruiz, 1996: 290, Fig. 16). It is not completely clear because of the imperfect preservation whether these structures were truly situated on the septum or rather on the inside of the outer shell wall.

4.7 Septal Myoadhesive Band or Pseudosuture?

Another type of structure superficially resembles a combination of drag bands and pseudosutures. They are marked by pairs of faint narrow dark lines within the lobes of *Ceratites* (SMNS 65537 and 65533, Figs. 10.3–10.5). In most cases, these lines are preserved only in the last two to ten chambers. These pairs of lines run more or less parallel to each other and more or less parallel to the flanks of the lobes. They

Fig. 10.6 (continued) *(C) Lateral view of the left side. x 0.3. (D) Detail of E showing the tension wrinkles. x 4. (E) Detail of C, displaying the pseudosutures, drag bands and tension wrinkles. x 0.5. (F) Detail of B, note the drag bands. x 1. (G) Detail of A, showing incompletely filled chamber with serrate winnowing pattern. For the formation of internal moulds of ceratites see Seilacher (1966, 1968, 1971). x 3.*

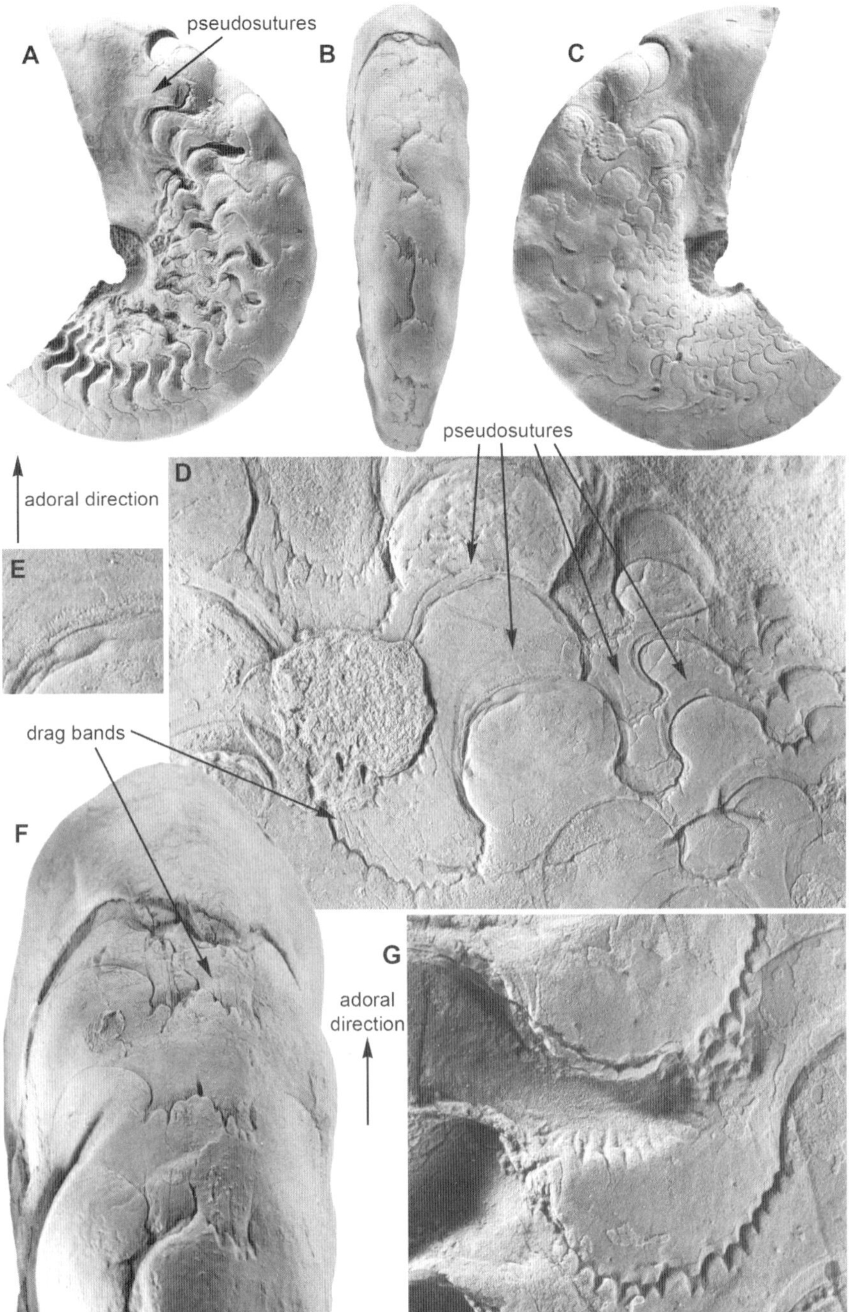

Fig. 10.6 Ceratites dorsoplanus *Philippi, 1901, SMNS 18996–4, col. G. Wagner,* semipartitus *Zone, early Langobardian, early Ladinian, Middle Triassic, speculatively from northern Baden-Württemberg, Germany. (A) Lateral view of the right side. x 0.3.(B) Ventral view. x 0.3.*

form very shallow elevations on the internal moulds. Their color differs from that of the sediment filling of the internal mould. Consequently, it appeared interesting to perform an element analysis again.

For this purpose, BSEI (back scattered electron imagery) pictures were taken of one specimen (SMNS 65533, Fig. 10.3). These images clearly show differences in the chemical composition of the dark lines and the sediment. The subsequent EDAX analyses, however, showed that, in contrast to our expectations, phosphorous is not present. The differences in composition between the dark lines and the surrounding sediment consisted in the higher content of the elements silicon, aluminum, and potassium, in addition to the expected elements oxygen and calcium. This probably reflects the predominance of one or several, possibly authigenic clay minerals which are present in a higher concentration than in the marly limestone of the internal mould. It is assumed that this difference in composition originated from the presence of organic matter.

Finally, the question arises whether this structure was formed during forward movements of the soft body or whether it represents some other kind of soft-tissue attachment structure. Since it is combined with dark-colored preseptal fields and with the scars of the unpaired ventral muscle in two of our specimens, it comes to mind that these structures might rather represent some kind of soft-tissue attachment, especially because only one set of these lines is preserved between two septa. Because the septal myoadhesive band probably runs more or less straight

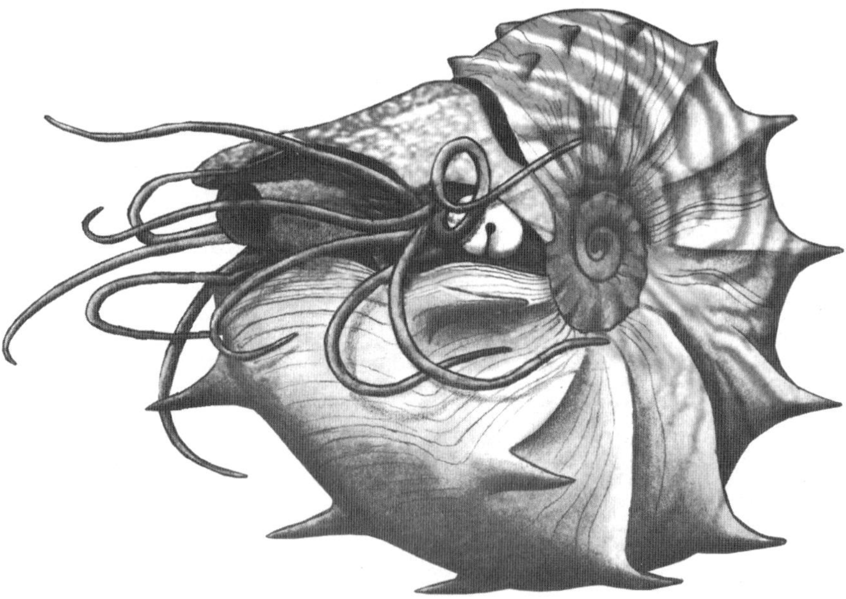

Fig. 10.7 *Reconstruction of* Ceratites spinosus *Philippi, 1901. Modified after Klug et al. (2004). Note the black band. The presence of a hood and the shape and number of arms are largely speculative.*

across the flank (see 4.3), this structure cannot represent a part of the septal myoadhesive band. Therefore, these lines may rather present a pseudosuture.

4.8 Black Layer

The "black layer" represents the attachment structure of the dorsal mantle in Recent *Nautilus* (see Ward, 1987). In this genus, the hood is directly connected with the dorsal mantle and therefore, the presence of the black layer is here considered to be an indication for the possible presence of a non-mineralized hood. In contrast to representatives of the ammonoids *Paraceratites* and *Ceratites* from the Upper Muschelkalk (Middle Triassic; Fig. 10.7), remains of *Germanonautilus* only rarely show remains of the "black layer" (i.e., the attachment structure of the dorsal mantle; Klug et al., 2004). Triassic ammonoids are frequently preserved with remains of the dorsal shell. The organic part of the dorsal shell was documented by Klug et al. (2004) and the mineralized part of the dorsal shell was published by a number of authors (Kummel and Steele, 1962; Kulicki et al., 2001; Keupp, 2000). Among Triassic ammonoids, the dorsal shell or parts of it are reported to be preserved in Triassic deposits of Spitzbergen, South China, Nevada, Germany, Greece, Austria, Siberia, and various other sites. For the forms of the Germanic Basin we refer here to the details shown by Klug et al. (2004).

5 Conclusions

Although rare, soft-tissue attachment structures can be preserved and hence used for a detailed study of the taphonomy and taxonomy of ceratites from the Germanic Basin. In this article, we describe structures visible on the internal moulds of *Ceratites* which probably are homologous to

(1) the "large scar" of the cephalic retractor
(2) the scar of the mantle myoadhesive band
(3) the scar of the septal myoadhesive band
(4) the scar of one of the paired dorsal muscles
(5) the scar of the unpaired ventral muscle
(6) the black layer

Additionally, we figure pseudosutures which consist of pairs of dark stripes on the internal mould. These run subparallel to the flanks of the lobes and apparently are indicated by elements pointing to the post-depositional formation of a clay mineral. The preseptal field, a dark field surrounding the scar of the unpaired ventral muscle, displays a similar preservation. In contrast, the remains of the annular elevation containing the structures listed above except (6) are at least partially phosphatized.

Acknowledgments

We are deeply indebted to H. Hagdorn (Ingelfingen), who generously put an important specimen at our disposal and by doing so, he initiated this study. Some of the photographs were taken by W. Gerber (Tübingen), G. Schweigert, and M. Urlichs (both Stuttgart) helped to find some of the additional material. Hugo Bucher (Zürich), two anonymous reviewers and the editors Royal Mapes (Ohio) as well as Neil Landman (New York) helped to improve the manuscript with their helpful annotations, corrections, and fair comments.

References

Bachmann, G. H., G. Beutler, H. Hagdorn, and N. Hauschke. 1999. Stratigraphie der Germanischen Trias. *In* N. Hauschke, and V. Wilde (editors), *Trias, eine ganz andere Welt: Mitteleuropa im frühen Erdmittelalter*, pp. 81–104. München: Pfeil Verlag.

Brack, P., H. Rieber, and M. Urlichs. 1999. Pelagic successions in the Southern Alps and their correlation with the Germanic Middle Triassic. *Zentralblatt für Geologie und Paläontologie Teil I*, **1998**(7–8): 835–876.

Checa, A. 1996. Origin of intracameral sheets in ammonoids. *Lethaia* **29**: 61–75.

Checa, A., and J. M. Garcia-Ruiz. 1996. Morphogenesis of the septum in ammonoids. *In* N. H. Landman, K. Tanabe, and R. A. Davis (editors), *Ammonoid Paleobiology*, pp. 253–296. New York: Plenum Press.

Crick, G. C. 1898. On the muscular attachment of the animal to its shell in some fossil Cephalopoda (Ammonoidea). *Transactions of the Linnean Society of London, Zoology* **2**(7): 71–113.

Doguzhaeva, L. A., and G. K. Kabanov. 1988. Muscle scars in ammonoids. *Doklady Akademii Nauk SSSR* **301**: 210–212. [in Russian].

Doguzhaeva, L. A., and H. Mutvei. 1991. Organization of the soft body in *Aconeceras* (Ammonitina), interpreted on the basis of shell-morphology and muscle-scars. *Palaeontographica A* **218**: 17–33.

Doguzhaeva, L. A., and H. Mutvei. 1996. Attachment of the body to the shell in ammonoids. *In* N. H. Landman, K. Tanabe, and R. A. Davis (editors), *Ammonoid Paleobiology*, pp. 44–64. New York: Plenum Press.

Griffin, L. E. 1900. The anatomy of *Nautilus pompilius*. *Memoirs of the National Academy of Sciences* **8**: 101–203.

Hagdorn, H., and R. Mundlos. 1983. Aspekte der Taphonomie von Muschelkalk-Cephalopoden. Teil 1: Siphozerfall und Füllmechanismus. *Neues Jahrbuch für Geologie und Paläontologie, Abhandlungen* **16**: 369–403.

Hewitt, R. A., A. Checa, G. E. G. Westermann, and P. M. Zaborski. 1991. Chamber growth in ammonites inferred from color markings and naturally etched surfaces of Cretaceous vascoceratids from Nigeria. *Lethaia* **24**: 271–287.

Isaji, S., T. Kase, K. Tanabe, and K. Uchiyama. 2002. Ultrastructure of muscle-shell attachment in *Nautilus pompilius* Linnaeus (Mollusca: Cephalopoda). *The Veliger* **45**(4): 316–330.

Jones, D. L. 1961. Muscle attachment impressions in a Cretaceous ammonite. *Journal of Paleontology* **35**: 502–504.

Jordan, R. 1968. Zur Anatomie mesozoischer Ammoniten nach den Strukturelementen der Gehäuse-Innenwand. *Beihefte Geologisches Jahrbuch* **77**: 1–64.

Keupp, H. 2000. *Ammoniten – paläobiologische Erfolgsspiralen*. Stuttgart: Thorbecke.

Klug, C. 2001. Functional morphology and taphonomy of nautiloid beaks from the Middle Triassic of southern Germany. *Acta Palaeontologica Polonica* **46**(1): 43–68.

Klug, C. 2004. Mature modifications, the black band, the black aperture, the black stripe, and the periostracum in cephalopods from the Upper Muschelkalk (Middle Triassic, Germany).

Mitteilungen aus dem Geologisch-Paläontologischen Institut der Universität Hamburg **88**: 63–78.

Klug, C., H. Hagdorn, and M. Montenari. in press. Phosphatised soft-tissue in Triassic bivalves from Germany. *Palaeontology* **48**(1): ca. 20 pp.

Klug, C., D. Korn, U. Richter, and M. Urlichs. 2004. The black layer in cephalopods from the German Muschelkalk (Middle Triassic). *Palaeontology* **47**(6): 1407–1425.

Klug, C., and A. Lehmkuhl. 2004. Soft-tissue attachment and taphonomy of the Middle Triassic nautiloid *Germanonautilus*. *Acta Palaeontologica Polonica* **49**(2): 243–258.

Kulicki, C., K. Tanabe, N. H. Landman, and R. H. Mapes. 2001. Dorsal shell wall in ammonoids. *Acta Palaeontologica Polonica* **46**: 23–42.

Kummel, B., and G. Steele. 1962. Ammonites from the *Meekoceras gracilitatus* Zone at Crittenden Springs, Elko County, Nevada. *Journal of Paleontology* **36**: 638–703.

Landman, N. H., K. Tanabe, R. H. Mapes, S. M. Klofak, and J. Whitehill. 1993. Pseudosutures in Paleozoic ammonoids. *Lethaia* **26**: 99–100.

Lehmann, U. 1985. On the dietary habits and locomotion of fossil cephalopods. *In* J. Wiedmann, and J. Kullmann (editors), pp. 633–640. Stuttgart: Schweizerbart.Lominadze, T., M. Sharikadze, and I. Kvantaliani. 1993. On the mechanism of soft body movement within the body chamber in ammonites. *Geobios* **15**: 267–273.

Mapes, R. H., K. Tanabe, and N. H. Landman. 1999. Siphuncular membranes in Upper Paleozoic prolecanitid ammonoids from Nevada, USA. *In* K. Histon (editor), *Cephalopods – present and past. Abstracts volume*, p. 83. Wien: Geologische Bundesanstalt.

Menning, M., and Deutsche Stratigraphische Kommission. 2002. *In* Deutsche Stratigraphische Kommission (editor), *Stratigraphische Tabelle von Deutschland; mit Beiheft, eine geologische Zeitskala*, pp. 1–16. Potsdam: GeoForschungsZentrum Deutschland.

Müller, A. H. 1963. Über Conchorhynchen (Nautil.) aus dem Oberen Muschelkalk des germanischen Triasbeckens. *Freiberger Forschungsheft C* **164**: 5–32.

Müller, A. H. 1969. Nautiliden-Kiefer (Cephalopoda) mit Resten des Cephalopodiums aus dem Muschelkalk des germanischen Triasbeckens. *Monatsberichte der deutschen Akademie der Wissenschaften, mathematisch-naturwissenschaftlichen Klasse* **4**: 308–315.

Müller, A. H. 1974. Über den Kieferapparat fossiler und rezenter Nautiliden (Cephalopoda) mit Bemerkungen zur Ökologie, Funktionsweise und Phylogenie. *Freiberger Forschungsheft C* **298**: 7–17.

Mutvei, H. 1957. On the relations of the principal muscles to the shell in *Nautilus* and some fossil nautiloids. *Arkiv för Mineralogi och Geologi* **10**: 219–253.

Mutvei, H. 1964. Remarks on the anatomy of recent and fossil Cephalopoda. *Stockholm Contributions in Geology* **11**(4): 79–112.

Mutvei, H., J. M. Arnold, and N. H. Landman. 1993. Muscles and attachment of the body to the shell in embryos and adults of *Nautilus belauensis* (Cephalopoda). *American Museum Novitates* **3059**: 1–15.

Philippi, E. 1901. Die Ceratiten des deutschen Muschelkalkes. *Paläontologische Abhandlungen, Neue Folge* **4**: 347–458.

Rakús, M. 1978. Sur l'éxistence de deux types distincts d'empreintes de muscles retracteurs chez les ammonites. *Bulletin de la Société Vaudoise Sciences Naturelles* **354**(74): 139–145.

Rein, S. 1993. Eine Platte mit Kauapparaten der germanischen Ceratiten. *Veröffentlichungen Naturhistorisches Museum Schleusingen* **7/8**: 3–8.

Rein, S. 1995. Organische Lamellen in Steinkernphragmokonen der germanischen Ceratiten. *Veröffentlichungen Naturhistorisches Museum Schleusingen* **14**: 173–184.

Richter, U. 2002. Gewebeansatz-Strukturen auf pyritisierten Steinkernen von Ammonoideen. *Geologische Beiträge Hannover* **4**: 1–113.

Riedel, A. 1916. Beiträge zur Paläontologie und Stratigraphie der Ceratiten des deutschen Oberen Muschelkalkes. *Jahrbuch der Königlich Preuβischen Geologischen Landesanstalt* **37**(1): 1–116.

Seilacher, A. 1966. Lobenlibellen und Füllstruktur bei Ceratiten. *Neues Jahrbuch für Geologie und Paläontologie, Monatshefte* **125** (Festband Schindewolf): 480–488.

Seilacher, A. 1968. Sedimentationsprozesse in Ammonitengehäusen. *Akademie der Wissen-schaften und Literatur, Abhandlungen der Mathematisch-Naturwissenschaftlichen Klasse* **9**: 191–203.

Seilacher, A. 1988. Why are nautiloid and ammonoid sutures so different? *Neues Jahrbuch für Geologie und Paläontologie, Abhandlungen* **177**(1): 41–69.

Tanabe, K., N. H. Landman, and R. H. Mapes. 1998. Muscle attachment scars in a Carboniferous goniatite. *Paleontological Research* **2**(2): 130–136.

Urlichs, M., and R. Mundlos. 1988. Zur Stratigraphie des Oberen Trochitenkalks (Oberer Muschelkalk, beranis) bei Crailsheim. *In* H. Hagdorn (editor), *Neue Forschungen zur Erdgeschichte von Crailsheim. Sonderbände der Gesellschaft für Naturkunde in Württemberg* **1**: 70–84.

Vogel, K. P. 1959. Zwergwuchs bei Polyptychiten (Ammonoidea). *Geologisches Jahrbuch* **76**: 469–540.

Wang, Y., and G. E. G. Westermann. 1993. Paleoecology of Triassic Ammonoids. *Geobios* **15**: 373–392.

Ward, P. D.1987. *The Natural History of Nautilus*. Boston: Allen & Unwin.

Weitschat, W. 1986. Phosphatisierte Ammonoideen aus der Mittleren Trias von Central-Spitzbergen. *Mitteilungen des Geologisch-Paläontologischen Instituts der Universität Hamburg* **61**: 249–279.

Weitschat, W., and K. Bandel. 1991. Organic components in phragmocones of boreal Triassic ammonoids: implications for ammonoid biology. *Paläontologische Zeitschrift* **65**: 269–303.

Weitschat, W., and K. Bandel. 1992. Formation and function of suspended organic cameral sheets in Triassic ammonoids: reply. *Paläontologische Zeitschrift* **66**: 443–444.

Wenger, R. 1957. Die Ceratiten der germanischen Trias. *Palaeontographica A* **108**: 57–129.

Westermann, G. E. G. 1992. Formation and function of suspended organic cameral sheets in Triassic ammonoids – discussion. *Paläontologische Zeitschrift* **66**: 437–441.

Zaborski, P. M. P. 1986. Internal mould markings in a Cretaceous ammonite from Nigeria. *Palaeontology* **29**: 725–738.

Chapter 11
The Preservation of Body Tissues, Shell, and Mandibles in the Ceratitid Ammonoid *Austrotrachyceras* (Late Triassic), Austria

Larisa A. Doguzhaeva,[1] Royal H. Mapes,[2] Herbert Summesberger,[3] and Harry Mutvei[4]

[1] Paleontological Institute, Russian Academy of Sciences, Moscow, Russia, ldoguzhaeva@rambler.ru;

[2] Department of Geological Sciences, Ohio University, Athens, OH 45701, USA, mapes@ohio.edu;

[3] Geologisch-Paläontologische Abteilung, Museum of Natural History, Vienna, Austria, herbert.summesberger@nhm-wien.ac.at;

[4] Department of Palaeozoology, Swedish Museum of Natural History, Stockholm, Sweden, harry.mutvei@nrm.se

Keywords: ammonoids, soft tissues, *Austrotrachyceras*, Late Triassic, ink

1 Introduction

For many years, the discovery of mandibles associated with the shells of ammonoids was considered important to paleobiology, and these occurrences were reported to the paleontological community. Such an association was reported by Trauth (1935b), who described and illustrated mandibles associated with the Late Triassic ceratitid ammonoid *Austrotrachyceras*. The specimens were discovered from the Carnian beds in the northern Calcareous Alps near Lunz (Lower Austria). Observation of his described and illustrated specimens and especially a detailed analysis of previously unstudied specimens from this locality has provided new insight into the preservation of ammonoids.

N. H. Landman et al. (eds.), *Cephalopods Present and Past: New Insights and Fresh Perspectives*, 221–238.

For convenience in the past, organs within the body of a fossil cephalopod including the mantle, stomach, ink sac, mandibles, and even arm hooks have been placed within the concept of "soft tissues" for ease of describing the fossil remains (Stenzel, 1964). In other cases the mandibles and the associated parts of jaws have been considered as "organic remains" (Teichert et al., 1964). Recognizing that different organs may have different ultrastructures, chemical compositions, and preservation potential, we believe that separation of these organs into categories is desirable for a better understanding of the paleobiology of fossil cephalopods.

We recognize three different categories: these are the "soft tissues," the "hard parts," and the "skeleton." We define "soft tissues" as the musculature of the buccal mass and mantle, the crop, stomach and gut, intestines, siphuncular cord, gills, ink sac, and other organs and tissue attachments that are generally within the head and body region of the animal. We separate from the "soft tissues" the category of "hard parts" that includes those structures that are all or partly chitin and some of which may be partly mineralized with calcium carbonate. These structures include the mandibles, the radula, and arm hooks. The "skeleton" includes such structures as the shell, rostrum, the guard, pro-ostracum, and gladius. We and many others have observed that some "hard parts" like the mandibles, have a much greater potential for fossilization than the "soft tissues" such as the internal organs and the tentacles, and that the "skeleton" material has the best fossilization potential of all.

2 Previous Work on Soft Tissues and Hard Parts

2.1 Fossil Mandible and Radula Overview

There is a voluminous literature on cephalopod hard parts, especially the mandibles, which we will not attempt to completely survey here. As a brief overview, Biguet described the first fossil mandible as a rhyncholite in 1819. From that time till the 1960s it was common for paleontologists to describe these parts of cephalopods as new genera and species (see, e.g., Trauth, 1927, 1935a, b) and in the Treatise of Invertebrate Paleontology, Volume K, Teichert et al. (1964) discussed these body parts extensively and provided a comprehensive compilation of the then known genera. More recent reports by Closs and Gordon (1966), Lehman (1987, 1990), Mapes (1987), Dagys et al. (1989), Doguzhaeva et al. (1997) and many others have not followed this naming trend and have generally used simpler terminology with the goal of investigating the modes of preservation, function – especially the operculum versus mandible problem (see Dagys and Dagys, 1975; Morton, 1981; Lehmann, 1970, 1972 a, b, 1987, 1990, and others), and paleobiology of both isolated and in situ occurrences (Closs and Gordon, 1966; Mapes, 1987; Tanabe and Mapes, 1997, and numerous other reports). To our knowledge, this report is the first study of the ultrastructure of fossil mandibles and their chemical makeup where the mandibles have not been mineralogically replaced (for phosphate and other mineral replacements and alteration examples, see Mapes, 1987;

Dagys and Weitschat, 1988). Indeed, as far as we are aware, the only report of the chemical makeup of the mandibles of an externally shelled cephalopod is that of modern *Nautilus* by Lowenstam et al. (1984).

Reports of fossil cephalopod radula are much fewer in number, and most occurrences are in association with mandibles (for examples, see Closs and Gordon, 1966; Lehmann, 1979; Tanabe and Mapes, 1995; Tanabe and Fukuda, 1983; Doguzhaeva and Mutvei, 1992). Many reported radula occurrences are in situ within the body chambers of fossil cephalopods, but some are not (for examples of both, see Saunders and Richardson, 1979). Most radula reports are descriptive, and the research focus has usually been to determine the morphological changes that reflect the evolutionary changes of this important structure.

2.2 Soft Body Tissues Overview

Coleoid soft body tissues, including ink-filled ink sacs and mantle tissue, and hard parts, including mandibles and arm hooks, have been known and reported for over 150 years (Buckland, 1829, in Huxley, 1864). These kinds of descriptive reports have been the research focus over the past decades (see, e.g., Naef, 1922); however, in the last decade new discoveries and interpretations have been made possible by using new technologies on older reposited material when the SEM was employed to elucidate the soft tissues, hard parts, and skeleton structures of different coleoids (see, e.g., Kear et al., 1995, and the numerous works of Doguzhaeva and her colleagues over the last decade).

Soft body tissues in ammonoids have only been reported in a few instances for the Ammonoidea. Lehman (1964) reported the presence of an ink sac with ink in a Jurassic ammonite from Germany; however, he retracted the report in Lehmann (1981). Additionally, he reported the preservation of a stomach and its contents in an ammonite from the same unit. In 2004b, Doguzhaeva and her colleagues described the remains of muscular mantle tissue and possible ink in specimens of the Late Triassic ceratitid *Austrotrachyceras* from the Lower Carnian-Austriacum Zone of Lunz (Schindelberg locality), Lower Austria. Additional material from this locality is now available including specimens with the mandibles near the aperture of the shell, and by using both standard and newer technology, new details are presented herein, as well as information on the ammonoid mandibles and shell material from the two collecting sites.

3 Locality and Material

The specimens studied are shells of the Late Triassic ceratitid *Austrotrachyceras* from the Lower Carnian–Austriacum Zone at Lunz (Schindelberg locality), Lower Austria. More than 100 years ago, the specimens were excavated with the permission of Mr. Haberfelner, the director of the coal mine at Lunz, for the Geological Survey of Austria in 1885 and the Museum of Natural History in Vienna in 1905. Both sites are now inaccessible for collecting (Doguzhaeva et al., 2004b).

Fifteen relatively complete (30–75 mm in diameter) shells in a medium gray shale were available for study. All the ammonoid shells are crushed by compaction with the left and right body chamber walls in more-or-less contact with each other. Shiny, black, asphaltic-like material is located between the body chamber shell walls in places. Despite the fact that the shell of the animals is crushed, the pieces of the shell remain together in a fractured mosaic pattern. Five mandibles are associated with the ammonoid specimens. Three are located in front of or partly within the aperture; a similar condition was described by Krystyn (1991). The body chamber length (bcl) varies depending on the stage of maturity; on specimen 5 the bcl is interpreted to be about 230° at a crushed diameter of 32 mm, and on specimen 6 the bcl is about 270° at a crushed diameter of 48 mm.

4 Purpose of this Study

The purpose of this research is to study the unusual shiny, black, asphaltic-like material that occurs within the body chambers of the shells of *Austrotrachyceras* and the material of the mandibles to determine the composition and ultrastructure. In addition, the black material is compared to the mandibles to determine similarities and differences. The shell ultrastructure is evaluated to determine the overall quality of preservation. To accomplish these goals, (1) energy dispersive spectrometry (EDS), (2) scanning electron microscopy (SEM), and (3) light microscopy were used.

5 Ultrastructure and Preservation of the Soft Tissue, Hard Parts, and Skeleton in *Austrotrachyceras*

5.1 The Black Bituminous Substance in the Body Chamber

The black, bituminous substance in the body chamber (Fig. 11.1A, C–E) looks like shiny pitch that has solidified after being squeezed between the crushed shell walls of the body chamber. The black substance (1) is missing in the chambers of the phragmocone, (2) is only present orad of the last septum, (3) has a somewhat variable thickness and extent, and (4) is restricted to the body chamber and is not present in the surrounding sediment. The black substance is best exposed where the wall of the body chamber is removed when the enclosing shale is split. The black material separates easily from the underlying shell wall and is lighter than the shell material.

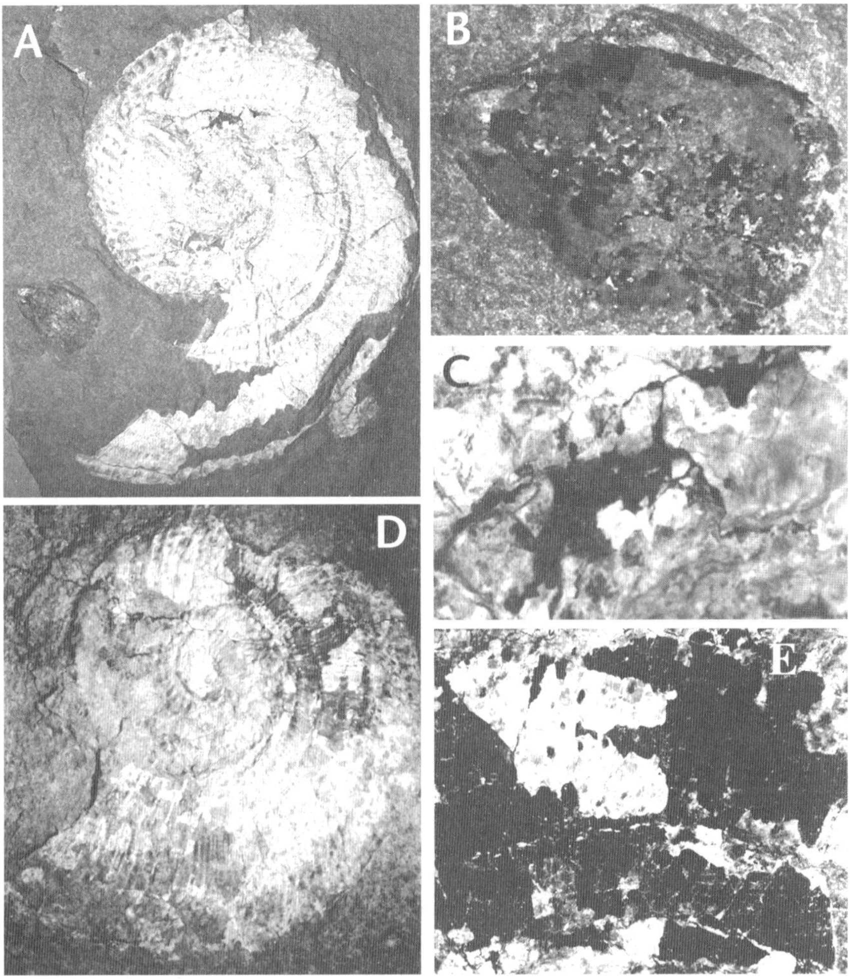

Fig. 11.1A–E Austrotrachyceras *sp. A–C. NHMW 2005z/0006/0006. D,E. NHMW 2005z/0006/0001. A. Shell and mandibles near the aperture. Black material in the shell is exposed where the body chamber shell wall has been removed. x 1.5. B. Enlarged view of the upper and lower mandibles; the lower mandible is poorly exposed around the periphery of the upper mandible. x 5.5 C. Enlarged view of a fragment of the black material in the body chamber. x 5.2, D, E. Lateral view of the compressed shell with black material squeezed in the body chamber and exposed in its middle part where the shell wall is removed. x 1.0. E. Enlarged view of the black material showing its pitch-like appearance. x 3.5.*

The SEM study shows that the black substance forms two sheets that are indistinctly separated by an uneven interspace (Fig. 11.4A, B). These sheets are laminated (Figs. 11.2A, C, 4C), and the laminae are often broken into irregular plates and patches (Figs. 11.4E, F, 6C). Due to differences in preservation, the sheets have a varying ultrastructure: a granular porous ultrastructure (Fig. 11.3B, C), a globular ultrastructure consisting either of irregularly sized globules (Fig. 11.6A) or regularly sized globules (Fig. 11.5C), or an irregular rodlike interconnected ultrastructure (Fig. 11.5D–F). In places, the laminae show a fibrous ultrastructure with the fibers consisting of numerous globules (Figs. 11.2C, 3A) or granules (Figs. 11.4E, F, 6C; see also Doguzhaeva et al., 2004b: Fig. 11.2A). The interspace between the two sheets contain debris that was probably originally organic material and organ pieces, which has been plastically deformed and sometimes folded (Fig. 11.4D). The outer surface of the sheets facing the wall of the body chamber shows a regular honeycomb-like pattern (Doguzhaeva et al., 2004b: Fig. 11.3A–C) with cells the size of which corresponds to that of nacreous tablets of the shell wall (ca. 0.3 μm). The microlamination of the two sheets of black substance, and the fibrous structure of each lamina (Fig. 11.3A), indicate that the sheets were originally a muscular tissue.

The SEM observations on the dispersed material preserved within the interspace between the two sheets of black substance reveal agglomerations of tiny globules (diameter is ca. 0.1–0.4 μm) each of which consists of smaller particles (Fig. 11.6A, B). The globules lack laminations, fibrous patterns, or any other structure. This material with a globular ultrastructure between both sheets of the black substance is comparable to described occurrences of fossil and modern coleoid ink (Doguzhaeva et al., 2002, 2003, 2004a). This globular ultrastructure of the ink in cephalopods has been demonstrated in *Loligo* and in several other undetermined living squids, sepiids, and octopuses and in the fossil ink of Carboniferous, Jurassic, and Cretaceous coleoids (Doguzhaeva et al., 2002; Doguzhaeva and Mutvei, 2003; Doguzhaeva et al., 2003, 2004a). The solidified ink seems to be the result of rapid coagulation of the melanin particles (the main constituent of ink) during precipitation. The ink solidification requires an acid or neutral environment (Fox, 1966). Such a pH environment could be produced by either bacterial decomposition activity on the dead animal or by chemical alteration of the ocean water at the water/sediment interface by bacterial activity or both.

The ultrastructure of the black material in the body chamber was compared to the muscle tissue of the buccal mass in a specimen of the modern squid *Loligo*, which was air dried for one year, and in a specimen of *Nautilus* muscle mantle tissue, which was preserved in alcohol for 20 years. The muscle tissue in the buccal mass of the modern squid shows that a globular ultrastructure similar to the globular ultrastructure seen in the black material had formed. The longitudinal muscles have an almost smooth appearance with a fine granular surface, whereas the transverse muscles have a more robust globular size (Fig.11.7C–D). In *Nautilus*, the muscular mantle tissue is well preserved and has a globular ultrastructure (Fig. 11.7E, F).

The black substance that forms two sheets in the body chamber in *Austrotrachyceras* was compared by SEM with isolated blobs of black substances in the shale that surround the ammonoid shells. These blobs lack a fibrous, globular, or laminar ultrastructure, and thus, they are not related to the black material in the body chambers of the ammonoids.

EDS analysis on one specimen demonstrates that the black substance has the following chemical composition in percents of the total weight: C (60–65%); O (30%); S (2–6%); Si (1–2%); Cd (0.5–1.8%); Fe and K (1%), Al and Zn (each less than 1%) (Doguzhaeva et al., 2004b). The lack of significant amounts of iron

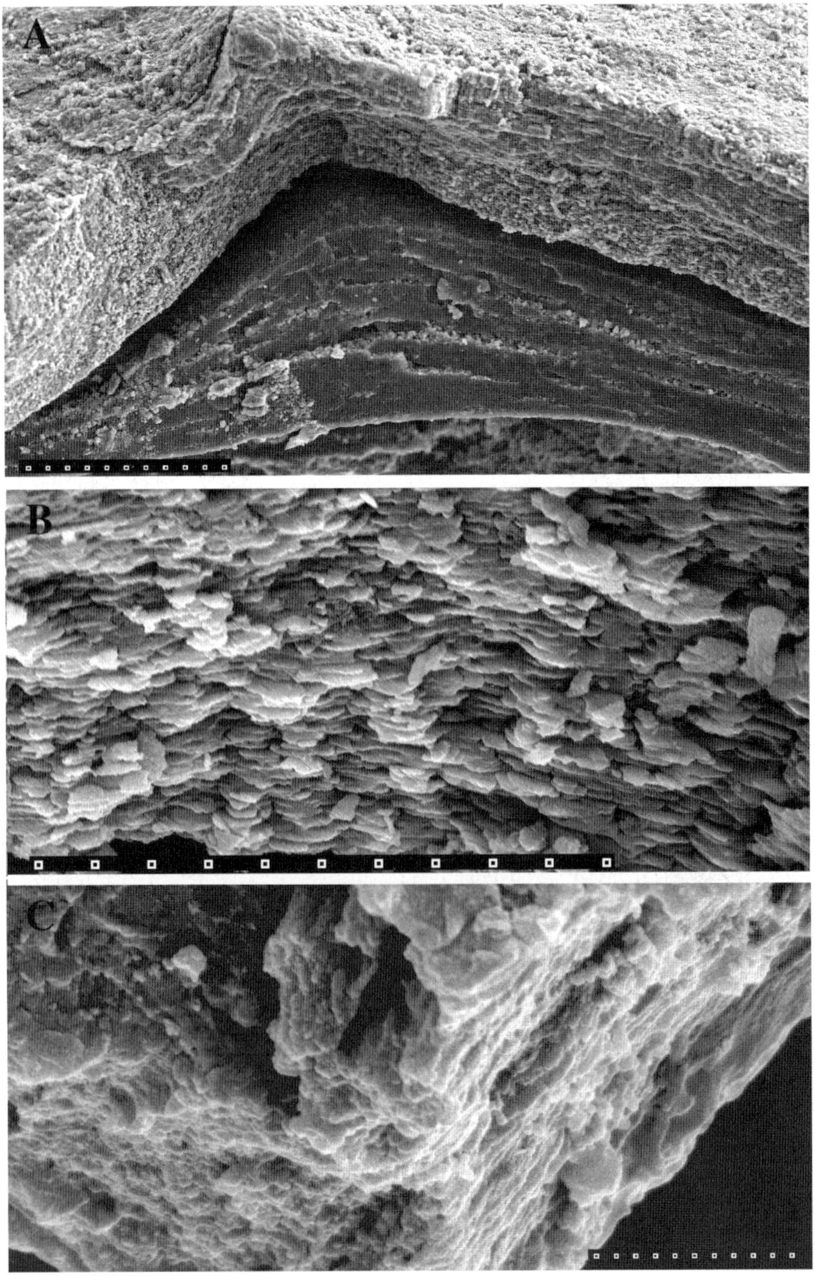

Fig. 11.2A–C Austrotrachyceras *sp. A, B. NHMW 2005z/0006/0002. C. NHMW 2005z/0006/0003. A. Fracture of the shell showing the squeezed black material (bottom) in compressed body chamber (top). The lamination in the middle of the black material is supposed to represent the contact between the left and right sides of the fossilized mantle preserved in the squeezed body chamber. Scale bar = 60 μm. B. Enlargement of the body chamber shell wall composed mostly of the nacreous layer. Scale bar = 10 μm; C. Enlarged view of A showing the microlaminations in the black material. Scale bar = 1.2 μm.*

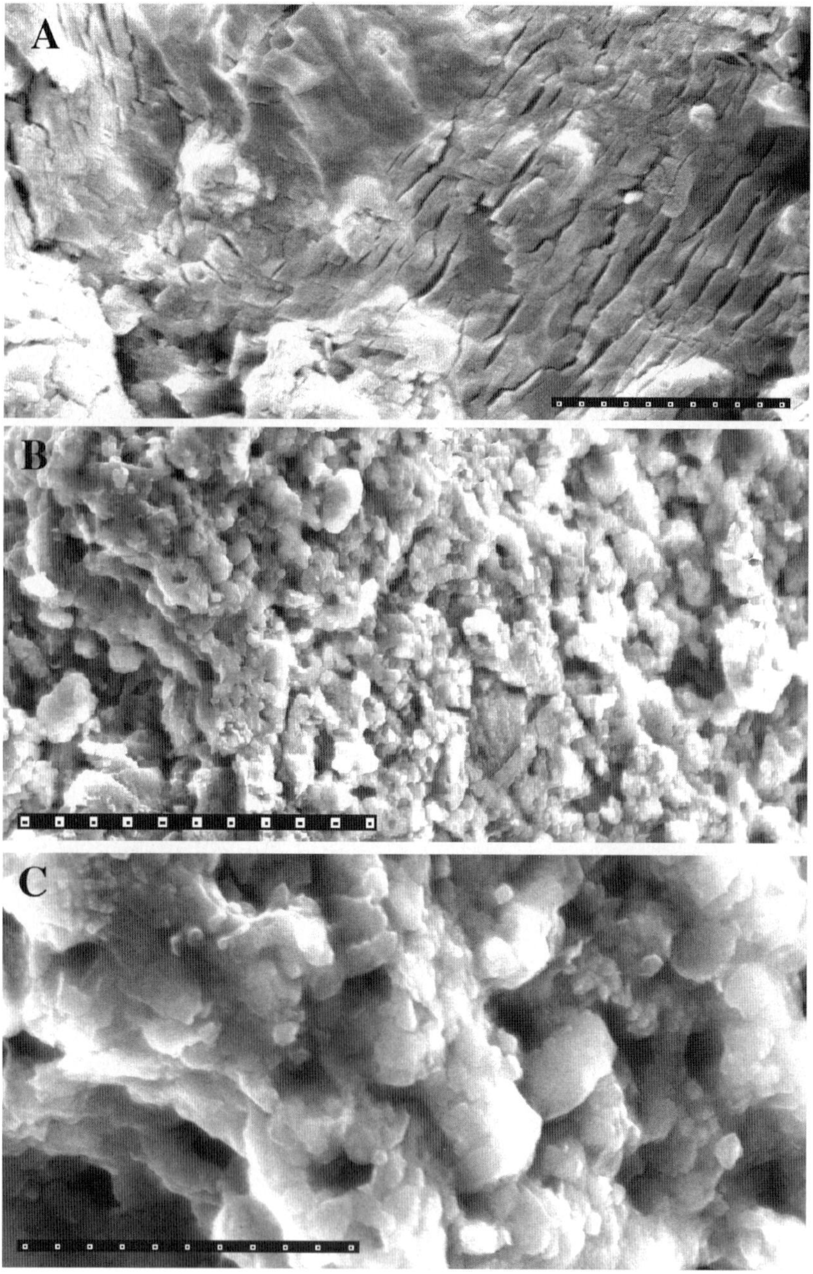

Fig. 11.3A–C Austrotrachyceras *sp., NHMW 2005z/0006/0004. A. Surface view of the black material (supposed fossilized mantle) showing microfractures possibly parallel to the muscle fibers. Scale bar = 3 µm. B. Piece of the mantle tissue showing porous surface. Scale bar = 12.0 µm. C. Enlarged view of B to show tubes perforated by pores. Scale bar = 3.0 µm.*

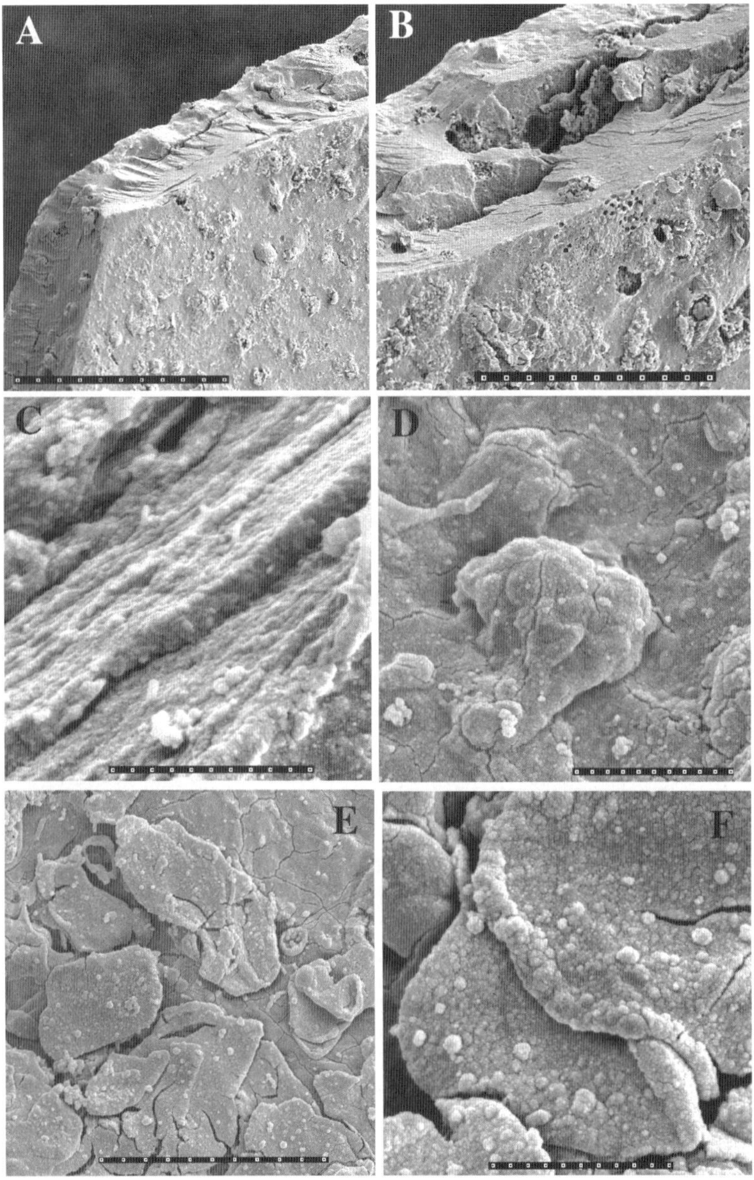

Fig. 11.4A–F Austrotrachyceras *sp., NHMW 2005z/0006/0001. A. Fragment of the black material from the body chamber showing its surface and fractures. Scale bar = 60.0 μm. B. Enlargement of A to show that the black material consists of the left and right portions separated by an interspace about midway between them. The interspace contains possible soft tissue debris. Scale bar = 30.0 μm. C. Fractured surface of the black material to show longitudinal microlamination. Scale bar = 1.2 μm. D. Irregularly shaped swollen structures on the surface of the black material supposed to be deformed soft tissue debris. Scale bar = 1.5 μm. E. Fragment of the black material (supposed fossilized mantle) to show that it was fractured into small broken pieces giving an impression of being originally composed of plastic organic material. Scale bar = 6.0 μm. F. Fractured and irregular plates as seen in E above. The plates have a granular surface. Scale bar = 1.2 μm.*

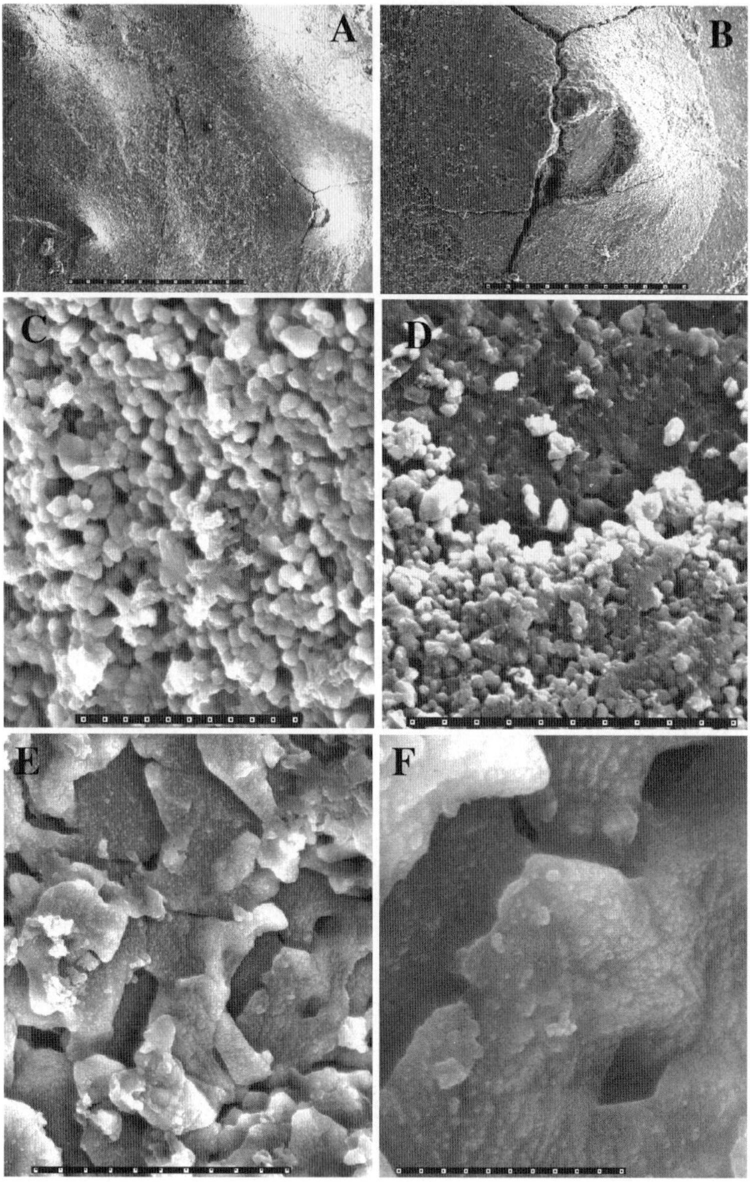

Fig. 11.5A–F Austrotrachyceras *sp., NHMW 2005z/0006/0009. A. Surface view on the exposed black material with the underlying shell wall of the body chamber bearing tubercles. Scale bar = 1.2 µm. B. Enlarged view of the black material coating the inner surface of the tubercle and forming the circular outline around its base. Scale bar = 0.3 mm. C. Porous globular ultrastructure of the black material on the conical surface of the tubercle. Scale bar = 6.0 µm. D. Enlargement of B to show the ultrastructural differences between the black material infilling the tubercle and lining the rest of the inner surface of the body chamber. The lighter material at the base of the photograph has a spherical ball-like globular structure while the upper darkened material has an irregular rodlike interconnected structure. Scale bar = 12.0 µm. E. Close up of the interconnected rodlike structures seen in D. Scale bar =6.0 µm. F. Detailed enlargement showing parts of the irregular rodlike structures and their granular ultrastructure. Scale bar = 1.5 µm.*

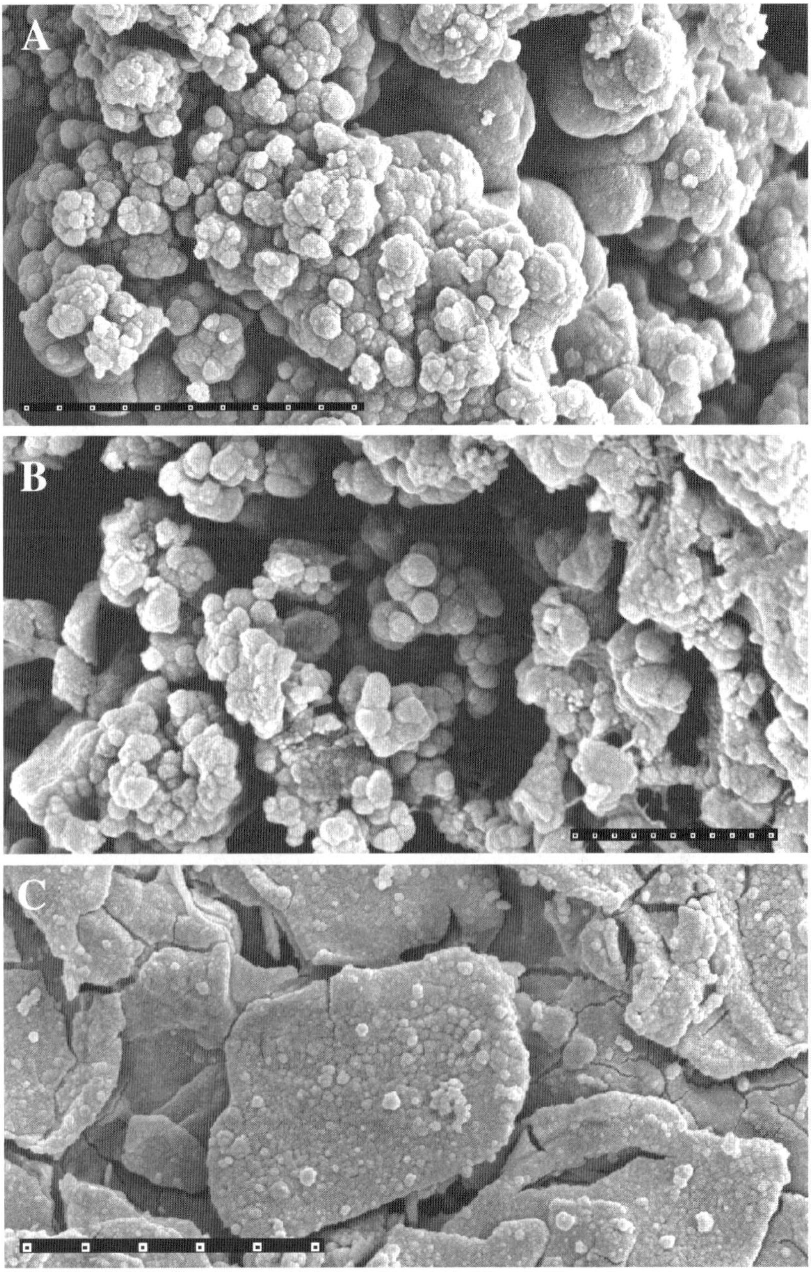

Fig. 11.6A–C Austrotrachyceras *sp. A. NHMW 2005z/0006/0002. View of the fractured black material in places where it consists of larger and smaller irregularly shaped and sized globular particles (supposed to be a mixture of organic debris preserved on the inner surface of the mantle). Scale bar = 3.0 µm. B. NHMW 2005z/0006/0001. View of the fractured black material in places where it consists of dispersed globules (central and left parts of the photo) that are similar in shape to described fossil and Recent ink of coleoids (also supposed to be an ink). Scale bar = 1.2 µm. C. Enlarged detail of Fig. 11.4E to show the granular ultrastructure of the fractured pieces of black material (supposed mantle). Scale bar = 3.0 µm.*

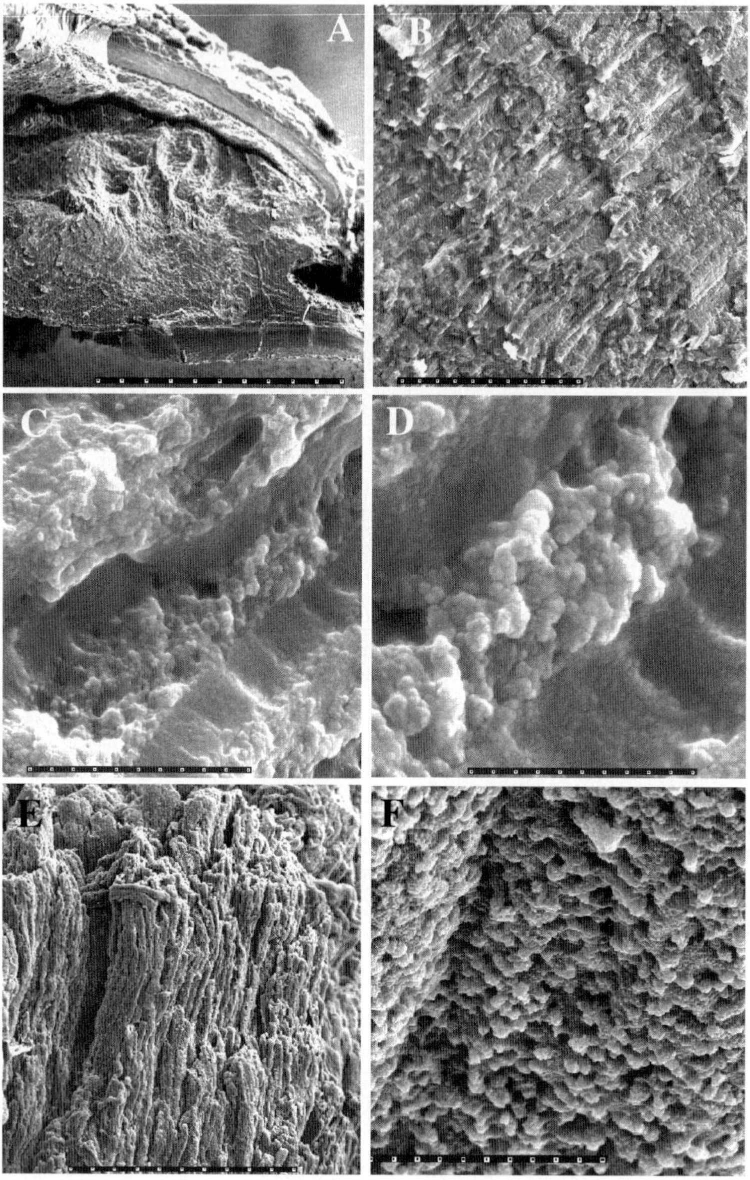

Fig. 11.7A–D *Modern squid* (Loligo) *after one year drying. A. Fracture of the buccal mass showing mandible and muscle preservation. Scale bar = 0.6 mm. B. Fibrous pattern of the buccal mass muscle seen in A, above. Scale bar = 30.0 μm. C. Enlarged view of B to show alternation of longitudinal and transverse muscle. Scale bar = 6.0 μm. D. Enlargement of C showing the transverse muscles with a globular surface bordered by longitudinal muscles with an almost smooth appearance consisting of smaller and less pronounced grains. Scale bar = 1.2 μm. E, F. Recent* Nautilus *after 20 years in 95% ethyl alcohol. Fibrous pattern of the muscular mantle. Scale bar = 120 μm. F. Enlarged view of the muscular mantle showing the interconnected globular ultrastructure with each globule having a granular surface. Scale bar = 12.0 μm.*

and calcium indicates that neither pyrite nor calcite is of importance as preservational elements in this substance. Even more significant is the lack of phosphorus. Many soft tissues in fossil coleoids are commonly replaced by phosphorus in the form of apatite. The lack of this element indicates that the preservation of the muscular mantle tissue and possible ink is not controlled by a phosphorus rich medium.

To summarize, the black substance (1) is restricted to the body chamber and is absent in the chambers of the phragmocone, (2) consists of the left and right sheets that are almost fused but separated by an indistinct interspace containing dispersed organic material, (3) exhibits in places both fine laminations, (4) a fibrous ultrastructure in each lamina, and (5) consists predominantly of carbon. The interspace between the two sheets of the black substance contains isolated agglomerations of tiny globules (ca. 0.1–0.4 μm in diameter); each globule consists of smaller particles. The globule agglomerations lack laminations, fibrous patterns, or any other discernable organized arrangement and are interpreted to be fossil ink. The fine laminations and fibrous ultrastructure of the black sheets, in combination with the high carbon content, are known to be the principal characters of fossilized muscular mantle material. The morphological, ultrastructural, and chemical features listed above allow the interpretation that the sheets of black substance are the bituminous remnants of the mantle squeezed within the compressed body chamber so that their left and right sides became nearly fused during compaction. Due to the fusion of the two portions of the mantle, the dispersed fragments of soft body tissues and organs seem to be partly preserved in the interspace between the sheets. It is likely that the ink sac was ruptured, and ink was dispersed throughout the body cavity and into the mantle tissues. This would explain the dispersed nature of the isolated agglomerations of tiny globules, which are typical of the ultrastructure of ink in other fossil and modern cephalopods. Our observations and conclusions are based on a limited number of specimens. The possible occurrence of ink in ammonoids is an important paleobiological discovery with many implications for the life mode and biology of this extinct group of animals. Previous suggestions that ammonoids had ink have been rejected as inconclusive. However, based on the ultrastructural evidence presented above, we suggest with reservations that it is possible that some ammonoids may have had ink and that it was used as a defensive mechanism. Thus, Lehmann's (1967) report of the occurrence of ink in some ammonoids should not be rejected, and additional specimens should be sought to confirm or reject this possibility in at least some ammonoids.

The replacement of the soft tissues in *Austrotrachyceras* by carbon is probably the result of the metabolism of carbon-accumulating anaerobic bacteria that replaced the organic fibers in the mantle by the globular granules of carbon during the slow fiber decay. The case for slow decay can be supported since the bottom environment in the sediment at and below the sediment/water interface at that time is assumed to be of low oxygen content (Griffith, 1977), and if this was the case, anaerobic bacteria would have been the dominant biological breakdown agent. Additionally, the removal of other chemical constituents other than carbon was probably promoted by liquefying the body tissues and organs by bacterial action, hydrostatic and lithostatic pressures with pore fluid movement, and chemical reactions within the mud surrounding the ammonoid body and shell during diagenesis.

The depositional environment of the Lower Carnian Trachyceras Shale in the Lower Austrian Alps and that of the Posidonia Shale at Holzmaden (both have yielded numerous coleoids with preserved soft parts) was previously believed to have been similar (Seilacher, 1982). However, the EDS analysis of the soft tissues of the cephalopods from both localities does not completely support this depositional environmental interpretation (Doguzhaeva et al., 2004b). In contrast to the carbon-dominated preservation of the soft tissues in the ammonoid *Austrotrachyceras* from the Trachyceras Shale, the soft tissues in the coleoids preserved in the Holzmaden Shale are replaced by phosphorus in the form of phosphate minerals. The geochemical conditions that supported soft tissue preservation by carbon coating and/or carbon concentration by distillization versus phosphate replacement and/or coating have not been precisely determined. However, even though both environments are interpreted to have been a low oxygen environment, the pH of the sediment and pore water in which the dead cephalopods were encased was probably very different in the two shales.

5.2 The Mandible Ultrastructure

EDS analysis on one specimen demonstrates that the mandibles have the following chemical composition in percents of the total weight: C (58–53%); O (30%); S (2%); Si (1–5%); Fe (1–2%), and Ca (less than 1%). The lack of significant amounts of iron and calcium indicates that neither pyrite nor calcite is an important preservational element in these fossils. Even more significant is the lack of phosphorus. Many soft tissues in fossil coleoids and the chitinous mandibles of all fossil cephalopods are commonly replaced by phosphorus in the form of apatite (for examples, see Dagys and Dagys, 1975; Mapes, 1987). However, the EDS data indicate that replacement by apatite or any other phosphorus mineral has not taken place in *Austrotrachyceras*. Thus, the preservation of the mandibles in these ammonoids is by concentrated carbon. For the ultrastructural comparison, the mandibles of modern squid (*Loligo*) were studied with SEM (Fig. 11. 8E, F).

Most mandible studies have utilized standard light microscopy for the descriptive analysis. The only study we are aware of utilizing SEM analysis on

Fig. 11.8A–F (continued) *after one year drying. E. Broken surface of a longitudinal fracture showing the steplike conchoidal fracture surface; faint laminae are present. Scale bar = 60.0 μm. F. Enlargement of the fracture surface of the mandible showing a faint granular texture. Blob with fractures near the center is a SEM "burn" showing that the organic material is very sensitive to even short term exposure of the electron beam. Scale bar = 6.0 μm.*

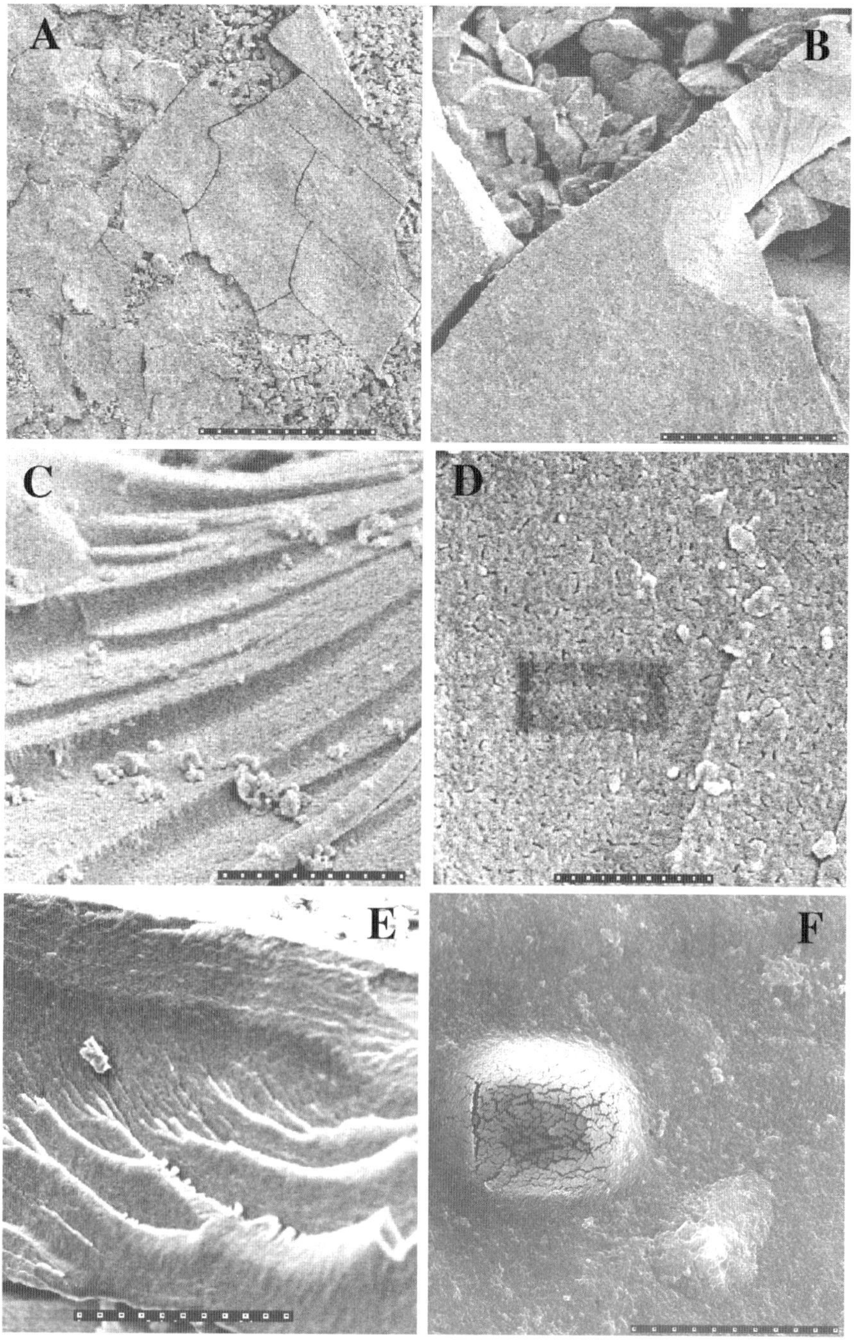

Fig. 11.8A–F *A–D.* Austrotrachyceras *sp. NHMW 2005z/0006/0014, mandibles. A. Mosaic fracture pattern of the mandible. Scale bar = 0.3 mm. B. Enlarged detail of A showing the conchoidal fracture surface. Scale bar = 30.0 µm. C. Enlargement of the fracture seen in B showing the stepped fracture surface. Scale bar = 3.0 µm. D. Fracture surface showing the ultrastructure of microporosity of small microslits. Scale bar = 1.5 µm. E, F. Modern squid* (Loligo) *mandibles*

ammonoid mandibles is the present report. It could be concluded that the original material was organic and that the carbonization had taken place to preserve the mandibles in situ.

5.3 The Shell Ultrastructure

The nacreous layer forms the main bulk of the shell wall in the body chamber of this genus of ammonoid and these layers give a bright iridescent luster to the shell. SEM analysis on one specimen demonstrates that the shell material is composed of nacreous plates and that the shell is well preserved and retains its original ultrastructure despite having been extensively crushed by diagenetic forces (Fig. 11.1 A, D).

EDS analysis demonstrates that the body chamber wall has the following chemical composition in percents of the total weight: C (6–8%); O (37–41%); and Ca (42–48%). In places, no trace elements were detected. This indicates that the shell is extremely pure calcium carbonate, which is most probably in the form of aragonite. In addition, the lack of trace elements indicates that the specimen has undergone virtually no replacement during fossilization. This then supports the conclusion that the entire animal, including the shell, mandibles, muscular mantle tissue, and the possible ink are not produced by chemical introduction from outside the specimen.

6 Conclusions

In summary, despite the diagenetic crushing of the shells, the preservational condition of the specimens is outstanding. The presence of the mandibles in life position indicates that the soft body was partly retained within the body chamber of some ammonoid specimens when the animal fell to the ocean bottom after death. Burial must have been rapid to preserve the body and jaws of the ammonoid within and in front of the shell, respectively. Preservation of the mantle, possible ink, and fragments of the internal organs between the two sides of the preserved muscular mantle began immediately after burial. Carbonization of the tissues by bacteria had already begun before the ammonoid shell was crushed. Crushing brought the mantle coatings on the left and right sides of the body chamber in contact. The fossilization process of carbonization continued to completion after crushing. Based on the pristine condition of the shell nacre in the body chamber and the EDS data, these ammonoid specimens with their preserved soft and hard tissues remained sealed from additional alteration by bacteria, pore water migration, weathering, and other geologic events until they were excavated more than 100 years ago. Given the preservation of the soft tissues in an externally shelled cephalopod and the rarity of such discoveries over the past 200 years of cephalopod fossil research, one must consider the preservation of muscular mantle tissue, possible ink, and mandibles that are carbonized and not replaced, as a truly extraordinary occurrence.

References

Biguet, F. 1819. Caractère du Rhyncholyte: considérations sur les Belemnites. *suivies d'un essai de Belemnitologie Synoptique*. Lyon, France: Kindelem.

Closs, D., and M. Gordon, Jr. 1966. An Upper Paleozoic goniatite radula. *Rio Grande do Sul, Universidade, Escola de Geologia, Notas e Estudos* **1, 2**: 79–80.

Dagys, A. S., and A. A. Dagys. 1975. The morphology and functional significance of anaptychi. *Paleontological Journal* **9**: 180–192.

Dagys, A., U. Lehmann, K. Bandel, K. Tanabe, and W. Weitschat. 1989. The jaw apparati of ectocochleate cephalopods. *Paläontologische Zeitschrift* **63**: 41–53.

Dagys, A. S., and W. Weitschat. 1988. Ammonoid jaws from the Boreal Triassic realm (Svalbard and Siberia). *Mitteilungen aus dem Geologisch-Paläontologischen Institut der Universität Hamburg* **67**: 53–71.

Doguzhaeva, L. A., R. H. Mapes, and H. Mutvei. 2003. The shell and ink sac morphology and ultrastructure of the Late Pennsylvanian cephalopod *Donovaniconus* and its phylogenetic significance. *Berliner Paläobiologische Abhandlungen* **3**: 61–78.

Doguzhaeva, L. A., R. H. Mapes, and H. Mutvei. 1997. Beaks and radulae of Early Carboniferous goniatites. *Lethaia* **30**: 305–313.

Doguzhaeva, L. A., R. H. Mapes, and H. Mutvei. 2004a. Occurrence of ink in Paleozoic and Mesozoic coleoids (Cephalopoda). *Mitteilungen aus dem Geologisch-Paläontologischen Institut der Universität Hamburg* **88**: 145–155.

Doguzhaeva, L. A., R. H. Mapes, H. Mutvei, and R. K. Pabian. 2002. The Late Carboniferous phragmocone-bearing orthoconic coleoids with ink sacs: their environment and mode of life. *First International Palaeontological Congress*, *Geological Society of Australia, Abstracts* **68**: 200.

Doguzhaeva, L. A., and H. Mutvei. 1992. Radula of the Early Cretaceous ammonite *Aconeceras* (Mollusca: Cephalopoda). *Palaeontographica A* **223**: 167–177.

Doguzhaeva, L. A., and H. Mutvei. 2003. Gladius composition and ultrastructure in extinct squid-like coleoids: *Loligosepia, Trachyteuthis* and *Teudopsis*. *Revue de Paléobiologie* **22**: 877–894.

Doguzhaeva, L. A., H. Mutvei, H. Summesberger, and E. Dunca. 2004b. Bituminous soft body tissues in Late Triassic ceratitid *Austrotrachyceras*. *Mitteilungen aus dem Geologisch-Paläontologischen Institut der Universität Hamburg* **88**: 37–50.

Fox, D. L. 1966. Pigmentation of molluscs. *In* K. M. Wilber, and C. M. Young (editors), *Physiology of Mollusca,* pp. 249–274. New York: Academic Press.

Griffith, J. 1977. The Upper Triassic fishes from Polzberg bei Lunz. *Zoological Journal of the Linnean Society* **60**: 1–93.

Huxley, T. H. 1864. On the structure of the Belemnitidae; with a description of a more complete specimen of *Belemnites* than any hitherto known, and an account of a new genus of Belemnitidae *Xiphoteuthis*. *Memoir of the Geological Survey of the United Kingdom*, Monograph II: 1–22.

Kear, A. J., D. E. G. Briggs, and D. T. Donovan. 1995. Decay and fossilization of non-mineralized tissue in coleoid cephalopods. *Palaeontology* **38**: 105–131.

Krystyn, L. 1991. Die Fossillagerstätten der alpinen Trias. *In* W. Vasicek, L. Krysten, D. Nagel, and G. Rabeder (editors), *Exkursionen im Jungpaläozoikum und Mesozoikum Österreichs. Österreichischen Paläontologischen Gesellschaft*: 23–78.

Lehmann, U. 1967. Ammoniten mit Tintenbeutel. *Paläontologische Zeitschrift* **41**: 132–136.

Lehmann, U. 1970. Lias-Anaptychen als Kieferelemente (Ammonoidea). *Paläontologische Zeitschrift* **44**: 25–31.

Lehmann, U. 1972a. Jaws, radula, and crop of *Arnioceras* (Ammonoidea). *Palaeontology* **14**: 338–341.

Lehmann, U. 1972b. Aptychen als Kieferelemente der Ammoniten. *Paläontologische Zeitschrift* **46**: 34–48.

Lehmann, U. 1979. The jaws and radula of the Jurassic ammonite *Dactylioceras*. *Palaeontology* **22**: 265–271.

Lehmann, U. 1981. *The Ammonites: Their Life and Their World*. London: Cambridge University Press.

Lehmann, U. 1987. *Ammoniten, ihr Leben und ihre Umwelt, 2*. Stuttgart: Enke-Verlag.

Lehmann, U. 1990. *Ammonoideen*. Stuttgart: Enke-Verlag.

Lowenstam, H. A., W. Traub, and S. Weiner. 1984. *Nautilus* hard parts: a study of the mineral and organic constituents. *Paleobiology* **10**: 268–279.

Mapes, R. H. 1987. Upper Paleozoic cephalopod mandibles: frequency of occurrence, modes of preservation, and paleoecological implications. *Journal of Paleontology* **61**: 521–538.

Morton, N. 1981. Aptychi: the myth of the ammonite operculum. *Lethaia* **14**: 57–61.

Naef, A. 1922. *Die Fossilen Tintenfische*. Jena: Gustav Fischer.

Saunders, W. B., and E. S. Richardson. 1979. Middle Pennsylvanian (Desmoinesean) Cephalopoda of the Mazon Creek fauna, northeastern Illinois. *In* M. H. Nitecki (editor), *Mazon Creek Fossils*, pp. 333–359. NewYork: Academic Press.

Seilacher, A. 1982. Posidonia Shale (Toarcian, S. Germany) – stagnant basin model revalidated. *In* E. Montanaro-Gallitelli (editor), *Palaeontology: Essentials of Historical Geology*, pp. 25–55. Modena: S.T.E.M. Mucchi.

Stenzel, H. B. 1964. Living *Nautilus*. *In* R. C. Moore (editor), *Treatise on Invertebrate Paleontology, Volume K, Mollusca 3*, pp. K59–K93. New York and Lawrence, KS: Geological Society of America and University of Kansas Press.

Tanabe, K., and Y. Fukuda. 1983. Buccal mass structure of the Cretaceous ammonite *Gaudryceras*. *Lethaia* **16**: 249–256.

Tanabe, K., and R. H. Mapes. 1995. Jaws and radula of the Carboniferous ammonoid *Cravenoceras*. *Journal of Paleontology* **69**: 703–707.

Teichert, C., R. C. Moore, and D. E. Nodine Zeller. 1964. Rhyncholites. *In* R. C. Moore (editor), *Treatise on Invertebrate Paleontology, Volume K, Mollusca 3*, pp. K467–K484. New York and Lawrence, KS: Geological Society of America and University of Kansas Press.

Trauth, F. 1927. Über die Aptychen im Allgemeinen. *Annalen des Naturhistorischen Museums in Wien* **41**: 171–259.

Trauth, F. 1935a. Die Aptychen des Palaozoicums. *Jahrbuch der Preussischen Geologischen Landesanstalt* **55**: 44–83.

Trauth, F. 1935b. Die Aptychen der Trias. *Akademie der Wissenschaft Wien Sitzungsberichte mathemathisch-naturwissenschaft* **144**: 455–482.

Chapter 12
Connecting Ring Ultrastructure in the Jurassic Ammonoid *Quenstedtoceras* with Discussion on Mode of Life of Ammonoids

Harry Mutvei[1] and Elena Dunca[1]

[1]Department of Palaeozoology, Swedish Museum of Natural History, S-10405, Stockholm, Sweden, harry.mutvei@nrm.se, elena.dunca@nrm.se

Keywords: ammonoids, connecting ring, ultrastructure, mode of life

1 Introduction

It was generally assumed (e.g., Mutvei, 1967, 1975; Grégoire, 1984; Westermann, 1971, 1982, 1993, 1996; Bandel, 1981; Obata et al., 1980; Tanabe, 1977, 1979; Hewitt et al., 1993; Kulicki, 1994; Hewitt, 1996) that the connecting ring in prosiphonate ammonoids, in which the septal necks are directed adorally, had similar structure and composition as that in the retrosiphonate Recent *Nautilus,* in which the septal necks are directed adapically, and in which the bathymetry is well known. Based on this assumption, Westermann (1971) calculated the mechanical strength of the ammonoid siphuncle against hydrostatic pressure (siphuncular strength index) from the siphuncular wall thickness, multiplied by 100 and divided by the siphuncle inner radius. This index was thought to indicate the maximum water depth where ammonoid shell imploded and, hence, the maximum habitat depth of the animal. In addition,

N. H. Landman et al. (eds.), *Cephalopods Present and Past: New Insights and Fresh Perspectives*, 239–256.

the postmortem distribution of ammonoids, and the water depth of the marine sediments where the ammonoid shells accumulated were calculated from this index.

However, as recently demonstrated (Mutvei et al., 2004), the connecting ring structure and composition in two Mesozoic genera, *Aconeceras* and *Grammoceras*, are considerably different from those in Recent *Nautilus*. The connecting ring in these ammonoids is composed of glycoprotein fibers that probably were hardened by phenolic tanning (=sclerotization) in a similar way as reported in crustacean exoskeletons (Dennell, 1960; Florkin, 1960; Passano, 1960; Dalingwater and Mutvei, 1990).

In order to make the connecting ring permeable for cameral liquid, it was traversed by numerous narrow pore canals that probably housed microvillous-like extensions from the siphuncular epithelium (Mutvei et al., 2004). The ammonoid connecting ring is therefore often preserved in fossil environments where the untanned, organic matter in the connecting rings of fossil nautiloids and coleoids is destroyed during diagenesis.

In the present paper, the connecting ring structure is described in the Jurassic ammonoid *Quenstedtoceras* spp. that were recovered from erratic boulders at Luckow, Poland. This structure is compared with that in nautiloid and coleoid cephalopods. The mechanical strength of the connecting ring and the mode of life in ammonoids, particularly swimming by jet propulsion, are discussed.

2 Material and Methods

Shells of *Quenstedtoceras* spp. (Cardioceratidae) were collected by the senior author from calcareous concretions of an erratic boulder at a brick factory near Luckow, eastern Poland (Makowski, 1962). The concretions are of Middle Jurassic, Late Callovian age, and come from the *Quenstedtoceras lamberti* Zone.

Shells of *Aconeceras trautscholdi* were collected by L. A. Doguzhaeva from sideritic concretions in the Lower Cretaceous at the Volga River, near the town Uljnovsk, Russia.

Numerous shells, partially preserved with empty chambers, were mechanically fractured. The exposed siphuncle segments were placed on specimen stubs, gold coated, and studied with a Hitachi S 4300 scanning electron microscope at the Swedish Museum of Natural History, Stockholm. The elemental composition in the connecting ring and shell was analyzed with Energy Dispersive X-Ray Microanalysis (EDAX).

3 Description

3.1 Connecting Ring Structure

Obata et al. (1980), Tanabe et al. (1982), Westermann (1982), and Grégoire (1984) pointed out that the ammonoid connecting ring is composed of glycoprotein fibers that form thin lamellae. These fibers are similar to those in the inner layer of the connecting ring in Recent *Nautilus*, but in ammonoids they seem to have been hardened by tanning and form a compact, semielastic structure that is resistant against diagenesis (Mutvei et al., 2004). The glycoprotein fibers in ammonoids are diagenetically transformed into chain-like aggregates of tiny granules (Fig. 12.1) or globular aggregates (Doguzhaeva and Mutvei, 2003; Mutvei et al., 2004). Structurally similar transformations were observed in muscular fibers of fossil coleoids, and in fibrous organic material of fossil coleoid gladii and arm hooks (Doguzhaeva and Mutvei, 2003).

Two structurally different sublayers can often be distinguished on the vertical fracture planes of the connecting rings in *Quenstedtoceras* (Fig. 12.2E). The outer sublayer differs from the inner sublayer by its more compact structure, probably because the glycoprotein fibers are thinner and more closely packed than those in the inner sublayer (Fig. 12.2E, F). The boundary between the two sublayers is distinct and often forms a narrow interspace. Two structurally similar sublayers occur in the connecting ring of *Grammoceras quadratum* (Fig. 12.2C; Mutvei et al., 2004: Fig. 12.3A). In most connecting rings the glycoprotein fibers in the inner sublayer are diagenetically transformed into chains of large-sized granules, 0.2–0.3 µm in diameter. The fibrous-lamellar structure of the connecting ring is indistinctly visible. Only in well-preserved connecting rings where the glycoprotein fibers consist of smaller granules, about 0.1 µm in diameter or less, is the fibrous-lamellar structure distinctly recognizable (Figs. 12.1B, C, 12.2F, G, 12.3C).

Both sublayers of the connecting ring are traversed by numerous narrow pore canals (Figs. 12.1B, C, 12.2F, G, 12.3C). They can only be observed in the best preserved connecting rings, but even then, the granular structure of the glycoprotein fibers more or less obscures their outlines. The pore canals have a diameter of about 0.1 µm and the distances between adjacent canals are about 1.0 µm. The density of the pore canals in *Quenstedtoceras* (Figs. 12.1B, C, 12.3C) seems to be somewhat higher than that in *Aconeceras* (Figs. 12.2A, B, 12.3A) and *Grammoceras* (Fig. 12.2C, D) in which about 500,000 canals occur per mm^2. The density of the pore canals in *Aconeceras* and *Grammoceras* is much higher than previously estimated by Mutvei et al. (2004).

As previously pointed out (Mutvei et al., 2004: Fig.12.4A, B), the diameter and density of the pore canals in an ammonoid connecting ring roughly correspond to those of the pore canals in arthropod exoskeletons (compare Fig. 12.3A, B). The pore canals in arthropod exoskeletons house microvillous-like epithelial extensions (Mutvei et al., 2004: Fig. 12.4B; Compére and Goffinet, 1987). It is reasonable to conclude that similar pore canals in ammonoid connecting rings also contained epithelial extensions (Mutvei et al., 2004: Fig. 12.4A). These extensions participated in the hardening processes of the connecting ring by tanning and created a considerably enlarged, physiologically active surface of the siphuncular epithelium for efficient removal of cameral liquid.

3.2 Elemental Composition

Energy Dispersive X-Ray Microanalysis (EDAX) showed that the siphonal tube of *Quenstedtoceras* is rich in phosphorus, while the shell and septa are made of

Fig. 12.1 Quenstedtoceras *sp. A – General view of connecting ring. B – Vertical fracture plane of the connecting ring. C – Detail of B in higher magnification to show the granular structure of the glycoprotein fibers and the pore canals. Scale bars: A – 0.1 mm; B, C – 6 μm.*

calcium carbonate. A total of 25 points were analyzed as follows: 8 points on the outer surface of the shell and inner surface of the septa, 16 points on the connecting ring and 1 point on the septal neck. The analysis results are shown in Table 12.1.

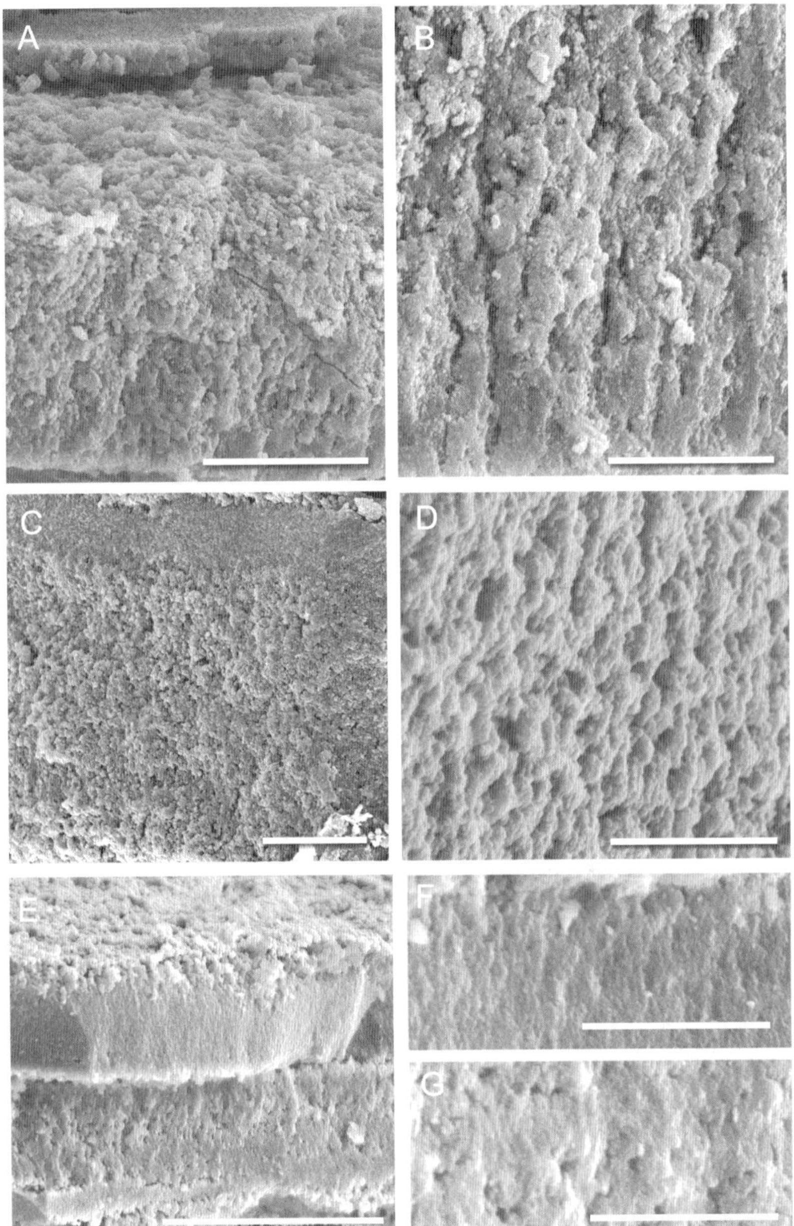

Fig. 12.2 *A, B.* Aconeceras trautscholdi. *A – Vertical fracture plane of connecting ring to show pore canals. B – Detail of A in higher magnification. C, D.* Grammoceras quadratum. *C – Vertical fracture plane of connecting ring to show the outer sublayer with compact structure and the inner sublayer with less compact structure. D – Detail of C in higher magnification to show the pore canals. E–G.* Quenstedtoceras *sp. E – Vertical fracture plane of connecting ring to show the two sublayers. F, G – Outer and inner sublayer, respectively, in higher magnification to show the granular structure of glycoprotein fibers and pore canals. Scale bars: A, C, E – 12 μm; B, D – 6 μm; F, G – 3 μm.*

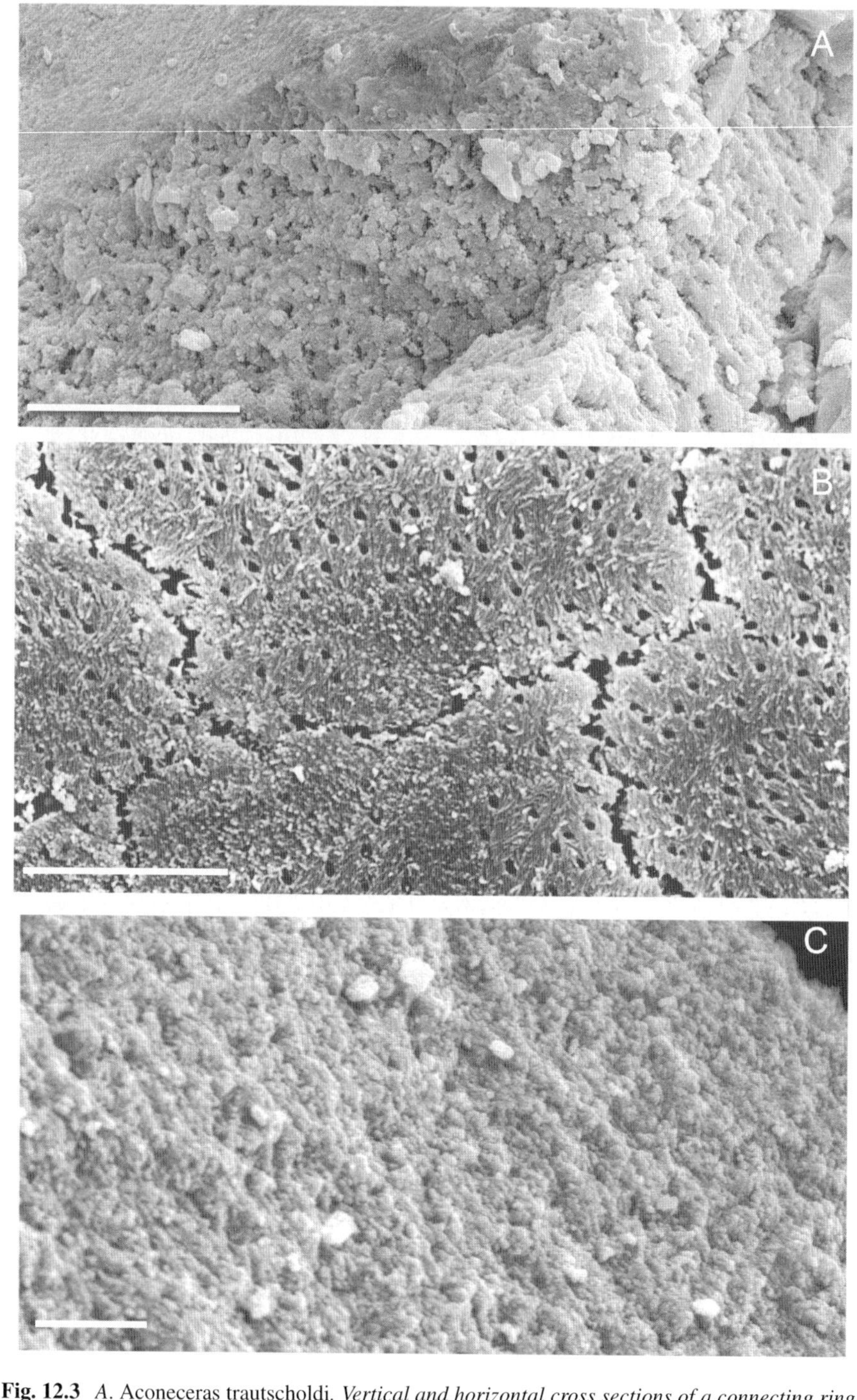

Fig. 12.3 *A*. Aconeceras trautscholdi. *Vertical and horizontal cross sections of a connecting ring to show the pore canals. B*. Carcinus maenas (*Crustacea*). *Horizontal fracture plane of the exoskeleton to show the pore canals and glycoprotein fibers. C*. Quenstedtoceras *sp. Oblique vertical fracture plane of the connecting ring to show granular structure and pore canals. Scale bars: A, B – 10 μm; C – 1 μm.*

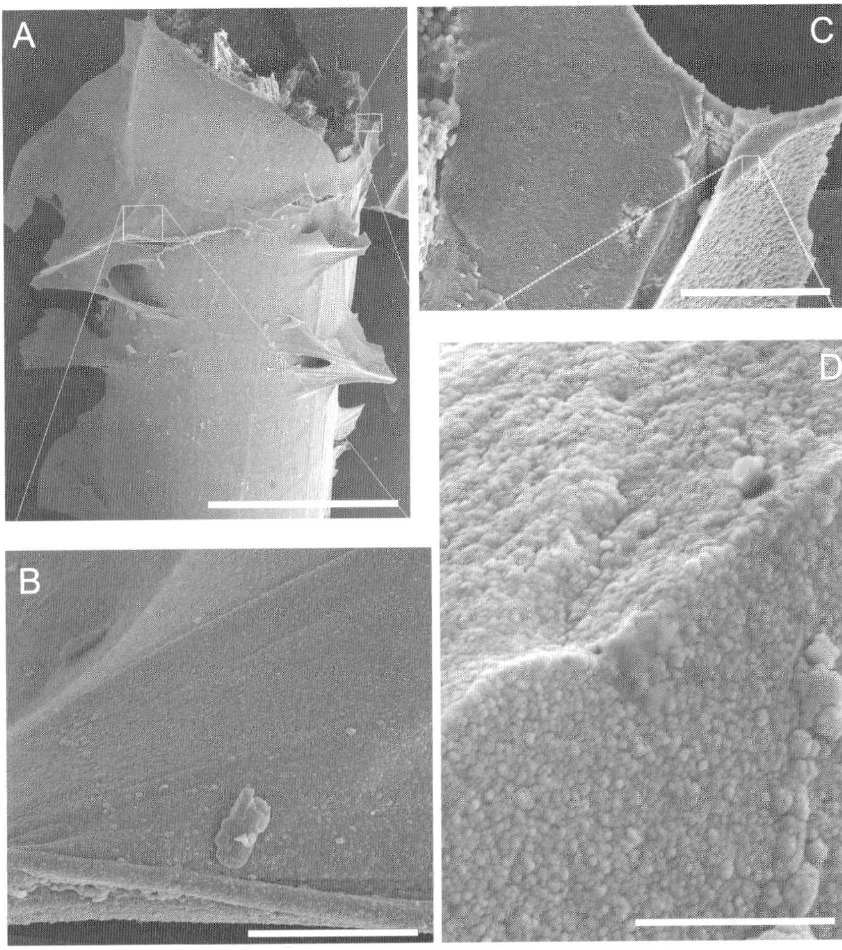

Fig. 12.4 A–D. Quenstedtoceras *sp. A – General view of connecting ring with several siphuncular sheets. B – Granular surface of siphuncular sheet. C – Detail of the vertical fracture plane of the connecting ring and siphuncular sheet. D – Siphuncular sheet in higher magnification. Scale bars: A – 300 μm; B – 30 μm; C – 12 μm; D – 3 μm.*

4 Discussion

4.1 Connecting Ring in Ammonoids

As described in a previous paper (Mutvei et al., 2004) and herein, the connecting rings in three Mesozoic ammonoid taxa *Grammoceras*, *Aconeceras*, and *Quenstedtoceras* form a semisolid, probably tanned layer of glycoprotein (conchiolin)

Table 12.1 *The results (in atomic weight %) of Energy Dispersive X-Ray Microanalysis.*

Points	Calcium	Carbon	Oxygen	Phosphorus	Sulphur
Shell and septa (8)	19–63	16–38	17–63	–	–
Septal neck (1)	18	17	64	1	–
Connecting ring (16)	6–23	19–46	18–63	3–18	1

fibers. In order to make the layer permeable for cameral liquid, it is perforated by numerous, narrow pore canals. The number and diameter of the pore canals in the ammonoid connecting ring roughly match the number and diameter of the pore canals in a crustacean exoskeleton that is composed of tanned glycoprotein fibers (Compére and Goffinet, 1987). In the crustacean exoskeleton the pore canals house thin filaments from the body epithelium representing elongated microvilli. It is probable that the pore canals in ammonoids also housed thin epithelial filaments that participated in the transport of the cameral liquid between the shell chambers and the siphuncular cord (Mutvei et al., 2004: Fig. 12.4A).

4.2 Connecting Ring in Nautiloids

The connecting ring in living *Nautilus* is composed of an inner glycoprotein (conchiolin, horny) layer and an outer, calcified spherulitic-prismatic layer (Mutvei, 1972, 2002a; Grégoire, 1984). The inner layer is an uncalcified and structurally modified continuation of the nacreous layer of the septal neck. It is composed of a thin lamella of glycoprotein fibers similar to those in the connecting ring of ammonoids (Obata et al., 1980; Westermann, 1982; Grégoire, 1984). In contrast to that in ammonoids, this layer is elastic and lacks pore canals. It can stand against hydrostatic pressures of 50–80 atm, but has a low permeability for cameral liquid. The outer, calcified, spherulitic-prismatic layer is a continuation of a structurally similar layer on the outer surface of the septal neck. The latter layer is porous without mechanical strength.

As reported by Mutvei (2002a), the fossil nautiloid taxa Nautilida and Tarphycerida have a similar connecting ring structure to that seen in *Nautilus* and, hence fundamentally different from that in ammonoids. The fibrous, glycoprotein layer of the connecting ring in the Nautilida and Tarphycerida is practically always destroyed during fossilization. However, the direction of the growth lines in the distal end of the septal neck suggests that an inner glycoprotein layer of the connecting ring was present during the animal's lifetime.

In Paleozoic orthoceratid and actinoceratid nautiloids the connecting ring has a different structure. The outer spherulitic-prismatic layer is present but is usually thinner than that in other nautiloids. The inner layer of the connecting ring is calcified and perforated by numerous pore canals, usually about 0.1–0.5 mm in diameter (Fig. 12.5A, B; Mutvei, 1997, 1998, 2002a, b). Thus, the pore canals occur in the connecting rings of actinocerid and orthocerid nautiloids, but they have a considerably larger diameter than those in ammonoids. It is highly probable that the pore canals in orthocerids and actinocerids also housed cellular extensions from the

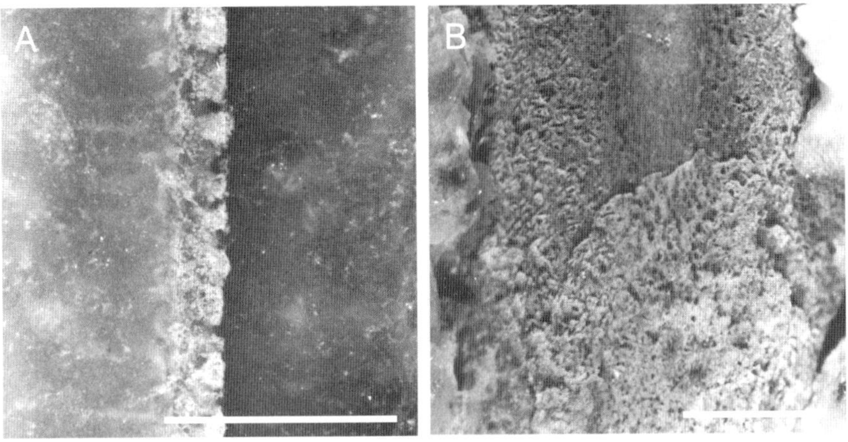

Fig. 12.5 *Connecting rings in Ordovician nautiloids. A –* Orthoceras scabridum. *Longitudinal section of connecting ring to show the calcified layer with pore canals. B –* Lituites *sp. Paramedian section of connecting ring to show the calcified layer with pore canals. Scale bar – 1 mm.*

epithelium of the siphuncular cord, and that these extensions considerably increased the surface of the physiologically active epithelium, thereby increasing the capacity and speed of the "osmotic pumping" function of this epithelium.

4.3 Connecting Ring in Coleoids

Three structural types have been distinguished in coleoids (Mutvei and Donovan, 2006, in press).

In Recent *Spirula* the connecting ring has a structure similar to that in Recent *Nautilus*. It is composed of an outer, calcified, spherulitic-prismatic layer and an inner, fibrous, glycoprotein layer. Because the septal necks are long, the siphuncular wall is mechanically strong to withstand high hydrostatic pressures. The connecting ring is exposed only in a narrow interspace between consecutive necks where the exchange of the cameral liquid takes place. When *Spirula* is in its normal swimming position this liquid is almost completely decoupled from the permeable, exposed region of the connecting ring (Denton et al., 1967; Denton and Gilpin-Brown, 1971). *Spirula* is able to make extensive daily vertical migrations (Clarke, 1969), but the osmotic exchange mechanism to remove cameral liquid is still unknown.

In the Jurassic phragmoteuthid, *Phragmoteuthis huxley*, the connecting ring has a primitive uncomplicated structure, unknown in other cephalopods. Each connecting ring has a calcified, porous, prismatic structure. It is extremely long and extends through 5–6 chambers. The siphuncular wall in each chamber is therefore composed of 5–6 superimposed connecting rings from consecutive septal necks (Mutvei and Donovan, 2006, in press).

The third structural type of connecting ring occurs in the Jurassic belemnoid *Megateuthis gigantea*, and in the aulacocerid *Mojsisovicteuthis* The permeable siphuncular wall in each chamber consists of two superimposed connecting rings from two consecutive septal necks. The connecting ring is composed of calcified lamellae that are perforated by numerous pore canals (Mutvei and Donovan, 2006, in press). Thus, the pore canals in the connecting rings occur not only in belemnoid and aulacocerid coleoids, but also in actinocerid and orthocerid nautiloids, and in ammonoids (Fig. 12.5).

4.4 Ammonoid Paleobathymetry and Paleobiogeography

Westermann (1971) calculated the siphuncular strength index in ammonoids from the siphuncular wall thickness, multiplied by 100 and divided by the inner radius of the siphuncle. This calculation was based on the assumption that the ammonoid connecting ring had the same structure and composition as the inner glycoprotein (cochiolin, horny) layer of the connecting ring in living *Nautilus*. From the siphuncular strength index, Westermann (1971, 1982, 1987) and Hewitt (1996) calculated the habitat limits of several ammonoid taxa. However, as demonstrated in the present paper the siphuncular strength index can only be used as an indicator of relative siphuncular strength between different ammonoid taxa.

The composition and structure of the ammonoid connecting ring is also important for estimations of the postmortem drift of ammonoid shells and hence, for paleobiogeography. Chamberlain et al. (1981) stressed that the postmortem drift in ammonoids was a relatively rare event because "the rate of water influx into the phragmocone due to ambient hydrostatic pressure is sufficiently rapid in most cases to overcome positive buoyancy before the shell reaches the surface" (Chamberlain et al., 1981: 494). The filling of the phragmocone with water is facilitated by postmortem loss of the body and the siphuncular cord. "If the fill rate is rapid compared to the ascent rate, the shell will not reach the surface but will sink back down to the sea floor" (Chamberlain et al., 1981: 497). Consequently, the structure of the connecting ring, its permeability, and the living depth of the animal regulate the postmortem biogeographic distribution of ammonoids (see also Maeda and Seilacher, 1996). As demonstrated here, the permeabilility coefficient of the connecting ring in ammonoids was probably much higher than that in *Nautilus* and this has to be taken into consideration in estimations of the paleobiogeographic distribution of ammonoids.

4.5 Ammonoid Septa

The function of the peripheral folding of ammonoid septa has generally been explained as providing increased mechanical strength for the shell wall from implosion by hydrostatic pressure (e.g., Westermann, 1975). In addition to the mechanical function, the septal folding probably also had a physiological function (Mutvei, 1967;

Kulicki, 1979; Kulicki and Mutvei, 1988). As in living *Nautilus*, the septal surface was covered by a wettable glycoprotein (conchiolin) sheet, but due to the considerably increased septal surface, by septal folding, this sheet attained a considerably larger surface in ammonoids. Also the inner surface of each chamber wall in ammonoids was coated by an organic sheet (Weitschat, 1986; Weitschat and Bandel, 1991). In many ammonoids (summarized in Weitschat and Bandel, 1991), the chambers contained siphuncular sheets that extended from the septal sheets and chamber-wall sheets to the siphuncular surface and became attached to it (Weitschat, 1986; Weitschat and Bandel, 1991; Tanabe and Landman, 1996). The siphuncular sheets created circumsiphonal reservoirs around the siphuncle. Checa (1996) interpreted some of the siphuncular sheets as being formed by desiccation of mucus in cameral liquid, but that interpretation is not supported by our observations. As illustrated in Fig. 12.4A–D, the structure of the siphuncular sheets is identical to that of the connecting ring. Additional sheets, transverse and horizontal, may also occur in the chambers of some ammonoids (Weitschat and Bandel, 1991). As pointed out by the latter writers, all these sheets were interconnected and they facilitated the transport of the cameral liquid through the siphuncular wall (Mapes et al., 1999).

As experimentally demonstrated by Mutvei and Reyment (1973), the amount of cameral liquid in ammonoids was relatively much greater than that in adult *Nautilus* (see also Heptonstall, 1970; Weitschat and Bandel, 1991; Kröger, 2002). Kulicki and Mutvei (1988) showed that spaces between the shell wall and the septal folds in ammonoids were capable of keeping a large volume of cameral liquid by surface tension. The higher amount of cameral liquid seems to have been an adaptive feature for vertical movements. Rapid buoyancy regulation was made possible by the highly porous but mechanically strong connecting ring, and by the epithelial extensions within the pore canals that provided a considerably enlarged surface area for the osmotically pumping, siphuncular epithelium.

4.6 Ammonoid Anatomy

As pointed out by several authors (e.g., Mutvei, 1967; Mutvei and Reyment, 1973; Weitschat and Bandel, 1991; Donovan, 1993; Jacobs and Landman, 1993; Saunders and Ward, 1994; Doguzhaeva and Mutvei, 1996; Richter and Fischer, 2002), the anatomy of the ammonoid body is different from that in living *Nautilus*. (a) In many ammonoid taxa the body was longer than that in *Nautilus* and more or less wormlike in shape; this indicates that the ventral mantle cavity was long and narrow, and contained only a small volume of water, not sufficient for jet-powered swimming (see Mutvei and Reyment, 1973: Text-fig. 12.8B). In Recent *Nautilus*, an animal that swims by jet propulsion, the volume of the mantle cavity is about one half of the total body volume (Ward, 1987). (b) The ammonoid shell aperture often has a ventral keel instead of a hyponomic sinus. This is in sharp contrast to *Nautilus* and fossil nautiloids. Also living coleoids have a hyponomic sinus in the ventral mantle margin. (c) The mantle-shell attachment in ammonoids was considerably different from that in living *Nautilus* (Crick, 1898; Jordan, 1968; Mutvei, 1967, 1975; Mutvei and Reyment, 1973; Weitschat and Bandel, 1991; Doguzhaeva and Mutvei, 1996; Richter and Fischer, 2002). In *Nautilus* jet propulsion is created by powerful, laterally attached, cephalic muscles that pull the body into the shell, accompanied by simultaneous contractions of the hyponome muscles. The water is thereby forcibly expelled from the mantle cavity through the hyponome. Except in a few taxa, in which the annular area forms a lateral lobe, the lateral muscle attachments are not recognized in ammonoids (Doguzhaeva and Mutvei, 1996; Kennedy et al., 2002; Richter and Fischer, 2002).

The following hypotheses have been generated to explain a possible jet-powered swimming mechanism in ammonoids: (1) The head and a highly muscular hyponome were outside the body chamber and the contractions of the hyponome alone created the jet propulsion for swimming (Doguzhaeva and Mutvei, 1991; Chamberlain, 1993; Saunders and Ward, 1994; Richter and Fischer, 2002). (2) Ammonoids had a muscular, coleoid-like mantle in front of the shell aperture and the contractions of this mantle created the propulsive power (Jacobs and Landman, 1993). However, neither hypothesis can be supported by the morphology and structure of the ammonoid shell.

4.7 Ammonoid Shell Morphology

In Recent *Nautilus*, the intra- and interspecific variations of shell morphology are very small. In fully matured shells the apertural edge is somewhat thickened, the apertural width is slightly contracted, and the ocular sinuses are somewhat deeper. There is also a small change of shell coiling and an increase in the length of the living chamber during growth (Collins and Ward, 1987). Sexual dimorphism differences are also very small and are only expressed by larger shells with somewhat wider shell apertures in males (Saunders and Ward, 1987; Ward, 1987). At early growth stages, before the nepionic constriction, the shell is ornamented by a reticulate pattern that is replaced by fine parallel ridges across the transverse growth lines in some species at later growth stages.

Shells in many ammonoid taxa differ considerably from *Nautilus* shells by their large scale intraspecific variation and more pronounced sexual dimorphism. As summarized by Davis et al. (1996), the late ontogenetic shell modifications comprise: (1) changes in coiling, whorl cross section, and ornamentation, and (2) development of an apertural shell thickening, apertural constriction, and formation of apertural rostrum (keel) or lappets. Sexual dimorphism is expressed in large differences in shell size of dimorphs (micro- and macroconch). At the terminal growth stages, the microconch (probably male) forms a prominent apertural rostrum or lappets whereas the whorls in the macroconch (probably female) become broad and inflated, and the surface of the living chamber becomes smooth (Makowski, 1962; Sarti, 1999).

Makowski (1962) described a considerable range of sexual dimorphism characters in *Quenstedtoceras*. The microconchs and macroconchs are identical up to the 5–5.5 whorls stage. The surface of these shells is ornamented by distinct ribs and the shell is strongly compressed in cross section. During further growth the macroconch attains a large diameter, and the fully grown shell consists of at least 7 whorls. The shell becomes more involute, and after 6–7 whorls, the ribbing gradually disappears, and the shell surface becomes smooth. In cross section the shell becomes broad and inflated. The microconch grows only to 6 whorls and has therefore a much smaller diameter than the macroconch. It maintains a compressed cross section and distinctly ribbed ornamentation. At the terminal growth stage the aperture forms a long ventral keel. The shell surface becomes smooth only immediately behind this keel.

Some taxa show extreme intraspecific variation of shell shape and ornamentation. As reported by Dagys and Weitschat (1993a, b), the adult shells of the Triassic ammonoid *Czekanowskites rieberi* range morphologically from "keeled smooth suboxycones with narrow umbilicus, through feebly ribbed platycones with a little wider umblicus, to subcadicones with relatively wide umbilicus and straight ribs with bullae." Consequently, the "streamlined" oxycones and "subbenthic" cadicones lived together within the same biotope. Based on these observations, Dagys and Weitschat came to the conclusion that "streamlining of shell cannot play any important role among such slow-swimming animals as ammonoids" (Dagys and Weitschat, 1993b: 26). Those writers also pointed out that high intraspecific variation also occurs in several other groups of Triassic ammonoids, and that it has created serious problems in taxonomy.

4.8 Ammonoid Mode of Life

Several authors (e.g., Westermann, 1987, 1996; Jacobs, 1992; Seki et al., 2000; Klug and Korn, 2004) were of the opinion that many ammonoids utilized jet-powered swimming. Westermann (1996) and Jacobs (1992) even calculated the swimming speeds in several ammonoid taxa.

On the other hand, some writers (Mutvei and Reyment, 1973; Mutvei, 1975; Weitschat and Bandel, 1991; Dagys and Weitschat, 1993b) have been more or less critical of the idea of jet-powered swimming in ammonoids. These writers pointed out the great differences between most ammonoids and living *Nautilus* in anatomical design and in intraspecific variation of shell morphology. If ammonoids were jet-powered swimmers, their populations had to have been subdivided into several subpopulations characterized by different shell design and hence, different swimming ability and mode of life. This interpretation is not in agreement with the conditions either in living *Nautilus* or in living coleoid cephalopods. Furthermore, ammonoids do not show any sign of natural selection toward streamlined shell shape during their evolution.

Instead of evolutionary adaptation to jet-powered swimming, most ammonoids show a specialization to vertical, probably diurnal migrations by aid of changes of buoyancy. This is evident from the occurrence of numerous pore canals in the connecting ring that probably contained epithelial extensions from the siphuncular epithelium, and by the complex system of wettable glycoprotein sheets that extended from each chamber to the wall of the connecting ring and functioned for rapid transport of cameral liquid from the chambers.

5 Conclusions

(1) The ammonoid connecting ring was structurally and mechanically different from that in living *Nautilus*: it is composed of glycoprotein (conchiolin) fibers that probably were tanned and formed a semielastic, mechanically strong structure. This made it highly resistant to diagenesis. In order to make the connecting ring permeable for cameral liquid, it was perforated by a great number of very fine pore canals that probably housed microvilli-like epithelial extensions. The epithelium of the siphuncular cord consequently acquired a considerably enlarged surface area for an efficient osmotic pump function.

(2) Connecting rings, perforated by pore canals, occur in Paleozoic orthocerid and actinocerid nautiloids, and in belemnoid and aulacocerid coleoids. This similarity may indicate phylogenetic relationships between these taxa and ammonoids.

(3) The siphuncular strength index in ammonoids cannot be used to estimate implosion depth of the shells and hence the maximum depth habitat, nor can be used to estimate postmortem paleobiogeographic distribution of the shells, and depth of the marine environment where the ammonoid shells were postmortally deposited. However, it can be used to compare the mechanical strength of the connecting ring between different ammonoid taxa.

(4) Ammonoids lack indications that they were jet-powered swimmers. (a) The muscle attachment to the shell differs considerably from that in living *Nautilus*. (b) Contrary to *Nautilus*, many ammonoids had a long, wormlike body that only contained a small, narrow, ventral mantle cavity; the water volume in this cavity was insufficient for jet propulsion. (c) Many ammonoids lack a hyponomic sinus. (d) The shell morphology and ornamentation in many ammonoid taxa show pronounced intraspecific variation, and great differences between juveniles, females, and males. If ammonoids were adapted for swimming by jet propulsion, these variations would indicate that every single population was subdivided into several subpopulations with different modes of life. In contrast, *Nautilus* shows very small intraspecific variation of their shell morphology and ornamentation. This is necessary to maintain a streamlined shell shape for the entire population for jet-powered swimming.

(5) Previously, the idea of vertical, diurnal migrations of ammonoids by means of buoyancy changes was refuted because in living *Nautilus*, the emptying of chambers of cameral liquid is a slow process. However, as shown in the present

paper, the occurrence of a great number of pore canals in the connecting ring indicates that ammonoids were capable of a rapid exchange of cameral liquid and gas between the siphuncle and the chambers. In addition, the physiologically active siphuncular epithelium extended into the pore canals and acquired a larger surface area for osmotic "pumping." Furthermore, the exchange of cameral liquid through the wall of the porous connecting ring was facilitated by a complex system of wettable glycoprotein (conchiolin) sheets that extended from each chamber to the wall of the connecting ring; these sheets were interconnected and facilitated the rapid transport of cameral liquid. All these features strongly indicate that the majority of ammonoids had acquired an efficient adaptation for diurnal vertical migrations by means of buoyancy changes.

Acknowledgments

We thank Dr. L. A. Doguzhaeva for permission to study her material of *Aconeceras trautscholdi* and for her valuable comments.

References

Bandel, K. 1981. The structure and formation of the siphuncular tube of *Quenstedtoceras* compared with that of *Nautilus*. *Neues Jahrbuch für Geologie und Paläontologie, Abhandlungen* **161**: 153–171.

Chamberlain, J. A. 1993. Locomotion in ancient seas: constraint and opportunity in cephalopod adaptive design. *Geobios* **15**: 49–61.

Chamberlain, J. A., P. D. Ward, and J. S. Weaver. 1981. Post-mortem ascent of nautilus shells: implications for cephalopod paleobiogeography. *Paleobiology* **7**: 494–550.

Checa, A. 1996. Origin of intracameral sheets in ammonoids. *Lethaia* **29**: 61–75.

Clarke, M. R. 1969. Cephalopoda collected on the SOND cruise 1965. *Journal of the Marine Biological Association of the United Kingdom* **50**: 961–1000.

Collins, D., and P. D. Ward. 1987. Adolescent growth and maturity in *Nautilus*. *In* W. B. Saunders, and N. H. Landman (editors), *Nautilus, the Biology and Paleobiology of a Living Fossil*, pp. 421–432. New York: Plenum Press.

Compére, P., and G. Goffinet. 1987. Ultrastructural shape and three dimensional organization of the intracuticular canal system in the mineralized cuticle of the green crab *Carcinus maenas*. *Tissue and Cell* **19**: 839–858.

Crick, G. C. 1898. On the muscular attachment of the animal to its shell in some Cephalopoda (Ammonoidea). *Transactions of the Linnean Society of London, Zoology* **7**: 71–113.

Dagys, A. S., and W. Weitschat. 1993a. Intraspecific variation in Boreal Triassic ammonoids. *Geobios* **15**: 107–109.

Dagys, A. S., and W. Weitschat. 1993b. Extensive intraspecific variation in a Triassic ammonoid from Siberia. *Lethaia* **26**: 113–121.

Dalingwater, J. E., and H. Mutvei. 1990. Arthropod exoskeletons. *In* J. G. Carter (editor), *Skeletal Biomineralization. Patterns, Processes and Evolutionary Trends*. Volume I, pp. 83–96. New York: Van Nostrand Reinhold.

Davis, R. A., N. H. Landman, J. L. Dommergues, D. Marchand, and H. Bucher. 1996. Mature modifications and dimorphism in ammonoid cephalopods. *In* N. H. Landman, K. Tanabe, and R. A. Davis (editors), *Ammonoid paleobiology*, pp. 463–539. New York: Plenum Press.

Dennell, R. 1960. Integument and exoskeleton. *In* T. H. Waterman (editor), *The Physiology of Crustacea*. Volume I, pp. 449–472. New YorK: Academic Press.

Denton, E. J., J. B. Gilpin-Brown, and J. V. Howarth. 1967. On the buoyancy of *Spirula spirula*. *Association of the United Kingdom* **47**: 181–191.

Denton, E. J., and J. B. Gilpin-Brown. 1971. Further observations on the buoyancy of *Spirula*. *Journal of the Marine Biological Association of the United Kingdom* **51**: 363–373.

Doguzhaeva, L. A., and Mutvei, H. 1991. Organization of the soft body in *Aconeceras* (Ammonitina), interpreted on the basis of shell morphology and muscle scars. *Palaeontographica A*. **218**: 17–33.

Doguzhaeva, L. A., and H. Mutvei. 1996. Attachment of the body to the shell in ammonoids. *In* N. H. Landman, K. Tanabe, and R. A. Davis (editors), *Ammonoid Paleobiology*, pp. 43–63. New York: Plenum Press.

Doguzhaeva, L. A., and H. Mutvei. 2003. Gladius composition and ultrastructure in extinct squid-like coleoids: *Loligosepia, Trachyteuthis* and *Teudopsis*. *Revue de Paléobiologie* **22**: 877–894.

Donovan, D. T. 1993. Ammonites in 1991. *In* M. R. House (editor), *The Ammonoidea: Environment, Ecology, and Evolutionary Change*, pp. 85–97. Oxford: Clarendon Press.

Florkin, M. 1960. Blood chemistry. *In* T. H. Waterman (editor), *The Physiology of Crustacea*. Volume I, pp. 141–160. New York: Academic Press.

Grégoire, C. 1984. Remains of organic components in the siphonal tube and in the brown membrane of ammonoids and fossil nautiloids. *Akademie der Wissenschaften und der Litteratur, Abhhandlungen der matematisch-naturwissenschaftlichen Klasse, Mainz* **1984**: 4–56.

Heptonstall, W. B. 1970. Buoyancy control in ammonoids. *Lethaia* **3**: 317–328.

Hewitt, R. A. 1996. Architecture and strength of the ammonoid shell. *In* N. H. Landman, K. Tanabe, and R. A. Davis (editors), *Ammonoid Paleobiology*, pp. 297–339. New York: Plenum Press.

Hewitt, R. A., U. A. Abdelsalam, M. A. Dokainish, and G. E. G. Westermann. 1993. Comparision of the relative strength of siphuncles with prochoanitic and retrochoanitic septal necks by finite-element analysis. *In* M. R. House (editor), *The Ammonoidea: Environment, Ecology, and Evolutionary Change*, pp. 85–97. Oxford: Clarendon Press.

Jacobs, D. K. 1992. Shape, drag, and power in ammonoid swimming. *Paleobiology* **18**: 203–220.

Jacobs, D. K., and N. H. Landman. 1993. *Nautilus* – a poor model for the function and behavior of ammonoids? *Lethaia* **26**: 101–111.

Jordan, R. 1968. Zur Anatomie mesozoischen Ammoniten nach den Strukturelementen der Gehäuse-Innerwand. *Beihefte des Geologischen Jahrbuch* **77**: 1–64.

Kennedy, W. J., W. A. Cobban, and H. C. Klinger. 2002. Muscle attachment and mantle related features in Upper Cretaceous *Baculites* from the United States Western interior. *Abhandlungen der Geologischen Bundesanstalt* **57**: 89–112.

Klug, C., and D. Korn. 2004. The origin of ammonoid locomotion. *Acta Palaeontologica Polonica* **49**: 235–242.

Kröger, B. 2002. On the efficiency of the buoyancy apparatus in ammonoids: evidences from sublethal shell injuries. *Lethaia* **35**: 61–70.

Kulicki, C. 1979. The ammonite shell: its structure, development and biological significance. *Acta Palaeontologica Polonica* **39**: 97–142.

Kulicki, C. 1994. Septal neck-siphuncular complex in *Stolleyites* (Ammonoidea), Triassic, Svalbard. *Polish Polar Research* **15**: 37–49.

Kulicki, C., and H. Mutvei. 1988. Functional interpretation of ammonoid septa. *In* J. Wiedmann, and J. Kullmann (editors), *Cephalopods – Present and Past*, pp. 713–718. Stuttgart: Schweitzerbart'sche Verlagbuchhandlung.

Maeda, H., and A. Seilacher. 1996. Ammonoid taphonomy. *In* N. H. Landman, K. Tanabe, and R. A. Davis (editors), *Ammonoid Paleobiology*, pp. 544–578. New York: Plenum Press.

Makowski, H. 1962. Problem of sexual dimrphism in ammonites. *Palaeontologia Polonica* **12**: 1–92.

Mapes, R. H., K. Tanabe, and N. H. Landman. 1999. Siphuncular membranes in Upper Paleozoic prolecanitid ammonoids from Nevada, USA. *In* K. Histon (editor), *V International Symposium Cephalopods – Present and Past, Vienna 6–7 September 1999. Berichte der Geologischen Bundesanstalt*. Abstracts Volume, **46**: 83.

Mutvei, H. 1967. On the microscopic shell structure in some Jurassic ammonoids. *Neues Jahrbuch für Geologie und Paläontologie, Abhandlungen* **129**: 157–166.

Mutvei, H. 1972. Ultrastrucural studies on cephalopod shells, Part I. The septa and siphonal tube in *Nautilus. Bulletin of the Geological Institutions of the University of Uppsala*. New Series **3**: 237–261.

Mutvei, H. 1975. The mode of life in ammonoids. *Paläontologische Zeitschrift* **49**: 196–202.

Mutvei, H. 1997. Characterization of actinoceratoid cephalopods by their siphuncular structure. *Lethaia* **29**: 339–348.

Mutvei, H. 1998. Siphuncular structure in a Silurian narthecoceratid nautiloid cephalopod from the Island of Gotland. *Geologiska Föreningens i Stockholm Förhandlingar* **120**: 375–378.

Mutvei, H. 2002a. Nautiloid systematics based on siphuncular structure and position of muscle scars. *Abhandlungen der Geologischen Bundesanstalt* **57**: 379–392.

Mutvei, H. 2002b. Connecting ring structure and its significance for classification of the orthoceratid cephalopods. *Acta Palaeontologica Polonica* **47**: 157–168.

Mutvei, H., and D. D. Donovan. 2006. Siphuncular structure in some fossil coleoids and Recent *Spirula. Palaeontology* (in press).

Mutvei, H., and R. A. Reyment. 1973. Buoyancy control and siphuncle function in ammonoids. *Palaeontology* **16**: 623–636.

Mutvei, H., W. Weitschat, L. A. Doguzhaeva, and E. Dunca. 2004. Connecting ring with pore canals in two genera of Mesozoic ammonoids. *Mitteilungen aus dem Geologisch-Paläontologischen Institut der Universität Hamburg* **88**:135–144.

Obata, I., K. Tanabe, and Y. Fukuda. 1980. The ammonite siphuncular wall: its microstructure and functional significance. *Bulletin of the National Science Museum Tokyo*. Series C **6**: 59–72.

Passano, L. M. 1960. Molting and its control. *In*. T. H. Waterman (editor), *The Physiology of Crustacea*. Volume I, pp. 473–536. New York: Academic Press.

Richter, U., and R. Fischer. 2002. Soft tissue attachment structures on pyritized internal molds of ammonites. *Abhandlungen der Geologischen Bundesanstalt* **57**: 139–149.

Sarti, C. 1999. Whorl width in the body chamber of ammonites as a sign of dimorphism. *In* F. Oloriz, and J. Rodriguez-Towar (editors), *Advancing Research on Living and Fossil Cephalopods*, pp. 315–332. New York: Kluwer Academic/Plenum Publishers.

Saunders, W. B., and P. D. Ward. 1987. Ecology, distribution and population characteristics of *Nautilus. In* W. B. Saunders, and N. H. Landman (editors), *Nautilus, the Biology and Paleobiology of a Living Fossil*, pp. 137–162. New York: Plenum Press.

Saunders, W. B., and P. D. Ward. 1994. *Nautilus* is not a model for the function and behavior of ammonoids. *Lethaia* **27**: 47–48.

Seki, K., K. Tanabe, N. H. Landman, and D. K. Jacobs. 2000. Hydrodynamic analysis of Late Cretaceous desmoceratine ammonites. *Revue de Paléobiologie*. Volume Spécial **8**: 141–155.

Tanabe, K. 1977. Functional evolution of *Otoscaphites puerculus* (Jimbo) and *Scaphites planus* (Yabe), Upper Cretaceous ammonites. *Memoirs of the Faculty of Science, Kyushu University*. Series D, *Geology* **23**: 367–407.

Tanabe, K. 1979. Palaeoecological analysis of ammonoid assemblages in the Turonian *Scaphites* facies of Hokkaido, Japan. *Palaeontology* **22**: 609–630.

Tanabe, K., Y. Fukuda, and I. Obata. 1982. Formation and function of the siphuncle-septal neck structures in two Mesozoic ammonites. *Transactions and Proceedings of the Palaeontological Society of Japan* **128**: 433–443.

Tanabe, K., and N. H. Landman. 1996. Septal neck-siphuncular complex of ammonoids. *In* N. H. Landman, K. Tanabe, and R. A. Davis (editors), *Ammonoid Paleobiology*, pp. 129–165. New York: Plenum Press.

Ward, P. D. 1987. *The Natural History of Nautilus*. Boston, MA: Allen & Unwin.

Weitschat, W. 1986. Phosphatisierte Ammonoideen aus der Mittleren Trias von Central-Spitzbergen. *Mitteilungen aus dem Geologisch-Paläontologischen Institut der Universität Hamburg* **61**: 249–279.

Weitschat, W., and K. Bandel. 1991. Organic components in phragmocones of Boreal Triassic ammonoids: implications for ammonoid biology. *Paläontologische Zeitschrift* **65**: 269–303.

Westermann, G. E. G. 1971. Form, structure and function of shell and siphuncle in coiled Mesozoic ammonoids. *Life Science Contributions of the Royal Ontario Museum* **78**: 1–39.

Westermann, G. E. G. 1975. Model for origin, function and fabrication of fluted cephalopod septa. *Paläontologische Zeitschrift* **49**: 235–253.

Westermann, G. E. G. 1982. The connecting rings in *Nautilus* and Mesozoic ammonoids: implications for ammonoid bathymetry. *Lethaia* **15**: 373–384.

Westermann, G. E. G. 1987. New developments in ecology of Jurassic-Cretaceous ammonoids. *Atti del secondo convegno internazionale Fossili, Evoluzione, Ambiante. Pergola* 25–30 ottobre 1987, pp. 459–478.

Westermann, G. E. G. 1993. Hydrostatics and hydrodynamics of cephalopod shells: form, structure, and function. *Anales Academia Nacional de Ciencias Exactas, Fisicas y Naturales de Buenos Aires* **45**: 183–204.

Westermann, G. E. G. 1996. Ammonoid life and habitat. *In* N. H. Landman, K. Tanabe, and R. A. Davis (editors), *Ammonoid Paleobiology*, pp. 607–707. New York: Plenum Press.

Chapter 13
Jaws and Radula of *Baculites* from the Upper Cretaceous (Campanian) of North America

Neil H. Landman,[1] Neal L. Larson,[2] and William A. Cobban[1,3]

[1]Division of Paleontology (Invertebrates), American Museum of Natural History, New York, NY 10024, USA, landman@amnh.org;

[2]Black Hills Museum of Natural History, Hill City, SD 57745, USA, ammoniteguy@bhigr.com;

[3]70 Estes Street, Lakewood, CO 80226, USA

Keywords: *Baculites*, *Pseudobaculites*, jaws, *Rugaptychus*, radula, Campanian, Upper Cretaceous, Cody Shale, Mooreville Chalk, Pierre Shale, Lewis Shale

1 Introduction

The seminal discovery by Meek and Hayden (1865) of an aptychus preserved inside the body chamber of a Late Cretaceous scaphite in close association with what is unmistakably an upper jaw demonstrated that the aptychus is part of the lower jaw. Lower jaws are known from many ammonites, but, surprisingly, these structures have rarely been reported in baculites from North America, even though these ammonites are extraordinarily abundant in Upper Cretaceous strata of this

N. H. Landman et al. (eds.), *Cephalopods Present and Past: New Insights and Fresh Perspectives*, 257–298.

continent. In contrast, baculite lower jaws are relatively common in upper Campanian chalks of northern Europe where they are preserved as isolated elements, and much more rarely, inside the body chamber. This paradox has led to speculation about the taxonomic distribution of baculite lower jaws, the degree of variation in their morphology, and the extent to which the vagaries of preservation have altered their appearance.

We report the discovery of lower jaws inside the body chambers of *Baculites* sp. (smooth) and *Baculites* sp. (weak flank ribs) from the lower Campanian Pierre Shale of South Dakota, and the Cody Shale of Wyoming, respectively. We also describe aptychi of presumably one or both of these same species from the lower Campanian Mooreville Chalk of Alabama and the Smoky Hill Chalk Member of the Niobrara Formation of Kansas. Isolated lower jaws are also present in the upper Campanian Pierre Shale of South Dakota, where they are attributed to *Baculites compressus*, Say, 1821, and *Baculites cuneatus*, Cobban, 1962, and the upper Campanian Lewis Shale of Wyoming, where they are attributed to *Pseudobaculites natosini* (Robinson, 1945).

In addition to jaws, we describe a radula preserved inside the body chamber of a specimen of *Baculites* sp. (smooth). Radulae have been reported from fewer than a dozen ammonite species worldwide and the discovery of a new specimen substantially adds to our knowledge of these structures.

2 Previous Work

There are numerous reports of baculite aptychi from the Campanian Boreal Chalk facies of northern Europe (England, France, Germany, Belgium, and Sweden). Trauth (1927: 245) assigned these aptychi to the form genus *Rugaptychus*. The aptychi generally occur as isolated elements loose in the matrix (Hébert, 1856: 367, pl. 28, Fig. 6; Sharpe, 1857: 57, 58, pl. 24, Figs. 8a, b, 9, 10a, b; Binckhorst, 1861: 33 (said to be Maastrichtian); Lundgren, 1874: 70, pl. 3, Fig. 14; Schlüter, 1876: 144, 145, pl. 40, Fig. 8; Moberg, 1885: 41–43, pl. 1, Figs. 14–19; pl. 6, Figs. 25, 26; Blackmore, 1896: 532–533, pl. 16, Fig. 16; Ravn, 1902: 259; de Grossouvre, 1908: 39, pl. 10, Figs. 7–13; Diener, 1925: 40; Trauth, 1927: 245; Trauth, 1928: 122–130, pl. 2, Figs. 1–9; Picard, 1929: 436; Arkell, 1957: L440, Figs. 557.3, 558.5; Moore and Sylvester-Bradley, 1957: L469; Kennedy, 1986: 192, pl. 16, Figs. 1–22; Kennedy, 1993: 114, pl. 7, Fig. 25; Kennedy and Christensen, 1997: 112, 114, Fig. 31a–h; Klinger and Kennedy, 2001: 72–79, Fig. 55). Outside of Europe, isolated baculite aptychi have been reported from the Smoky Hill Chalk Member of the Niobrara Formation of Kansas (Miller, 1968: 48, 49, pl. 8, Fig. 14).

There are fewer reports of baculite jaws preserved inside the body chamber. Schlüter (1876: 147, pl. 39, Fig. 16) described and illustrated a lower jaw inside the body chamber of a large specimen of *Baculites knorrianus* Desmarest, 1817, from the upper Campanian Mukronatenkreide of Lüneburg, Germany. Nowak (1908: 339, pl. 14, Fig. 11) illustrated a lower jaw in the body chamber of

Baculites leopoliensis Nowak, 1908, from the same locality. Giers (1964: 256) described a jaw in a specimen of *Baculites vertebralis* Giers, 1964, Lamarck, 1801, from the lower Campanian of northwest Germany, but did not illustrate it. He also figured a structure in another body chamber of the same species but W. J. Kennedy (in Klinger and Kennedy, 2001: 79) concluded that the structure did not belong to the ammonite. Tanabe and Landman (2002: 161, pl. 1, Fig. 8; see also Matsumoto and Obata, 1963: 55) illustrated a lower jaw in the body chamber of *Baculites* cf. *princeps* Matsumoto and Obata, 1963, from the Campanian of western Japan.

Baculite jaws are not as common in older strata. Fritsch (1893: 80, Fig. 63c–e) described and illustrated a lower jaw inside the body chamber of what he called *Baculites faujassi* var. *bohemica* von Priesen, from the Coniacian of Bohemia, but the drawing is stylized and the description difficult to follow. Breitkreutz et al. (1991: 42, Figs. 6, 7; reillustrated in Klinger and Kennedy, 2001: 72, Fig. 56a, b) reported a lower jaw inside the body chamber of *Sciponoceras bohemicum anterius*? Wright and Kennedy, 1984, from the upper Cenomanian of Germany. Outside of Europe, Matsumoto and Obata (1963: 13, pl. 4; reillustrated in Tanabe and Landman, 2002: 161, pl. 1, Fig. 9) reported a lower jaw inside the body chamber of *Sciponoceras kossmati* (Nowak, 1908) from the lower Turonian of Hokkaido.

3 List of Localities

Numbers refer to Figs. 13.1 and 13.2.

1. AMNH loc. 3274: Upper Campanian *Baculites compressus–B. cuneatus* Zones, Pierre Shale, SE¼ sec. 6, T. 3N, R. 14E, Elk Creek, Meade County, South Dakota.
2. AMNH locs. 3280, 3281, 3294, and 3296: Lower Campanian *Baculites* sp. (smooth) Zone, Gammon Ferruginous Member, Pierre Shale, SW¼ sec. 21, east center sec. 27, SE¼ sec. 17, N½ sec. 27, T. 11N, R. 2E, respectively, Butte County, South Dakota.
3. Lower Campanian *Baculites* sp. (weak flank ribs) Zone, Cody Shale, sec. 6, T. 40N, R. 79W, Natrona County, Wyoming.
4. Upper Campanian *Baculites eliasi* Zone, Lewis Shale, sec. 30, T. 35N, R. 84W, Natrona County, Wyoming.
5. Lower Campanian *Baculites* sp. (smooth)-*Baculites* sp. (weak flank ribs) Zones, upper part of Mooreville Chalk, sec. 10, T. 22N, R. 1W, Greene County, Alabama.
6. Lower Campanian *Baculites* sp. (smooth)-*Baculites* sp. (weak flank ribs) Zones, upper part of Mooreville Chalk, sec. 26, T. 22N, R. 1W, Dallas County, Alabama.
7. Lower Campanian *Hesperornis* Zone, Smoky Hill Chalk Member of the Niobrara Formation, NW corner, Rooks County, Kansas.

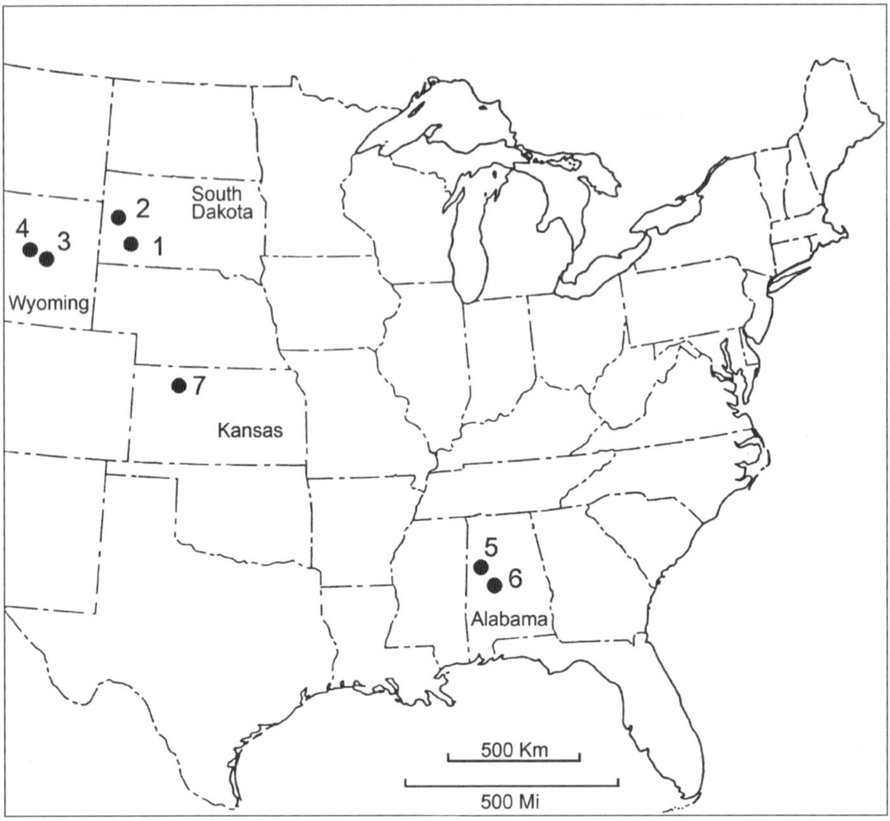

Fig. 13.1 *Map of part of the USA showing the localities mentioned in the text.*

4 Geologic Setting

A total of 12 specimens of *Baculites* sp. (smooth) with jaws preserved inside the body chamber were discovered in the upper part of the Gammon Ferruginous Member of the Pierre Shale, Butte County, South Dakota [lower Campanian Zone of *Baculites* sp. (smooth)] (Figs. 13.1, 2). Most of the specimens were collected by N.L.L. This locality was described by Gill and Cobban (1961), Robinson et al. (1964), and Bishop (1985). Baculites are extremely abundant, comprising thousands to tens of thousands of specimens. They occur loose and in concretions, in a silty shaly matrix, forming what Bishop (1985: 607) called a "Baculite Epibole." The baculites are preserved as steinkerns with pieces of aragonitic outer shell still attached (= composite internal molds) and bear traces of muscle scars and other imprints (Klinger and Kennedy, 2001; Henderson et al., 2002; Kennedy et al., 2002). Associated fossils include other ammonites, bivalves, gastropods, scaphopods, decapod crustaceans, bony fish, sharks, and mosasaurs. The sheer abundance of baculites, nearly all of which are large and consist of macroconchs and

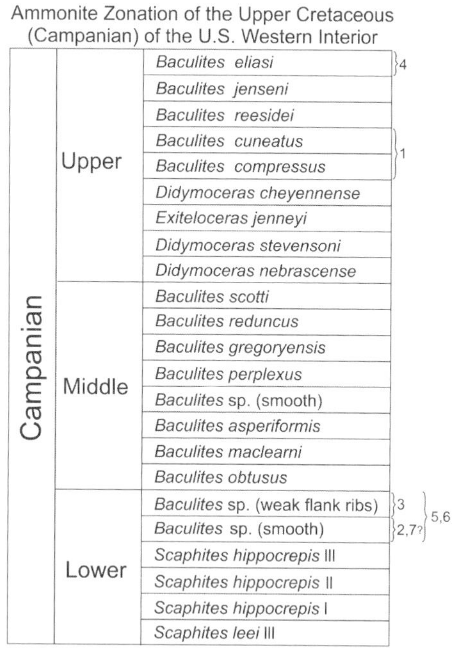

Fig. 13.2 *Ammonite zonation of the Campanian (Upper Cretaceous) of the US Western Interior (Cobban et al., 2006), with the biostratigraphic positions of the baculite localities indicated on the right.*

microconchs, suggests massive die-offs, perhaps following mating or spawning. Surprisingly, however, the incidence of jaws is very low. Based on our collecting, we estimate a recovery rate of 1 jaw for every 4,000 specimens.

A single specimen of *Baculites* sp. (weak flank ribs) with the jaw preserved inside the body chamber was discovered from a baculite-rich interval in the Cody Shale of Natrona County, Wyoming (Figs. 13.1, 2). This interval represents the lower Campanian Zone of *Baculites* sp. (weak flank ribs).

Isolated aptychi were recovered from near the top of the Mooreville Chalk below the Arcola Limestone Member in Dallas and Greene counties, Alabama (Figs. 13.1, 2). No aragonitic shells are preserved in this formation. The most common fossils are the originally calcitic bivalves *Exogyra* and *Cadeceramus*, and the worm tube *Hamulus*. The aptychi are relatively abundant; for example, N. Larson and T. Rust collected a total of 146 aptychi in a period of 7.5 h at five small (1–3 acre) outcrops.

The aptychi co-occur with baculite steinkerns, but these specimens are not well enough preserved for identification, thus precluding a definitive attribution of the aptychi. However, Kennedy et al. (1997) recorded *Scaphites hippocrepis* DeKay, 1827, form II Cobban, 1969, from the upper part of the Mooreville Chalk, which is below the aptychi-bearing interval. They also recorded *Menabites* (*Delawarella*) *danei* (Young, 1963), indicative of the Zone of *Baculites obtusus*, from strata above the aptychi-bearing interval. Therefore, the aptychi-bearing interval spans the zones of *Scaphites hippocrepis II* to *Baculites obtusus*, including the zones of *Baculites* sp. (smooth) and *Baculites* sp. (weak flank ribs), and is thus approximately equivalent to the baculite occurrences at localities 2 and 3 in South Dakota and Wyoming. Therefore, the aptychi in the Mooreville Chalk are tentatively attributed to *Baculites* sp. (smooth) and *Baculites* sp. (weak flank ribs).

Rugaptychi were previously reported from the Smoky Hill Chalk Member of the Niobrara Formation of Kansas (Miller, 1968). The rugaptychi occur as isolated elements in the upper part of this member, corresponding to the lower Campanian Zone of *Hesperornis* (see Stewart, 1990; Everhart, 2005: 36, Table 2.1). The associated baculites are preserved as smooth steinkerns and are referred to as *Baculites* sp. by Everhart (2005). It is possible that this species is synonymous with *Baculites* sp. (smooth) and, therefore, by implication, this site is equivalent in age to site 2.

A baculite jaw was discovered in a concretion in the upper Campanian *Baculites compressus–B. cuneatus* Zones in the Pierre Shale of Meade County, South Dakota (Figs. 13.1, 2). The concretion contains benthic and nektic organisms, all of which are preserved with their original aragonitic shells. Associated ammonites include *Hoploscaphites landesi* (Riccardi, 1983), *Jeletzkytes nodosus* (Owen, 1852), *J. brevis* (Meek, 1876), *Placenticeras meeki* (Böhm, 1898), and *P. costatum* (Hyatt, 1903). The jaws of these other ammonites differ from those of *Baculites* and have been described elsewhere (for *Hoploscaphites* and *Jeletzkytes*, see Landman and Waage, 1993; for *Placenticeras*, see Landman et al., 2006). Although the baculite jaw is not preserved inside a body chamber, it is attributed to *B. compressus* or *B. cuneatus* because these are the only two species of *Baculites* at this site. Additional baculite jaws may be present in other concretions from this same horizon but require further study.

Several isolated jaws were discovered in the upper Campanian *Baculites eliasi* Zone of the Lewis Shale of Natrona County, Wyoming (Figs. 13.1, 2). The jaws occur in a sandstone concretion associated with pieces of *Pseudobaculites natosini* (Robinson, 1945). This species is the largest baculite known from the Upper Cretaceous of the Western Interior of North America (Cobban and Kennedy, 1994). Other ammonites in these strata include *Jeletzkytes plenus* (Meek, 1876) and *Baculites eliasi* (Cobban, 1958). The large size of the jaws matches the large size of *P. natosini* and, therefore, the jaws are attributed to this species.

5 Conventions

The terminology used to describe the lower jaws of ammonites is that of Kanie (1982), Tanabe (1983), and Tanabe and Fukuda (1987), and illustrated in Fig. 13.3. We employ the terms anterior, posterior, ventral, dorsal, left, and right, to refer to the jaws as they were oriented in life. The jaws are illustrated with the anterior end on top. The most recently formed portion of the jaw is at the posterior end. The two symmetric halves of the lower jaw are called the wings, with the hinge line (symphysis = commissure) along the midline.

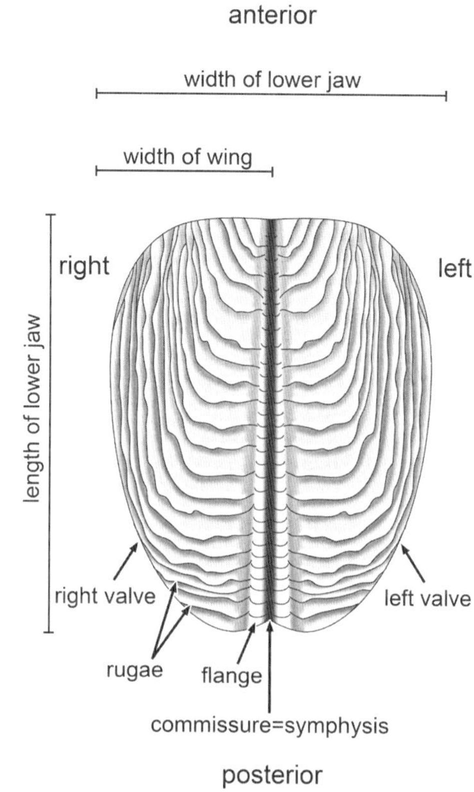

Fig. 13.3 *Terminology and measurements of the lower jaw of baculites, showing the ventral surface covered with an aptychus.*

Most of the jaws we describe are incomplete. We measured the width and length of each wing of the jaw, irrespective of curvature, following the approach of Kanie (1982) and Tanabe and Fukuda (1987). These measurements represent the maximum dimensions of the wing perpendicular and parallel to the symphysis, respectively. Width is a more reliable indicator of size than length, because the long end of the jaw is commonly broken, especially in isolated elements. The width of the jaw equals twice the width of the wings. The ratio of jaw width to length provides an approximation of jaw shape.

Landman et al. (2006) and Landman and Grebneff (2006) discussed the terminology surrounding aptychi. They defined an aptychus type lower jaw as a lower jaw bearing an aptychus. An aptychus is the entire calcareous layer covering the ventral surface of the lower jaw. It consists of a pair of left and right valves. (For additional discussions about the terminology used to describe aptychi, see Trauth, 1927–1936, 1938; Arkell, 1957; Moore and Slyvester-Bradley, 1957; and Farinacci et al., 1976).

Trauth (1927) assigned the coarsely ornamented aptychi of *Baculites* to the form genus *Rugaptychus*, and erected parataxa with binomial names written in italics, to describe variation within this morphotype. Moore and Sylvester-Bradley (1957:

L469) characterized *Rugaptychus* as follows: "Elongate diaptychi [pair of valves] with strong sharp ridges on outer surface, ridges characteristically arranged with angulated bend; inner surface with growth lines, nearly smooth." We follow the practice of Engeser and Keupp (2002) in employing the term rugaptychus as a common name without italics to refer to this kind of aptychus.

Specimens are reposited in the American Museum of Natural History (AMNH), New York; the Black Hills Museum of Natural History (BHMNH), Hill City, South Dakota; and the Sternberg Museum of Natural History (FHSM), Hays, Kansas; and the US National Museum (USNM), Washington, DC.

6 Description of Jaws

6.1 *Baculites* sp. (smooth), Pierre Shale, South Dakota

The body chambers containing lower jaws are fragmentary and range from 43.2 mm to 120.9 mm in length (Figs. 13.4–14, Table 13.1). All of the body chambers are steinkerns but several retain pieces of the aragonitic outer shell (= composite internal molds). The body chambers consist of four stout and eight slender individuals, presumably representing mature macroconchs and microconchs, respectively.

In specimens that preserve the ultimate septum or in which the shell expands enough to detect the adoral direction, the jaw is located in the adoral part of the body chamber. In most specimens, the long axis of the jaw lies at an angle of 45°–90° to the long axis of the body chamber, with the anterior end of the jaw on the dorsal side of the shell, and the posterior end of the jaw on the ventral side of the shell (Figs. 13.5–7, 8D–F, 9, 10, 11, 12A–C, 13). The jaws are nearly perpendicular to the whorl section except in BHMNH 5494a, in which the jaw is displaced to the left hand side of the shell (Fig. 13.7). In BHMNH 5147 and 5497 (Figs. 13.8A–C, 12D–F), in contrast, the long axis of the jaw is approximately parallel to the long axis of the body chamber, with the jaw lodged in the dorsal and ventral half of the shell, respectively.

The jaw consists of two wings with the symphysis along the midline. Both wings are preserved in AMNH 47109, BHMNH 5494a, 5147, 5143, 5491, and 5496

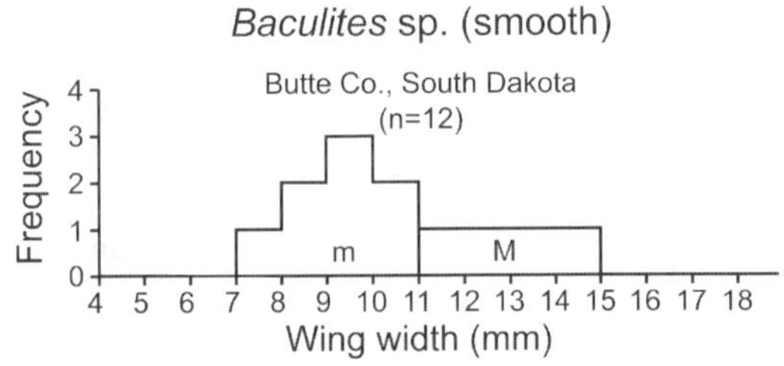

Fig. 13.4 *Size–frequency histogram of the wing width of the lower jaw of* Baculites *sp. (smooth) from the Gammon Ferruginous Member of the Pierre Shale, South Dakota. M = macroconch; m = microconch.*

Fig. 13.5 *Steinkern of part of the body chamber of* Baculites *sp. (smooth), macroconch, AMNH 47109, Pierre Shale, South Dakota, with lower jaw inside. A. Right lateral view. X1. B. Ventral view. X1. C. Lower jaw with parts of the aptychus preserved, dorsal side of the shell on top, coated. X1.5.*

(Figs. 13.5, 7, 8A–C, 9, 11, 12A–C). The wings are folded in a U-shape in BHMNH 5147, 5143, and 5491 (Figs. 13.8A–C, 9, 11). In contrast, they are nearly flattened out in AMNH 47109 and BHMNH 5494a (Figs. 13.5, 7) and folded slightly outward in BHMNH 5496 (Fig. 13.12A–C). In the other six specimens, only one wing or one wing and part of the other wing is preserved. In BHMNH 5148 and 5495 (Figs. 13.8D–F, 13D–F), the partially preserved wing is oriented nearly perpendicular to the plane of the other wing.

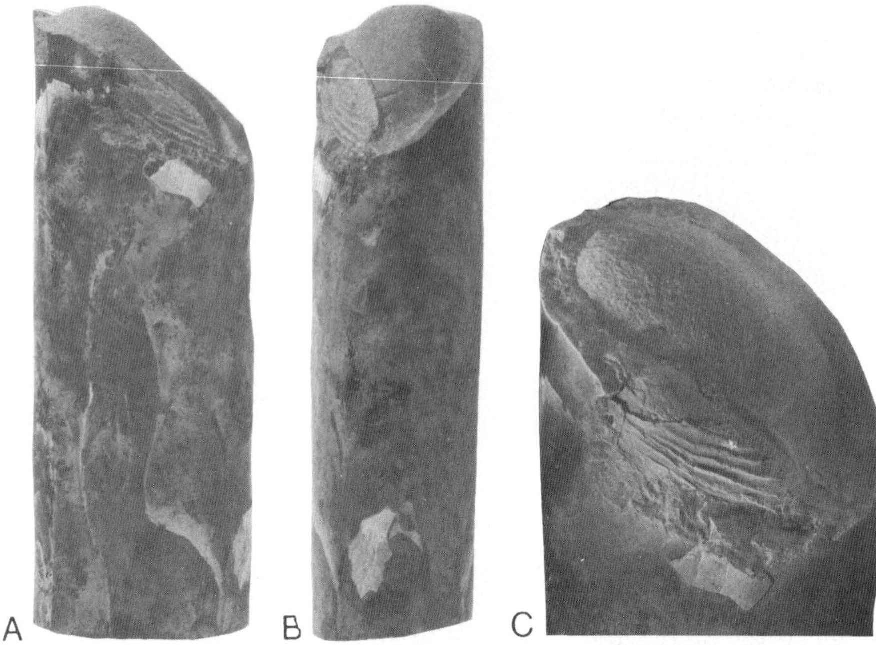

Fig. 13.6 *Steinkern of part of the body chamber of* Baculites *sp. (smooth), microconch, AMNH 51329, Pierre Shale, South Dakota, with the lower jaw inside. A. Right lateral view. X1. B. Ventral view. X1. C. Lower jaw with part of the right side of the aptychus preserved, coated. X1.5.*

The jaw measurements are reported in Table 13.1. Jaw length ranges from 20.8 mm to 37.2 mm. Wing width ranges from 7.8 mm to 14.4 mm. Jaw width, which equals twice the wing width, ranges from 15.6 mm to 28.8 mm. The ratio of jaw width to jaw length ranges from 0.68 to 0.87 and averages 0.77, indicating an elongate shape.

A comparison of the shape of the jaws with that of the whorl cross section reveals that the jaws fit snugly into the body chamber (Table 13.1). The ratio of jaw length to whorl height ranges from 0.72 to 1.03, and averages 0.85. The ratio of jaw width to whorl width ranges from 0.81 to 1.04, and averages 0.94. The ratio of jaw width to jaw length is the same or slightly higher than the ratio of whorl width to whorl height. When one or both wings are splayed out across the whorl section, the sides of the jaw touch the sides of the shell, e.g., AMNH 47109, BHMNH 5148, 5144, and 5496 (Figs. 13.5, 8D–F, 10, 12A–C).

The variation in jaw size with respect to sexual dimorphism is illustrated in Fig. 13.4. The baculites were divided into macroconchs and microconchs based on the size and robustness of the shell. Wing width (= 1/2 aptychus width) was used as a measure of jaw size. The histogram of wing width shows no overlap between dimorphs.

The jaw is covered with a pair of calcareous plates or valves (= the aptychus). Each valve has a broadly rounded lateral margin, a straight symphysal margin, and

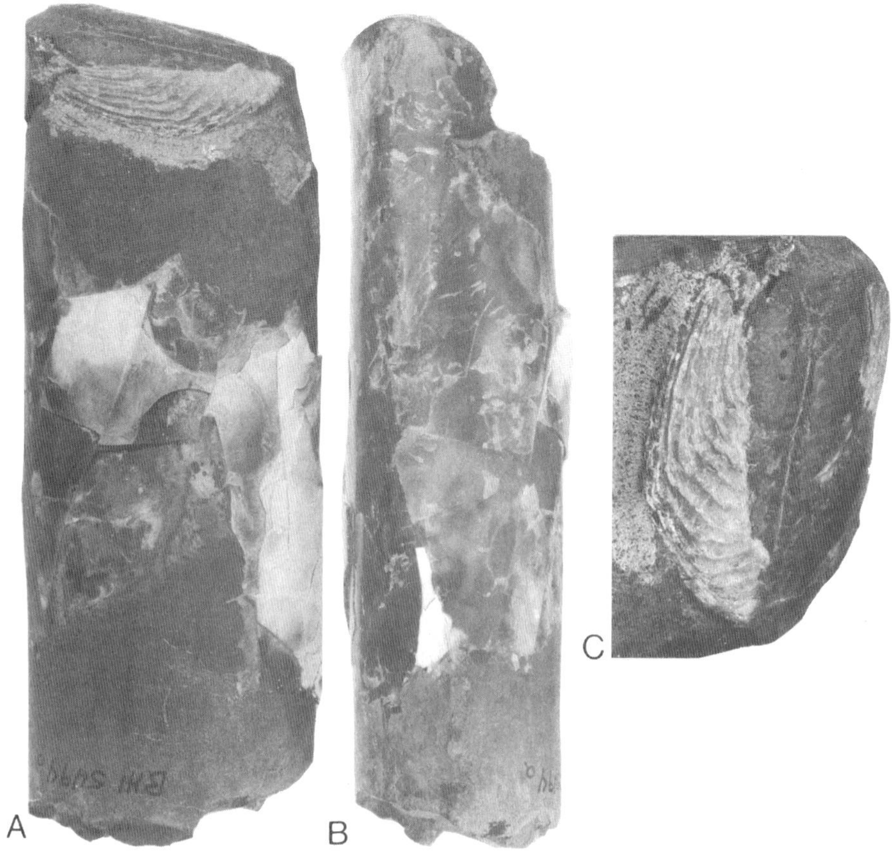

Fig. 13.7 *Steinkern of part of the body chamber of* Baculites *sp. (smooth), macroconch, BHMNH 5494a, Pierre Shale, South Dakota, with lower jaw inside. A. Lateral view. X1. B. Dorsal view. X1. C. Lower jaw with part of the right side of the aptychus preserved, dorsal side of the shell on top. X1.5.*

a narrowly rounded posterior margin. The anterior margin is missing or obscured in all specimens. The symphysis is bordered by a flange with a sharp crest, as shown in BHMNH 5144 and 5491 (Figs. 13.10, 11).

The aptychi are composed of calcite, as indicated by X-ray diffraction analysis of a sample from BHMNH 5491. The calcite in this specimen is 0.50 mm thick at a ridge and 0.17 mm thick near the symphysis. The aptychi are covered with coarse, irregular rugae characteristic of rugaptychi. The rugae parallel the lateral and posterior margins and approach the symphysis at nearly a right angle. In BHMNH 5144 (Fig. 13.10), with the best preserved aptychus, there are 13 rugae, although additional ones are undoubtedly obscured in the anterior region. The rugae are more broadly spaced in the middle of the aptychus than near the posterior margin.

In some specimens, parts of the aptychus are broken off exposing large portions of the underlying chitinous layer. The chitinous layer bears a midline groove

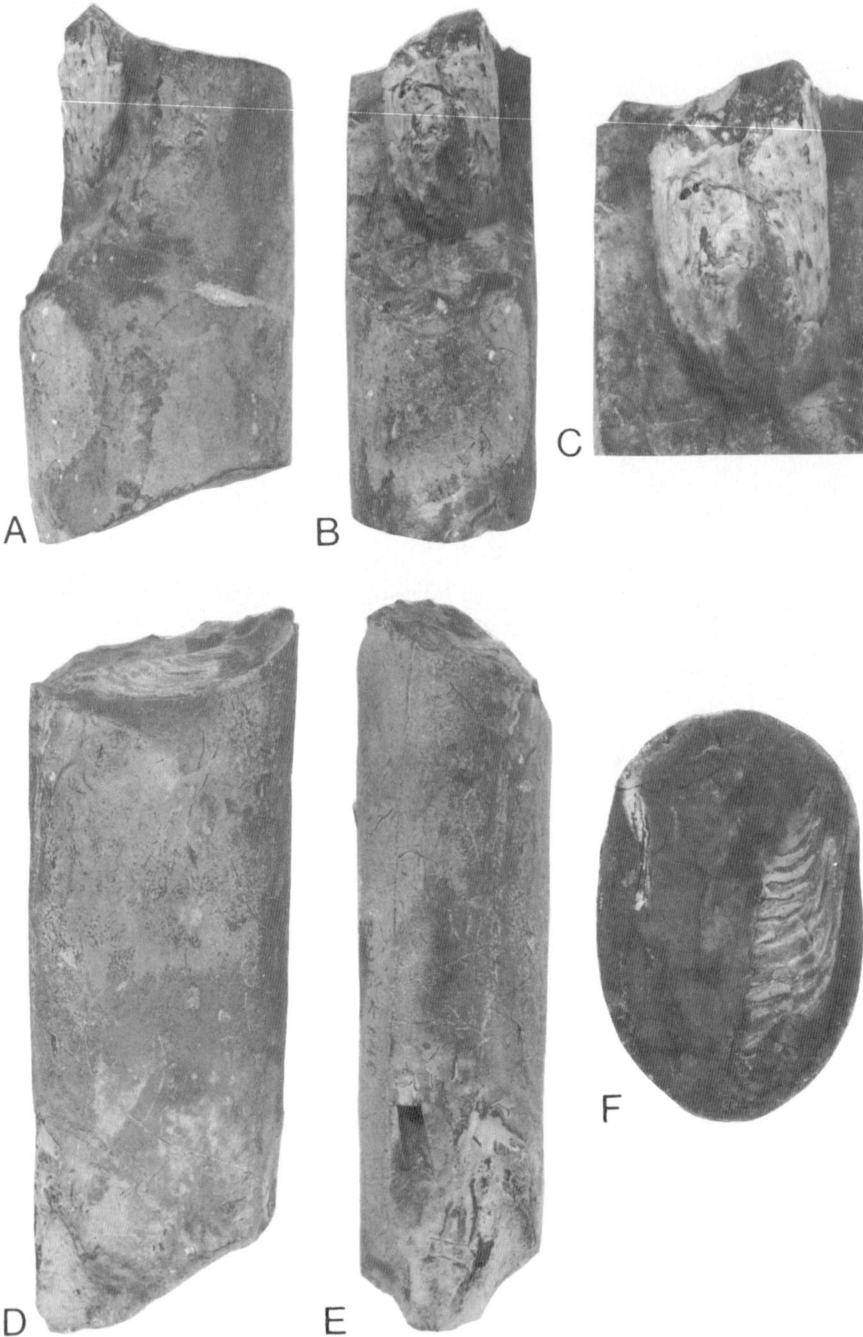

Fig. 13.8 *Steinkerns of parts of the body chambers of* Baculites *sp. (smooth), Pierre Shale, South Dakota, with lower jaws inside. A–C. BHMNH 5147, microconch. A. Right lateral view. X1. B. Dorsal view. X1. C. Close-up of the lower jaw folded in a U-shape. X1.5. D-F. BHMNH 5148, microconch. D. Left lateral view. X1. E. Ventral view. X1. F. Lower jaw with part of the left side of the aptychus preserved, dorsal side of the shell on top. X1.5.*

Fig. 13.9 *Steinkern of part of the body chamber of* Baculites *sp. (smooth), BHMNH 5143, macroconch, Pierre Shale, South Dakota, with the lower jaw inside. A. Right lateral view. X1. B. Ventral view. X1. C. Lower jaw, dorsal side of the shell on top. X1.5.*

bordered by flanges. In BHMNH 5494a (Fig. 13.7), this layer is smooth with subdued undulations. The two wings in this specimen diverge at the anterior end and the symphysis disappears, so that there is a triangular gap between the wings.

Two layers are visible below the calcitic aptychus in BHMNH 5491: a black crystalline layer (80 μm thick near the symphysis) and an underlying tan layer (60 μm thick near the symphysis). X-ray diffraction analysis of these two layers indicates that they consist of an amorphous material and magnesium enriched calcite, respectively.

A radula is present in BHMNH 5496 (Figs. 13.12A–C, 14). It occurs in the middle of the jaw at the anterior end. The aptychus is missing in this area and the surface is eroded away. The radula is approximately 4.1 mm long and 2.0 mm wide. The teeth

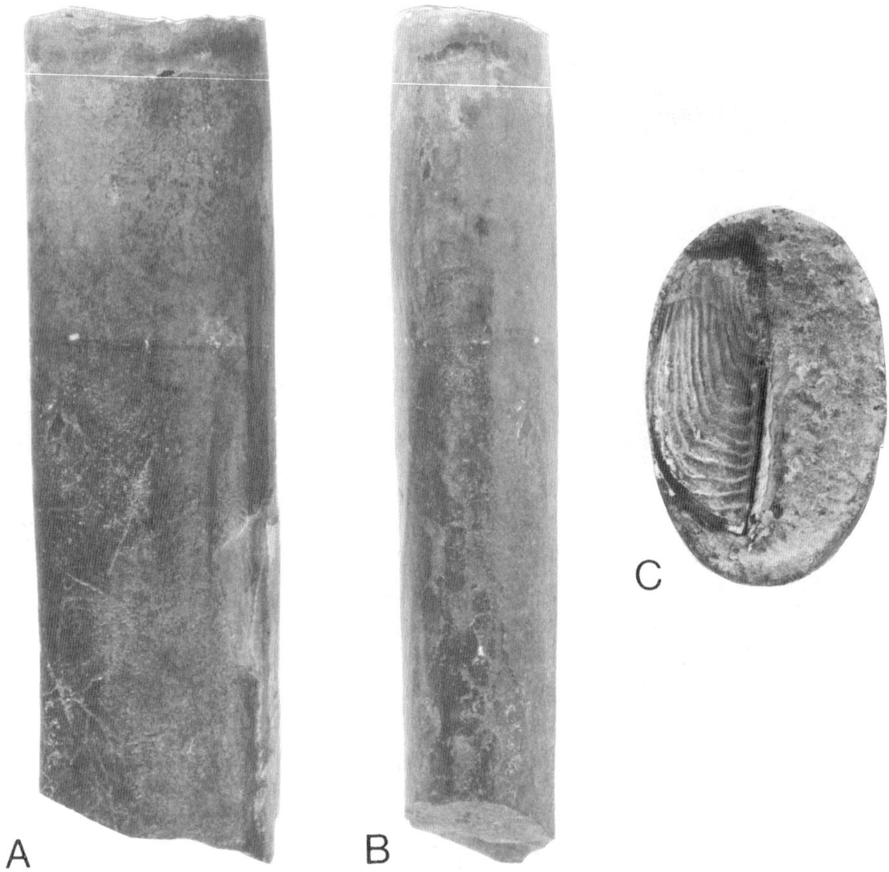

Fig. 13.10 *Steinkern of part of the body chamber of* Baculites *sp. (smooth), BHMNH 5144, microconch, Pierre Shale, South Dakota, with the lower jaw inside. A. Left lateral view. X1. B. Ventral view. X1. C. Close-up of the lower jaw with the right side of the aptychus preserved, dorsal side of the shell on top. X1.5.*

are dark brown with reddish overgrowths. Three elongate, slightly curved elements are visible in the upper right side of the radula and represent the marginal teeth (Fig. 13.14B). Each tooth is approximately 990 μm long and lies on a curved sheet. The teeth point forward and inward. They are ornamented with ridges spaced at equal intervals of approximately 70 μm (Fig. 13.14C).

Three blunter, rectangular elements are visible just to the right of the marginal teeth, and represent the marginal plates. They are aligned almost perpendicular to the marginal teeth and are spaced between them.

Additional teeth are present on the lower right side of the radular complex but are indistinct (Fig. 13.14A). The middle of the radula is recrystallized. A small triangular tooth is visible at the lower end flanked by longer teeth on either side, which point inward. The left side of the complex also contains elongate teeth like those in the upper right side, but they are broken and embedded in the matrix.

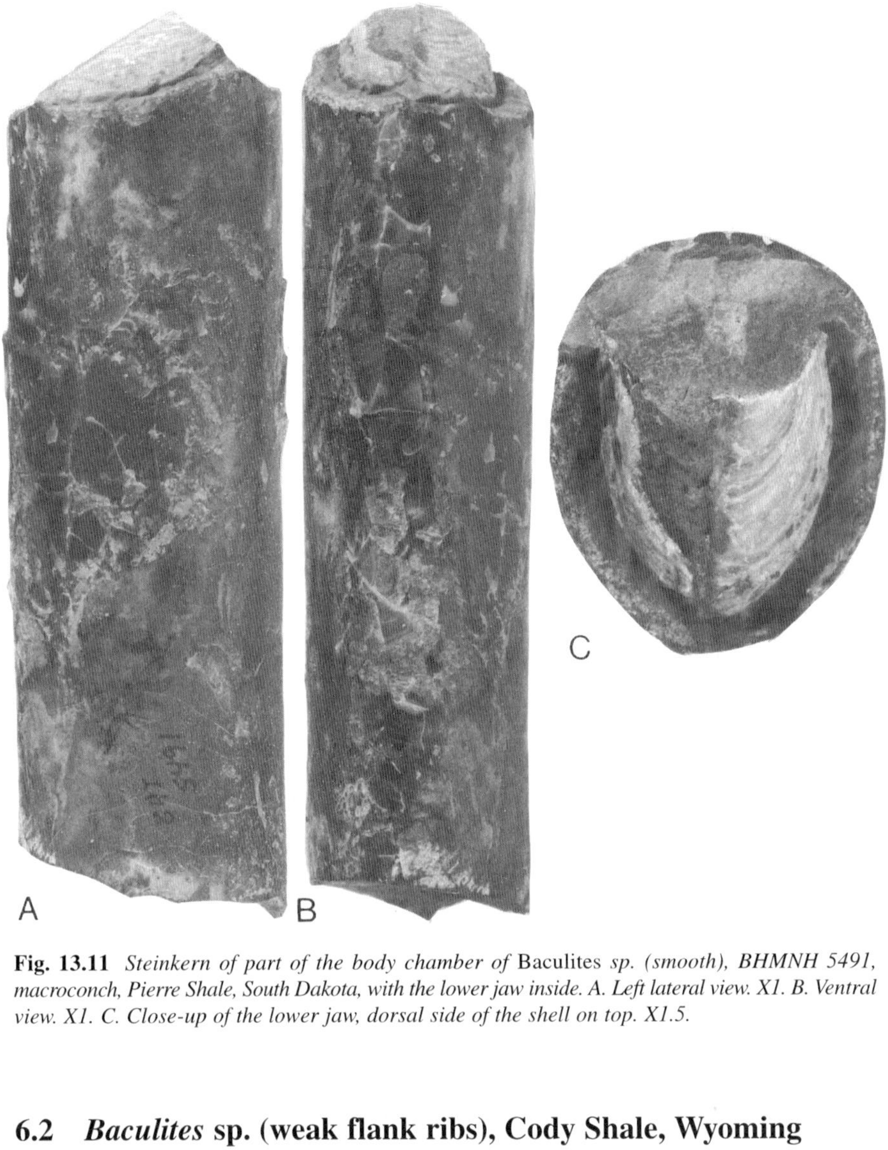

Fig. 13.11 *Steinkern of part of the body chamber of* Baculites *sp. (smooth), BHMNH 5491, macroconch, Pierre Shale, South Dakota, with the lower jaw inside. A. Left lateral view. X1. B. Ventral view. X1. C. Close-up of the lower jaw, dorsal side of the shell on top. X1.5.*

6.2 *Baculites* sp. (weak flank ribs), Cody Shale, Wyoming

A lower jaw (BHMNH 5331) occurs inside the adoral end of a weathered fragment of a body chamber of *Baculites* sp. (weak flank ribs) (Fig. 13.15, Table 13.1). The body chamber fragment is 111.0 mm long with a maximum whorl height of 33.8 mm. The long axis of the jaw is oriented at an angle of 70° with the long axis of the body chamber. The anterior end of the jaw points slightly backward against the ventral (?) side of the shell. The jaw is folded along the symphysis so that the two wings are perpendicular to each other, with the right wing splayed out across the opening.

Fig. 13.12 *Steinkerns of parts of the body chambers of* Baculites *sp. (smooth), Pierre Shale, South Dakota, with lower jaws inside. A-C. BHMNH 5496, microconch. A. Right lateral view. X1. B. Dorsal view. X1. C. Close-up of the lower jaw, dorsal side of the shell on top. The radula (arrow) occurs at the anterior end of the jaw between the two wing tips. X1.5. D-F. BHMNH 5497, microconch. D. Left lateral view. X1. E. Ventral view. X1. F. Close-up of the lower jaw (arrow), with the left side of the aptychus preserved in the ventral half of the shell, coated. X1.5.*

Fig. 13.13 *Steinkerns of parts of the body chambers of* Baculites *sp. (smooth), Pierre Shale, South Dakota, with lower jaws inside. A–C. BHMNH 5146, microconch. A. Right lateral view. X1. B. Dorsal view. X1. C. Close-up of the lower jaw with the right side of the aptychus preserved, dorsal side of the shell on top. X1.5. D-F. BHMNH 5495, microconch. D. Left lateral view. X1. E. Ventral view. X1. F. Close-up of the lower jaw with the right side of the aptychus preserved, dorsal side of the shell on top, coated. X1.5.*

Table 13.1 *Measurements of the whorl cross section and lower jaws of* Baculites/Pseudobaculites *from the Pierre Shale, Cody Shale, Lewis Shale, Mooreville Chalk, and Niobrara Formation. Abbreviations: M/m = macroconch/microconch; WW = whorl width; WH = whorl height; W = width of wing as measured perpendicular to the symphysis, irrespective of convexity; L = length of wing measured parallel to the symphysis; 2W/L = ratio of jaw width to jaw length; 2W/WW = ratio of jaw width to whorl width; L/WH = ratio of jaw length to whorl height; *=estimate due to poor or incomplete preservation. All measurements in millimeters.*

Specimen M/m	WW	WH	WW/WH	W	L	2W/L	2W/WW	L/WH
Baculites sp. (smooth), Pierre Shale, South Dakota								
BHMNH 5143 M	25.8*	33.2	0.78*	11.3	26.0*	0.87*	0.88	0.78*
5144 m	17.8	27.0	0.66	8.4	20.8	0.81	0.94	0.72
5146 m	18.8	28.4	0.66	9.8	25.0*	0.78*	1.04	0.88*
5147 m	22.6	32.4	0.70	10.2	25.2*	0.81*	0.90	0.75*
5148 m	21.9	31.4	0.70	10.2	26.3	0.78	0.93	0.86
5491 M	30.1	37.8	0.80	14.4	–	–	0.96	–
5494a M	28.1	41.1	0.68	13.2	34.5	0.76	0.94	0.87
5496 m	19.1	27.4	0.70	9.2	27.0*	0.68*	0.96	0.98
5497 m	16.0	22.2	0.72	7.8	–	–	0.98	–
5495 m	17.2	25.3	0.68	8.5	21.2	0.80	0.99	0.81
AMNH 47109 M	26.9	36.2	0.74	12.9	37.2	0.69	0.96	1.03
51329 m	22.8	30.4	0.75	9.2*	25.1*	0.73*	0.81*	0.82*
Baculites sp. (weak flanks ribs), Cody Shale, Wyoming								
BHMNH 5331	26.6	33.8	0.79	12.5	33.2	0.75	0.94	0.98
Baculites sp. (smooth and weak flank ribs), Mooreville Chalk, Alabama (selected specimens)								
BHMNH 5406				15.8	38.6	0.82		
5407				11.7	27.5*	0.85*		
5408				9.9	21.5*	0.92*		
5409				13.3	31.5*	0.84*		
5410				12.1	27.2*	0.89*		
5484				11.8	30.6	0.77		
5450				6.4	15.2*	0.84*		
5150				17.2	–	–		
AMNH 51293				13.3	30.6*	0.87*		
51316				13.8	–	–		
Baculites sp., Niobrara Formation, Kansas								
FHSM 968:1				10.2	20.1*	0.51*		
Baculites compressus or *B. cuneatus*, Pierre Shale, South Dakota								
BHMNH 5498				7.9	19.5	0.81		
Baculites compressus, Pierre Shale, South Dakota								
AMNH 51885	17.2	31.0	0.55					
51880	13.9	24.9	0.56					
51865	15.5	27.1	0.57					
51879	19.6	40.4	0.48					
51871	15.7	26.9	0.58					
51872	16.3	30.0	0.54					
51886	19.3	36.1	0.53					
51887	16.1	30.3	0.53					

Specimen M/m	WW	WH	WW/WH	W	L	2W/L	2W/WW	L/WH
Baculites cuneatus, Pierre Shale, South Dakota								
AMNH 51866	24.6	50.2	0.49					
51873	16.8	32.4	0.52					
50422	15.2	26.2	0.58					
BHMNH 5498	25.2	40.6	0.62					
Pseudobaculites natosini, Lewis Shale, Wyoming								
BHMNH 5501				25.8	50.1*	1.03*		
AMNH 51328				26.3	59.0	0.89		
BHMNH 5502				23.6	52.8	0.89		
BHMNH 5500				28.1	–	–		
USNM 458243	31.9	87.3	0.36					
458241	60.0	141.0	0.43					
458239	72.0	152.0	0.48					

Some of the rugose ornament of the aptychus is preserved on the left wing. The calcite is approximately 0.4 mm thick. Most of the aptychus is missing on the right wing and there is a dark brown surface comprising the underlying chitinous layer. This layer is broadly convex and is covered with fine ridges spaced at intervals of 300 μm and very fine lines spaced at intervals of 80 μm, both of which parallel the lateral and posterior margins.

6.3 *Baculites* sp. (smooth and weak flank ribs), Mooreville Chalk, Alabama

There are approximately 150 isolated aptychi. They are attributed to *Baculites* sp. (smooth) and *Baculites* sp. (weak flank ribs) by comparison with the jaws inside the body chambers of these species and the fact that these two species occur in age-equivalent strata elsewhere. In terms of measurable specimens, the collection contains 8 pairs of valves found in close association, 19 left valves, and 11 right valves. The valves range from 15.2 mm to 38.6 mm in length and 6.4 mm to 17.2 mm in width, although nearly all of the valves are slightly broken (Table 13.1). The ratio of jaw width (=twice the width of the valve) to jaw length (=length of the valve) ranges from 0.77 to 0.92.

The size distribution of 38 valves in our collection is illustrated in Fig. 13.16, using valve width rather than length as an indicator of size, because the specimens are more commonly broken in length than width. Only one valve of each pair was counted. As a cautionary note, the widths of valves in a pair do not always match, presumably due to breakage of one of the valves, but the difference is usually trivial. In case of a discrepancy, the larger value was used. The resultant histogram shows a bimodal distribution of valve width with modes at 9–10 mm and 13–14 mm.

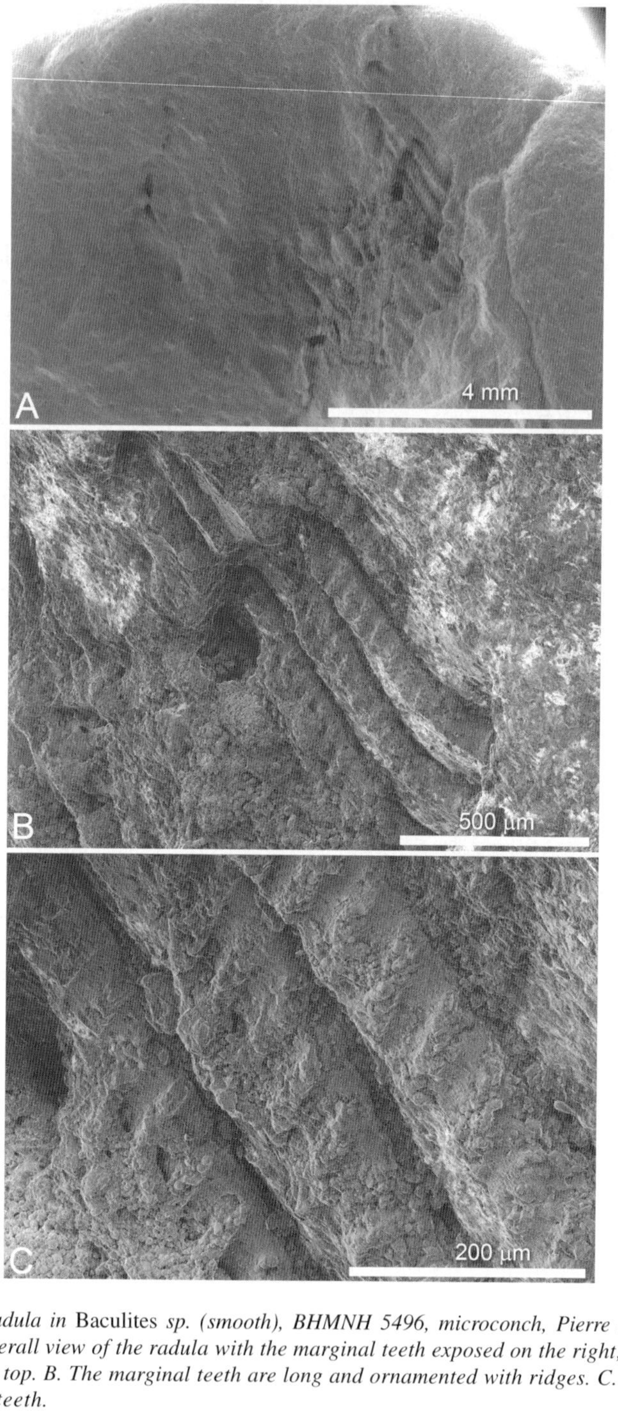

Fig. 13.14 *Radula in* Baculites *sp. (smooth), BHMNH 5496, microconch, Pierre Shale, South Dakota. A. Overall view of the radula with the marginal teeth exposed on the right; dorsal side of the shell on top. B. The marginal teeth are long and ornamented with ridges. C. Close-up of the marginal teeth.*

Fig. 13.15 *Steinkern of part of the body chamber of* Baculites *sp. (weak flank ribs), BHMNH 5331, macroconch, Cody Shale, Wyoming, with the lower jaw inside. A. Right lateral (?) view. X1. B. Ventral (?) view. X1. C. Close-up of the lower jaw with part of the aptychus and underlying chitinous layer preserved, ventral (?) side of the shell on top. X1.5.*

Because some of the isolated valves may belong to the same jaw and are thus counted twice, we reevaluated the data by constructing histograms using only left valves (including the left valve of each pair), and histograms using only right valves (including the right valve of each pair). The histogram of left valves retains the mode at 13–14 mm but the other mode disappears. The histogram of right valves yields the original two modes at 9–10 mm and 13–14 mm.

This bimodal distribution may reflect sexual dimorphism. However, it is difficult to demonstrate that all of the aptychi were derived from adults. One indication of ontogenetic stage is the degree of convexity of the aptychus, with flatter aptychi characteristic of later ontogenetic stages (see below). According to this criterion, most of these aptychi are mature, with the exception of the two smallest specimens.

The shape of each valve is semilunate (Figs. 13.17–20). The lateral margin is broadly rounded and the posterior margin is more sharply rounded. The anterior

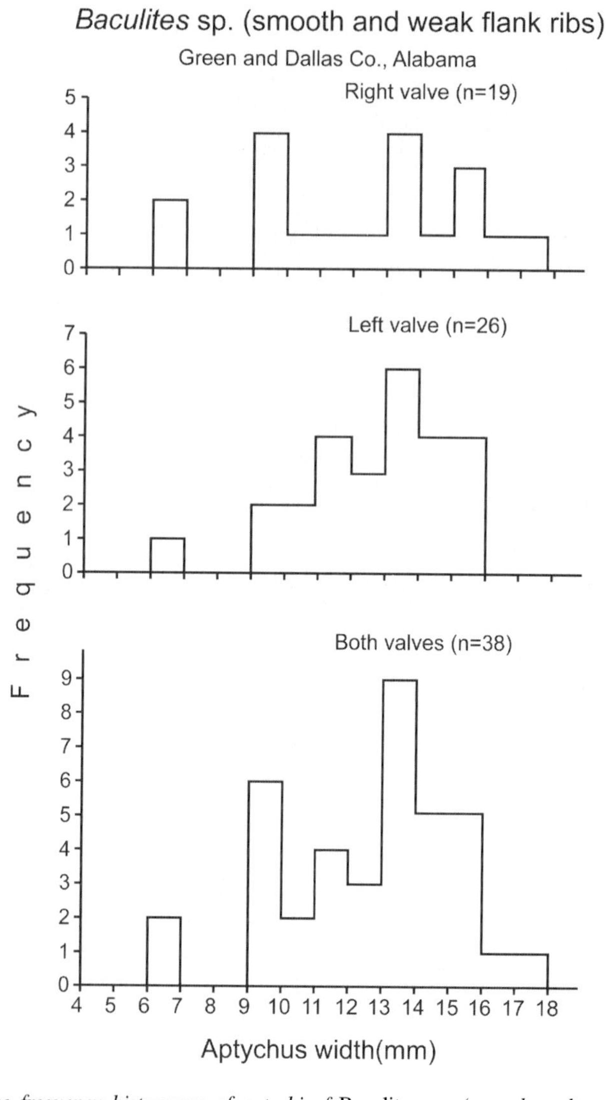

Fig. 13.16 *Size–frequency histograms of aptychi of* Baculites *sp. (smooth and weak flank ribs), Mooreville Chalk, Greene and Dallas counties, Alabama. The bimodal distribution with peaks at 9–10 mm and 13–14 mm may reflect sexual dimorphism, provided that the specimens are all mature.*

margin usually ends in a notch that parallels the rugae and is probably the result of breakage. The symphysal edge is bordered by a flange, reflecting the morphology of the underlying chitinous layer.

The aptychi are weakly convex on the ventral side and weakly concave on the dorsal side. Smaller specimens are more strongly curved than larger specimens, which probably reflects a change in shape during ontogenetic development (compare

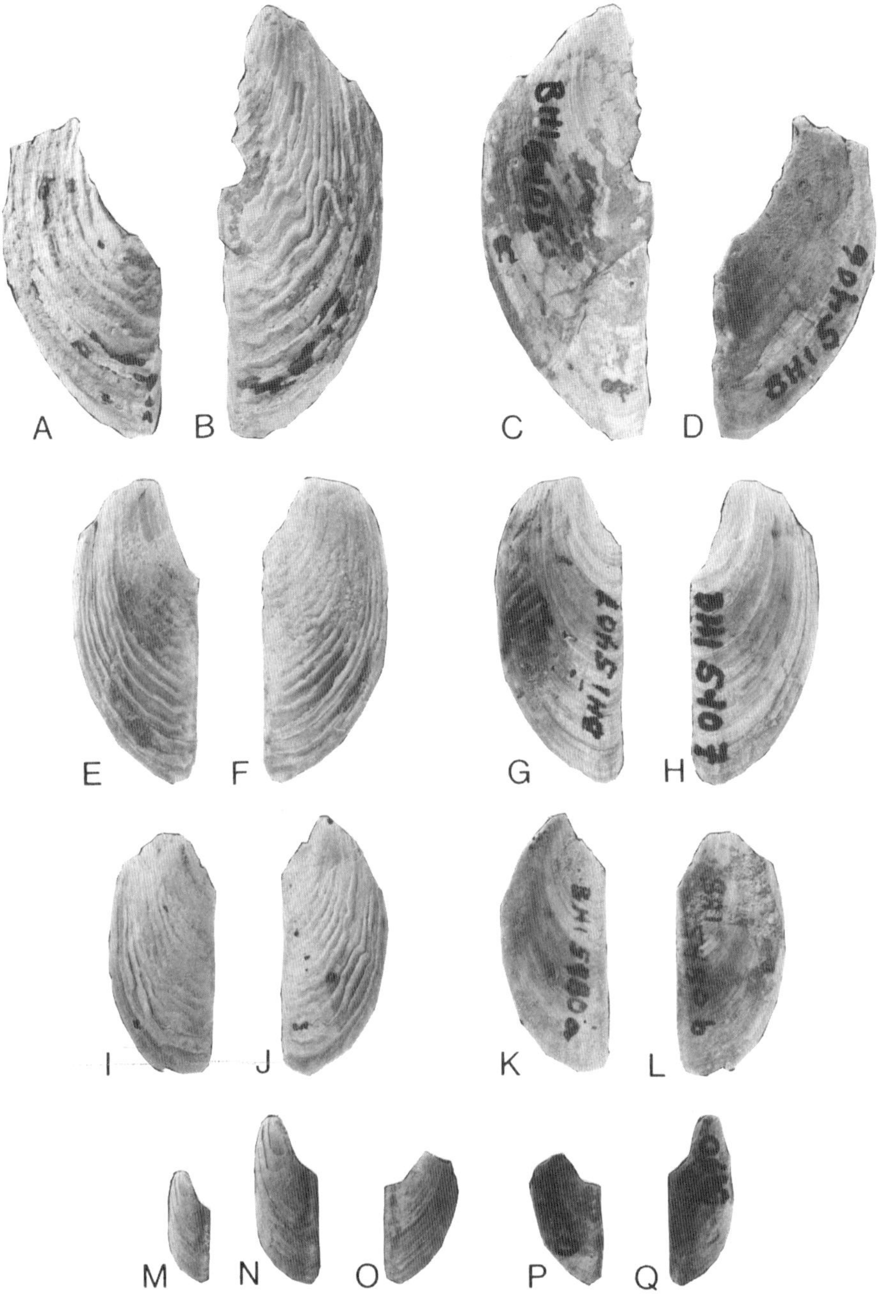

Fig. 13.17 *Aptychi of* Baculites *sp. (smooth and weak flank ribs), Mooreville Chalk, Greene and Dallas counties, Alabama. A–D. Aptychus, BHMNH 5406. A. Right valve, ventral view. B. Left valve, ventral view. C. Left valve, dorsal view. D. Right valve, dorsal view. E-H. Aptychus, BHMNH 5407. E. Right valve, ventral view. F. Left valve, ventral view. G. Left valve, dorsal view. H. Right valve, dorsal view. I-L. Aptychus, BHMNH 5480. I. Right valve, ventral view. J. Left valve, ventral view. K. Left valve, dorsal view. L. Right valve, dorsal view. M–Q. Aptychus, BHMNH 5450. M, N. Right valve, ventral view. O. Left valve, ventral view. P. Left valve, dorsal view. Q. Right valve, dorsal view. M is X1; all other figures X1.5.*

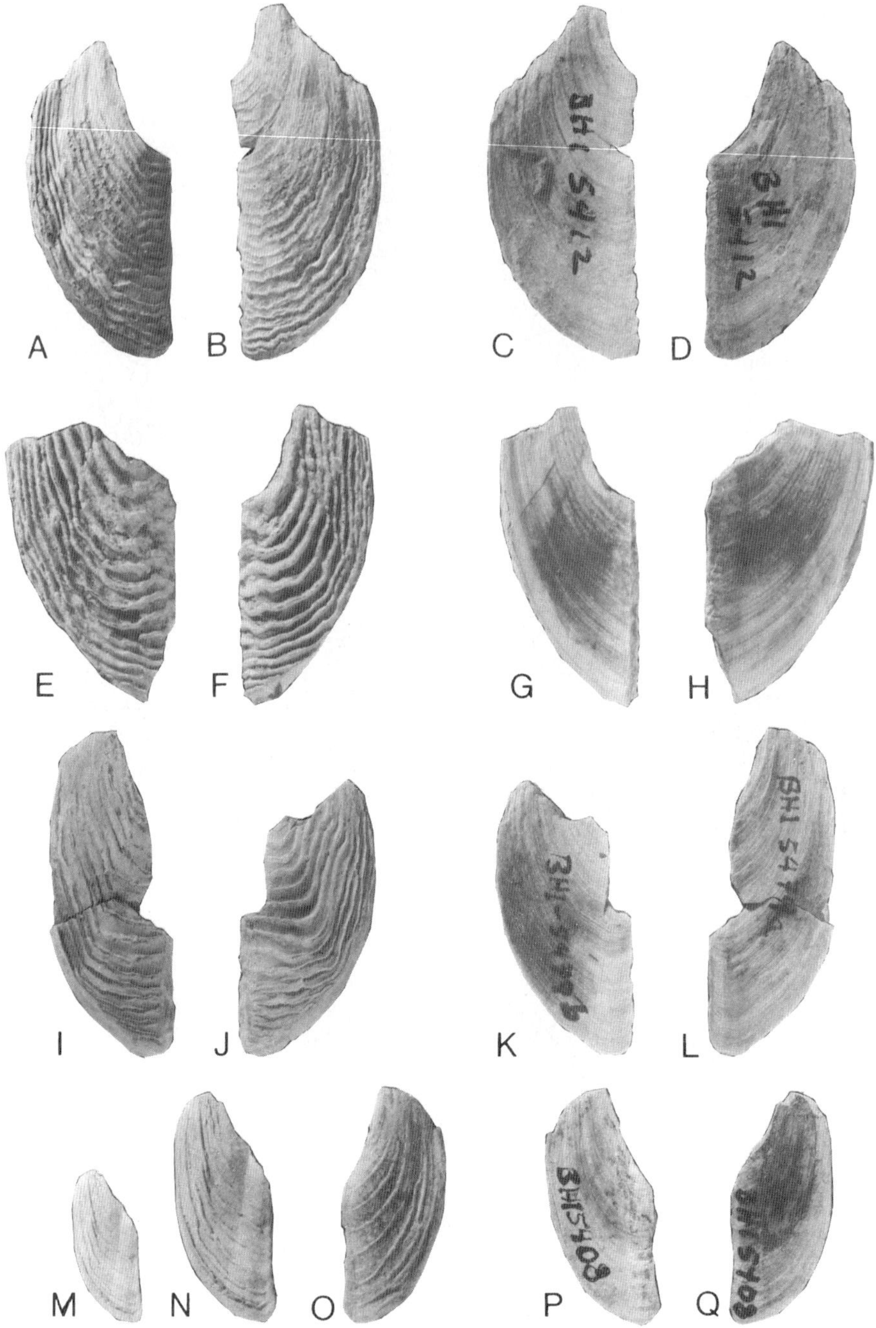

Fig. 13.18 *Aptychi of* Baculites *sp. (smooth and weak flank ribs), Mooreville Chalk, Greene and Dallas counties, Alabama. A–D. Aptychus, BHMNH 5412. A. Right valve, ventral view. B. Left valve, ventral view. C. Left valve, dorsal view. D. Right valve, dorsal view. E–H. Aptychus, BHMNH 5149. E. Right valve, ventral view. F. Left valve, ventral view. G. Left valve, dorsal view. H. Right valve, dorsal view. I–L. Aptychus, BHMNH 5479. I. Right valve, ventral view. J. Left valve, ventral view. K. Left valve, dorsal view. L. Right valve, dorsal view. M–Q. Aptychus, BHMNH 5408. M, N. Right valve, ventral view. O. Left valve, ventral view. P. Left valve, dorsal view. Q. Right valve, dorsal view. M is X1; all other figures X1.5.*

Fig. 13.19 *Aptychi of* Baculites *sp. (smooth and weak flank ribs), Mooreville Chalk, Greene and Dallas counties, Alabama. A–C. Right valve, BHMNH 5156. A, B. Ventral view. C. Dorsal view. D, E. Left valve, BHMNH 5409. D. Ventral view. E. Dorsal view. F, G. Right valve, BHMNH 5152. F. Ventral view. G. Dorsal view. H, I. Right valve, BHMNH 5157. H. Ventral view. I. Dorsal view. J, K. Left valve, AMNH 51342. J. Ventral view. K. Dorsal view. A is X1; all other figures X1.5.*

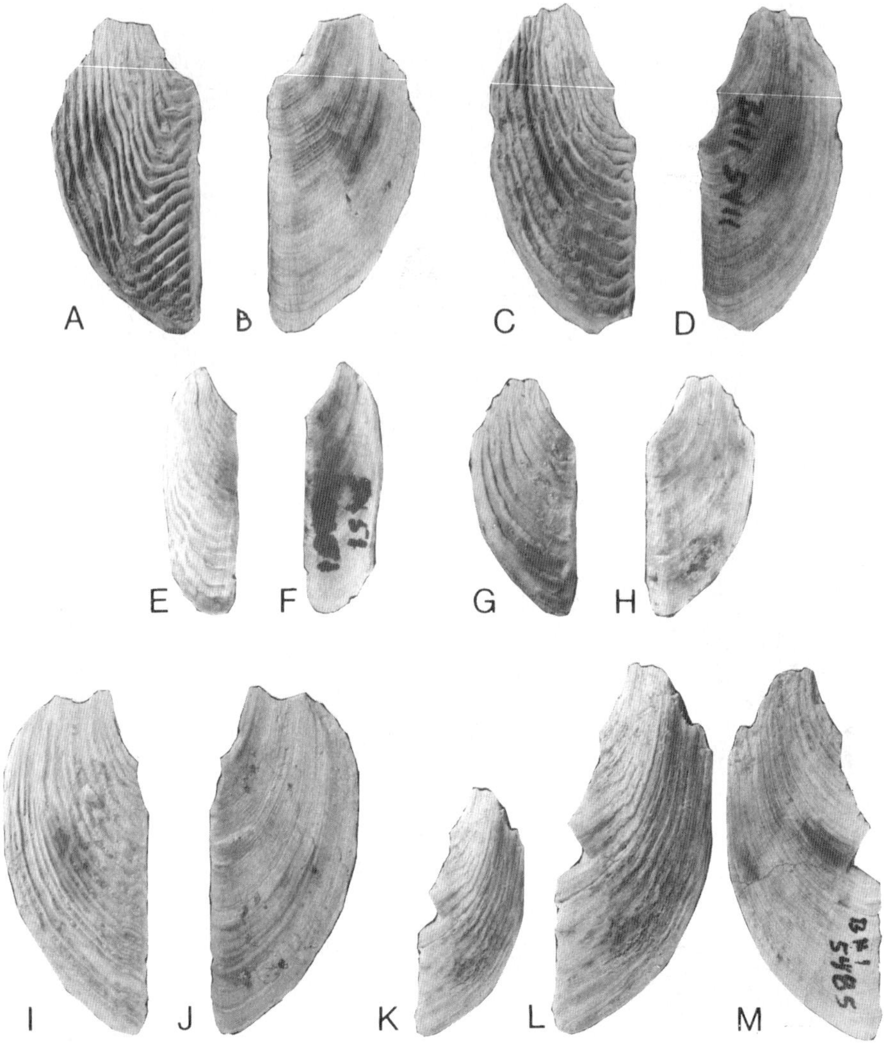

Fig. 13.20 *Aptychi of* Baculites *sp. (smooth and weak flank ribs), Mooreville Chalk, Greene and Dallas counties, Alabama. A, B. Right valve, AMNH 51316. A. Ventral view. B. Dorsal view. C, D. Right valve, BHMNH 5411. C. Ventral view. D. Dorsal view. E, F. Right valve, BHMNH 5451. E. Ventral view. F. Dorsal view. G, H. Right valve, BHMNH 5151. G. Ventral view. H. Dorsal view. I, J. Right valve, AMNH 51293. I. Ventral view. J. Dorsal view. K–M. Left valve, BHMNH 5485. K, L. Ventral view. M. Dorsal view. K is X1; all other figures X1.5.*

Figs. 13.17M–Q and 20K–M). Some valves, such as BHMNH 5152, also show longitudinal flexures, associated with fractures on the dorsal side (Fig. 13.19F, G).

The aptychi vary in thickness. For example, the thickness at the posterior margin ranges from 2.8 mm in BHMNH 5149 to 0.8 mm in BHMNH 5485. In any one specimen, the valve is thickest in the center and thinnest at the symphysis, which

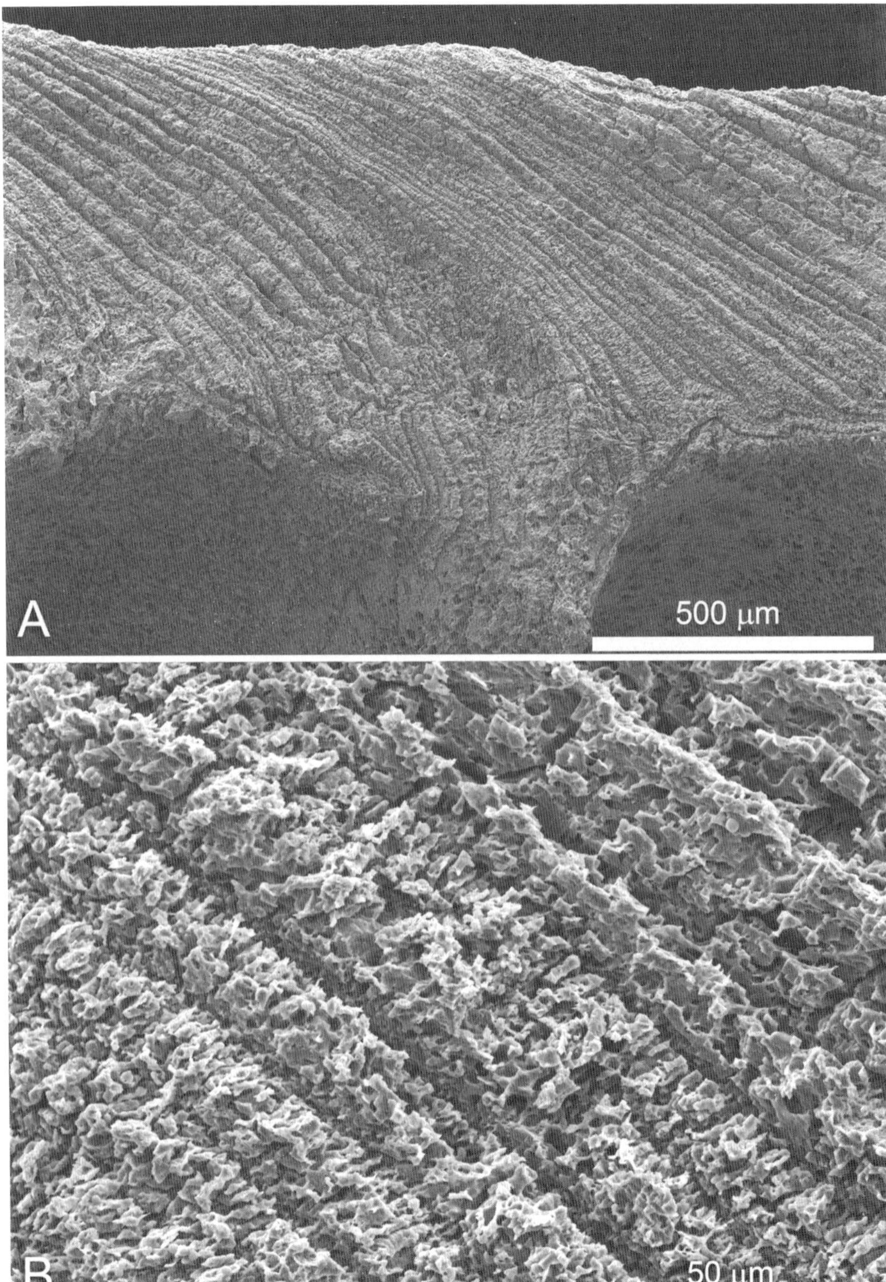

Fig. 13.21 *Cross section through the long axis of an aptychus of* Baculites *sp. (smooth or weak flank ribs), AMNH 54277, Mooreville Chalk, Alabama. A. The aptychus is composed of thin calcitic increments secreted at the posterior end. Dorsal surface is on the top, ventral surface on the bottom. B. Each increment is approximately 20 um thick.*

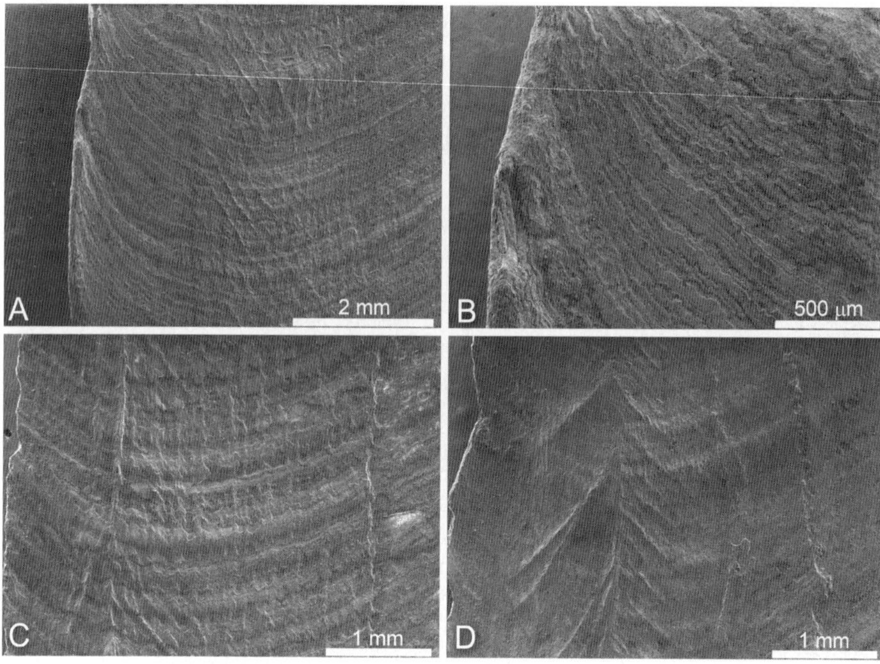

Fig. 13.22 *Dorsal surface of aptychi of* Baculites *sp. (smooth and weak flank ribs), Mooreville Chalk, Alabama. The symphysal edge is on the left side of each photo and the anterior direction is toward the top. A. BHMNH 5479. View of the symphysal edge showing delicate diagonal striations. B. BHMNH 5479. The increments composing the aptychus bend forward as they approach the symphysis. C. BHMNH 5152. Thin longitudinal striations are common near the symphysis. D. BHMNH 5151. The area near the symphysis is covered with chevrons in this specimen, which may be the result of an injury or growth pathology.*

ends in a flat, beveled edge (e.g., 2.8 mm versus 0.6 mm, respectively, in BHMNH 5154). Variations in thickness are also associated with injuries and pathologies. The thick ridge and accompanying fold at the posterior margin in BHMNH 5406 is such an example (Fig. 13.17A–D).

X-ray diffraction analysis indicates that the aptychi are composed of calcite. A cross section through the long axis of a valve reveals a series of thin increments, each approximately 20 μm thick (Fig. 13.21). The increments are stacked shingle-style with more recent increments lying underneath older increments, forming an angle of 30° with the dorsal edge. The increments seem to be bundled into clusters, but this requires further research.

The dorsal surface of the aptychus is relatively smooth. There are broad undulations covered with fine lines that parallel the lateral and posterior margins. The lines bend forward at the symphysis, which they approach at an angle of 45° (Fig. 13.22B). It is unclear if the fine lines become more closely spaced toward the posterior end. Many specimens also show delicate longitudinal or diagonal striations, especially near the symphysis (Fig. 13.22A, C). In BHMNH 5151 (Fig. 13.22D), this area is covered with chevrons, which may be the result of a repaired injury.

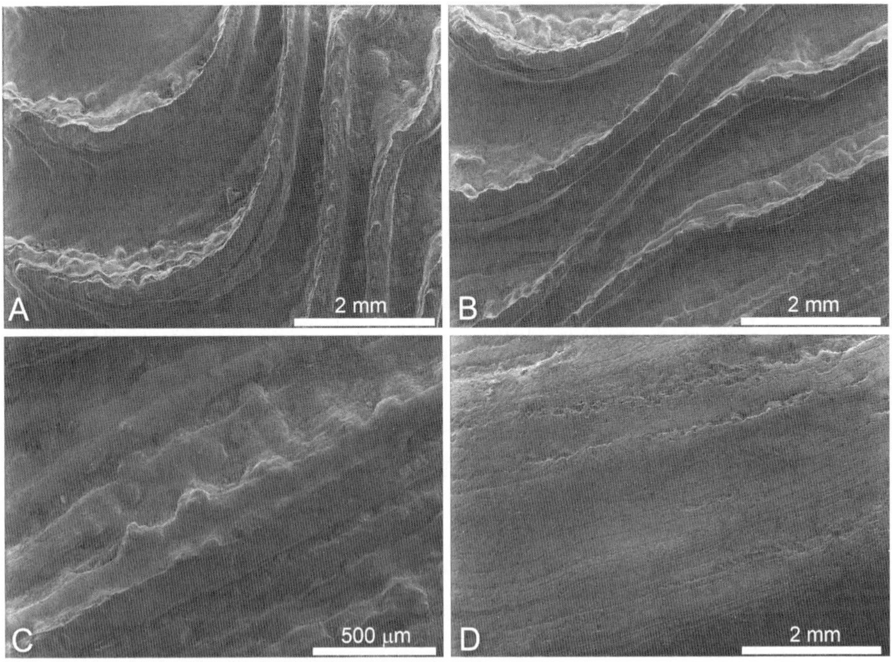

Fig. 13.23 *Ventral surface of aptychi of* Baculites *sp. (smooth and weak flank ribs), Mooreville Chalk, Alabama. The symphysal edge is toward the left of each photo and the anterior direction is toward the top. A–C. AMNH 51342. A. The rugae turn almost perpendicularly (left) toward the symphysis, forming a geniculation. B. The rugae, especially on the lateral margin, are twisted over, resembling flexible sheets of metal. C. Rugae with botryoidal excrescences. D. BHMNH 5485. The increments composing the aptychus are visible where the rugae have eroded away.*

The ventral surface of the aptychus is ornamented with coarse ridges or rugae that parallel the lateral and posterior margins. The rugae usually attain their widest spacing at midlength (Figs. 13.18E, F, I, J, M–O, 19B, D, J), and sometimes bunch up at the posterior end (compare Figs. 13.18E, F, 20C). In general, the rugae are closely spaced along the lateral margins (Fig. 13.17A, B, E, F). They turn almost perpendicularly toward the symphysis, forming a geniculation (Figs. 13.18I, J, 20A, 23A), but do not continue onto the symphysal fold, which is covered instead with fine ridges that are convex toward the posterior end.

In addition to the geniculation, the rugae sometimes develop a series of kinks (zigzags) that are aligned longitudinally (Fig. 13.19A, B, H). These kinks may or may not correspond to fractures on the dorsal side. Conversely, fractures on the dorsal side may or may not correspond to kinks on the ventral side.

The rugae vary in their degree of coarseness and waviness, which may reflect ontogenetic development. For example, the rugae in BHMNH 5450 (Fig. 13.17M–O), the smallest specimen in our collection, and presumably from a juvenile, are fine and relatively straight, whereas those in BHMNH 5149 (Fig. 13.18E, F), one of the larger specimens in our collection, and presumably from an adult, are coarse and wavy. However, specimens of nearly the same size also show wide variation in coarseness and sinuosity (compare Fig. 13.20A, I), which may be related to different states of preservation.

The rugae are asymmetric in profile, with one side more steeply sloping than the other. Commonly, the crests of the rugae bend backward (anteriorly) at an acute angle, e.g., AMNH 51316 (Fig. 13.20A). In AMNH 51342, the rugae on the lateral margin are twisted over, resembling flexible sheets of metal (Fig. 13.23B). In some specimens, the edges of the rugae are very sharp, whereas in others, they are worn down. In general, the rugae are sharper on the posterior end, suggesting erosion of earlier formed rugae during ontogeny, but this pattern requires further documentation.

Close examination of the ventral side of the aptychi reveals several additional features: a thin, chalky outer layer covering the surface; botryoidal excrescences (Fig. 13.23C), which may be accentuated by weathering; fine ridges that cross the sides of the rugae diagonally and disappear in the troughs (e.g., AMNH 51316); and thin increments in areas where the rugae are worn down (Fig. 13.23D).

6.4 *Baculites* sp., Niobrara Formation, Kansas

FHSM 968:1 was previously described and illustrated by Miller (1968: 87, pl. 8, Fig. 14) and consists of an aptychus with part of the anterior portion of the left valve missing (Fig. 13.24G). According to Miller (1968: 87), the specimen was closely associated with the mold of a smooth baculite. The aptychus is 10.2 mm wide and 20.1 mm long, with a ratio of aptychus width to aptychus length of 0.51, which is probably an underestimate. The ventral surface of the aptychus is ornamented with coarse rugae that parallel the lateral and posterior margins.

6.5 *Baculites compressus/B. cuneatus*, Pierre Shale, South Dakota

BHMNH 5498 is a right wing of a lower jaw (Fig. 13.24F, Table 13.1). It is 7.9 mm wide and 19.5 mm long. The estimated ratio of jaw width to jaw length is 0.81. The wing is convex with a prominent flange along the symphysal edge. The anterior margin of the wing is broadly rounded and the posterior margin is more sharply rounded; the lateral margin is obscured.

The jaw consists of a dark brown to black crystalline layer approximately 200 μm thick. X-ray diffraction analysis indicates that it consists of magnesium enriched calcite. This layer shows broad folds that parallel the posterior margin. In areas where the black layer is broken away, it exposes a smooth brown layer bearing finer ridges, spaced at approximately equal distances of 0.5 mm.

An analysis of the relationship between the shape of the jaw and the whorl cross section is complicated because the jaw is not preserved inside the body chamber,

Fig. 13.24 *A–E. Lower jaws of* Pseudobaculites natosini *(Robinson, 1945), Lewis Shale, Natrona County, Wyoming. A, B. BHMNH 5501. A. Right wing, anterior direction toward the right. B. Left wing, anterior direction toward the left. C. Right wing, BHMNH 5500, anterior direction toward the right. D, E. AMNH 51328. D. Left wing, anterior direction toward the left. E. Right wing, anterior direction toward the right. F. Right wing of a lower jaw attributed to* Baculites compressus *Say, 1820, or* Baculites cuneatus *Cobban, 1962, BHMNH 5498, Pierre Shale, Meade County, South Dakota. Anterior direction toward the right. G. Aptychus attributed to* Baculites *sp., FHSM 968:1, Smoky Hill Chalk Member, Niobrara Formation, Rooks Co., Kansas. G is X2; all other figures are X1.*

and, therefore, no measurements are available for the corresponding whorl cross section. Instead, we measured specimens of *Baculites compressus/B. cuneatus* from the same concretion as the jaws, as well as specimens from other concretions in the same stratigraphic interval. Our calculations indicate that the shape of the jaw is much broader than the whorl cross section (Table 13.1).

6.6 *Pseudobaculites natosini,* Lewis Shale, Wyoming

BHMNH 5501 is 50.1 mm long but is incomplete (Fig. 13.24A, B; Table 13.1). Each wing is approximately 25.8 mm wide. The specimen is folded in half along the symphysis. A flange is present on the middle one-third of the jaw. The posterior margin is broken but the lateral margin is intact and broadly curved. The apex is more strongly convex than the rest of the wing.

The jaw is preserved as a fine-grained, tan layer covered by a dark brown to black, coarsely crystalline layer 160 µm thick. This crystalline layer bears broad ridges that parallel the posterior margin. X-ray diffraction analysis indicates that it consists of calcite. This layer is covered near the symphysal margin of the left wing by patches of a thin (100 µm thick) translucent layer bearing very fine lirae spaced at approximately equal distances of 60 µm. X-ray diffraction analysis indicates that this layer consists of gypsum, which is undoubtedly diagenetic.

AMNH 51328 is 59.0 mm long with a wing width of 26.3 mm (Fig. 13.24D, E; Table 13.1). The estimated ratio of jaw width to jaw length is 0.89. The jaw is folded together along the midline and the flange is visible on the middle two-thirds of the jaw. The left apical tip is bluntly rounded and points away from the midline. The apex is more strongly convex than the rest of the wing. The specimen is preserved as a fine grained tan layer. The counterpart is covered with a dense, dark brown crystalline layer 220 µm thick, similar to that in BHMNH 5501. X-ray diffraction analysis of this layer indicates that it is composed of calcite. It bears low folds and very fine, evenly spaced lirae, which parallel the posterior margin.

BHMNH 5502 is a negative of the right wing of a lower jaw with a well-defined lateral margin (not illustrated). The width and length of the wing are 23.6 mm and 52.8 mm, respectively. The estimated ratio of jaw width to jaw length is 0.89. The anterior portion is composed of tan grainy material and the rest of the jaw is composed of dark brown to black crystalline material with a thickness of approximately 200 µm. X-ray diffraction analysis indicates that this crystalline material consists of calcite.

BHMNH 5500 is a partially preserved right wing (Fig. 13.24C). The width of the wing is 28.1 mm. The flange is visible on most of the specimen. The jaw is covered with a coarsely crystalline brown to black material overlain in parts with a platy yellowish material. The flange is ornamented with thin ridges that intersect the symphysis at an angle of 60°.

An analysis of the relationship between the shape of the jaw and the whorl cross section is again complicated by the fact that none of the jaws is preserved inside the body chamber and, therefore, no measurements are available for the corresponding whorl cross section. To circumvent this problem, we relied on measurements reported by Cobban and Kennedy (1994) for specimens of this species from the Lewis Shale of Wyoming (Table 13.1). Based on these measurements, the shape of the jaw is broader than the whorl cross section.

7 Discussion

7.1 Preservation

Differences in the preservation of jaws can obscure similarities in morphology (Landman et al., 2006). Therefore, in comparing jaws, it is important to take into account their state of preservation. In some instances, only the chitinous layer of the lower jaw, now altered, is preserved, whereas in other instances, only the calcitic layer (aptychus) is preserved. However, the absence of one or the other of these layers does not necessarily imply that it was absent during life.

We have documented three modes of occurrence of baculite jaws in North America:

(1) In the Gammon Ferruginous Member of the Pierre Shale and in the Cody Shale, the jaws occur inside the body chamber and retain both the chitinous layer, now altered, and the calcitic aptychus. The outer aragonitic shell of the body chamber is also usually present.
(2) In the Mooreville Chalk, only the aptychus is present, not the underlying chitinous layer. The aragonitic outer shells of the baculites are absent. Likewise, in the Smoky Hill Chalk Member of the Niobrara Formation, only the aptychi are present, not the outer shells of the baculites. This mode of occurrence is similar to that of baculite jaws in the Boreal chalks of northern Europe and has traditionally been interpreted as resulting from differential dissolution in which the aragonite of the outer shell dissolves away leaving the calcitic aptychi behind (Morton and Nixon, 1987; Barthel et al., 1990; Sanders, 2003).
(3) In the Lewis Shale and the upper part of the Pierre Shale, the jaws occur as isolated elements in concretions containing aragonite-preserved fossils. The chitinous layer of the lower jaw, now altered, is present, but the outer calcitic layer (aptychus) is absent, fragmentary, or recrystallized. The aptychi may have partially or completely disintegrated before or during the formation of the concretion. A similar loss of aptychi has been hypothesized in scaphites (Landman and Waage, 1993) and, possibly, placenticeratids (Landman et al., 2005).

In instances in which the chitinous layer of the lower jaw is still present, it consists of black crystalline material, similar to that reported in other ammonite jaws (Landman et al., 2006). X-ray diffraction analysis of samples of this material from five specimens (AMNH 5138, BHMNH 5496, 5498, 5501, and 5502) reveals the presence of calcite, calcite enriched in magnesium, and amorphous material. However, it is possible that part of this black crystalline material actually represents a recrystallized portion of the aptychus. In order to better understand these relationships, it is important to more thoroughly study the diagenetic history of the chitinous and calcitic parts of the ammonite jaw.

The rarity of baculite jaws inside the body chamber indicates the low preservation potential of these structures. After death, the soft body must have easily fallen out of the orthoconic body chamber and become separated from the shell. It required unusual circumstances for both the shell and soft body to fall to the sea bottom and be preserved together.

7.2 Comparative Jaw Morphology

The lower jaws of baculites from North America are elongate. The ratio of jaw width to jaw length in *Baculites* sp. (smooth) from South Dakota averages 0.77. The ratio in two specimens of *Pseudobaculites natosini* averages 0.89. These differences in proportions may be related to differences in the whorl cross section. However, it is notable that the proportions of the jaws of the early Campanian species more closely match those of the corresponding whorl cross section than in the late Campanian species (see below).

The outer lamella of the lower jaw consists of two wings with a midline groove bordered by flanges. The inner layer, which was originally composed of chitin, is ornamented with broad undulations and fine ridges, which parallel the posterior margin, as shown in BHMNH 5331 (Fig. 13.15). This layer is covered by a pair of calcitic valves, although these are poorly preserved, if present at all, in the two late Campanian species.

In the material from the lower Campanian Pierre Shale, Cody Shale, Mooreville Chalk, and Nobrara Formation, representing *Baculites* sp. (smooth and weak flank ribs), the aptychi conform to the description of rugaptychi. They are elongate with a flange at the symphysal edge, reflecting the morphology of the underlying chitinous layer. The aptychi from Alabama commonly show a notch at the anterior end, which was also noted in material from the Campanian of England by Sharpe (1857: 57), and which may be the result of breakage.

The aptychi are ornamented on the dorsal side with broad undulations covered with fine lines that parallel the lateral and posterior margins. The aptychi from the Campanian of Europe also show this same pattern (Sharpe, 1857: pl. 24, Figs. 8b, 10a; Moberg, 1885: pl. 1, Figs. 14, 15; reillustrated by Kennedy and Christensen, 1997: Fig. 31E, H).

The aptychi are ornamented on the ventral side with coarse rugae that parallel the lateral and posterior margins, and approach the symphysis almost perpendicularly. In many aptychi, the rugae show a well-developed geniculation as they turn toward the symphysis, forming a series of chevrons pointing posteriorly (Figs. 13.18I, J, 20A). This geniculation is characteristic of rugaptychi and has been widely documented in European material (Hébert, 1854: pl. 28, Fig. 6; Sharpe, 1857: pl. 26, Fig. 9; reillustrated by Blackmore, 1896: pl. 16, Fig. 16; Moore and Sylvester-Bradley, 1957).

The aptychi from the lower Campanian Pierre Shale, Cody Shale, Mooreville Chalk, and Niobrara Formation are nearly identical to those from the Campanian of northern Europe. For example, the North American specimens closely match those from

Belgium illustrated by de Grossouvre (1908: 39, pl. 10, Figs. 7–13; reillustrated with other specimens from the same locality by Kennedy, 1986: 192, pl. 16, Figs. 1–22). These similarities imply that rugaptychi are common features of many baculite species.

The aptychi in *Baculites compressus/B. cuneatus* and *Pseudobaculites natosini* are not well preserved, making comparisons with other species difficult. One specimen of *P. natosini* (BHMNH 5501) retains a thin layer of clear material with very fine lirae, which may represent part of the aptychus. If so, it is thinner and more finely ornamented than the aptychi of the early Campanian species, but this requires further study.

The radula in *Baculites* sp. (smooth) features long marginal teeth and blunt marginal plates. Although the other teeth of the radula are not well preserved, enough of the radula is present to suggest its similarity to those reported in other ammonites (Nixon, 1996). The radular teeth are distinct from the hooklike elements reported in scaphites (Landman and Waage, 1993; Kennedy et al., 2002; Landman and Klofak, 2004), and variously interpreted as radular teeth or tentacular hooks. The ridges on the marginal teeth of the baculite radula resemble those on the marginal teeth of the radula of *Vampyroteuthis infernalis* Chun, 1903, as illustrated by Solem and Roper (1975).

7.3 Function of the Aptychus Type Jaw

The aptychus type jaw has generally been interpreted as an adaptation for feeding on small prey (microphagy) (Morton and Nixon, 1987; Lehmann and Kulicki, 1990; Seilacher, 1993; Engeser and Keupp, 2002). Several morphological features of the jaw have been cited in support of this interpretation: (1) the presence of a blunt anterior margin with the two wing tips diverging at the apex, (2) the general absence of a thickened calcareous deposit at the apical end, (3) the presence of a groove, rather than a thickening, along the middle of the jaw, which creates a line of weakness incompatible with a crushing function (Seilacher, 1993), (4) the much larger size of the lower jaw relative to the upper jaw, with a concomitant loss of the gliding joint between the upper and lower jaws (Seilacher, 1993), and (5) the reduction in size of the inner lamella of the lower jaw, thus reducing the surface area required for muscular attachment to facilitate biting (Lehmann and Kulicki, 1990). A microphagous feeding habit is also consistent with rare finds of ammonite stomach contents (Lehmann, 1981; Jäger and Fraaye, 1997), although such finds have not yet been documented in baculites.

In contrast, the morphological features of lower jaws that are adapted for biting and crushing are very different. The apical end of the jaw is beaklike and usually reinforced with a mineralized deposit [e.g., *Nautilus*, *Gaudryceras*, and some species of Ammonitina with thin aptychi such as *Aconeceras* (see Kulicki et al., 1988; Doguzhaeva and Mutvei, 1992) and, possibly, *Placenticeras* (see Landman et al.,

2006)]. In addition, the lower jaw is smaller or equal in size to the upper jaw (Lehmann, 1981; Morton and Nixon, 1987).

Although the morphology of the lower jaws of baculites is consistent with a microphagous feeding habit, the exact feeding mechanism is unknown. Several hypotheses have been proposed for other members of the Aptychophora, including scooping up soft-bodied organisms off the seafloor (Lehmann, 1981), straining out plankton from seawater (Morton and Nixon, 1987), and sucking in prey through a pump-like action of the lower jaw (Seilacher, 1993). Of these, the first scenario is the least likely to apply to baculites, because these animals probably lived several tens of meters above the bottom, based on their lithologic and faunal associations (Westermann, 1996), and oxygen isotopic analyses of their shells (Landman et al., in prep.).

In addition to its role in food capture, the aptychus type jaw has also been interpreted as an operculum (Trauth, 1927–1936; Schindewolf, 1958), largely because of the close correspondence between the shape and size of the jaw and the whorl cross section. Lehmann and Kulicki (1990) developed a model to accommodate a dual function – both as an operculum and as a feeding device. According to these authors, the aptychus type jaw was capable of rotating into a nearly vertical position to serve as an operculum. Seilacher (1993) also advocated a dual function, and detailed the evolutionary steps culminating in this innovation. This interpretation was also favored in the most recent treatment of the subject by Engeser and Keupp (2002).

With respect to baculites, the evidence for an opercular function of the lower jaw is based on the following criteria:

(1) The correspondence between the shape and size of the lower jaw and the whorl cross section. In *Baculites* sp. (smooth and weak flank ribs) from the lower Campanian Pierre Shale and Cody Shale, the correspondence between the shape and size of the lower jaw and the whorl cross section is excellent. When the jaw is splayed out, it touches the sides of the shell. However, the jaw may not have been splayed out during life. Indeed in some specimens where the jaw is preserved in the body chamber, it forms a U-shaped structure (Fig. 13.8A–C). In addition, in *Baculites compressus/B. cuneatus* and *Pseudobaculites natosini*, the fit is much poorer. The shape of the jaw in these species is much broader than the whorl cross section.

(2) The mode of occurrence of the lower jaw in the body chamber. In *Baculites* sp. (smooth and weak flank ribs) from the lower Campanian Pierre Shale and Cody Shale, most of the jaws are aligned at an angle of 45°–90° relative to the long axis of the shell, with the anterior end of the jaw near the dorsal side of the shell, and the posterior end near the ventral side of the shell, effectively occluding the opening (Figs. 13.5–7, 8D–F, 9, 10, 11, 12, 13; see also Nowak, 1908: pl. 14, Fig. 13.11). However, this position may simply represent an artifact of preservation related to how the jaw settled in the body chamber after death [see Morton (1975), Seilacher (1990), and Landman and Waage (1993) for a discussion of taphonomic biases affecting the preservation of ammonite jaws].

(3) The presence of healed injuries on the aptychus. Many of the aptychi from the Mooreville Chalk show fractures on the dorsal side, which may represent healed injuries. However, healed injuries are also present on the jaws of *Nautilus*, which are used for biting, not as opercula (Landman and Kruta, 2006). The injuries in *Nautilus* may have been inflicted during feeding, while subjugating prey.

(4) The composition and ornament of the aptychus. In *Baculites* sp. (smooth and weak flank ribs), the aptychi are composed of hard calcite and covered with coarse, rugose ornament. These features are usually interpreted as adaptations against predation. However, Seilacher (1993) has argued that the ornament on some aptychi may simply represent fabricational noise associated with biomineralization, with no functional value at all.

In conclusion, the evidence for an opercular function in baculites is ambiguous, with as many arguments as counterarguments. We suggest an alternative interpretation that is consistent with many of the morphological features described above. The jaws of baculites may have protruded out from the aperture to facilitate food capture. Even if the jaws were exposed to the outside temporarily, they would have been vulnerable to predation. The thick, coarsely ornamented aptychus could have served as protection. Indeed, in some modern coleoids, the jaws can extend out from the oral cavity during feeding (Kasugai, 2001; S. v. Boletzky, 2004, personal communication). Such flexibility may also have characterized baculites and other ammonites.

7.4 Rate of Growth

The jaws provide some insights into the rate of growth of baculites. Analysis of the microstructure of the aptychi from the Mooreville Chalk reveals that they are composed of thin increments, each approximately 20 um thick (Fig. 13.21). The rate of formation of these increments is unknown, but studies of aptychi from other ammonites suggest that such increments were deposited daily (Hewitt et al., 1993: 207).

If we assume a daily rate of deposition, we can calculate the approximate age of a baculite by dividing the length of its aptychus by the thickness of a single increment. The size–frequency histogram of valve width of aptychi from the Mooreville Chalk reveals two modes at 9–10 mm and 13–14 mm. These modes presumably represent the modal size classes of mature microconchs and macroconchs, respectively. The corresponding aptychus lengths are approximately 21 and 31 mm. Using these values and dividing by the thickness of a single increment yields an average age at maturity of approximately 3 and 4 years for microconchs and macroconchs, respectively. This estimate is consistent with estimates of the age at maturity of other shallow-water ammonites (Bucher et al., 1996). However, this result must be confirmed with oxygen isotopic analyses of the aptychi and examination of the variation in increment thickness through ontogeny.

8 Conclusions

The lower jaws of baculites are longer than wide and consist of two subquadrate wings. The inner chitinous layer of the outer lamella shows a groove along the midline bordered by flanges. This layer bears a pair of calcitic plates (=the aptychus), comparable to those in all other members of the Aptychophora.

In the early Campanian species in which the aptychi are well preserved, they conform to the rugaptychus morphotype. The dorsal side is characterized by fine lines and broad undulations that parallel the posterior margin. The ventral side is characterized by irregular rugae that parallel the lateral and posterior margins and approach the symphysis almost perpendicularly. In contrast, the aptychi of the late Campanian species seem thinner and less coarsely ornamented, but this difference may be due to preservation and requires further research.

Jaws are a fundamental feature of all cephalopods and were undoubtedly present in all members of the Baculitidae. This family ranges from the Albian to the Maastrichtian, but so far jaws have only been reported from the Cenomanian, Turonian, and Campanian. We suspect that it is only a matter of time before they are discovered in the rest of the Upper Cretaceous.

Acknowledgments

George Harlow and Jamie Newman (AMNH) analyzed the mineral composition of the jaws using the Rigaku DMAX/RAPID X-ray microdiffraction unit at the AMNH, funded by National Science Foundation Grant EAR-0418088. The authors thank the property owners Chance and Cindy Davis, Sonny and Lois Jolly, David Cameron, Jim and Myrna Smith, and Tom and Sheila Trask for permission to collect on their land. We especially thank Henry (Tad) Rust, Kraig Derstler, Glenn and Barbara Rockers, Jim and Linda Schmeidt, and Mike Ross who collected and donated critical specimens for our study. We also thank Susan Klofak, Kathy Sarg, Niles Eldredge, David Weinreb, Leif and Elisha Larson, and Steven Jorgensen for their help in the field. R. J. Zakrzewski kindly arranged for the loan of specimens from the Sternberg Museum of Natural History. Kazushige Tanabe (University of Tokyo), W. James Kennedy (Oxford University), and Royal H. Mapes (Ohio University) read an earlier draft of this manuscript and made many valuable suggestions. This paper also benefited from discussions about ammonite jaws with Isabella Kruta (Muséum National d'Histoire Naturelle, Paris). At the AMNH, Landman thanks Jacob Mey and Angela Klaus for help in the SEM, and Steve Thurston, Kathy Sarg, and Stephanie Crooms for help in preparing the figures and manuscript. Larson thanks the Black Hills Institute of Geological Research for financial and technical support in the collection, preparation, and storage of specimens in the Black Hills Museum of Natural History, and Larry Shaffer for his help in preparing the poster on the subject of this paper for the VI International Symposium on Cephalopods – Present and Past in September, 2004. The US Geological Survey provided facilities for examining the extensive collection of baculites housed at the Denver Federal Center, Colorado.

References

Arkell, W. J. 1957. Aptychi. *In* R. C. Moore (editor), *Treatise on Invertebrate Paleontology, Part L. Mollusca* **4**: L437–L440. New York: Geological Society of America; Lawrence, KS: University of Kansas Press.

Barthel, K. W., N. H. M. Swinburne, and S. C. Morris. 1990. *Solnhofen: A Study in Mesozoic Palaeontology*. Cambridge: Cambridge University Press.

Binckhorst van den, J. T. 1861. *Monographie des Gastropodes et des Céphalopodes de la Craie Supérieure du Limbourg*. vi + 83pp. Brussels: Muquardt; Maastricht: Muller Frères.

Bishop, G. A. 1985. Fossil decapod crustaceans from the Gammon Ferruginous Member, Pierre Shale (Early Campanian), Black Hills, South Dakota. *Journal of Paleontology* **59**: 605–624.

Blackmore, H. P. 1896. Some notes on the aptychi from the Upper Chalk. *Geological Magazine* **3**: 529–533.

Böhm, J. 1898. Über *Ammonites pedernalis* v. Buch. *Zeitschrift der Deutschen Geologischen Gesellschaft* **50**: 183–201.

Breitkreutz, H., R. Diedrich, and R. Metzdorf. 1991. Fossilfunde aus der Schwartz-Bunten Wechselfolge (Ob. Cenoman bis Unter Turon) des Ostwestfalendammes bei Bielefeld. *Berichte des Naturwissenschaftlichen Vereins für Bielefeld und Umgegend* **32**: 37–48.

Bucher, H., N. H. Landman, S. M. Klofak, and J. Guex. 1996. Mode and rate of growth in ammonoids. *In* N. H. Landman, K. Tanabe, and R. A. Davis (editors), *Ammonoid Paleobiology*, pp. 407–461. New York: Plenum Press.

Chun, K. 1903. *Aus den tiefen des weltmeeres*. Jena: G. Fischer.

Cobban, W. A. 1958. Two new species of *Baculites* from the Western Interior region. *Journal of Paleontology* **32**(4): 660–665.

Cobban, W. A. 1962. New baculites from the Bearpaw Shale and equivalent rocks of the Western Interior. *Journal of Paleontology* **36**(1): 125–135.

Cobban, W. A. 1969. The Late Cretaceous ammonites *Scaphites leei* Reeside and *Scaphites hippocrepis* (DeKay) in the Western Interior of the United States. *U.S. Geological Survey Professional Paper* **619**: 1–29.

Cobban, W. A., and W. J. Kennedy. 1994. A giant baculite from the Upper Campanian and Lower Maastrichtian of the Western Interior. *U.S. Geological Survey Bulletin* **2073C**: C1–C4.

Cobban, W. A., I. Walaszczyk, J. D. Obradovich, and K. C. McKinney. 2006. A USGS zonal table for the Upper Cretaceous middle Cenomanian-Maastrichtian of the Western Interior of the United States based on ammonites, inoceramids, and radiometric ages. *U.S. Geological Survey Open – File Report* 2006–1250, 45p.

Desmarest, A. G. 1817. Mémoire sur deux genres coquilles fossiles cloisonnées et à siphon. *Journal de Physique, de Chimie, d'Histoire naturelle et des Arts* **85**: 42–51.

DeKay, J. E. 1827. Report on several multilocular shells from the State of Delaware; with observations of a second specimen of the new fossil genus *Eurypterus*. *Annals of the Lyceum of Natural History of New York* **2**: 273–279.

Diener, C. 1925. Ammonoidea neocretacea. *Fossilium Catalogus (1: Animalia)* **29**: 1–244.

Doguzhaeva, L. A., and H. Mutvei. 1992. Radula of the Early Cretaceous ammonite *Aconeceras* (Mollusca: Cephalopoda). *Palaeontographica Abteilung A* **223**: 167–177.

Engeser, T., and H. Keupp. 2002. Phylogeny of aptychi-possessing Neoammonoidea (Aptychophora nov., Cephalopoda). *Lethaia* **24**: 79–96.

Everhart, M. J. 2005. *Oceans of Kansas: A Natural History of the Western Interior Sea*. Bloomington, IN: Indiana University Press.

Farinacci, A., N. Mariotti, R. Matteucci, U. Nicosia, and G. Pallini. 1976. Structural features of some Jurassic and Early Cretaceous Aptychi. Bollettino della Società Paleontologica Italiana **15**: 111–143.

Fritsch, A. 1893. Studien im Gebiete der böhmischen Kreideformation, Palaeontologische Untersuchungen der einzelnen Schichten. V. Die Priesener Schichten. *Archiv für naturwissenschaftliche Landesdurchforschung von Böhmen* **9**: 1–135.

Giers, R. 1964. Die Grossfauna der Mukronatenkreide (unteres Obercampan) im östlichen Münsterland. *Fortschritte in der Geologie von Rhineland und Westfalen* **7**: 213–294.

Gill, J. R., and W. A. Cobban. 1961. Stratigraphy of lower and middle parts of the Pierre Shale, northern Great Plains. *U.S. Geological Survey Professional Paper* **424D**: D185–D191.

Grossouvre, A. de. 1908. Description des ammonitides du Crétacé supérieur du Limbourg belge et hollandais et du Hainaut. *Mémoires du Musée royal d'histoire naturelle de Belgique* **4**: 1–39.

Hébert, E. 1856. Tableau des fossiles de craie de Meudon et description de quelques espèces nouvelles. *Mémoires de la Société Géologique de France* **5**(2): 345–374.

Henderson, R. A., W. J. Kennedy, and W. A. Cobban. 2002. Perspectives of ammonite paleobiology from shell abnormalities in the genus *Baculites. Lethaia* **35**: 215–230.

Hewitt, R. A., G. E. G. Westermann, and A. Checa. 1993. Growth rates of ammonites estimated from aptychi. *Geobios Mémoire Spécial* **15**: 203–208.

Hyatt, A. 1903. Pseudoceratites of the Cretaceous. *Monograph. U.S. Geological Survey* **44**: 1–351.

Jäger, M. and R. Fraaye. 1997. The diet of the Early Toarcian ammonite *Harpoceras falciferum. Palaeontology* **40**(2): 557–574.

Kanie, Y. 1982. Cretaceous tetragonitid ammonite jaws: a comparison with modern *Nautilus* jaws. *Transactions and Proceedings of the Palaeontological Society of Japan*, New Series **125**: 239–258.

Kasugai, T. 2001. Feeding behavior of the Japanese pygmy cuttlefish *Idiosepius paradoxus* (Cephalopoda: Idiosepiidae) in captivity: evidence for external digestion? *Journal of the Marine Biological Association of the United Kingdom* **81**: 979–981.

Kennedy, W. J. 1986. The ammonite fauna of the type Maastrichtian with a revision of *Ammonites colligatus* Binkhorst, 1861. *Bulletin de l'Institut royal des Sciences naturelles de Belgique (Sciences de la Terre)* **56**: 151–267.

Kennedy, W. J. 1993. Campanian and Maastrichtian ammonites from the Mons Basin and adjacent areas, Belgium. *Bulletin de l'Institut royal des Sciences naturelles de Belgique (Sciences de la Terre)* **63**: 99–131.

Kennedy, W. J., and W. A. Christensen. 1997. Santonian to Maastrictian ammonites from Scania, southern Sweden. *Fossils and Strata* **44**: 75–128.

Kennedy, W. J., W. A. Cobban, and N. H. Landman. 1997. Campanian ammonites from the Tombigbee Sand Member of the Eutaw Formation, the Mooreville Formation, and the basal part of the Demopolis Formation in Mississippi and Alabama. *American Museum Novitates* **3201**: 1–44.

Kennedy, W. J., W. A. Cobban, and H. C. Klinger. 2002. Muscle attachment and mantle-related features in Upper Cretaceous *Baculites* from the United States Western Interior. *In* H. Summesburger, K. Histon, and A. Daurer (editors.), *Cephalopods-Present and Past. Abhandlungen der Geologischen Bundesanstalt* **57**: 89–112.

Klinger, H. C., and W. J. Kennedy. 2001. Stratigraphic and geographic distribution, phylogenetic trends and general comments on the ammonite family Baculitidae Gill, 1871 (with an annotated list of species referred to the family). *Annals of the South African Museum* **107**(1): 1–290.

Kulicki, C., L. A. Doguzhaeva, and G. K. Kabanov. 1988. *Nautilus*-like jaw elements of a juvenile ammonite. *In* J. Wiedmann, and J. Kullmann (editors), *Cephalopods – Present and Past*, pp. 679–686. Stuttgart: Schweizerbart'sche Verlagsbuchhandlung.

Lamarck, J. P. B. A. de M. de. 1801. *Système des Animaux sans vertebrès*. Deterville, Paris: The Author.

Landman, N. H., and A. Grebneff. 2006. Jaws of Triassic ammonoids from New Zealand. *New Zealand Journal of Geology and Geophysics* 49: 121–129.

Landman, N. H., and S. M. Klofak. 2004. Radulae and red herrings. *Sixth International Symposium, Cephalopods – Present and Past, Abstracts Volume*, p. 99.

Landman, N. H., and I. Kruta. 2006. Injuries on the jaws of modern *Nautilus* as clues to the function of ammonite aptychi. *A Symposium on the Paleontology, Geology, and Stratigraphy of the*

Late Cretaceous Western Interior Seaway. A Tribute to the Life of Willian Aubrey "Bill" Cobban, 26 & 27 August 2006, *Abstracts Volume*, p. 31.

Landman, N. H., C. J. Tsujita, W. A. Cobban, N. L. Larson, K. Tanabe, and R. L. Flemming. 2006. Jaws of Late Cretaceous placenticeratid ammonites: how preservation affects the interpretation of morphology. *American Museum Novitates* 3500: 1–48.

Landman, N. H., and K. M. Waage. 1993. Scaphitid ammonites of the Upper Cretaceous (Maastrichtian) Fox Hills Formation in South Dakota and Wyoming. Bulletin of the American Museum of Natural History **215**: 1–257.

Lehmann, U. 1981. *The Ammonites: Their life and Their World*. Cambridge: Cambridge University Press.

Lehmann, U., and C. Kulicki. 1990. Double function of aptychi (Ammonoidea) as jaw elements and opercula. *Lethaia* **23**: 325–331.

Lundgren, B. 1874. Om en Comaster och en Aptychus frän Köpinge. *Öfversigt af Kongl. Vetenskaps- Akademiens Förhandlingar* **3**: 61–73.

Matsumoto, T., and I. Obata. 1963. A monograph of the Baculitidae from Japan. *Memoirs of the Faculty of Science, Kyushu University* (series D, Geology) **13**(1): 1–116.

Meek, F. B., and F. V. Hayden. 1865. Paleontology of the Upper Missouri: a report upon collections made principally by the expeditions under command of Lieut. G. K. Warren, US. Top. Engrs., in 1855 and 1856, Invertebrates. Part I. *Smithsonian Contributions to Knowledge* **14**(Art. 5): 1–136.

Meek, F. B. 1876. A report on the invertebrate Cretaceous and Tertiary fossils of the upper Missouri country. *U.S. Geological Survey of the Terri*tories (Hayden) **9**: 1–629.

Miller, H. W. 1968. Invertebrate fauna and environment of deposition of the Niobrara Formation (Cretaceous) of Kansas. *Fort Hays Studies* series no. 8: 1–90.

Moberg, J. C. 1885. Cephalopoderna i Sveriges Kritsystem. II. Artbeskrifning. *Sveriges geologiska undersökning. Afhandlingar och uppsatser* (series C) **73**: 1–63.

Moore, R. C., and P. C. Sylvester-Bradley. 1957. Taxonomy and nomenclature of aptychi. *In* R. C. Moore (editor), *Treatise on Invertebrate Paleontology, Part L, Mollusca 4*: L464–L471. New York: Geological Society of America; Lawrence, KS: University of Kansas Press.

Morton, N. 1975. The position of the aptychus in some Jurassic ammonites. *Neues Jahrbuch für Geologie und Paläontologie Abhandlungen* **1975**(7): 409–411.

Morton, N., and M. Nixon. 1987. Size and function of ammonite aptychi in comparison with buccal masses of modern cephalopods. *Lethaia* **20**: 231–238.

Nixon, M. 1996. Morphology of the jaws and radula in ammonoids. *In* N. H. Landman, K. Tanabe, and R. A. Davis (editors), *Ammonoid Paleobiology*, pp. 23–42. New York: Plenum Press.

Nowak, J. 1908. Untersuchungen über die Cephalopoden der oberen Kreide in Polen, I. Teil, Genus *Baculites* Lamarck. *Bulletin international de l'Académie des Sciences de Cracovie (Classe des Sciences mathématiques et naturelles)* **1908**(4): 326–353.

Owen, D. D. 1852. Description of new and imperfectly known genera and species of organic remains, collected during the geological surveys of Wisconsin, Iowa, and Minnesota, by D. D. Owen. *Report of a Geological Survey of Wisconsin, Iowa, and Minnesota*, 638p.

Picard, L. 1929. On Upper Cretaceous (chiefly Maestrichtian) Ammonoidea from Palestine. *Annals and Magazine of Natural History* **3**(10): 433–456.

Ravn, J. P. J. 1902–1903. Molluskerne i Danmarks Kridtaflejringer. *Kongelige Danske Videnskabernes Selkskabs Skrifter* (raekke 6) **11**, 2, Scaphopoder, Gastropoder g Cephalopoder, 209(5)–269(65), pl. 1–5 (1902); 3, Stratigrafisker undersøgelser, 339(5)–443(99), 1 pl. (1903).

Riccardi, A. C. 1983. Scaphitids from the upper Campanian-lower Maastrichtian Bearpaw Formation of the Western Interior of Canada. *Bulletin Geological Survey of Canada* **354**: 1–103.

Robinson, C. S., W. J. Mapel, and M. H. Bergendahl. 1964. Stratigraphy and structure of the northern and western flanks of the Black Hills Uplift, Wyoming, Montana, and South Dakota. *U.S. Geological Survey Professional Paper* **404**: 1–134.

Robinson, H. R. 1945. New baculites from the Cretaceous Bearpaw Formation of southwestern Saskatchewan. *Transactions of the Royal Society of Canada*, 3rd series **39**(4): 51–54.

Sanders, D. 2003. Syndepositional dissolution of calcium carbonate in neritic carbonate environments: Geological recognition, processes, potential significance. *Journal of African Earth Sciences and the Middle East* **36**: 99–134.

Say, T. 1821. Art. IV. Observations on some species of zoophytes, shells, & tc., principally fossil. *American Journal of Science and Arts* **2**: 34–45.

Schindewolf, O. H. 1958. Über Aptychen (Ammonoidea). *Palaeontographica Abteilung A* **111**: 1–46.

Schlüter, C., 1871–1876, Cephalopoden der oberen deutschen Kreide. *Palaeontographica* **21**: 1–24, pl. 1–8 (1871); **21**: 25–120, pl. 9–35 (1872); **24**: 1–144, pl. 36–55 (1876).

Seilacher, A. 1993. Ammonite aptychi: how to transform a jaw into an operculum. *American Journal of Science* **293A**: 20–32.

Sharpe, D. 1853–1857. Description of the fossil remains of Mollusca found in the Chalk of England. 1. Cephalopoda. *Palaeontographical Society Monographs* 1–26(1853); 27–36(1855); 37–68(1857).

Solem, A., and C. F. E. Roper. 1975. Structures of Recent Cephalopod radulae. *The Veliger* **18**(2): 127–133.

Stewart, J. D. 1990. Niobrara Formation vertebrate stratigraphy. *In* S. C. Bennett (editor), *Niobrara Chalk Excursion Guidebook*, pp. 19–39. Lawrence, KS: University of Kansas Museum of Natural History and Kansas Geological Survey.

Tanabe, K. 1983. The jaw apparatuses of Cretaceous desmoceratid ammonites. *Palaeontology* **26**(3): 677–686.

Tanabe, K., and Y. Fukuda. 1987. The jaw apparatus of the Cretaceous ammonite *Reesidites. Lethaia* **20**: 41–48.

Tanabe, K., and N. H. Landman. 2002. Morphological diversity of the jaws of Cretaceous Ammonoidea. *In*: H. Summesberger, K. Histon, and A. Daurer (editors), *Cephalopods – Present and Past, Abhandlungen der Geologischen Bundesanstalt* **57**: 157–165.

Trauth, F. 1927–1936. Aptychenstudien I–VIII. *Annalen des Naturhistorischen Museums in Wien* **41**: 171–259 (1927); **42**: 121–193 (1928); **44**: 329–411 (1930); **45**: 17–136 (1931); **47**: 127–145 (1936).

Trauth, F. 1938. Die Lamellaptychi des Oberjura und der Unterkreide. *Palaeontographica Abteilung A* **88**: 115–229.

Westermann, G. E. G. 1996. Ammonoid life and habitat. *In* N. H. Landman, K. Tanabe, and R. A. Davis (editors), *Ammonoid Paleobiology*, pp. 607–707. New York: Plenum Press.

Wright, C. W., and W. J. Kennedy. 1984. The Ammonoidea of the Lower Chalk. I. *Palaeontographical Society Monographs (London)* **137**: 1–126.

Young, K. 1963. Upper Cretaceous ammonites from the Gulf Coast of the United States. *University of Texas Publications* **6304**: ix + 367pp.

Chapter 14
Ultrastructural Analyses on the Conotheca of the Genus *Belemnotheutis* (Belemnitida: Coleoidea)

Dirk Fuchs,[1] Helmut Keupp,[1] Vasilij Mitta,[2] and Theo Engeser[1]

[1]Institute of Geological Sciences, Paleontology, Freie Universität Berlin, Malteserstrasse 74–100. D-12249 Berlin, Germany, drig@zedat.fu-berlin.de;

[2]Russian Academy of Sciences, Palaeontological Institute, Profsojuznaja 123, Moscow 117997, Russia, vmitta@mail.ru

Keywords: *Belemnotheutis*, ultrastructure, conotheca, proostracum, periostracum

1 Introduction

The internal skeleton of extinct belemnoids consists of the phragmocone (Fig. 14.1) and the rostrum. The rostrum covers the posterior part of the phragmocone and acts as a counterweight that brings the animal into a horizontal swimming position. Structural elements of the phragmocone are the septal-siphonal complex and the conotheca (phragmocone wall). If the proostracum (Fig. 14.1), the anterior dorsal projection of the phragmocone, is also part of the phragmocone or an independent development, is currently under discussion. The proostracum is usually considered to be the dorsal remnant of the body chamber inherited from ectocochlean ancestors (Crick, 1896; Naef, 1922; Jeletzky, 1966; Doyle and Shakides, 2004). Rostral layers (primordial rostrum + rostrum) and the mural parts of the septa do not belong to the conotheca proper.

N. H. Landman et al. (eds.), *Cephalopods Present and Past: New Insights and Fresh Perspectives*, 299–314.

Although the conothecal ultrastructure of belemnoids has been studied for a long time (Jeletzky, 1966; Barskov, 1972; Bandel et al., 1984; Doguzhaeva et al., 1999; Doguzhaeva et al., 2002; Doguzhaeva et al., 2003a, b), morphological knowledge of this very important feature is still incomplete.

This is partly due to the bad preservational potential of these fragile aragonitic structures. In most cases, only empty alveola without remains of the phragmocone are available. Only in a few cases phragmocones are preserved, but usually their ultrastructure has been recrystallized. Especially Bandel et al. (1984) and Bandel (1989) warned of diagenetic alterations and mentioned that ultrastructural analyses on the conotheca require unaltered material.

Additionally, data about the conothecal wall ultrastructure is problematic to compare as important methodological details such as preparation, magnification, ontogenetic stage, or morphological orientation (ventral, lateral, dorsal) given in previous publications are often insufficient. This resulted in an inconsistent terminology in the literature. Today, it is difficult to correlate conothecal layers observed in previous investigations. In order to overcome these obstacles, a consistent method and terminology are necessary.

New material of *Belemnotheutis* provided by one of us (Dr. V. Mitta, Moscow, Russia) exhibits the ultrastructure of well-preserved unaltered conothecal layers. Comparisons and correlations with previous studies on belemnoid ultrastructure were possible and could give additional information about the poorly understood belemnoid morphology.

It is the purpose of the present study to reinvestigate the conotheca of *Belemnotheutis* and to correlate the conothecal layers with those of other belemnoid taxa described in previous studies.

2 Previous Studies

In Table 1, we have summarized important ultrastructural analyses on the belemnoid conotheca. It seems that the sequence of layers within the conotheca is highly variable.

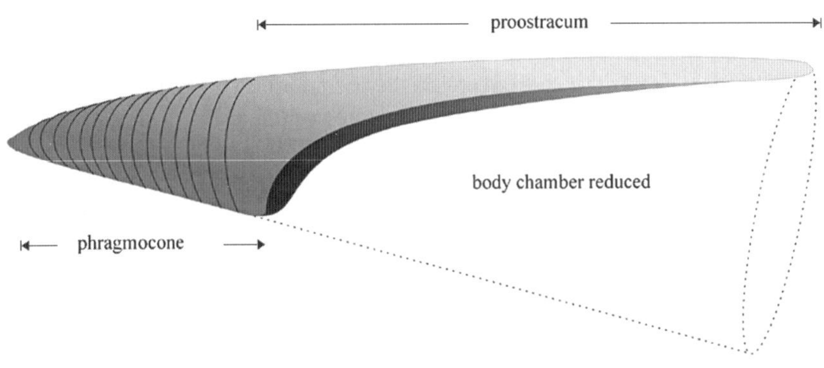

Fig. 14.1 *General morphology of the internal skeleton of belemnoids (rostrum not drawn).*

Interests in the ultrastructure of the belemnoid conotheca started with Christensen (1925) and Müller-Stoll (1936). Jeletzky (1966) revised older literature and postulated a belemnoid conotheca consisting of a thin inner prismatic layer, a thin central mineralized layer, and a thick outer prismatic layer. Jeletzky (1966: 110) also added: "Our ideas about microscopic structure of phragmocone and conotheca of the Belemnitida are now in a state of confusion." Since microscopic magnifications have proved to be insufficient for ultrastructural analyses, these older investigations are negligible. Modern SEM analyses with adequate magnifications started with Barskov (1972). In *Conobelus, Pachyteuthis,* and *Mesohibolithes*, Barskov observed an inner prismatic and an outer nacreous layer in the conotheca. Observations are in agreement with those of Jeletzky (1966), at least concerning the presence of an innermost prismatic conothecal layer.

Especially concerning the presence of an outer prismatic layer, observations are inconsistent. As is well documented by Bandel et al. (1984) and Doguzhaeva et al. (1999, 2003b), the layer forming the primordial rostrum continues in a layer along the outside of the conotheca. It is possible that this layer was sometimes interpreted as the outer prismatic layer of the conotheca (Table 14.1), because Bandel and Kulicki (1988) identified a comparably thin outer prismatic layer within the conotheca of *Belemnotheutis* not belonging to the primordial rostrum.

Likewise it is difficult to correlate intermediate layers. Interpretations vary between a thick nacreous layer (Mutvei, 1964; Barskov, 1972; Hewitt and Pinkney, 1982; Bandel and Kulicki, 1988; Doguzhaeva et al., 1999, 2002) and thin organic sheets (Doguzhaeva et al., 2003b).

3 Material and Methods

Several well-preserved phragmocones from the Upper Callovian (– *lamberti* – zone) of Dubki near Saratov, Russia (provided by Dr. V. Mitta) were investigated (for the stratigraphy of Dubki, see Keupp and Mitta, 2004).

Our phragmocones might be easily assigned to *Belemnotheutis polonica* Makowski, 1952, which occurs in the – *lamberti* – zone of Lukow (Poland). But as there are no distinctive characters for a morphological differentiation between our specimens and the type material of *B. antiquus* Pearce, 1847 (Donovan and Crane, 1992; Doyle and Shakides, 2004) from the Callovian (– *athleta* – zone) of Christian Malford, England, we consider these taxa in synonymy.

Six uncrushed specimens (MC-1–MC-6) are between 22–31mm in maximum length and 12–22 mm in maximum diameter (Fig. 14.2). The cross section is almost circular. The apical angle is always 20°. The earliest chambers including the protoconch are not preserved. Another dorsoventrally flattened specimen (MC-7), which is 100 mm in maximum length, might be considered as fully grown. In the same specimen the most apical part of the rostrum is preserved but the earliest camerae including the conotheca are replaced by pyrite. None of the studied specimens retains a proostracum in situ.

Table 1. *Comparison of previous ultrastructural observations on belemnoid conotheca.*

Order	Authors	Sequence of layers			Ontogenetic	Investigated	
		External	Intermediate	Investigated	taxa	stage	plane
	Christensen, 1925	Prismatic (= primordial rostrum?)	–	Prismatic	*Megateuthis*	embryonic	?
	Müller-Stoll, 1936	Thick Prismatic (=primordial rostrum?)	Thin organic	Thin prismatic	*Chitinoteuthis*	embryonic	?
	Mutvei 1964 (p. 97)	Periostracum?	Nacreous	Thin prismatic	*Megateuthis*	?	longitudinal?
	Jeletzky 1966 (p. 124)	Thick Prismatic (Primordial rostrum?)	Very thin, mineralized	Thick prismatic	*Megateuthis*	Embryonic	Medial (ventral)
	Barskov, 1972	–	Nacreous	Prismatic	*Conobelus, Pachyteuthis, Mesohibolithes*	Embryonic	?
	Hewitt and Pinkney, 1982 (p. 144)	?	Thick nacreous	Thin prismatic	*Acroteuthis*	Juvenile-adult	Transversal (dorsal?)
	Bandel et al., 1984 (p. 276, 285)	Primordial rostrum	Organic	Thin prismatic	*Hibolithes*	Embryonic-postembryonic	Medial
	Bandel and Kulicki, 1988 (p. 310)	Thin Prismatic	Nacreous	Thin prismatic	*Belemnotheutis Megateuthis*	Adult?	?
	Doguzhaeva et al., 1999 (p. 255)	Primordial rostrum	Nacreous	–	*Conobelus*	Embryonic-postembryonic	Medial
	Doguzhaeva et al. 2002 (p. 324)	Thin proostracal layers	Nacreous	Prismatic	*Megateuthis*	Adult	Longitudinal (dorsal)
	Doguzhaeva et al. 2003b (p. 83)	Primordial rostrum	Thin proostracal layers	Prismatic	*Passalotheutis?*	Embryonic-postembryonic	Medial + Transversal (dorsal + ventral)

Aulaco-cerida	Dauphin, 1983	–	–	Lamellar	Prismatic	*Aulacoceras*	? ?	
	Bandel, 1985 (p. 233)	Periostracum	Thin prismatic	Nacreous	Prismatic	*Dictyoconites*	Adult	?
	Cuif and Dauphin, 1979	–	Organic	Nacreous	Prismatic	*Atractites*	? ?	
	Doguzhaeva et al., 2002	–	–	Nacreous	Prismatic	*Mutveiconites*	Juvenile	?
Hematitida	Doguzhaeva et al., 2002, 2003a	5–6 prismatic layers without nacre?				*Hematites*	Postembryonic	?
Phragmo-teuthida	Doguzhaeva et al., 2003a (p. 67)	Irregulary mineralised	Proostracal layers	Lamello-fibrillar	Prismatic	*Donovaniconus*	Juvenile-adult	Medial (dorsal?)

The thin sheath-like rostrum surrounding the phragmocone has a brownish color. X-ray diffraction analyses revealed that it is typically composed of aragonite. On the mid-dorsal line of the rostrum, characteristic rounded ridges (Fig. 14.2) are present. As demonstrated by Makowski (1952), they run parallel in the earliest ontogenetic stages but diverge at later stages until they disappear at a phragmocone diameter of approximately 18 mm. Specimen MC-7 demonstrates that the dorsal ridges are already developed at the earliest ontogenetic stages.

Conotheca and septal sutures are visible where the thin rostrum has flaked off. When this is the case in the dorsolateral region, a sharp and colored separation is evident on the outer conothecal surface (Fig. 14.2). Toward the dorsum, the surface is black. Weak imprints of proostracal growth lines are visible. They are similar to those described by Doguzhaeva et al. (2002) in *Megateuthis*. Toward the venter, the surface exhibits an iridescence that takes up two-thirds of the ventral phragmocone circumference.

Two specimens selected for ultrastructural studies (MC-8, MC-9) were fractured. Cross and longitudinal fractures were produced between a phragmocone diameter of 4 mm and 10 mm. This region corresponds presumably to the 10th–36th camera, i.e., juvenile, adolescent, and presumably adult stages.

Opened camerae were completely filled with sediment. Due to diagenetic dissolution, the septal and siphonal morphology of *Belemnotheutis* could not be investigated.

Shell fractures were only cleaned, coated with gold, and studied with SEM. All specimens (MC-1–MC-9) are stored in the Institute of Geological Sciences, Paleontology, Freie Universität Berlin.

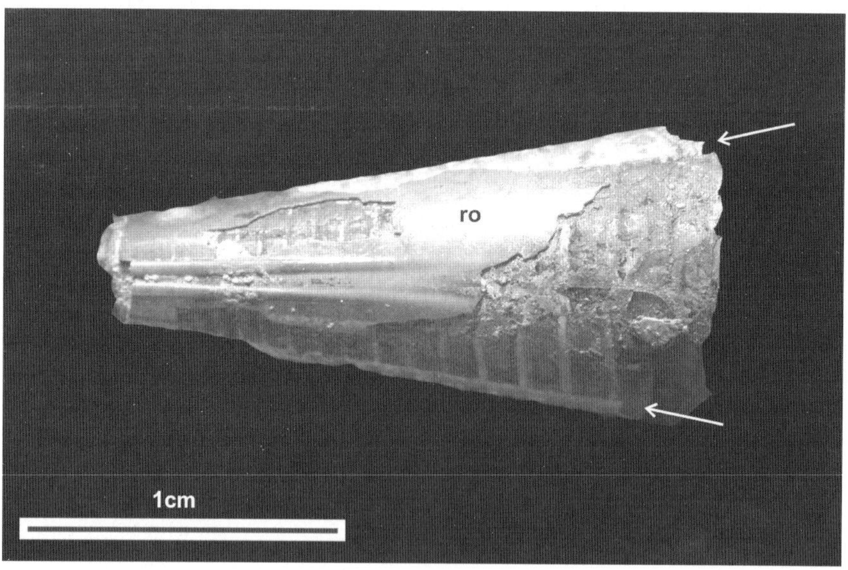

Fig. 14.2 Belemnotheutis antiquus (=polonica). *Dorsal view of specimen MC-5. The paired ridges are visible. Where the rostrum (ro) is flaked off on the dorsolateral part, an abrupt change in color is visible (arrows).*

4 Ultrastructural Observations on the Conotheca of *Belemnotheutis*

4.1 Ventral Conotheca (Phragmocone Diameter = 10 mm)

Ventrally to ventrolaterally, the conotheca consists of (from inside out) an inner prismatic, a nacreous, a thin outer prismatic, and a lamellar layer (Fig. 14.3). The rostrum overlies the outermost lamellar layer. The inner prismatic layer is between 17 μm and 25 μm thick. Thickness increases lateralward. The outer prismatic layer is constantly 3–4 μm thick (Fig. 14.3B). A well-developed 20–40 μm thick nacreous layer (platelets-nacre, *Nautilus*-Type, Type 1) separates the inner and outer prismatic layer. Longitudinal fractures between a phragmocone diameter of 4 mm and 10 mm show that the thickness of nacre is correlated with the ontogenetic stage, i.e., the thickness of nacre increases with the phragmocone diameter. Prisms of the inner and outer layer are clearly delimited from the intermediate nacreous layer. There are absolutely no interspaces between them.

The outside of the outer prismatic layer is covered by a 4–5 μm thick lamellar layer (Figs. 14.3B, 14.4A). Only the rostrum occurs outside this lamellar layer.

4.2 Lateral Conotheca (Phragmocone Diameter = 10 mm)

Laterally, the conotheca is composed of (from inside out) an inner prismatic, a nacreous, a thin outer prismatic, and a lamellar layer. The only difference between the ventral and the lateral conotheca is the comparatively thin nacreous layer (10 μm). Figure 14.4B shows that sometimes there might be a hollow gap between the outer prismatic layer and the rostrum. The lamellar layer must have disintegrated.

4.3 Dorsolateral Conotheca (Phragmocone Diameter = 10 mm)

Dorsolaterally (between 10 o'clock and 2 o'clock respectively on both sides of the conotheca), the most remarkable observation is the wedging out of the nacreous layer (Fig. 14.5A, B). At a distance of 200 μm, the nacreous layer thins out and disappears completely (Fig. 14.5B). Consequently, both inner and outer prisms merge together (25 μm) and are no longer distinguishable. The laminar layer

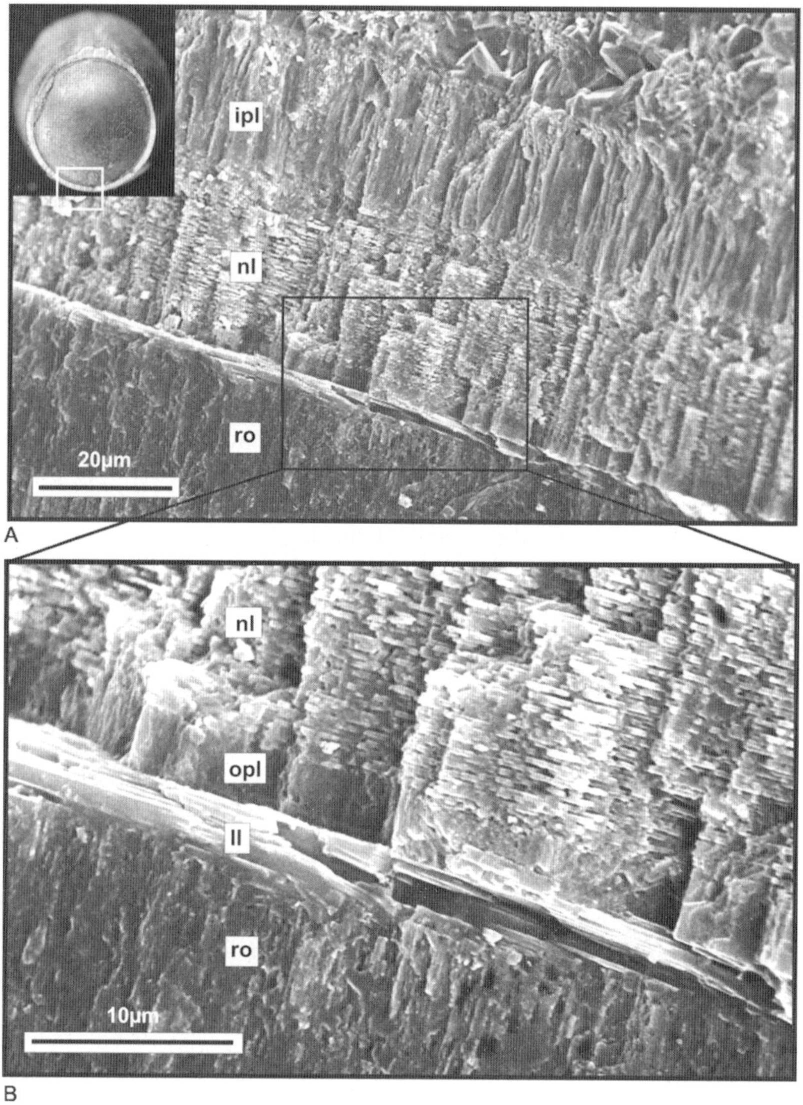

Fig. 14.3 *(A) Ventral cross-fracture of specimen MC-8 (phragmocone diameter = 10 mm) to show inner prismatic layer (ipl), nacreous layer (nl), and rostrum (ro); 1,000× enlarged. (B) Detail of (A) to show nacreous layer (nl), outer prismatic layer (opl), lamellar layer (ll), and rostrum (ro); 3,000× enlarged.*

between the rostrum and the now single inner prismatic layer is still present (2–3 μm). In some places (Fig. 14.5A), the inner surface of the rostrum displays similar borings as described in Doguzhaeva et al. (2003b), which indicate a high organic content. In other places, gaps are filled with secondary calcite (Fig. 14.5B).

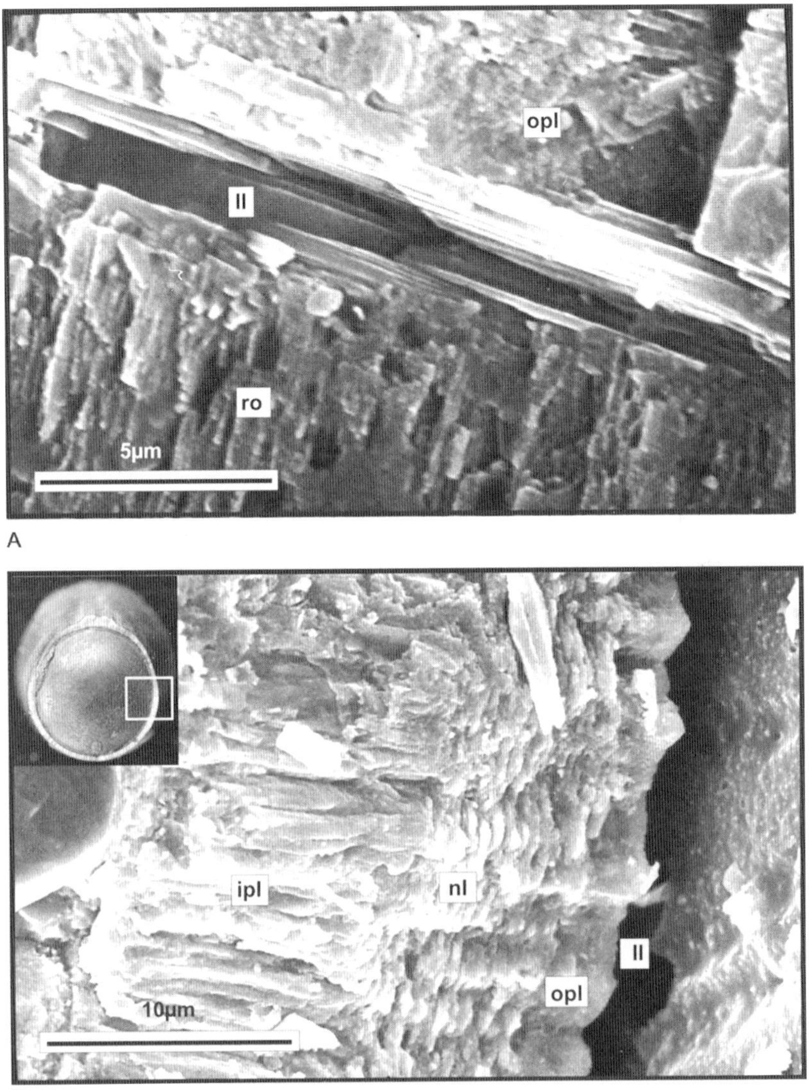

Fig. 14.4 *(A) Detail of Fig. 14.3B to show outer prismatic layer (opl), lamellar layer (ll), and rostrum (ro); 7,000× enlarged. (B) Lateral cross-fracture of specimen MC-8 (phragmocone diameter = 10 mm) to show inner prismatic layer (ipl), nacreous layer (nl), and outer prismatic layer (opl). The lamellar layer (ll) is disintegrated; 3000× enlarged.*

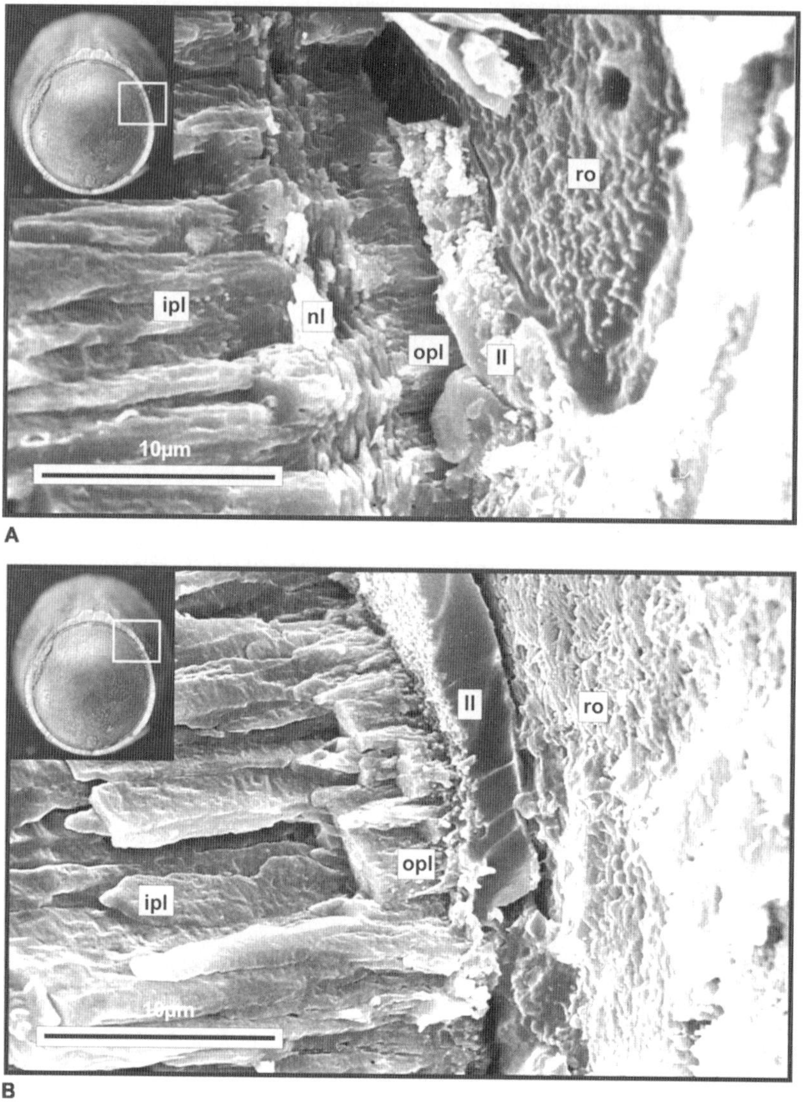

Fig. 14.5 *(A) Dorsolateral cross-fracture of specimen MC-8 (phragmocone diameter = 10 mm) to show inner prismatic layer (ipl), thin nacreous layer (nl), outer prismatic layer, lamellar layer (ll), and inner surface of rostrum (ro). The lamellar layer is bored; 3,000× enlarged. (B) Dorsolateral cross-fracture (only 50 µm dorsal from A). The nacreous layer wedged out. The lamellar layer is replaced by secondary calcite; 3,000× enlarged.*

4.4 Dorsal Conotheca (Phragmocone Diameter = 10 mm)

Dorsally, the conotheca is composed of an inner prismatic layer and a layer that corresponds to the lamellar layer. No nacreous layer is observable either in cross fractures or in longitudinal fractures.

Sideward from the paired ridges on the outer surface of the rostrum, the lamellar layer is 4–5 μm thick. Laminae were sometimes replaced by presumably phosphorus granules similar to those described by Doguzhaeva and Mutvei (2003). They are nearly 1 μm in diameter.

Mid-dorsally below the paired ridges the layer under discussion swells up to a spindle-like bulge of 40 μm in thickness (Fig. 14.6A). The spindle has a length of 500 μm. Here the lamellar layer is preserved as secondary calcite (Fig. 14.6B).

5 Discussion

Our results partly confirm earlier observations by Bandel and Kulicki (1988) in which the conotheca of *Belemnotheutis* consists of an inner and an outer prismatic layer separated by a nacreous layer (Fig. 14.7). In contrast to Bandel and Kulicki (1988), we found a fourth lamellar layer, which is external to the outer prismatic layer. It occurs all around the phragmocone. Most probably, the laminae of this layer were primarily organic because in some places they are penetrated by borings (fungi? algae?), disintegrated or replaced by secondary calcite.

However, a four-layered construction of the conotheca does not occur all around the whole phragmocone circumference (Fig. 14.7). Whereas Bandel and Kulicki (1988) assumed a persistent conotheca, we conclude from our observations that in the genus *Belemnotheutis* only two-thirds of the phragmocone circumference consist of four layers. The remaining dorsal one-third of the circumference is built of only two layers. Ventrolaterally to dorsolaterally the thickness of the nacreous layer decreases continuously until it wedges out. As a result, inner and outer prismatic layers merge together. Consequently, only a single inner prismatic layer and an outer lamello-organic layer constitute the dorsal conotheca.

Because of its outermost position, it seems reasonable to interpret the lamello-organic layer as the periostracum. As characteristic for most shelled molluscs the periostracum covers the inner mineralized layers. Dorsally, where the nacreous layer is absent, the rather thin periostracum is modified into a comparably thick layer. Where the sheath-like rostrum has been removed, this sudden shift from a nacre-dominated conotheca to a periostracum-dominated conotheca is even visible macroscopically (Fig. 14.2). Weak, forwardly curved growth lines on the dorsal black surface of the periostracal layer are typical for a belemnoid proostracum. This led us to assume that the periostracum forms the proostracum.

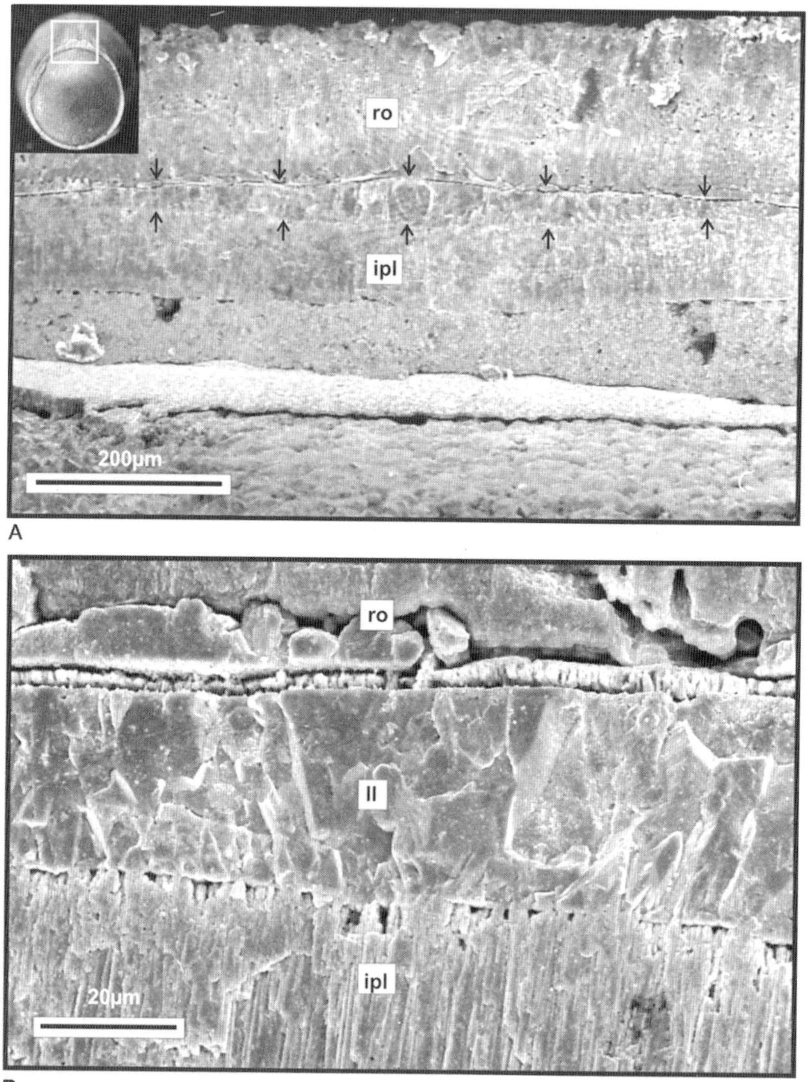

Fig. 14.6 *(A) Dorsal cross-fracture of specimen MC-8 (phragmocone diameter = 1 cm) to show inner prismatic layer (ipl), spindle-like swelling of the lamellar layer (arrows), and rostrum (ro); 100× enlarged. (B) Detail of (A). The lamellar layer (ll) between the inner prismatic layer (ipl) and the rostrum (ro) is comparatively thick and replaced by secondary calcite; 1,000 × enlarged.*

6 Conclusions

It was previously unknown that nacre does not occur along the entire phragmocone circumference in belemnoid shells. Certainly, the presence of nacre depends on the ontogenetic stage. According to our material, nacre is already present after completion of the tenth camera (phragmocone diameter = 4mm). The absence of nacre in

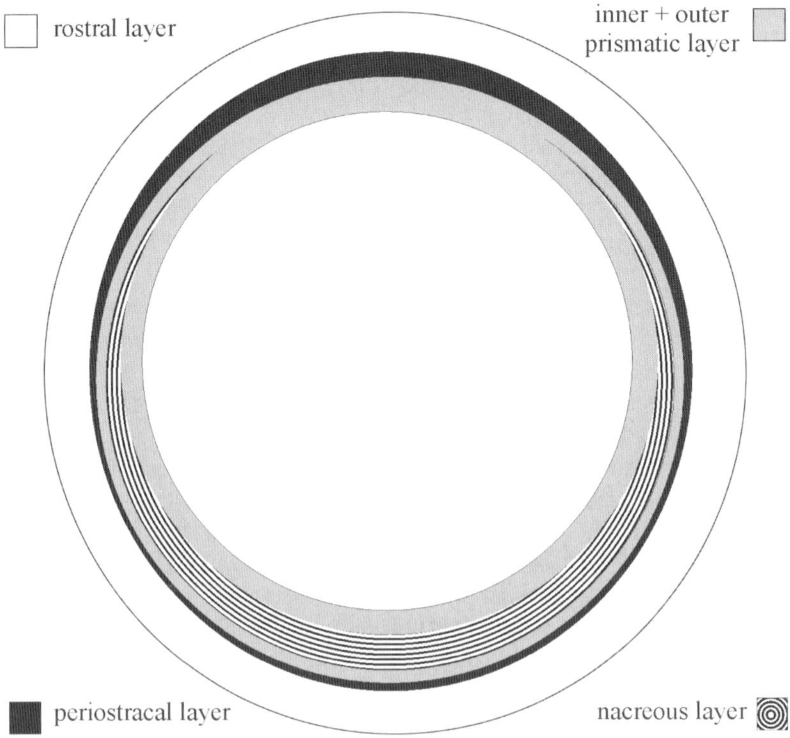

Figure 14.7. *Schematic cross section through the conotheca and the rostrum of* Belemnotheutis. *Ventrally the conotheca is four-layered. Dorsolaterally the nacreous layer wedges out. As a result, the inner and the outer prismatic layer merge together. Dorsally the periostracal layer increases in thickness.*

the dorsal part can be observed in these presumably juvenile and in later stages. MC-8, a specimen with approximately 36 chambers and a phragmocone diameter of 10 mm, is probably still an adolescent. Nevertheless, our conclusions can also be applied to adult forms, because specimen MC-7 (phragmocone length = 100 mm) shows the distinct color difference between the dorsum and the remaining ventral side. We are aware that *Belemnotheutis* is an unusual representative of the Belemnitida and that dorsal reduction of nacre is not necessarily a feature of all Belemnoidea. Nevertheless, previously analyzed specimens should be reinvestigated in the light of this new ultrastructural pattern. Our observations might explain why nacre sometimes was not observed in earlier studies on belemnoid shells (Table 14.1). It is likely that if just dorsal fragments of individuals are investigated, no nacre is detectable. Older statements that the nacreous layer forms the bulk of the belemnoid conotheca must be revised, if future observations on other taxa support our observations.

So far only Bandel (1985) and Bandel and Kulicki (1988) observed a thin outer prismatic layer in the belemnoid contheca. In the present study, it was difficult to confirm this thin outer prismatic layer. We suppose that most workers who did not find this outer prismatic layer either overlooked it or investigated dorsal parts of the conotheca where only a single prismatic layer occurs (Fig. 14.7). Consequently, we conclude that a thin outer prismatic layer is present at least in the ventral to dorsolateral parts of the belemnoid contheca.

An irregularly mineralized or predominantly organic layer external to a nacreous layer was reported by Mutvei (1964) from presumably adult *Megateuthis*, by Doguzhaeva et al. (2002) from adult *Megateuthis*, by Doguzhaeva et al. (2003b) from adult *Passaloteuthis*, and by Doguzhaeva et al. (2003a) from adult *Donovaniconus*. Bandel et al. (1984) described an organic layer ("innere organische Zwischenschicht") between the inner prismatic protoconch wall and the primordial rostrum, which continues without interruption into the postembryonic stages. Judging by the difficulty of detecting the thin outer prismatic layer, these predominantly organic layers most probably correspond to our outermost lamello-organic layer. Doguzhaeva et al. (2003b) called their irregularly mineralized layer with a high organic content "proostracal layer." They believed that the layer under discussion forms the proostracum, but cannot be considered as a continuation of the conotheca, because it is situated between the conotheca and the rostrum. However, Doguzhaeva et al. (2003a, b) did not refer to the possibility that their organic layer might correspond to the periostracum, which is a typical feature of all mollusc shells. Already Bandel (1985) stated that belemnoids display exactly the same sequence of conothecal layers (including a periostracal layer) as nautilids and ammonoids (Fig. 14.8). In the conotheca of spirulid and sepiids, in contrast, a nacreous layer is completely absent (Doguzhaeva, 1996; Fuchs, 2006).

Independently if the lamello-organic layer corresponds to the periostracum or not, we follow Doguzhaeva et al. (2003a, b) in considering this layer as the layer that forms the proostracum. (1) The lamello-organic layer in the belemnoid conotheca seems to possess the ability to accomplish remarkable swellings in the dorsal region. (2) Strongly forwardly curved growth lines on the dorsolateral surface of the layer

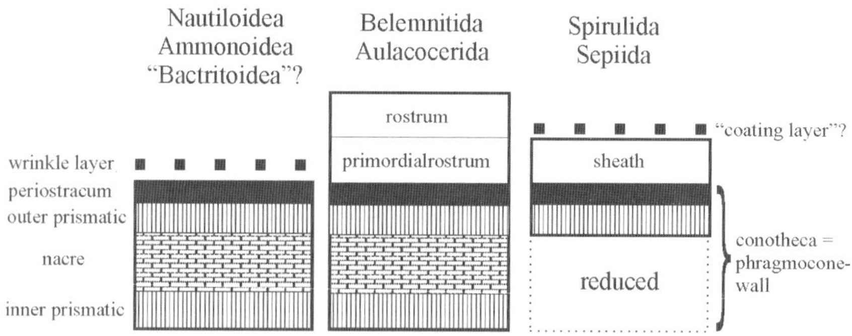

Fig. 14.8 *Homologous conothecal layers of Cephalopoda. Structures such as the wrinkle layer, rostrum, primordial rostrum, and sheath do not belong to the conotheca. Belemnoids exhibit the same sequence of conothecal layers as ectocochleates, whereas in spirulids and sepiids the nacreous layer is reduced. The periostracum and an inner prismatic layer are present in each taxon.*

under discussion (i.e., the phragmocone surface) are characteristic for a belemnitid proostracum (Fig. 14.1). (3) The comparatively abrupt change from a nacre-dominated conotheca into a more organic conotheca accomplished by a colored separation on the dorsolateral phragmocone surface appears where the proostracal growth lines bend forward. It remains unclear whether the inner prismatic layer also helps to form the projecting part of the proostracum. So far reliable data on cross sections through a proostracum are poor (Hewitt and Pinckney, 1982).

According to our interpretation regarding the proostracum as a derivation of the periostracum, the proostracum is part of the conotheca. This idea contradicts Doguzhaeva et al. (2002, 2003a, b) in that the belemnoid proostracum is independent and not involved in the composition of the conotheca. Doguzhaeva et al. (2002, 2003) argued that the possession of a proostracum is unique and therefore apomorphic within the Coleoidea, because ectocochleate cephalopods lack a comparable layer. They further concluded that their idea does not support the widely accepted hypothesis by which the proostracum is a dorsal remnant of the body chamber (Naef, 1922; Jeletzky, 1966). The development of the coleoid proostracum through the progressive reduction of the ventral part of a formerly closed body chamber, hence, affords the involvement of the periostracum. Our conclusions therefore strongly support the ideas of Naef (1922) and Jeletzky (1966). The apomorphic status of the proostracum as a dorsal extension of the phragmocone within the Coleoidea is beyond any doubt, but the layer that builds up the proostracum, namely the periostracum, is plesiomorphic within the Coleoidea.

References

Bandel, K. 1985. Composition and ontogeny of *Dictyoconites* (Aulacocerida, Cephalopoda). *Paläontologische Zeitschrift* **59**: 223–244.

Bandel, K. 1989. Cephalopod shell structure and general mechanisms of shell formation. *In* Carter, J. G. (editor), *Skeleton Biomineralization: Patterns, Processes and Evolutionary Trends*, pp. 97–115. New York: Van Nostrand Reinhold.

Bandel, K., T. Engeser, and J. Reitner. 1984. Die Embryonalentwicklung von *Hibolithes* (Belemnitida, Cephalopoda). *Neues Jahrbuch für Geologie und Paläontologie, Abhandlungen* **167**: 275–303.

Bandel, K., and C. Kulicki. 1988. *Belemnoteuthis polonica*: a belemnite with an aragonitic rostrum. *In* J. Wiedmann, and J. Kullmann (editors), *Cephalopods – Present and Past*, pp. 303–316. Stuttgart: Schweitzerbart'sche Verlagsbuchhandlung.

Barskov, I. S. 1972. Microstructure of the skeletal layers of belemnites compared with the shell layers of other molluscs. *Paleontological Journal* (Translation of the Paleontologicheskiy Zhurnal) **4**: 492–500.

Christensen, E. 1925. Neue Beiträge zum Bau der Belemniten. *Neues Jahrbuch für Geologie und Paläontologie* **51**: 118–158.

Crick, G. C. 1896. On the Proöstracum of a belemnite from the upper Lias of Alderton, Gloucestershire. *Proceedings of the Malacological Society* **2**: 117–119.

Cuif, J.-P., and Y. Dauphin. 1979. Mineralogie et microstructures d'Aulacocerida (Mollusca-Coleoidea) du Trias de Turquie. *Biomineralisation* **10**: 70–79.

Dauphin, Y. 1983. Microstructure du phragmocone du genre triassique *Aulacoceras* (Cephalopoda-Coleoida) – remarques sur les homologies des tissues coquilliers chez les Cephalopodes. *Neues Jahrbuch für Geologie und Paläontologie, Abhandlungen* **165**: 418–437.

Doguzhaeva, L. A. 1996. Two early Cretaceous spirulid coleoids of the northwestern Caucasus: their shell ultrastructure and evolutionary implications. *Paleontology* **39**(3): 681–707.

Doguzhaeva, L. A., R. H. Mapes, and H. Mutvei. 2003a. The shell and inc sac morphology and ultrastructure of the Late Pennsylvanian cephalopod *Donovaniconus* and its phylogenetic significance. *Berliner Paläobiologische Abhandlungen* **3**: 61–78.

Doguzhaeva, L. A., and H. Mutvei. 2003. Gladius composition and ultrastructure in extinct squid-like coleoids *Loligosepia*, *Trachyteuthis* and *Teudopsis*. *Revue de Paleobiologie* **22**: 877–894.

Doguzhaeva, L. A., H. Mutvei, and D. T. Donovan. 2002. Pro-ostracum, muscular mantle and conotheca in the Middle Jurassic belemnite *Megateuthis*. *Abhandlungen der geologischen Bundesanstalt* **57**: 321–339.

Doguzhaeva, L. A., H. Mutvei, G. K. Kabanov, and D. T. Donovan. 1999. Conch ultrastructure and septal neck ontogeny of the belemnid *Conobelus* (Duvaliidae) from the Valangian of the Crimea (Black Sea). *In* F. Oloriz, and F. J. Rodriguez-Tovar (editors), *Advancing Research on Living and Fossil Cephalopods*, pp. 47–57. New York: Kluwer Academic/Plenum.

Doguzhaeva, L. A., H. Mutvei, and W. Weitschat. 2003b. The pro-ostracum and primordial rostrum at early ontogeny of Jurassic belemnites from north-western Germany. *Berliner Paläobiologische Abhandlungen* **3**: 79–89.

Donovan, D. T., and M. D. Crane. 1992. The type material of the Jurassic cephalopod *Belemnoteuthis*. *Palaeontology* **35**(2): 273–296.

Doyle, P. 1986. *Naefia* (Coleoidea) from late Cretaceous of southern India. *Bulletin of the British Museum of Natural History* **40**(4): 133–139.

Doyle, P., and E. V. Shakides. 2004. The jurassic belemnite suborder Belemnotheutina. *Palaeontology* **47**: 983–998.

Fuchs, D. 2006. Fossil erhaltungsfähige Merkmalskomplexe der Coleoidea (Cephalopoda) und ihre phylogenetische Bedeutung. *Berliner Paläobiologische Abhandlungen* **8**: 1–121.

Hewitt, R. A., and G. Pinckney. 1982. On the occurrence and the microstructure of the phragmocone and pro-ostracum of the belemnite Genus *Acroteuthis Stolley. Palaeontographica, Abteilung A* **179**: 142–147.

Jeletzky, J. A. 1966. Comparative morphology, phylogeny and classification of fossil Coleoidea. *Paleontological Contributions, University of Kansas, Mollusca* **7**:1–162.

Keupp, H., and V. V. Mitta. 2004. Septenbildung bei *Quenstedtoceras* (Ammonoidea) von Saratov (Russland) unter anomalen Kammerdruckbedingungen. *Mitteilungen des Geologisch-Paläontologischen Instituts, Universität Hamburg* **88**: 51–62.

Makowski, H. 1952. La faune callovienne de Lukow en Pologne, *Palaeontologia Polonica* **4**: 1–64.

Müller-Stoll, H. 1936. Beitraege zur Anatomie der Belemnoidea, *Nova Acta Leopoldina*. New Series **4**: 159–226.

Mutvei, H. 1964. Remarks on the anatomy of recent and fossil Cephalopoda – with description of the minute shell structure of belemnoids. *Stockholm Contributions to Geology* **11**: 79–102.

Naef, A. 1922. *Die fossilen Tintenfische – Eine paläozoologische Monographie*. Jena: Fischerverlag.

Pearce, J. C. 1847. On the fossil Cephalopoda from the Oxford Clay constituting the genus *Belemnoteuthis*, Pearce. *Proceedings of the Geological Society of London* **3**: 592–594.

Part III
Biogeography, Biostratigraphy, Ecology, and Taphonomy

Chapter 15
New Data on the Clymeniid Faunas of the Urals and Kazakhstan

Dedicated to the Memory of B. I. Bogoslovsky

Svetlana Nikolaeva

Paleontological Institute, Russian Academy of Sciences, Profsoyuznaya 123, Moscow 117997, Russia, 44svnikol@mtu-net.ru

Keywords: clymeniids, Urals, Kazakhstan, Bashkortostan, Famennian, biostratigraphy

1 Introduction

The clymeniid faunas of the Urals and Kazakhstan are renowned for their diversity and abundance. Thousands of excellently preserved shells represent all known families and most genera. Ammonoids are mostly found in carbonate

N. H. Landman et al. (eds.), *Cephalopods Present and Past: New Insights and Fresh Perspectives*, 317–343.

units containing almost exclusively cephalopods and conodonts with very little or no benthic fauna. The most comprehensive research on the fauna was by Bogoslovsky (1955, 1960, 1962, 1965, 1969, 1971, 1975, 1976, 1977, 1979a, b, 1981, 1983) although several previous workers (Tschernyshev, 1887; Perna, 1914; Kolotukhina, 1938; Kind, 1944; Nalivkina, 1953; among others) provided valuable data. Bogoslovsky's collection includes over 15,000 specimens collected by him and many mapping geologists. Unfortunately, most of Bogoslovsky's notes and field books were lost after his death in 1986; the huge collection remained poorly labeled and unusable until new fieldwork was conducted in this region in 2001. During research by the staff of the Paleontological Institute, Russian Academy of Sciences, Moscow, the most important localities were resampled and redescribed, and most horizons where Bogoslovsky's specimens were collected were found.

2 Geological Setting

The collection comes from a large territory of over 500,000 square miles in the Urals and Kazakhstan (Fig. 15.1) and covers the three Famennian ammonoid genozones: *Prolobites-Platyclymenia*, *Clymenia-Gonioclymenia*, and *Kalloclymenia-Wocklumeria* as defined by Bogoslovsky (1969, 1971) (Fig. 15.2). The genozone concept, definition, correlation, and distribution of Devonian genozones in the world are discussed in detail by Bogoslovsky (1969) and Becker and House (2000). For the purpose of this study, the subdivision of the Famennian into three stages is adopted in the sense of Sandberg et al. (2002) rather than into four substages, as ruled by the International Commission on Stratigraphy (2004).

The area studied includes several large regions each containing ammonoid assemblages of differing abundance [west to east direction – Bashkortostan, western Kazakhstan, Chelyabinsk Region (these three within the South Urals), and two regions in central Kazakhstan] (Fig. 15.3). The borders of these regions do not conform to political boundaries (e.g., the western Kazakhstan Region, as defined here, includes some localities that are politically in the Orenburg Region, Russia). The central Kazakhstan Region, as defined here, includes the southwestern margin of the Chingiz Mountain Range, which is politically in the former Semipalatinsk Region. The Uralian-Kazakhstanian Region includes several tectonic zones existing in the Famennian, i.e., the eastern slope of the Baltica Craton [the Volga-Ural Carbonate Platform on the eastern margin of the Euramerica (Laurussia) Continent], remains of the Magnitogorsk Island Arc, the slope of the Kazakhstania Continent, and that of the submarine Mugodzhar Microcontinent (Puchkov, 2000; Mizens, 2004; Veimarn et al., 2004). The complex tectonics of this region, with highly variable depths, facies, and currents are reflected in the variety of ammonoid communities, which included clymeniids and goniatitids. The differences in the abundances of specimens and in

Fig. 15.1 *Map showing ammonoid localities in the Uralian–Kazakhstanian Region. Major localities are at the southern termination of the Urals and Kazakhstan.*

the taxonomic structure of the assemblages are believed to reflect different habitats within the basin of the Uralian Ocean.

3 Facies and Taphonomy

The ammonoid-bearing Famennian sequences of the Uralian–Kazakhstanian Region are highly condensed in western Kazakhstan, but become thicker to the north (Bashkortostan) and are particularly thick in central Kazakhstan.

Stage	Genozone		Zones in the Rhenish Massif Becker & House, 2000; Becker et al., 2002	Genozones (Becker, 1993; Becker & House (2000)	Zones of the Rhenish Massif (Korn, 1981, 1986)	Zones of the Rhenish Massif (Korn, 1999; Korn& Ziegler, 2002)	Zones of the Urals and Kazakhstan
Famennian	Disappearance of clymeniids; VI Kalloclymenia-Wocklumeria	VI-F	Acutimitoceras (S.) prorsum	Acutimitoceras	prorsum	prorsum	
		VI-E	Cymaclymenia nigra	Cymaclymenia	Upper paradoxa	nigra	
		VI-D-2	Wocklumeria sphaeroides: Epiwocklum. applanata	Wocklumeria		sphaeroides	sphaeroides
		VI-D-1	Wocklumeria sphaeroides: Wocklum. sphaeroides				
		VI-C-2	Parawocklumeria paradoxa	Parawocklumeria	Lower paradoxa	paradoxa	nucleus
		VI-C-1	Kamptoclymenia endogona			endogona	
		VI-B	Balvia lens	Balvia	Upper subarmata	lens	
		VI-A	Kosmoclymenia (M.) sublaevis	Linguaclymenia	Lower subarmata	parundulata / sublaevis	frechi-corpulenta
	V Clymenia-Gonioclymenia	V-C	Piriclymenia piriformis	Kalloclymenia	piriformis	piriformis	
		V-B	Ornatoclymenia ornata	Ornatoclymenia	ornata	ornata	ornata
		V-A	Clymenia laevigata	Clymenia	acuticostata	laevigata	laevigata
	IV Prolobites-Platyclymenia	IV-C	Protoxyclymenia (F.) serpentina	Pachyclymenia	serpentina*	serpentina*	serpentina + dunkeri + annulata
		IV-B	Protoxyclymenia (P.) dunkeri	Protoxyclymenia	Upper annulata	dunkeri	
		IV-A	Platyclymenia (P.) annulata	Prionoceras	Lower annulata	annulata	
	III (Appearance of clymeniids)	III-C	Prolobites delphinus	Prolobites	delphinus	delphinus	delphinus
		III-B	Pseudoclymenia pseudogoniatites	Pseudoclymenia	sandberg.	pseudogon.	pseudogon. + dorsatum
		III-A	Pernoceras dorsatum	Pernoceras		dorsatum	
	Cheiloceras	II	Dimeroceras mamilliferum	Dimeroceras			

Fig. 15.2 *Correlation of the Famennian ammonoid zones proposed by various authors. The shaded area indicates the interval containing Famennian clymeniids in the Urals and Kazakhstan.*

Fig. 15.3 *Map showing five major areas with different Famennian ammonoid assemblages in the Uralian–Kazakhstanian Region. 1 – Western Kazakhstan, 2 – Chelyabinsk Region, 3 – Bashkortostan (these three within the South Urals), 4 – Karaganda Region, 5 – Semipalatinsk Region (these two in central Kazakhstan).*

The richest Famennian ammonoid occurrences are found in western Kazakhstan in settings that are interpreted as the eastern margin of the Volga-Ural Carbonate Platform. They occur in thin, relatively deepwater carbonates with abundant conodonts and nautiloids, but with scarce foraminifers, bryozoans, brachiopods, and other benthic organisms (Fig. 15.4). Usually the ammonoids are found as large accumulations of non-oriented, unsorted, and excellently preserved shells. Abundances differ, however, in the stratigraphic succession. In most sections, the richest assemblages are found in the Middle Famennian, and lower part of the Upper Famennian. The number and diversity of ammonoids decreases noticeably up section. The microfacies are mostly mudstone/wackestone, and less frequently homogeneous packstone. Bioclasts include thin-shelled planktonic ostracodes (varying quantity), rare foraminifers (e.g., *Parathurammina*), and rare, rounded, clear fragments of pelmatozoans and camenids (Akhmetshina et al., 2004; Nikolaeva et al., 2004).

The Famennian of western Kazakhstan includes most Upper Famennian ammonoid zones in uninterrupted succession. The total thickness of the Famennian does not exceed 20 m. The condensed nature of the sequences and the character of the microfacies suggest deepwater sedimentation far from the source of clastic material. In contrast, in Bashkortostan and central Kazakhstan, the beds with ammonoids are separated by relatively thick beds of barren rocks (sections up to 150 m). In Bashkortostan, the shell-bearing rocks are carbonates, the source of which was the Volga-Urals Carbonate Platform, and in central Kazakhstan these are shales and siltstones, of clastics supplied by washout from the Kazakhstania Paleocontinent (for a full list of ammonoid localities and detailed description of ammonoid-bearing sections, see Nikolaeva and Bogoslovsky, 2005a).

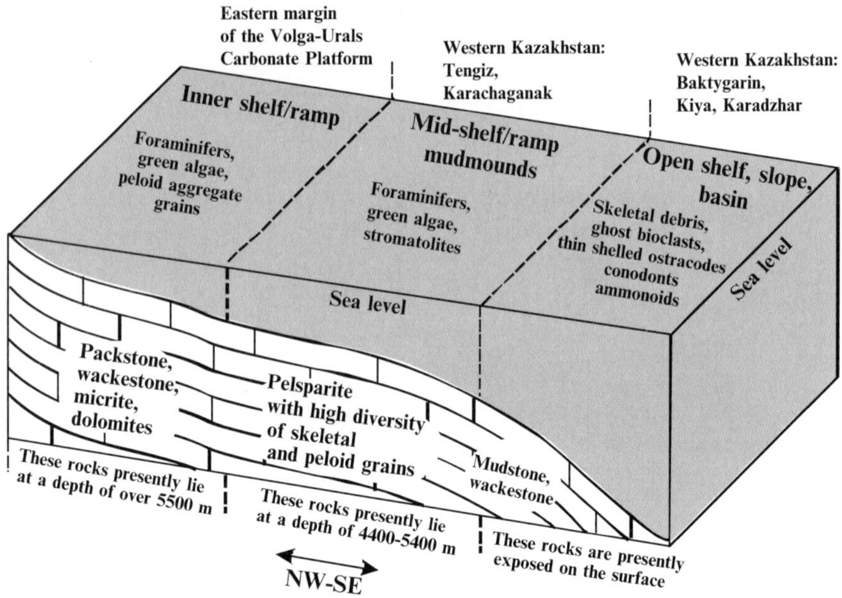

Fig. 15.4 *Reconstruction of the environmental settings in the Famennian basin of western Kazakhstan. Ammonoids are found only in the deep southeastern zone, where there is very little benthic fauna. This reconstruction is based on data from many sections and borings in the Peri-Caspian Region (Nikolaeva and Gibshman, in preparation). Mudmounds to the northwest of the zone with ammonoid-bearing sequences contain giant oil deposits.*

4 Ammonoid Assemblages

Clymeniids dominate all Middle–Late Famennian ammonoid assemblages of the Uralian–Kazakhstanian Region. The clymeniid faunas include two suborders, four superfamilies, 14 families, 41 genera, and some 130 species. Most taxa (over 70%) belong to the suborder Clymeniina, with Cyrtoclymeniidae, Rectoclymeniidae, and Cymaclymeniidae dominant in diversity and number of individuals (Fig. 15.5) (Bogoslovsky, 1981; Nikolaeva and Bogoslovsky, 2005a).

Four successive assemblages corresponding to ammonoid genozones are recognized in the Uralian–Kazakhstanian Region: *Cheiloceras, Prolobites-Platyclymenia, Clymenia-Gonioclymenia*, and *Kalloclymenia-Wocklumeria* (Bogoslovsky, 1971, 1981) (Fig. 15.2). The succession of species within these genozones is similar to that recognized in Germany and Morocco and internationally (Korn, 1999; Becker and House, 2000; Becker et al., 2002), although there are regional differences. Because the ammonoid succession in Germany has been studied in great detail, the species-based zones there are referred to for comparisons with distant regions of the world (see Korn, 1999; Becker, 2000; Becker et al., 2002; Korn and Klug, 2002). The comparisons are closest at the supraspecific level but there are clear regional and local differences at the species level.

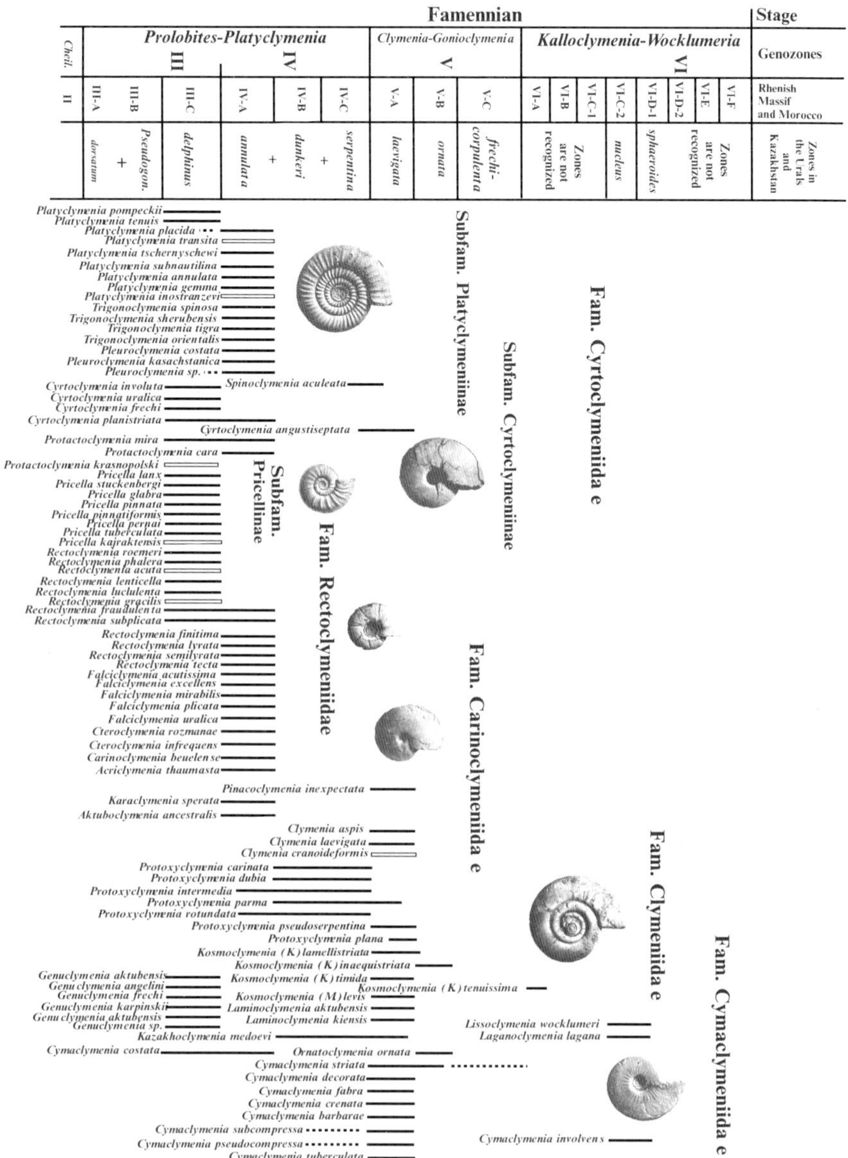

Fig. 15.5 *Distribution of the suborder Clymeniina in the Famennian of the Uralian–Kazakhstanian Region. This clymeniid suborder was the dominant component of the ammonoid faunas in this region. Solid black lines show the taxa recorded by Nikolaeva and Bogoslovsky (2005a). White lines are drawn based on records by other authors.*

In the South Urals and Kazakhstan, recognition of species zones has always been very problematic because the sections are highly condensed (Bogoslovsky, 1971; Simakov et al., 1983). However, it is evident that the equivalents of some species zones recognized in Germany can be recognized there too, although the precise location of the zonal boundaries will require more data. The zonal boundaries in some sections are drawn tentatively and are the subject of ongoing research.

4.1 *Cheiloceras* Genozone (UD-II)

The earliest Famennian ammonoids in the South Urals and Kazakhstan are known from many sections, but they are particularly abundant in the eastern slope of the Volga-Ural Platform in the periphery of the Mugodzhary Mountains and Chelyabinsk Region. This assemblage is diverse and contains many specimens, but all are goniatitids (mostly *Cheiloceras* and *Dimeroceras*).

4.2 *Prolobites–Platyclymenia* Genozone (UD-III-IV)

This genozone in the Urals and Kazakhstan includes the *dorsatum*, *pseudogoniatites, delphinus, annulata*, possibly *dunkeri*, and *serpentina* Zones, which is the complete sequence of zones identified at this level in the Rhenohercynian Region, with the *pseudogoniatites, delphinus*, and *annulata* zones best represented. For the composition of ammonoid assemblages of this age in the region studied, see Nikolaeva and Bogoslovsky (2005a).

4.2.1 *dorsatum* Zone (UD-III-A)

This zone is identified in the Chelyabinsk Region, South Urals, and in the Aktyubinsk Region, western Kazakhstan (both southwestern realms of the Uralian Ocean) where the assemblage includes *Pernoceras fundiferum* (Perna), *P. kochi* (Wedekind), *P. subtimidum* (Perna), *Sporadoceras* cf. *muensteri* (Buch), *S. equale* Becker, *S. clarkei* Wedekind, *Armatites planidorsatus* (Münster), and others.

4.2.2 *pseudogoniatites* Zone (UD-III-B)

This zone is identified in the western Kazakhstan and Chelyabinsk Region where assemblages are similar to the assemblages of the *pseudogoniatites* Zone in the Rhenohercynian Region where the zone is characterized by the wide distribution of the tornoceratid genus *Pseudoclymenia*. Localities with ammonoids of this age are also known in Germany (Sandberger, 1853a, 1853b; Gümbel, 1862; Frech, 1897, 1902; Born, 1912a, b; Schindewolf, 1923, 1937; Matern, 1931; Becker, 1993; Becker and House, 2000; Korn and Ziegler, 2002), Poland (Sobolev, 1914a, b; Dybczynski, 1913), Algeria and Morocco (Petter, 1960; Becker et al., 2002). The equivalents of the zone are also recorded in Australia (Jenkins, 1968; Petersen, 1975).

It is interesting that in Morocco *Pseudoclymenia* is rare at this level, whereas hundreds of *Pseudoclymenia* shells typically occur in assemblages in the South Urals and western Kazakhstan. Bogoslovsky (1971) suggested that the earliest species of *Prolobites* and the clymeniids could possibly have appeared within this zone, referring to records of *Prolobites insulcatus* by Lange (1929). Rzhonsnitskaya (2000) suggested the presence of clymeniids in the *pseudogoniatites* Zone. However, we found no *Prolobites* or clymeniids in this zone in western Kazakhstan. The *pseudogoniatites* Zone assemblages contain *Pseudoclymenia pseudogoniatites* (Sandberger), *Ps. dillensis* Drevermann, *Ps. aktubensis* Bogoslovsky, *Ps. elegans* Bogoslovsky, *Ps. plana* Bogoslovsky, *Sporadoceras* cf. *muensteri* (Buch), *S. clarkei* Wedekind, *S. equale* Becker, *S. lentiforme* Bogoslovsky, *Maeneceras rotundum* (Wedekind), *Araneites falcatus* Bogoslovsky, *Dimeroceras padbergense* Wedekind, *D. aktubense* Bogoslovsky, *Falcitornoceras bilobatum* (Wedekind), and others.

4.2.3 *delphinus* Zone (UD-III-C)

The *delphinus* Zone is recorded throughout the South Urals and Kazakhstan, but is especially well represented on the eastern slope of the Urals (Chelyabinsk Region) and in the Mugodzhary Mountains (western Kazakhstan) (Loewinson-Lessing, 1892; Perna, 1914; Kind, 1944; Nalivkina, 1953; Bogoslovsky, 1971, 1981) and in the north (Novaya Zemlya) (Nalivkina, 1936). This zone is recognized in Germany (Wedekind, 1908; Schindewolf, 1923; Lange, 1929; Matern, 1931; Price and Korn, 1989; Korn, 2002; Korn and Ziegler, 2002), Poland (Czarnocki, 1989), Austria (Frech, 1902, 1913), and Australia (Jenkins, 1968). Equivalents of this zone are recognized in Morocco (Becker et al., 2002) and in Montana (Miller, 1938; House, 1962; and others). The Uralian-Kazakhstanian assemblages of this age are dominated by *P. delphinus* (Sandberger), *P. auriformis* Bogoslovsky, *P. nanus* Perna, *P. aktubensis* Bogoslovsky, among others. The first clymeniids appeared at this level. The assemblage includes diverse *Rectoclymenia* (including *R. roemeri* Wedekind, *R. fraudulenta* Nikolaeva et Bogoslovsky, *R. acuta* (Perna), *Pricella* [including *P. stuckenbergi* (Tokarenko), *P. glabra* (Perna), *P. tuberculata* (Kind)], *Genuclymenia* [including *G. angelini* Wedekind, *G. frechi* Wedekind, *G. karpinskii* (Perna)], *Platyclymenia pompeckii* (Wedekind), *P. tenuis* Nikolaeva et Bogoslovsky, *Cyrtoclymenia frechi* (Tokarenko), *C. uralica* Nalivkina, and *C. involuta* (Wedekind). The assemblage also contains diverse *Sporadoceras* [e.g., *S.* cf. *muensteri* (Buch.), *S. clarkei* Wedekind, *S. lentiforme* Bogoslovsky, *S. biferum biferum* (Phillips)] and numerous *Posttornoceras contiguum* (Sobolev). On the whole, the assemblage is similar to those from Germany and Poland, but differs from the Moroccan and Algerian assemblages, where typical *Prolobites* are absent (Termier and Termier, 1950; Petter, 1959, 1960; Becker et al., 2002). We were not able to observe the succession of appearances of the earliest clymeniids similar to that reported from the Enken-Berg Section, Germany, where *Cyrtoclymenia frechi* (Tokarenko) and *Pricella stuckenbergi* (Tokarenko) are the first to appear and are followed by *Genuclymenia*, *Stenoclymenia*, and early *Platyclymenia* above (Korn and Ziegler, 2002).

4.2.4 *annulata* Zone (UD-IV-A)

In the South Urals and Kazakhstan this zone is widely represented (Perna, 1914; Kolotukhina, 1938; Kind, 1944; Nalivkina, 1953; Bogoslovsky, 1971, 1981; Nikolaeva and Bogoslovsky, 2005a.). This zone is also recognized in Germany (Wedekind, 1908; Lange, 1929; Matern, 1931; Price and Korn, 1989; Korn and Ziegler, 2002); Austria (Flügel and Kropfitsch-Flügel, 1965), Czech Republic (Rzehak, 1910), France (Frech, 1902), Great Britain (House, 1959, 1963), Spain (Schmidt, 1931; Montesinos and Arbizu, 1988), Morocco (Korn, 1999; Korn et al., 2000; Becker et al., 2002), Iran (Walliser, 1966), China (Ruan, 1981; Chang, 1960), Australia (Jenkins, 1968; Petersen, 1975; Becker, 2000), Alberta, Canada (House and Pedder, 1963). The assemblage of the Uralian–Kazakhstanian Region is similar to that of Germany, a reflection of the general worldwide similarity of ammonoid faunas of this age noted by many authors (House, 1985; Korn, 1999; Korn et al., 2000; Becker and House, 2000; Becker et al., 2002). The *annulata* Zone is recognized in the Aktyubinsk and Orenburg Regions (western Kazakhstan), Chelyabinsk Region (eastern South Urals), Bashkortostan (central South Urals), Karaganda and Semipalatinsk Regions (central Kazakhstan), and Komi Republic (North Urals).

The Uralian–Kazakhstanian clymeniid assemblages of the *annulata* Zone are dominated by *Platyclymenia* [*P. annulata annulata* (Münster), *P. annulata richteri* Wedekind, *P. intracostata* (Frech), *P. placida* (Perna), *P. tschernyschewi* (Rzehak), *P. gemma* Nikolaeva et Bogoslovsky] and *Rectoclymenia* [e.g., *R. acuta* (Perna), *R. tecta* Nikolaeva et Bogoslovsky, *R. lyrata* Nikolaeva et Bogoslovsky, *R. semilyrata* Nikolaeva et Bogoslovsky, *R. phalera* Nikolaeva et Bogoslovsky, *R. fraudulenta* Nikolaeva et Bogoslovsky, *R. subplicata* Nikolaeva et Bogoslovsky]. The assemblage also contains *Falciclymenia* (e.g., *F. plicata* Nikolaeva et Bogoslovsky, *F. uralica* Bogoslovsky, *F. acutissima* Nikolaeva et Bogoslovsky), *Protactoclymenia cara* Nikolaeva et Bogoslovsky, *Pleuroclymenia kasachstanica* (Kind), *Uraloclymenia volkovi* Bogoslovsky, and *Cymaclymenia costata* (Wedekind). Compared to the assemblages from other regions, *Rectoclymenia* and *Falciclymenia* are far more diverse here, whereas *Platyclymenia* is less abundant. Among goniatitids, *Prionoceras* is widespread [*P. divisum* (Münster), *P. frechi* (Wedekind)].

4.2.5 *dunkeri* Zone (UD-IV-B)

This zone or its equivalents are recognized in Germany, Morocco, and Poland, and possibly in the Uralian–Kazakhstanian Region. *Protoxyclymenia intermedia* Nikolaeva et Bogoslovsky is probably the earliest species of this genus in the Aktyubinsk Region, while *P. rotundata* Nikolaeva et Bogoslovsky is the earliest *Protoxyclymenia* in the Orenburg region. Some evidence exists on the presence of this zone in central Kazakhstan (Semipalatinsk Region).

4.2.6 *serpentina* Zone (UD-IV-C)

In western Kazakhstan, this zone is possibly represented in the Orenburg Region, in beds containing *Protoxyclymenia*, although the boundaries of this zone are uncertain. Possible equivalents of this zone are recognized in New Mexico (Percha Shale Formation) (Miller and Collinson, 1951) and Morocco (*Sporadoceras orbiculare* Zone) (Becker et al., 2002). The debatable position of the *serpentina* Zone at the base of the *Clymenia-Gonioclymenia* (Korn, 1999, 2002), or at the top of the *Prolobites-Platyclymenia* zone (Becker and House, 2000) cannot presently be resolved based on the Uralian material.

4.3 *Clymenia–Gonioclymenia* Genozone (UD-V)

In the Rhenohercynian Region this genozone includes the three species zones: *laevigata*, *ornata*, and *piriformis* Zones, the first two of which are recognized in the Uralian–Kazakhstanian Region. The topmost part of the genozone contains an assemblage transitional to the *Kalloclymenia–Wocklumeria* Genozone. The beds with this assemblage are recognized as the *frechi-corpulenta* Zone. For the composition of ammonoid assemblages of this age in the region studied, see Nikolaeva and Bogoslovsky (2005a).

4.3.1 *laevigata* Zone (UD V-A)

In the Uralian–Kazakhstanian Region, this Zone is best represented in the Orenburg Region (western Kazakhstan) and less so in the Chelyabinsk Region and central Kazakhstan. The *laevigata* Zone is also recognized in Britain, Spain, Poland, Austria, Czech Republic, and China. In Morocco, the equivalents of the *laevigata* Zone are recognized as the *Endosiphonites muensteri* and *Gonioclymenia subcarinata* subzones.

In the Uralian–Kazakhstanian Region, the assemblage contains numerous *Clymenia laevigata* (Münster), *C. aspis* Nikolaeva et Bogoslovsky, *Protoxyclymenia dubia* (Loewinson-Lessing), *Progonioclymenia* aff. *acuticostata* (Münster), *Gonioclymenia levis* Bogoslovsky, *Pachyclymenia intermedia* Bogoslovsky, *Kosmoclymenia lamellistriata* Nikolaeva et Bogoslovsky, *Cyrtoclymenia angustiseptata* (Münster), *Cymaclymenia striata* (Münster), *C. barbarae* (Loewinson-Lessing), *Costaclymenia binodosa* (Münster), *Discoclymenia cucullata* (Buch), *Falcitornoceras bilobatum* (Wedekind), *Prionoceras divisum* (Münster), *Alpinites kayseri* (Schindewolf), *Cymaclymenia subcompressa* Nikolaeva et Bogoslovsky, among others. Possibly, *Protoxyclymenia pseudoserpentina* Nikolaeva et Bogoslovsky also belongs to this level.

Ammonoids of this age are found in the Orenburg and Aktyubinsk Regions (both western Kazakhstan), Bashkortostan (central South Urals), Chelyabinsk Region (eastern South Urals), Karaganda, and the former Semipalatinsk Regions (central Kazakhstan). Further north, on the eastern slope of the North Urals, the assemblage contains *Posttornoceras contiguum* (Sobolew), *Maeneceras sulciferum* (Lange), *Uraloclymenia volkovi* Bogoslovsky, *Falciclymenia uralica* Bogoslovsky, *Protoxyclymenia dubia* (Loewinson-Lessing), and *Cymaclymenia costata* (Wedekind).

4.3.2 *ornata* Zone (UD-V-B)

In the Uralian–Kazakhstanian Region, this zone is possibly present in the sections of western Kazakhstan. An especially rich assemblage of this age is recorded from the Orenburg Region, where *Ornatoclymenia ornata* occurs with other typical taxa [*Gonioclymenia hoevelensis* Wedekind, *Sphenoclymenia maxima* (Münster), *Biloclymenia dubia* Bogoslovsky, *Progonioclymenia ventroplana* Bogoslovsky, *Protoxyclymenia carinata* Nikolaeva et Bogoslovsky, among others]. The zone is also recognized in Germany and Spain, possibly in Austria and Algeria, while in Morocco its equivalents are recognized as the *Gonioclymenia hoevelensis* Zone. Korn and Price (1987) suggested that *Kosmoclymenia inaequistriata* also appears at this level.

4.3.3 *frechi–corpulenta* Zone (UD-V-C–VI-A)

This zone is recognized in the Uralian–Kazakhstanian Region only. It corresponds to the *Kalloclymenia* Genozone (Becker and House, 2000), and possibly partly to the *piriformis* Zone (Korn, 1999) and *Piriclymenia piriformis* Zone (Becker and House, 2000; Korn, 2002). This correlation is only provisional because the *piriformis* Zone is only recognized in Germany (Korn, 1981; Price and Korn, 1989; Becker and House, 2000). Becker et al. (2002) recognized the equivalents of this zone in Morocco as the *Kalloclymenia subarmata* Zone. The shortcomings of this correlation are discussed in detail by Nikolaeva and Bogoslovsky (2005a). At Kiya, the genus *Kalloclymenia* first appears in the upper part of the *Clymenia-Gonioclymenia* Zone, a situation also observed in Morocco by Becker et al. (2002). The new *frechi-corpulenta* Zone in western Kazakhstan contains *Kalloclymenia frechi* Lange, *Pachyclymenia sinuconstricta* Bogoslovsky, *Kiaclymenia uralica* Bogoslovsky, *Kalloclymenia subarmata* (Münster), *K. pachydiscus* Bogoslovsky, *Otoclymenia* aff. *uhligi* (Frech), and *Gonioclymenia corpulenta* Bogoslovsky, which indicates that it partly corresponds with the lowest horizons of the *Kalloclymenia–Wocklumeria* Genozone (possibly the *laevis* Zone).

4.4 *Kalloclymenia–Wocklumeria* Genozone (UD-VI)

In Germany and Morocco, this genozone includes up to eight or nine species zones, most of which are not recognized elsewhere. Only two of these species zones *(nucleus* and *sphaeroides)* are positively identified in the Uralian–Kazakhstanian Region (Nikolaeva and Bogoslovsky, 2005a, b). So far ammonoids of this age have only been found in the western realms of the Uralian Ocean (Bashkortostan and western Kazakhstan). Faunas of the northern realms of the Uralian Ocean are as yet poorly dated. Bogoslovsky and Kuzina (1980) described a small assemblage from the Subpolar Urals (Kozhim River) that included *Rectimitoceras obsoletum* (Kuzina), *R. angustilobatum* (Kuzina), and *Kalloclymenia (Finiclymenia) kozhimensis* (Bogoslovsky). Bogoslovsky (1981) described *Kalloclymenia glabra* from the Subpolar Urals (Loz'va River). Probably these ammonoids belong to the upper part of the *frechi-corpulenta* Zone, i.e., to the lower part of the *Kalloclymenia–Wocklumeria* Genozone.

4.4.1 *nucleus* Zone (UD-VI-C2)

In the Uralian–Kazakhstanian Region, this zone is recognized by the presence of *Balvia (Mayeneoceras) nucleus* (Schmidt). It is possible that in western Kazakhstan this level also contains *Riphaeoclymenia canaliculata* Bogoslovsky and *Parawocklumeria* cf. *paprothae* Korn. This zone has a wide geographical distribution: it is also recognized in Germany, Morocco, Poland, Britain, and Austria (Schindewolf, 1937; Selwood, 1960; Czarnocki, 1989; Becker et al., 2002).

4.4.2 *sphaeroides* Zone (UD-VI-D1–2)

The *sphaeroides* Zone in western Kazakhstan contains *Wocklumeria sphaeroides* (Richter), *Glatziella glaucopis* (Renz), *Synwocklumeria kiensis* Bogoslovsky, *S. elata* Nikolaeva, *Tardewocklumeria perplexa* (Bogoslovsky), *Parawocklumeria laevigata* Selwood, and *Epiwocklumeria applanata* (Wedekind). This zone was originally established in Germany, and is recognized worldwide (Austria, possibly Britain,

Morocco, Poland, northern Caucasus, Bashkortostan, Orenburg Region, China, and USA) (Selwood, 1960; Becker, 1988; Czarnocki, 1989; Becker and House, 2000).

The upper part of the *sphaeroides* Zone in Germany is recognized as the *Epiwocklumeria applanata* Subzone (Becker and House, 2000). The same subdivision is possible in Poland (Czarnocki, 1989). The assemblage from western Kazakhstan contains *E. applanata*. However, it is not possible to establish whether it appears in the succession later than other ammonoids of the *sphaeroides* Zone, because the only section where it was found (Kiya River) is highly condensed, and any possible interval between the appearances would be no more than a few centimeters.

The uppermost Famennian of Germany and Morocco contains two zones that were not found in the Uralian–Kazakhstanian Region, although rocks of equivalent age are recorded in several sequences. These are the *nigra* Zone (UD-VI-E), corresponding to the *Cymaclymenia* Zone (Becker, 1993), or the *evoluta* Zone (Becker et al., 2002). The Famennian of Germany and Morocco also includes the lower half of the *prorsum* Zone (UD VI-E) (Korn et al., 2004), the equivalents of which may be present in Berchogur (western Kazakhstan) (Bogoslovsky, 1988).

5 Changes in Diversity

In the Famennian, the number and diversity of ammonoids (both goniatitids and clymeniids) in all sections studied in the Uralian–Kazakhstanian Region first increased during the transition from the *pseudogoniatites* to the *delphinus* to the *annulata* Zones, but then gradually decreased until clymeniids, the dominant group, totally disappeared in the Late Famennian. All the Middle–Late Famennian ammonoid assemblages in this region are dominated by clymeniids. The diversities of the goniatitids and clymeniids evolved differently in the Middle Famennian, but became more similar to each other toward the end of the Famennian (Fig. 15.6).

The total number of goniatitid genera in the Uralian Ocean sections halved from the end of the Middle Famennian to the end of the Famennian, while the number of clymeniid genera decreased by a third. However, the proportions of clymeniids and goniatitids in ammonoid assemblages in the Uralian Ocean did not change much during the transition from the Middle to the Late Famennian. Both groups showed a considerable decline. No indications of sudden extinction events are apparent.

At the same time, in the Rhenish Massif in the latest Devonian all clymeniid lineages but one became extinct immediately at the base of the Hangenberg Black Shale; "the extinction event appears to have been very sudden and did not only affect species but also entire evolutionary lineages" (Korn et al., 2004: 311). At the end of the Famennian, all clymeniid families (except Cymaclymeniidae) experienced an extinction crisis.

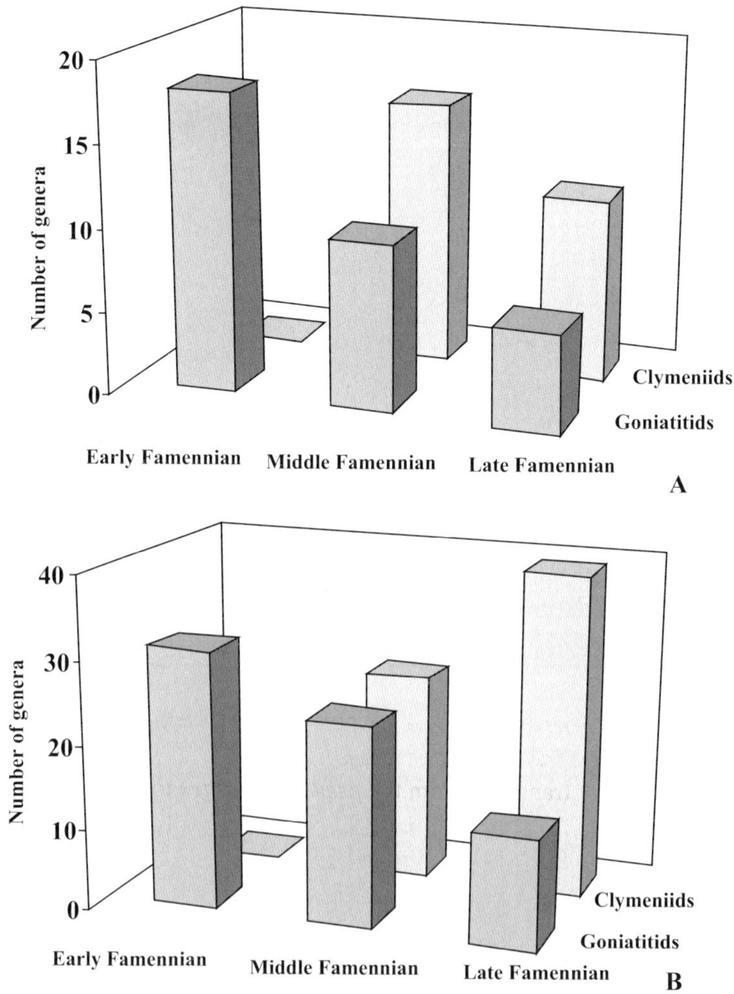

Fig. 15.6 *Diagrams showing changes in diversity of goniatitids and clymeniids. (A). Decrease in diversity of both groups from the Middle to Late Famennian in the sequences of the Uralian–Kazakhstanian Region. Note the sharp rise in the diversity of clymeniids shortly after their appearance in the Middle Famennian. (B). Decrease in diversity of goniatitids and increase in diversity of clymeniids in the Famennian in the sequences worldwide. Note the rise in the diversity of clymeniids just prior to the extinction of most clymeniid taxa in the Late Famennian.*

6 Distribution of Ammonoid Faunas in the Uralian Ocean

The first ammonoids in the Uralian Ocean appeared in the Emsian. In the Eifelian to Frasnian, the paleogeographic settings were different from those in the Famennian (Fig. 15.7). The ocean was more open toward the east with ammonoid

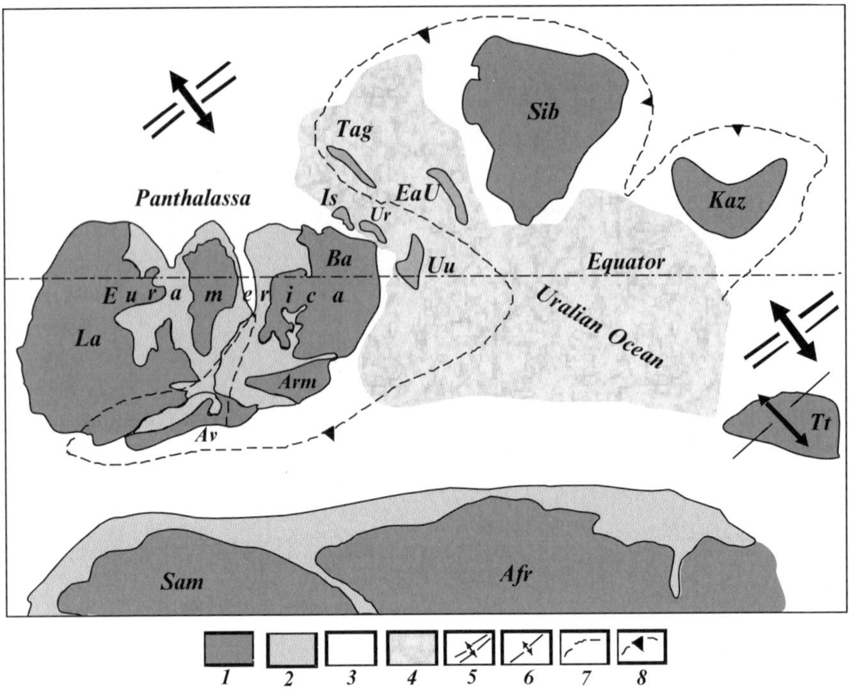

Fig. 15.7 *Paleotectonic reconstruction for the Early–Middle Devonian modified from Puchkov (2000); 1– continents, microcontinents, island arcs; 2 – other areas of continental crust; 3 – areas with oceanic crust; 4 – Uralian Ocean; 5 – rifts, 6 – continental rifts; 7 – zones of collision, 8 – zones of subduction; landmasses: Afr – Africa, Al – Alai Massif, Ar – Armorican Massif, Av – Avalonia, Ba – Baltica, Kaz – Kazakhstania, La – Laurentia, Sam – South America, Sib – Siberia, Tt – Tadjik-Tarim Massif, Uu – Ust-Urt, Ur – Uraltau.*

faunas inhabiting regions near Central Asia, eastern Kazakhstan, and northern Urals, whereas coastal areas were occupied by huge reefs. No ammonoid faunas of this age are found in western Kazakhstan.

In the Famennian, ammonoid habitats in the Uralian–Kazakhstanian Region were mostly concentrated in a relatively narrow zone of the deep open shelf on the eastern margin of the Volga-Ural Carbonate Platform in the far east of Euramerica (eastern Baltica), and the shallow western slope of Kazakhstania. The Famennian ammonoid habitats in western Kazakhstan were evidently not inherited from the Frasnian, being completely new developments (Fig. 15.8). The first Famennian communities of ammonoids (Lower Famennian preclymeniid *Cheiloceras* Zone) were abundant in the eastern South Urals and western Kazakhstan. A few contemporary ammonoid communities are known in the Middle Urals (to the north of western Kazakhstan). The patterns of distribution of communities of *Pernoceras*, *Pseudoclymenia*, and later clymeniids that followed *Cheiloceras* communities were the same, which suggests that the habitats in which clymeniids occurred in the Uralian Ocean in the Middle Famennian did not experience any major change from the Early Famennian.

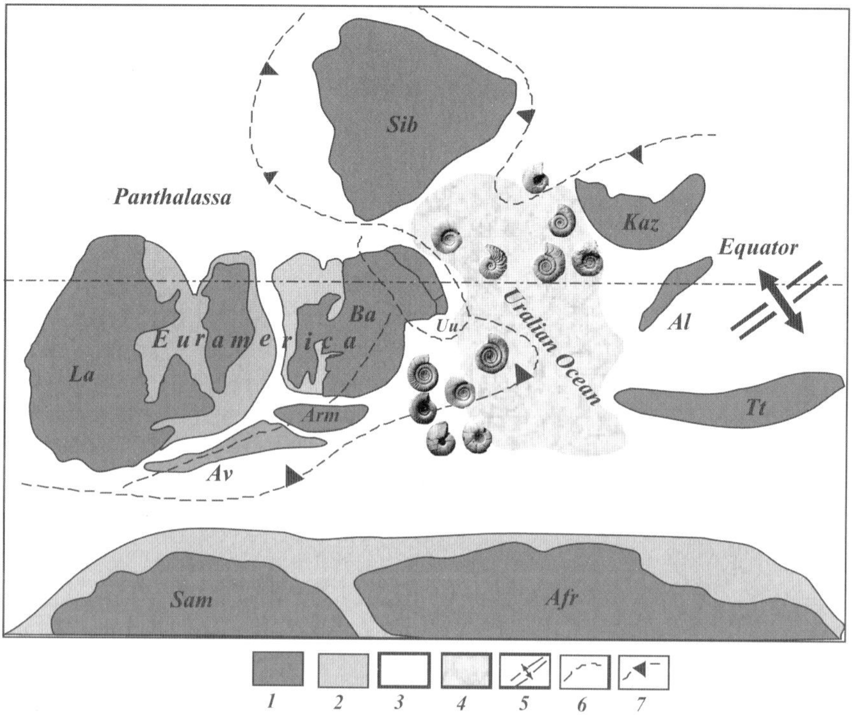

Fig. 15.8 *Paleotectonic reconstruction for the Famennian modified after Puchkov (2000). For explanations see Fig. 15.7.*

Of all ammonoid faunas known to date to have inhabited the Uralian Ocean, the fauna of western Kazakhstan stands out with 14 families and 49 genera. The basin of western Kazakhstan occupied an area presently within a large plain to the north-east of the Caspian Sea and further east to the Mugodzhary Mountains, which are the southern continuation of the Ural Mountains (Baktygarin, Kiya, Karadzhar). The basin was becoming shallower toward the west (median shelf/ramp and inner parts of the fore-slope) and was rimmed by organic buildups, the remains of which are found in the Peri-Caspian Region (Fig. 15.4). Further to the west, it was bounded by the Volga-Urals carbonate platform. In the south, the basin extended up to the South Emba area. Ammonoids are found in the eastern area of the basin, in the carbonate series terminating the carbonate–siltstone–cherty flysch sequences. The first clymeniids in the eastern area of the basin appeared in the Middle Famennian within assemblages that also included abundant goniatitids *Sporadoceras, Praeglyphioceras, Posttornoceras*, and the family Prolobitidae. The first clymeniid genera (*Platyclymenia, Genuclymenia, Cyrtoclymenia, Pricella*) appeared almost simultaneously, immediately after the disappearance of the goniatite genus *Pseudoclymenia*, which had "clymeniid-like" external shell morphology. Other goniatitid genera mentioned above continue into the clymeniid-bearing beds. The sedimentary situation in the basin at this time was not disrupted by any major

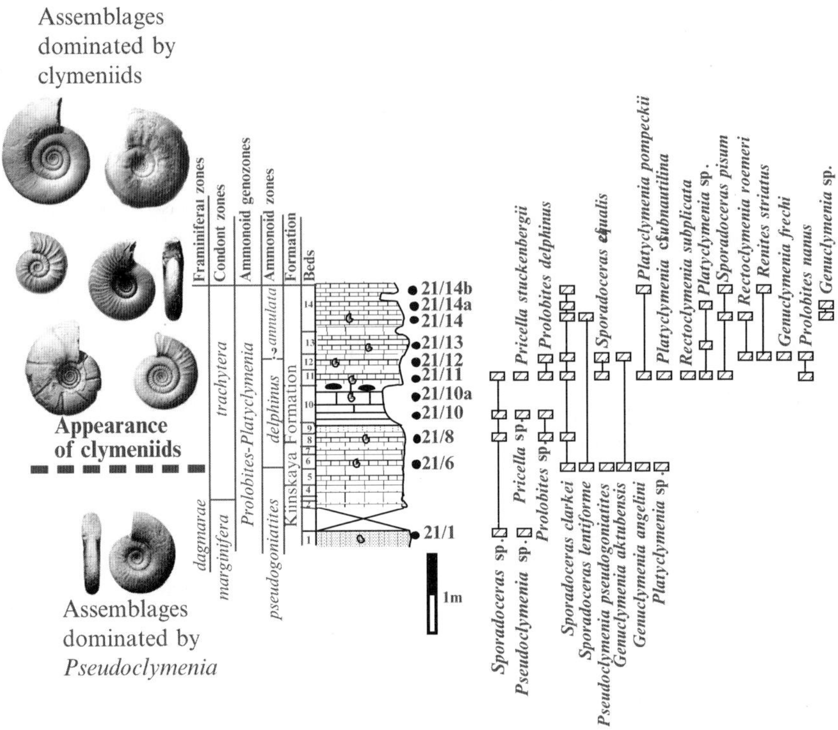

Fig. 15.9 *Upper Famennian succession in the Karadzhar Valley (Shiyli-Sai locality). Clymeniids appeared just above the level of the total disappearance of the goniatite* Pseudoclymenia. *Note the level with black shale and nodules.*

events, such as significant sea level fluctuations or essential changes in quality or quantity of sedimentary material.

Of the clymeniid assemblages studied, the earliest is that from Shiyli-Sai 2 and Ornektotas-Sai sections (western Kazakhstan) (Fig. 15.9). These carbonate sections were previously referred to as Karadzhar and are mostly in the *delphinus* Zone. The topmost part of these sections shows the transition to the *annulata* zone (UD-IV-A), which is marked by black shales and nodules and contains *Platyclymenia* cf. *subnautilina*. To the north of this region similar faunas inhabited the eastern margin of the East-Uralian flysch depression adjacent to the Magnitogorsk Island Arc Zone (Kirsa locality in the Chelyabinsk Region) (described by Perna, 1914; Nikolaeva and Bogoslovsky, 2005a). Both these basins were relatively deepwater and contained well-established communities of cephalopods and conodonts with no evidence of benthic fauna. Outside this deepwater area, the earliest ammonoid faunas appeared later, apparently in the *annulata* time. Ammonoid faunas of the *annulata* Zone were widespread throughout the Uralian Ocean. In the northwest, they reached the shallow-water carbonate basins of Bashkortostan (Ryauzyak and Terekly localities), and in the north, the shallow basins of the Polar Urals (localities

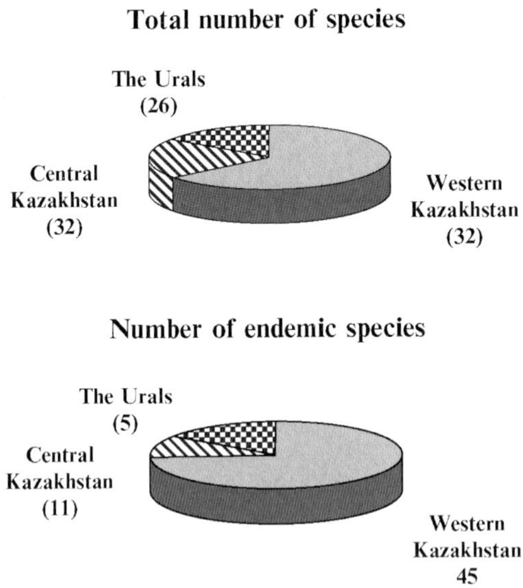

Fig. 15.10 *Diagram showing the low proportion of endemic genera in the clymeniid faunas of the Uralian–Kazakhstanian Region.*

on the Severnaya Sos'va River and its tributaries) (Bogoslovsky, 1971; Nikolaeva and Bogoslovsky, 2005a). This was the first time that clymeniids appeared in the shallow terrigeneous-carbonate basins of central Kazakhstan (Karaganda and Semipalatinsk Regions) (Nikolaeva and Bogoslovsky, 2005a).

Although few of the clymeniid genera of the *delphinus–annulata* interval were endemic to the Uralian–Kazakhstanian Region (e.g., *Spinoclymenia, Laminoclymenia, Riphaeoclymenia*), many were represented by endemic species (Fig. 15.10). For instance, in the family Cyrtoclymeniidae, four of eight *Platyclymenia*, three of four *Trigonoclymenia*, three of five *Cyrtoclymenia*, and five of six *Pricella* species are regional endemics. A similar pattern is observed in other families.

The faunas of Bashkortostan and central Kazakhstan are much less diverse than those of western Kazakhstan. However, the assemblages in these areas can be as high in number of individuals as those of western Kazakhstan, with hundreds of individuals of the same species, usually a species unknown in western Kazakhstan. Typical examples are numerous shells of *Rectoclymenia tecta* Nikolaeva and Bogoslovsky and *Trigonoclymenia sherubensis* Nikolaeva and Bogoslovsky in the *Prolobites–Platyclymenia* Genozone in central Kazakhstan (Fig. 15.11) and of *Platyclymenia tschernyschewi* (Rzehak) from the same genozone in Bashkortostan.

Sections in western Kazakhstan display an uninterrupted transition from the *Prolobites-Platyclymenia* to *Clymenia–Gonioclymenia* Genozones, which is marked by a gradual decline in diversity, although the proportion of the endemic taxa does not fall. The transition to the *Clymenia–Gonioclymenia* Genozone in central

Fig. 15.11 *Endemics of central Kazakhstan and Bashkortostan. (1a–1c).* Trigonoclymenia sherubensis *Nikolaeva et Bogoslovsky (x1): holotype PIN no. 1683/8488; central Kazakhstan, right bank of the Sherubai-Nura River, outcrop 6;* Prolobites–Platyclymenia *Genozone,* annulata *Zone. (2a–2c).* Rectoclymenia tecta *Nikolaeva et Bogoslovsky; holotype PIN no. 1683/7045 (x1); central Kazakhstan, right bank of the Sherubai-Nura River, village of Dzhalair, outcrop 10;* Prolobites–Platyclymenia *Genozone. (3a–3b).* Kazakhoclymenia medoevi *Bogoslovsky; holotype PIN no. 1683/6981 (x1); central Kazakhstan, right bank of the Sherubai-Nura River, southwest of the village of Dzhalair; transitional beds between the* Prolobites–Platyclymenia *and* Clymenia–Gonioclymenia *Genozones. (4a–4b).* Platyclymenia tschernyschewi *(Rzehak, 1910); specimen*

Kazakhstan is markedly different. For instance, *Clymenia laevigata* and the diverse assemblage of *Kosmoclymenia* species found in western Kazakhstan are not found there. The assemblages are impoverished, and contain only few taxa rooted in the lineages that were previously abundant locally. This strongly suggests that these areas were colonized anew in the *laevigata* Zone. A similar pattern is observed at the beginning of the *Kalloclymenia-Wocklumeria* Genozone in Bashkortostan.

The Famennian sections in the Urals and Kazakhstan contain very few horizons of black shale, a facies which is prominent in some other regions (e.g., the Rhenish Massif). Black shales are thought to indicate global eustatic events, which had major impacts on ammonoid evolution (Becker, 1993; Sandberg et al., 2002; and others). In the Uralian–Kazakhstanian Region, the decrease in diversity toward the end of the Famennian was gradual. Clymeniids with a compressed, evolute shell (families Gonioclymeniidae, Clymeniidae, Cymaclymeniidae) were the first to become extinct. The family Cymaclymeniidae, which in the Rhenish Massif is the only clymeniid family that is found in the Hangenberg Black Shale, and in Morocco even above the D–C boundary (Korn et al., 2004), decreased and disappeared in the Uralian–Kazakhstanian Region before the end of the Devonian (no cymaclymeniids are as yet recorded there above the *sphaeroides* Zone). The latest Famennian clymeniids are so far found only in western Kazakhstan, in assemblages with mostly small, involute, or semievolute, inflated clymeniids along with bizarre triangularly coiled wocklumeriids (Nikolaeva and Bogoslovsky, 2005b). The assemblages of wocklumeriids in western Kazakhstan are far more impoverished than those from Germany, Poland, or North Africa. The total number of wocklumeriids recovered from this region does not exceed 30–35 specimens.

The differences in the composition of ammonoid assemblages in the Uralian–Kazakhstanian and Rhenohercynian Regions prior to their extinction near the end of the Devonian are apparently related to a peculiar sedimentary setting in the west of the Uralian Ocean, particularly the reduction there of open-shelf habitats. Toward the end of the Devonian, the deep-shelf environments in this area sharply decreased due to the westward migration of Kazakhstania and progressive collision of the southeastern margin of Baltica with the island arcs accompanied by strong volcanism (Puchkov, 2000; Mizens, 2004; Veimarn et al., 2004). The tectonic activity inevitably influenced the biotopes of the cephalopods resulting in a decline in diversity.

The distribution of clymeniids in the Uralian Ocean is here interpreted as an expression of a marked degree of environmental sensitivity and facies-depth control, which may explain the differences in community structure, with higher diversity communities in the open-shelf areas and essentially monospecific communities in

Fig. 15.11 (continued) *PIN no. 3754/1001; South Urals, Bashkortostan, Terekly River;* Prolobites–Platyclymenia *Genozone;* annulata *Zone. (5a–5b).* Cymaclymenia subcompressa *Nikolaeva et Bogoslovsky; holotype PIN no. 1683/7008; central Kazakhstan, Sherubai-Nura River; outcrop 4, sample 4b;* Clymenia–Gonioclymenia *Genozone,* laevigata *Zone. (6a–6c).* Rectoclymenia lyrata *Nikolaeva et Bogoslovsky (x1); specimen PIN no. 1683/7043; central Kazakhstan, right bank of the Sherubai-Nura River, village of Dzhalair;* Prolobites–Platyclymenia *Genozone,* annulata *Zone.*

the more shallow shoreward areas of the ocean. It is possible that the communities in the shallower zones did not survive long enough to produce stable lineages of descendants.

The environmental factors responsible for these distribution patterns were not uncommon in the Phanerozoic, when epicontinental seas appeared as a result of major global-scale transgressions. It is not certain whether the basin of western Kazakhstan was the global center of the primary radiation of clymeniids, but it was most certainly a secondary center, i.e., an area with a favorable environment to which the newly appearing group migrated from its center of origin and primary distribution.

The place of origin of clymeniids, i.e., the primary center of distribution, is not yet known (the discussion of the possible locations of the primary center of clymeniid distribution is beyond the scope of this paper). However, it is normal for a group to be poorly represented in the place where it evolved from its immediate ancestor due to high competition pressure, and therefore the primary center is difficult to identify. Furthermore, the immediate descendant is usually not present at the site of occurrence of the ancestor. According to a typical model of the centers of a taxon's distribution, after migration and the discovery of a more favorable environment, the diversity and abundance of lower rank taxa increases. Later the group migrates widely in various directions and occupies new areas. The place where a group first reaches high abundance and diversity, which facilitates its migration, is the secondary center of a group's distribution. It is possible that a few secondary centers could occur simultaneously in different places. These patterns suggest that primitive forms, members of the oldest waves of immigrants were displaced. The difference in the fauna on different sides of the Uralian Ocean is greater with increasing distance from western Kazakhstan, which is consistent with the above model. The extent to which the model of centers of distribution can be applied to the case of ammonoids of the Uralian Ocean will be clearer when the Central Asian Famennian faunas (in the south of the ocean) are studied and compared with those in Kazakhstan.

The evolution of ammonoid faunas in the Uralian Ocean was subject both to global oceanic events such as major transgressions but more so to local tectonic restructuring. The constant decrease in the diversity of ammonoids in this region in the Famennian is interpreted as the aftermath of progressive reduction of the deep- and outer-shelf habitats resulting from the collision of Baltica, Kazakhstania, and the Magnitogorsk Island Arc.

7 Conclusions

1. The Famennian clymeniid faunas of the Uralian–Kazakhstanian Region are highly endemic, with evolutionary patterns different from those of contemporary clymeniid faunas from Germany, Poland, and Morocco.
2. Ammonoid assemblages are different in the three major zones of the Uralian–Kazakhstanian Region (Bashkortostan, western Kazakhstan, and central

Kazakhstan), which represented different sedimentary areas and environments of the Uralian Ocean.
3. The richest ammonoid occurrences belonged to the relatively deep zones of the open shelf in the western realms of the Uralian Ocean, where throughout the Famennian, diversity remained the highest compared to other zones of the ocean.
4. Faunas in the more shallow-water zones (Bashkortostan and central Kazakhstan) contain many short-lived endemic species that are not found in western Kazakhstan.
5. The deep basin of western Kazakhstan was probably a secondary center of clymeniid distribution in the Famennian, with faunas migrating northward and eastward to colonize outlying shallower areas on the edge of the platforms, following large-scale transgressions.
6. The evolution of ammonoid communities in the Uralian Ocean was likely to have been controlled by local tectonics, such as a collision of continents, micro-continents, and island arcs.

Acknowledgments

This study is a continuation of the research commenced by B. I. Bogoslovsky (Paleontological Institute, Moscow) on the clymeniids of the Urals and Kazakhstan. The shells in Bogoslovsky's collection, beautifully prepared by Boris Ivanovich Bogoslovsky himself, provided the basis of this paper. I am greatly indebted to N. B. Gibshman for valuable information on the microfacies and sedimentary settings in western Kazakhstan. M. S. Boiko, L. Z. Akhmetshina, and V. F. Korobkov helped to organize the field trips. I am grateful to T. B. Leonova, M. F. Bogoslovskaya, L. F. Kuzina, A. A. Shevyrev, L. A. Nevesskaya, and E. I. Kulagina for valuable discussions and help. J. Kullmann was very supportive and encouraging at all stages of this work. Photographs were produced at the Paleontological Institute, Moscow, by V. T. Antonova.

The study was supported by the Program of the Presidium of the Russian Academy of Sciences "Origin and Evolution of the Biosphere."

References

Akhmetshina, L. Z., N. B. Gibshman, and S. V. Nikolaeva. 2004. Famenskii neftegazonosnyi kompleks Zapadnogo Kazakhstana (Prikaspiskii Bassein) [Famennian Hydrocarbon-bearing complex of Western Kazakhstan (Caspian Basin)]. *Doklady kazakhstanskikh geologov na XXXII sessii MGK*, Kazakhstanskoe geologicheskoe obshchestvo, pp. 89–97. Almaty: Kazakhtsanskoe [in Russian].

Becker, R.T. 1988. Ammonoids from the Devonian-Carboniferous boundary in the Hasselbach Valley (Northern Rhenish Slate Mountains). *Courier Forschungsinstitut Senckenberg* **100**: 193–213.

Becker, R. T. 1993. Anoxia, eustatic changes, and Upper Devonian to lowermost Carboniferous global ammonoid diversity. *In* M. R. House (editor), *The Ammonoidea: Environment, Ecology, and Evolutionary Change*, pp. 115–163. Oxford: Clarendon Press.

Becker, R. T. 2000. Palaeobiogeographic relationships and diversity of Upper Devonian ammonoids from Western Australia. *Records of the Western Australian Museum Supplement* **58**: 385–401.

Becker, R. T., and M. R. House. 2000. Devonian ammonoid zones and their correlation with established series and stage boundaries. *Courier Forschungsinstitut Senckenberg* **220**: 113–151.

Becker, R. T., M. R. House, J. Bockwinckel, V. Ebbighausen, and Z. S. Aboussalam. 2002. Famennian ammonoid zones of the eastern Anti-Atlas (southern Morocco). *Münster Forschungen zur Geologie und Paläontologie* **93**: 159–205.

Bogoslovsky, B. I. 1955. O semeistve Biloclymeniidae, fam. nov. [On the family Biloclymeniidae, fam. nov.]. *Dokady Akademii Nauk SSSR* **104**(1): 134–137 [in Russian].

Bogoslovsky, B. I. 1960. Novye predstaviteli nekotorykh maloizvest'nyikh rodov devonskikh ammonoidei iz famenskikh otlozheniy Urala [New representatives of some poorly known genera of Devonian ammonoids from the Famennian beds of the Urals]. *Paleontologicheskii Zhurnal* **1960**(4): 69–73 [in Russian].

Bogoslovsky, B. I. 1962. Redkii tip skul'ptury u klimenii [Rare type of ornament in clymeniids]. *Paleontologicheskii Zhurnal* **1962**(1): 166–168 [in Russian].

Bogoslovsky, B. I. 1965. *Carinoclymenia* – novyi rod semeistva Rectoclymeniidae [*Carinoclymenia* – a new genus in the family Rectoclymeniidae]. *Paleontologicheskii Zhurnal* **1965**(4): 88–91 [in Russian].

Bogoslovsky, B. I. 1969. Devonskie ammonoidei. I. Agoniatity [Devonian ammonoids. I. Agoniatitida]. *Trudy Paleontologicheskogo Instituta Akademii Nauk SSSR* **124**: 1–341 [in Russian].

Bogoslovsky, B. I. 1971. Devonskie ammonoidei. II. Goniatity [Devonian ammonoids. I. Goniatitida]. *Trudy Paleontologicheskogo Instituta Akademii Nauk SSSR* **127**: 1–228 [in Russian].

Bogoslovsky, B. I. 1975. Novoe semeistvo klimenii [A new clymeniid family]. *Paleontologicheskii Zhurnal* **1975**(3): 35–41 [in Russian].

Bogoslovsky, B. I. 1976. Rannii ontogenez i proiskhozhdenie klimenii [Early ontogeny and origin of clymeniids]. *Paleontologicheskii Zhurnal* **1976**(2): 41–50 [in Russian].

Bogoslovsky, B. I. 1977. O semeistve Miroclymenidae Schindewolf, 1924 [On the family Miroclymenidae Schindewolf, 1924]. *Paleontologicheskii Zhurnal* **1977**(4): 47–58 [in Russian].

Bogoslovsky, B. I. 1979a. Sistematika i filogeniya klimeniin [Systematics and phylogeny of clymeniins]. *Paleontologicheskii Zhurnal* **1979**(2): 32–47[in Russian].

Bogoslovsky, B. I. 1979b. Novyi rod semeistva Rectoclymeniidae [A new genus in the family Rectoclymeniidae]. *Paleontologicheskii Zhurnal* **1979**(1): 140–143 [in Russian].

Bogoslovsky, B. I. 1981. Devonskie ammonoidei, III. Klimenii (Podotryad Gonioclymeniina) [Devonian ammonoids, III. Clymeniids (Suborder Gonioclymeniina)]. *Trudy Paleontologicheskogo Instituta Akademii Nauk SSSR* **191**: 1–122 [in Russian].

Bogoslovsky, B. I. 1983. Novyi rod semeistva Carinoclymenidae [A new genus in the family Carinoclymenidae]. *Paleontologicheskii Zhurnal* **1983**(3): 106–109 [in Russian].

Bogoslovsky, B. I. 1988. Evolyutsiya i biokhronologiya ammonoidei na rubezhe devona i karbona [Evolution and biochronology of ammonoids at the Devonian-Carboniferous boundary]. *Granitsa devona i karbona na territorii SSSR*, pp. 128–154. Minsk: Nauka i tekhnika [in Russian].

Bogoslovsky, B. I., and L. F. Kuzina. 1980. Pozdnedevonskiye ammonoidei basseina reki Kozhim na pripolyarnom Urale [Late Devonian ammonoids of the basin of the Kozhim River in the Subpolar Urals]. *Paleontologicheskii Zhurnal* **1980**(2): 67–73 [in Russian].

Born, A. 1912a. Die geologischen Verhältnisse des Oberdevons im Aeketal (Oberharz). *Neues Jahrbuch für Mineralogie, Geologie und Paläontologie, Beilage-Band* **34**: 553–632.

Born A. 1912b. Über eine Vergesellschaftung von Clymenien und Cheiloceren. *Zeitschrift der Deutschen Geologischen Gesellschaft Monatsberichte* **11**: 537–545.

Chang, A. 1960. New Late Upper Devonian ammonite-faunas of the Great Khingan and its biological classification. *Acta Palaeontologica Sinica* **8**(2): 180–193.

Czarnocki, J. 1989. Klimenie Gór Swietokrzyskich. *Prace Państwowego Instytutu Geologicznego* **127**: 1–91.

Dybczynski, T. 1913. Ammonity górnego Dewonu Kielc, Wiadomość tymczasowa. *Kosmos* **38**: 510–525.

Flügel H., and M. Kropfitsch-Flügel. 1965. Ammonoidea palaeozoica. *Fossilium Catalogus* **6**(1): 5–31.

Frech, F. 1897. Lethaea geognostica oder Beschreibung und Abbildung der für die Gebirgs-Formationen bezeichnendsten Versteinerungen. I. Theil. *Lethaea palaeozoica* 2. Band.

Frech, F. 1902. Über devonische Ammoneen. *Beiträge zur Paläontologie und Geologie des Österreich-Ungarns und des Orients* **14**: 27–112.

Frech, F. 1913. *Fossilium Catalogus. I. Animalia. Pars 1. Ammoneae Devonicae*. Berlin: Jungk.

Gümbel, C. W. 1862. Revision der Goniatiten des Fichtelgebirges. *Neues Jahrbuch für Mineralogie, Geologie und Paläontologie* **1862**: 284–326.

House, M. R. 1959. Upper Devonian ammonoids from North-West Dartmoor, Devonshire. *Proceedings of the Geological Association* **70**(4): 315–321.

House, M. R. 1962. Observations on the ammonoid succession of the North American Devonian. *Journal of Paleontolology* **36**: 247–284.

House, M. R. 1963. Devonian ammonoid successions and facies in Devon and Cornwall. *Quarterly Journal of the Geological Society of London* **119**(1): 1–27.

House, M. R. 1985. Class Cephalopoda. *In* J. W. Murray (editor), *Atlas of Invertebrate Macrofossils*, pp. 114–152. Essex: Longman, Harlow.

House, M. R., and A. E. H. Pedder. 1963. Devonian goniatites and stratigraphical correlations in Western Canada. *Palaeontology* **6**(3): 491–539.

International Union of Geological Sciences. International Commission on Stratigraphy (ICS). 2004. *Consolidated Annual Report for 2004*. F. M. Gradstein, and J. G. Ogg (editors), http://www.iugs.org/iugs/downloads/ annual_reports_04/ar04_tgggb.pdf.

Jenkins, T. B. H. 1968. Famennian ammonoids from New South Wales. *Palaeontology* **4**(4): 535–548.

Kind, N. V. 1944. Goniatity i klimenii zapadnogo sklona Mugodzharskikh gor [Goniatites and Clymenoidea in the Upper Devonian of the Western Mugodzhars]. *Uchenye zapiski Leningradskogo Universiteta, Seriya geologo-pochvennykh nauk* **11**: 137–166 [In Russian].

Kolotukhina, S. E. 1938. K voprosu o klimenievykh fatsiyakh neodevona Tsentral'nogo Kazakhstana [on the problem of the clymeniid facies of the Neodevonian of Central Kazakhstan]. *Izvestiya Akademii Nauk SSSR, otdelenie matematicheskih i estestvennykh nauk* **5–6**: 671–686 [in Russian].

Korn, D. 1981. *Cymaclymenia* – eine besonders langlebige Clymenien-Gattung (Ammonoidea, Cephalopoda). *Neues Jahrbuch für Geologie und Paläontologie* **161**(2): 172–208.

Korn, D. 1999. Famennian ammonoid stratigraphy of the Ma'der and Tafilalt (eastern Anti-Atlas, Morocco). *Abhandlungen der Geologischen Bundesanstalt* **54**: 147–179.

Korn, D. 2002. Die Ammonoideen-Fauna der *Platyclymenia annulata*-Zone vom Kattensiepen (Oberdevon, Rheinisches Schiefergebirge). *Senckenbergiana lethaea* **82**(2): 557–608.

Korn, D., and C. Klug. 2002. Ammoneae Devonicae. *Fossilium Catalogus. I: Animalia*, Pars 138. Leiden: Backhuys.

Korn, D., and J. Price. 1987. Taxonomy and Phylogeny of the Kosmoclymeniinae subfam. nov. (Cephalopoda, Ammonoidea, Clymeniida). *Courier Forschungsinstitut Senckenberg* **92**: 5–75.

Korn, D., and W. Ziegler. 2002. The ammonoid and conodont zonation at Enkenberg (Famennian, Late Devonian; Rhenish Mountains). *Senckenbergiana lethaea* **82**: 453–462.

Korn, D., C. Klug, and A. Reisdorf. 2000. Middle Famennian ammonoid stratigraphy in the Amessoui Syncline (Late Devonian; eastern Anti-Atlas, Morocco). *Travaux de l'Institut des Sciences, Rabat, Série Géologie et Géographie Physique* **20**: 69–77.

Korn, D., Z. Belka, S. Fröhlich, M. Rücklin, and J. Wendt. 2004. The youngest African clymeniids (Ammonoidea, Late Devonian) – failed survivors of the Hangenberg Event. *Lethaia* **37**: 307–315.

Lange, W. 1929. Zur Kenntnis des Oberdevons am Enkeberg und bei Balve (Sauerland). *Abhandlungen der Preußischen Geologischen Landesanstalt. Neue Folge* **119**: 1–132.

Loewinson-Lessing, F. 1892. Les ammonées de la zone à *Sporadoceras Münsteri* dans les monts Gouberlinskya Gory (gouv. d'Orenburg), Oural méridional. *Bulletin de la société belge de géologie de paléontologie et d'hydrologie* **6**: 15–25.

Matern, H. 1931. Das Oberdevon der Dill-Mulde. *Abhandlungen der Preußischen Geologischen Landesanstalt* **134**: 1–139.

Miller, A. K. 1938. Devonian ammonoids of America. *The Geological Society of America. Special Papers* **14**: 1–262.

Miller, A. K., and C. Collinson. 1951. A clymenoid ammonoid from New Mexico. *American Journal of Science* **249**: 600–603.

Mizens, G. A. 2004. Devonian palaeogeography of the southern Urals. *Geological Quarterly* **48**(3): 205–216.

Montesinos, J. R., and M. A. Arbizu. 1988. Ammonoideos y trilobites de la Formación Vidrieros (Horcada del Oro, Dominio Palentino, NO de España). *Cuadernos Laboratorio Xeoloxico Laxe Coruña* **12**: 93–98.

Nalivkina, A. K. 1936. O verkhnedevonskih goniatitakh Novoi Zemli [On the Upper Devonian goniatites of Novaya Zemlya]. *Trudy Arkticheskogo Instituta* **28**: 91–108 [in Russian].

Nalivkina, A. K. 1953. Verkhnedevonskie goniatity i klimenii Mugodzhar [Upper Devonian goniatitids and clymeniids of the Mugodzhars]. *Trudy Vsesoyuznogo neftyanogo nauchno-issledovatel'skogo geologicheskogo instituta, Novaya Seriya* **72**: 60–125 [in Russian].

Nikolaeva, S.V., and B. I.Bogoslovsky. 2005a. Devonskie ammonoidei, IV. Klimenii (Podotryad Clymeniina) [Devonian ammonoids, IV. Clymeniids (Suborder Clymeniina)]. *Trudy Paleontologicheskogo Instituta Rossiiskoi Akademii Nauk* **287**: 1–220 [in Russian].

Nikolaeva, S.V., and B. I. Bogoslovsky. 2005b. Late Famennian ammonoids from the upper part of the Kiya formation of the South Urals. *Paleontologicheskii Zhurnal* **39**(Suppl. 5): S527–S537.

Nikolaeva, S.V., and N. B. Gibshman. In preparation. *Ammonoid and Foraminiferal Habitats in the Famennian Basin of Western Kazakhstan.*

Nikolaeva S.V., L. Z. Akhmetshina, N. B. Gibshman, and A.N. Kan. 2004. Biostratigraphy and Microfacies of the Famennian Hydrocarbon-bearing complex in Western Kazakhstan (Peri-Caspian Basin). *Abstracts of the 32nd Geological Congress, 2004, Florence, Italy* **1**(1): 28–30.

Perna, E. Y. 1914. Ammonei verkhnego neodevona vostochnogo sklona Urala [Ammonoids of the Upper Neodevonian of the eastern slope of the Urals]. *Trudy Geologicheskogo Komiteta, novaya seriya* **99**: 1–114 [in Russian].

Petersen, M. S. 1975. Upper Devonian (Famennian) ammonoids from the Canning Basin, Western Australia. *Journal of Paleontology Memoirs* **8**: 1–55.

Petter, G. 1959. Goniatites dévoniennes du Sahara. *Publications du Service de la Carte géologique d'Algerie. Nouvelle série. Paléontologie* **2**: 1–313.

Petter, G. 1960. Clymènies du Sahara. *Publications du Service de la Carte géologique d'Algerie. Nouvelle série. Paléontologie* **6**: 1–58.

Price, J. D., and D. Korn. 1989. Stratigraphically important clymeniids (Ammonoidea) from the Famennian (Later Devonian) of the Rhenish Massif, West Germany. *Courier Forschungsinstitut Senckenberg* **110**: 257–294.

Puchkov, V. N. 2000. *Paleogeodynamics of the South and Middle Urals*. Ufa: Dauria.

Ruan, Y. 1981. Devonian and earliest Carboniferous ammonoids from Guangxi and Guizhou. *Memoirs of Nanjing Institute of Geology and Palaeontology, Academy Sinica* **15**: 1–152.

Rzehak, A. 1910. Der Brünner Clymenienkalk. *Zeitschrift des Mahrischen Landesmuseums* **10**(2): 149–216.

Rzhonsnitskaya, M. A. 2000. Devonian stage boundaries on the East European (Russian) platform. *Courier Forschungsinstitut Senckenberg* **225**: 227–237.

Sandberg, C. A., J. R. Morrow, and W. Ziegler. 2002. Late Devonian sea-level changes, catastrophic events, and mass extinctions. *In* C. Koeberl, and K. G. MacLeod (editors), *Catastrophic Events and Mass Extinctions: Impacts and Beyond* (*Geological Society of America Special Paper* **356**: 473–487.

Sandberger, G. 1853a. Einige Beobachtungen über Clymenien; mit besonderer Rücksicht auf die westphälischen Arten. *Verhandlungen des naturhistorischen Verein Rheinland und Westfalen* **10**: 171–216.

Sandberger, G. 1853b. Über Clymenien. *Neues Jahrbuch für Mineralogie, Geologie und Paläontologie* **1853**: 513–523.

Schindewolf, O. H. 1923. Beiträge zur Kenntnis des Paläozoikums in Oberfranken, Ostthüringen und dem Sächsischen Vogtlande. I. Stratigraphie und Ammoneenfauna des Oberdevons von Hof a. S. *Neues Jahrbuch für Mineralogie, Geologie und Paläontologie, Beilage-Band* **49**: 250–357, 393–509.

Schindewolf, O. H. 1937. Zwei neue, bemerkenswerte Goniatiten-Gattungen des Rheinischen Oberdevons. *Jahrbuch der Preussischen Geologischen Landesanstalt* **58**: 242–255.

Schmidt, H. 1931. Das Paläozoikum der spanischen Pyrenäen. *Abhandlungen der Gesellschaft der Wissenschaften zu Göttingen, Mathematisch-Physikalische Klasse* **3**(5): 1–85.

Selwood, E. B. 1960. Ammonoids and trilobites from the Upper Devonian and lowest Carboniferous of the Launceston area of Cornwall. *Palaeontology* **3**(2): 153–185.

Simakov, K. V., B. I. Bogoslovsky, M. K. Gagiev, L. I. Kononova, N. M. Kochetkova, L. F. Kuzina, E. I. Kulagina, U. I. Onoprienko, V. N. Pazukhin, E. P. Radionova, T. P. Razina, E. A. Reitlinger, L. V. Simakova, and M. G. Yanultanova. 1983. Biostratigrafiya pogranichnykh otlozhenii devona i karbona Mugodzhar [Biostratigraphy of the Devonian-Carboniferous beds in Mugodzhary]. *K kharakteristike pogranichnykh otlozhenii devona i karbona Mugodzhar*, 51pp. Magadan: Akademiya Nauk SSSR.

Sobolev, D. 1914a. Über Clymenien und Goniatiten. *Paläontologische Zeitschrift* **1**: 348–378.

Sobolev, D. 1914b. Nabroski po filogenii goniatitov [Synopsis of the goniatite phylogeny]. *Izvestia Varshavskogo politekhnicheskogo instituta* **1**(for 1914): 1–191.

Termier G., and H. Termier. 1950. Paléontologie Marocaine. 2. Invertébrés de l'ére Primaire. Fasc. 3. Mollusques. *Notes et memoires du Service géologique du Maroc* **78**: 1–246.

Tschernyshev, F. N. 1887. Fauna srednego i verkhnego devona zapadnago sklona Urala [Fauna from the Middle and Upper Devonian of the western slope of the Urals]. *Trudy geologicheskogo Komiteta* **3**(3): I–XII, 1–208 [in Russian].

Veimarn, A. B., V. I. Puchkov, A. A. Abramova, O. V. Artyushkova, V. N. Baryshev, K. E. Degtyaryov, L. I. Kononova, V. A. Maslov, V. M. Mosejchuk, V. N. Pazukhin, N. V. Pravikova, A. V. Tevelev, and A. V. Yarkova. 2004. Stratigraphy and geological events at the Frasnian-Famennian boundary in the southern Urals. *Geological Quarterly* **48**(3): 233–244.

Walliser, O. H. 1966. Preliminary notes on Devonian, Lower and Upper Carboniferous goniatites in Iran. *Report of the Geological Survey of Iran* **6:** 7–24.

Wedekind, R. 1908. Die Cephalopodenfauna des höheren Oberdevon am Enkeberge. *Neues Jahrbuch für Mineralogie, Geologie und Paläontologie, Beilage-Band* **26**: 565–633.

Chapter 16
Deformities in the Late Callovian (Late Middle Jurassic) Ammonite Fauna from Saratov, Russia

Neal L. Larson

Black Hills Institute of Geological Research, PO Box 643, 117 Main Street, Hill City, SD 57745, USA, ammoniteguy@bhigr.com

Keywords: ammonites, deformities, epizoa, injuries, Jurassic, Callovian

1 Introduction

Like other molluscs, every ammonite carries a record of its ontogeny and, commonly, its death, in its shell. Traumatic life events such as bites, diseases, epizoa, and diet all left evidence in the shell as scars, blisters, disfigurations, holes, nicks, crushing, attachments, and abnormalities, along with slow, rapid, stunted, or enlarged growth. If an ammonite survived bites due to predation, its shell documented the damage. Often, the cause and stage of life when that injury happened can, at times, be interpreted and diagnosed. If an ammonite did not survive the attack, its shell

N. H. Landman et al. (eds.), *Cephalopods Present and Past: New Insights and Fresh Perspectives*, 344–374.

commonly exhibits holes or missing portions. If most of the ammonite was consumed, only parts, pieces, or fragments of the shell will be found.

In the case of severe, sublethal bites, injuries to ammonite shells are expressed as conspicuous scars, blisters, or uneven growth. Deep bites to the body chamber sometimes resulted in the mantle rupturing out beyond the confines of the shell. Some of the fragmented shell that originally covered that area could be resorbed, and new shell was often secreted to patch the ruptured area. Deep bites to the mantle are expressed in the form of long scars, irregular apertures, deep-healed gouges, or displaced shell fragments molded back into the shell. If the ammonite was subjected to only minor attacks (nicks and bites to the aperture at some point in life), proof of these injuries are usually shown as interruptions or slight distortions in the normal shell structure. Sometimes, this shell damage can be distinguished as interruptions in the ribbing, rib displacement, irregular shell growth, small scars, or even as repeated "check marks" in the shell.

The presence of epizoa also can disfigure the shell. Commonly, these animals were serpulids, brachiopods, limpets, bryozoans, oysters, pelecypods, etc. Davis et al. (1999) noted that attachments of epizoa rarely occurred on living ammonites. Seilacher (1982) observed that it was quite common for epizoa to attach themselves to empty ammonite conchs as they floated in the sea or lay on the sea floor. Normally, epizoa did little to no obvious damage to living ammonites, merely etching the surface and perhaps causing little more than an inconvenience to the ammonite. In rare instances, epizoa were overgrown by the ammonite, and deformation occurred to the ammonite shell.

There have been numerous papers on the deformations of ammonite shells due to epizoa, most notably Hölder (1956), Hengsbach (1996), Maeda and Seilacher (1996), Davis et al. (1999), Checa et al. (2002), and Keupp (1996, 2000, 2005). Hölder (1956, 1973) assigned a complex series of "forma" names to identify the different types of pathologies present in ammonite shells and Hengsbach (1996) provided a nearly complete list of different types of "forma" names for each pathological deformation. Davis et al. (1999) tried to tackle the complex details and terminology, coining the new term "epicole" for any organism that spends its life attached to a hard surface, be it living, dead, or inorganic. Checa et al. (2002) described the complex deformities associated with epizoa attached to an ever-growing ammonite, and attempted to understand the growth patterns of the ammonite in response to these attached organisms. H. Keupp has spent the last several decades collecting and publishing on as many different ammonite pathologies as possible, while trying to establish the different ways that ammonites survived and adapted to other organisms that bit, damaged, and formed encrustations.

Large collections of ammonites from a single horizon or, perhaps, a single locality, can provide much information on the life habits, growth, predation, and even death of the ammonites. Individual growth, dimorphism, predation incidence, and interactions with other species can also be assessed. The study of large collections of a single ammonite species allows us to elucidate the life and death of these animals.

The upper Callovian *Quenstedtoceras (Lamberticeras) lamberti* Zone (Upper Jurassic), exposed in the Dubki Quarry near Saratov, Russia, has preserved a large number of unusually deformed ammonites. An unknown percentage of these ammonites were subjected to normal predation, with many of them preserving healed defects and scars in their shell. Along with healed predation, many of these ammonites also exhibit some unusual malformations. It seems that while the ammonites were living, tiny organisms attached themselves to some of the shells and began to grow. Probably, by the time the ammonites discovered something was growing on their shells, it was already too late to remove the epizoa. Because the ammonites continued to grow with these epizoa attached, their shell grew over and around their "guests." This commonly resulted in extreme shell disfigurement for the ammonites and death for the epizoa. Because this is such a large and unique collection of pathological specimens from one site, this paper will describe and illustrate most of the types of pathologic distortions in the species *Quenstedtoceras* (*Lamberticeras*) *lamberti* Sowerby, 1819, and make some interpretations as to why and how these abnormalities occurred.

2 Material

The *Quenstedtoceras* (L.) *lamberti* Zone lies in the uppermost upper Callovian of the Upper Jurassic. The fauna from this zone was deposited in clays and marls in the central portion of the Russian Platform (Meledina, 1988). The *Q*. (L.) *lamberti* Zone directly overlies the *Peltoceras athleta* Zone, corresponding to the western European ammonite zones (Meledina, 1988). The ammonite fauna in the *Q*. (L.) *lamberti* Zone is also quite similar to that of the stratotype described from Europe. In addition to *Quenstedtoceras*, other ammonite genera in the Dubki Quarry include *Eboraciceras*, *Grossouvria*, *Hecticoceras*, *Kosmoceras*, *Peltoceras*, *Prorsiceras*, *Cadoceras*, and *Rursiceras* (?). Further information on the geology of the area can be found in Aleksejev and Repin (1986).

Residing within the collections of the Black Hills Museum of Natural History and Black Hills Institute of Geological Research, Hill City, South Dakota, are approximately 1,100 ammonite specimens collected from the Dubki Quarry, a commercial clay quarry and brickyard, near Saratov, Russia. These collections contain nearly 1,000 specimens of *Quenstedtoceras* (Lamberticeras) *lamberti* and 100 mixed specimens of *Eboraciceras*, *Peltoceras*, *Kosmoceras*, *Grossouvria*, *Prorsiceras*, *Cadoceras*, and *Rursiceras* all collected from the same zone during 2001 and 2002. There are 167 specimens of *Q*. (*L*.) *lamberti* and 89 specimens of other ammonite genera that exhibit a wide range of healed injuries due mostly to predation. A total of 655 specimens of *Q*. (*L*.) *lamberti* display very unusual deformed growth of the shell (15 of these also exhibit healed predation scars and are included in the above count). There are 48 *Q*. (*L*.) *lamberti* with an unusual distortion, where the ammonite grew in a tilted manner due to an unknown cause. A total of 43 *Q*. (*L*.) *lamberti* exhibit small depressions in the venter, believed to be due to the attachment of

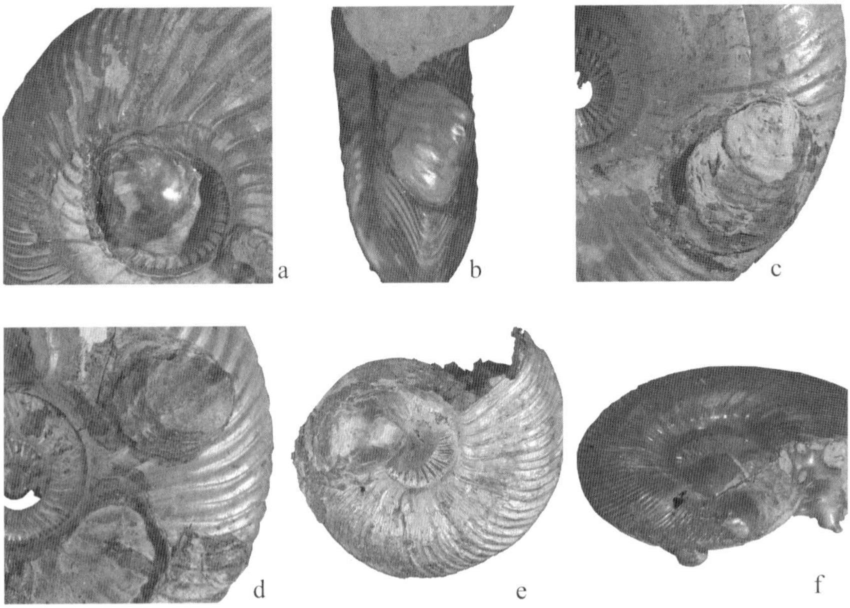

Fig. 16.1 *(a)* Placunopsis *(9 × 10 mm) on umbilicus of* Quenstedtoceras (L.) lamberti *(BHI-5360). (b)* Placunopsis *(7 × 9.7 mm) on venter of* Q. (L.) lamberti *(BHI-5345). (c) Ostrea (?) (15 × 19.8 mm) on flank of* Q. (L.) lamberti *(BHI-5336). (d) Ostrea (19.5 × 19.5 mm) on flank of* Q. (L.) lamberti *(BHI-5469). (e)* Placunopsis *(11.5 × 15 mm) on* Q. (L.) lamberti *(BHI-5306). (f) Four* Placunopsis *(largest 12.5 mm; smallest 8.3 × 10 mm) on flank, venter, and umbilicus of* Q. (L.) lamberti *(BHI-5340).*

epizoa on the venter or the result of a healed bite. Only 95 specimens of *Q.* (*L.*) *lamberti* in this collection exhibit little to no shell deformation.

While about 25% of these ammonites display healed, predatory scarring, 60% exhibit some very unusual and grotesque abnormalities from a nonpredatory source. These bizarre deformities appear to be the result of an infestation of immature bivalves attached to the shell of the ammonite. Originally, the author believed the epizoa were articulate brachiopods; however, H. Feldman (2005, personal communication) confirmed that they were not brachiopods, but rather bivalves. D. Seilacher finally identified the majority of the epizoa as the bivalve *Placunopsis* along with some minor *Ostrea* (2005, personal communication).

Although this collection contains eight different genera of ammonites with healed injuries, only *Quenstedtoceras* (*L.*) *lamberti* seems to have been utilized as a host for this infestation of epizoa. It appears that, in most cases, the epizoa were not removed by the ammonites and, as a result, the ammonites grew over and/or around their "guests." Some bivalves grew quite large relative to the size of their ammonite hosts, and this resulted in some very bizarre and erratic deformities to

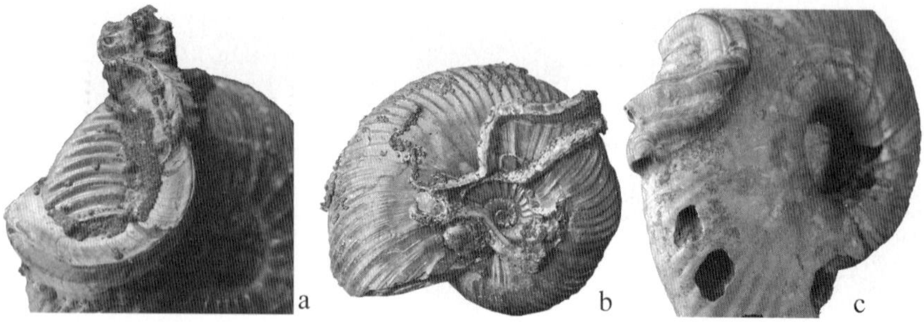

Fig. 16.2 *(a) Serpula on flank and venter of* Quenstedtoceras (L.) lamberti *(BHI-5440) (x1). (b) Serpulids lying across umbilicus, flank, and venter of* Q. (L.) lamberti *(BHI-5442) (x0.7). (c) Serpulids crossing over the venter of* Q. (L.) lamberti *(BHI-5597) (x1).*

the ammonites. The ammonite shell became further distorted when the animal tried to compensate for the additional weight and uneven distribution of the bivalves.

It is unknown why the bivalves attached themselves to the ammonites. It is possible that *Quenstedtoceras (L.) lamberti* were being used as carriers by the bivalves to migrate into other areas of the sea, as documented by Allen (1937) for extant articulate brachiopods attached on mobile gastropods and scallops. The ammonite fauna from the Dubki Quarry was collected in marl or clay (material suitable for making bricks). Deposition of these clay particles suggests a calm, deep marine environment (Reineck and Singh, 1980). The ammonite specimens contain a great deal of pyrite filling and replacement in the phragomocone, indicating that they were buried in a reducing or oxygen-deficient environment (Reineck and Singh, 1980). Because the ammonites would have been unable to survive in this low oxygen environment, they must have been living in the water column, well above the bottom, although (to my knowledge) there has not yet been any isotopic studies of *Quenstedtoceras* (L.) *lamberti* to determine what water depth they inhabited. They were most likely active swimmers, though it is not known how well they were able to swim.

As stated before, approximately one thousand ammonite specimens from the Dubki Quarry were used in this study. That is a small portion of the total number of ammonites collected from this site to date. S. Baskakov (personal communication, February, 2003) estimated that between 50,000 and 70,000 ammonite specimens were collected from this locality by the Spring of 2003, and thousands more since then. These numbers represent a very small percentage of the ammonites originally deposited in this particular stratum, as large excavations have previously taken place in the quarry to mine the clay (and the ammonites) for the production of industrial bricks. Of all the genera of ammonites identified from this locality, only *Quenstedtoceras (L.) lamberti* exhibits deformations due to attached epizoa, while all species from the locality exhibit sublethal pathological scarring due to bites or predation.

It is interesting to note that there are no complete or even nearly complete specimens of *Quenstedtoceras (L.) lamberti* in this entire collection. All of the specimens from this site consist primarily of phragmocones; partial body chambers

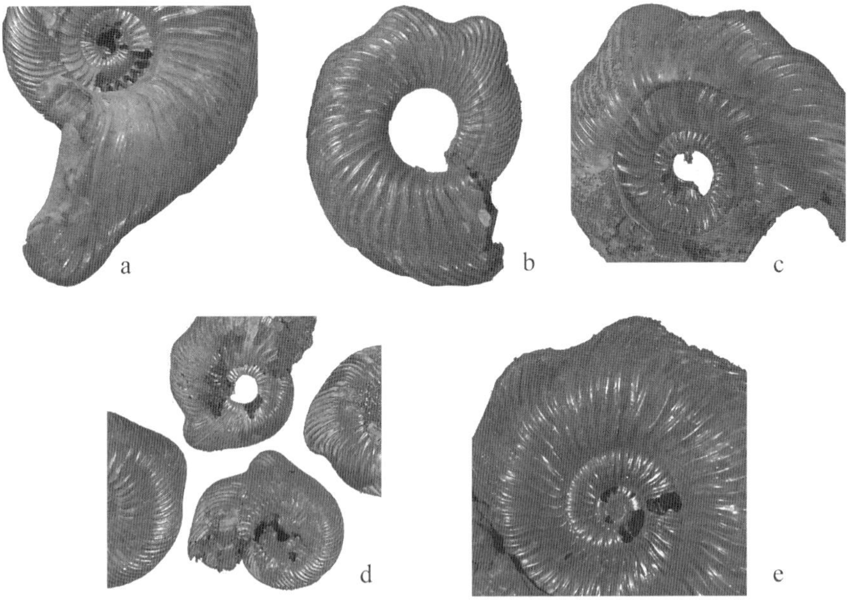

Fig. 16.3 *(a) Protuberance extruding from venter of* Quenstedtoceras (L.) lamberti *(BHI-5379) (x1). (b) Two small swellings on the venter of* Q. (L.) lamberti *(BHI-5380) (x0.9). (c) Protuberances extruding from the venter of* Q. (L.) lamberti *(BHI-5372) (x1). (d) Protuberances extruding from the venter of several* Q. (L.) lamberti *specimens (BHI-5366 left, BHI-5367 top, BHI-5335 right, BHI-5374 bottom) (x0.5). (e) Flattened swellings on the venter of* Q. (L.) lamberti *(BHI-5391) (x1.5).*

are rare. Apparently, either the body chambers were not filled with a stable enough host material, or postmortem predation did away with the body chambers before preservation. The large numbers of *Q. (L.) lamberti* from this site most likely represent mass spawning deaths over several years or generations. This is one of the largest known collections of a single ammonite species from a single site with such a large percentage of deformities resulting from the infestation of epizoa. All of the ammonites illustrated in this paper reside in the collection of the Black Hills Museum of Natural History, 117 Main St., Hill City, South Dakota and are identified by the prefix BHI.

3 Previous Reports of Epizoa on Ammonites

Reports of epizoa on ammonites are not something new; over the last 50 years there have been scores of papers. It is not rare or unusual to find epizoa attached to ammonite shells, but often they are overlooked. Discovering epizoa on ammonites is often dependant on the ammonite species as well as how much of the host rock is saved with the ammonite. Because certain ammonite species are rarely found with epizoa, we can assume that these species were either not suitable hosts or that they may have been able to keep themselves much cleaner than other species. In the case of saving the host rock with the fossil, many fossil preparators often remove all matrix from ammonites during preparation. Because epizoa (and the ammonite shell) are often weathered or textured on their outer surface, they stick to the rock when the ammonite is cleaned. In these cases, saving the host rock and looking at the outer surface of the ammonites can often result in the discovery of epizoa attached to the ammonite shells. Below (listed by date) are some cases of epizoa on ammonites that are important to this study.

Seilacher (1960) reported on the encrustation of a Late Cretaceous ammonite *Buchiceras* by oysters [for further discussion about this example, see Keupp et al. (1999) and Seilacher and Keupp (2000)]. The oysters attached to the ammonite during life, encrusting the flanks, the umbilicus, and the venter. The oysters on the lower flanks and venter grew quite large with respect to the ammonite, several nearly one-fourth to one-third the diameter of the ammonite. Seilacher used the oysters to determine the orientation of the swimming ammonite. Based on the encrustation, he concluded that *Buchiceras* was neither a crawler nor a rapid swimmer, but rather a slow floater that swam near the ocean floor.

Meischner (1968) described an adult *Ceratites* (Triassic) encrusted with *Placunopsis ostracina*, on the flanks, umbilicus, and venter. He determined that swarms of larval *Placunopsis* had settled on the ammonite several times during the life of the animal. Based on the different phases of *Placunopsis* settlements on the ammonite shell, he concluded that these attachments occurred over four annual periods, meaning that the last whorl of the ammonite was formed within four years.

Cope (1968) determined that the oyster distribution on some Kimmeridgian ammonites from Dorset was most likely a postmortem attachment. These ammonites were nearly completely encrusted on the side facing up, and relatively free from oysters on the side facing down, in contact with the sediments.

Seilacher (1982) analyzed the orientations and patterns of overgrowth of oysters, serpulids, bryozoans, and articulate brachiopods on ammonites from the Posidonia Shale of Holzmaden. The importance of this study is that it indicated that many of the epizoa were attached while the host (ammonite) was still alive.

Keupp (1996) illustrated some specimens of *Pavlovia* preserved without epizoa that exhibit distortions similar to some of the pathologies present in *Quenstedtoceras* from the Dubki Quarry. Keupp (1984, 2000: 126) later illustrated several extreme umbilical and dorsal distortions in *Pavlovia* and attributed them to parasites and brachiopods that had attached to the venter of the still-growing ammonites (the epizoa were no longer attached).

Kase et al. (1998) illustrated a specimen of *Placenticeras* from South Dakota with 102 limpets (*Acmaea occidentalis*) attached to both flanks of the ammonite. They concluded that the limpets either attached themselves to a floating shell or to a live ammonite. I have observed hundreds of specimens of *Placenticeras* with attached limpets and know that many of the limpets were attached during the lifetime of the ammonites because the limpets are frequently found under the overlying whorls. Shell distortion from infestation of the epizoa has not been recognized in any of these specimens. Limpet-like or oyster-like animals have been found under some layers of shell near the aperture of a large specimen of *Baculites grandis* (BHI 5502), indicating that they too attached during life.

Davis et al. (1999) tried to cover the many different forms of epizoa on shelled cephalopods. Their study surveyed the evidence of epizoa in many Paleozoic ammonoids from Morocco and Texas illustrating many of the different forms of attachment. They attempted to bring together the records of such epizoa along with the many different manners and places of attachment, whether on living or empty shells. They found that epizoa on Paleozoic ammonites was less common than on co-occurring nautiloids, but that differences could have been due to collector or preparator bias. They concluded that not enough data have been acquired yet to determine the distribution and evolution of epizoa on ammonites through time and that more data were needed.

Checa et al. (2002) illustrated many shell deformations caused by epizoa and listed this type of malformation in 16 Jurassic ammonite genera from Eurasia. They described the action taken by the ammonite to grow over the epizoa and listed two different coiling patterns that ammonites used as they continued their growth: zigzag and trochospiral. Zigzag was defined as a lateral deviation of the whorl with an epizoön more or less centered on the venter of the underlying whorl. Trochospiral was defined as a displacement in whorl growth to counterbalance the weight of the ever-enlarging epizoön upon the ammonite's center of gravity.

There has also been one paper written on the pathologies in *Quenstedtoceras* from the Dubki Quarry. Keupp (2005) published on some of the healed predations and deformations due to epizoa from this site. He noted that "parasites" caused much of these very bizarre deformities, and figured several specimens with these malformations. His publication showed many of the types and forms of grotesque deformities that this study also shows.

4 Terminology

Finding the correct terminology for the epizoa attached to living animals has been a difficult task. Following the presentation "Symbiotic deformities in the Late Callovian fauna from Saratov, Russia," at the Sixth International Cephalopod Symposium, held in Fayetteville, Arkansas, 2004, a discussion ensued on the proper use of terminology regarding the word "symbiont." G. Westermann (personal communication, September, 2004), argued that the term symbiont was improper, and should not be used to describe the attached organisms on *Quenstedtoceras* (*L.*) *lamberti*, while others argued in favor of this usage. There has been a host of terms that ammonite biologists have used when referring to attached "guests" on "host" ammonites, such as epibionts, epizoa, epizoans, epifauna, parasites, etc. (see Davis et al., 1999).

One of the goals of this study is to find the proper terminology to describe the ammonite and the attached bivalve (*Placunopsis*). Both were alive at the same time, neither feeding on the other nor needing the other to survive. The bivalves attached

Fig. 16.4 *(a)* Placunopsis *(7 × 9 mm) on the venter of* Quenstedtoceras (L.) lamberti *(BHI-5345). (b)* Placunopsis *(2.5 × 3 mm) on the flank and venter of* Q. (L.) lamberti *(BHI-5598). (c) Ventral swellings on* Q. (L.) lamberti *with* Placunopsis *(8 mm across) near the umbilicus (BHI-5354).*

themselves to only a small percentage of ammonites, yet the bivalves did not need to live on ammonites to survive. Both the host and the guest were ultimately harmed in different ways from this uncommon union, and it appears that neither one benefited, though one of them may have. This relationship has previously had little written about it.

Grier and Burk (1992) described symbiosis as "living together," along with "an interaction that brings animals of different species into close relationships throughout much or all of their lives, particularly commensalism, mutualism, and various forms of parasitism." The biological definition of symbiosis in the Oxford English Dictionary is an "Association of two different organisms (usually two plants, or an animal and a plant) which live attached to each other, or one as a tenant of the other, and contribute to each other's support. Also more widely, any intimate association of two or more different organisms, whether mutually beneficial or not." Parker (1994) described symbiosis as the interrelationship between two different organisms where the relationship is neither harmful nor beneficial. For the general definition of symbiosis, the term symbiont would work, but because both were ultimately harmed in the union, symbiosis is not the proper term.

Nor was this a commensal union. Parker (1994) described as commensal when animals or plants live with other animals or plants for support, or sometimes for mutual advantage, but not as parasites. Both the ammonite and the bivalves certainly lived together in a nonparasitic relationship, but it appears that there was neither any mutual advantage nor a single advantage for either animal.

Grier and Burk (1992) used the term amensalism for an encounter that results in harm or loss to one of the species while not affecting the other. Since both were ultimately harmed, by either deformation or death, this term does not seem to be quite correct either.

Among "ammonitologists," the terms epizoa (plural) and epizoön (single) (Davis et al., 1999) are frequently used, but for most epizoa, there is either no harm to the coexisting animals or the damage done to each other is usually negligible to nonexistent. Grier and Burk (1992) described epizoa as simply animals that live on other animals; but epizoa as described by Parker (1994) are any of various parasitic animals that live externally upon the bodies of other animals. Davis et al. (1999) proposed the term "epicole" for an organism that might live on a hard substrate or any empty shell or even a shell still occupied by its maker.

Grier and Burk (1992) described a defensive interaction as one "where one or both of the participants stand to lose, including amensalism and competition.... Both sides actually lose when one or both are defensive, even if they both survive." It appears that, in many ways, this was a defensive interaction between the two participants. The ammonites became greatly deformed, and most of the bivalves were buried under the ammonite shell; thus, it appears that neither one was a winner.

Normally, the choice for the correct terminology is simple, such as in the case for animals attached after death (epibionts in the act of epibiosis), or more commonly, if the attached animal was a parasite. R. A. Davis (2005, personal communication) noted how there seems to be a problem with finding the one word that best describes this relationship between the host and the "guest." The proper terminology is difficult to ascertain for this most unusual relationship because both organisms lost due to their ultimate connection. J. Grier (2005, personal communication) recommended against using the term symbiosis, and stated that the term epizoa are correct but it may carry inadvertent connotations. He suggested that maybe a new term should be coined, or maybe not use any terminology at all.

R. A. Davis (2005, personal communication) advised that the relationship should be called "epizoism" and the "guest" should be referred to as an epizoön (single) or epizoa (plural), but only if it could be demonstrated that the "guest" was attached to the "host" while the "host" was still alive. To prove that the "host" was alive while the "guest" was attached and to prove that the deformities were caused from this relationship is the basis for this study. Davis' suggested terminology will be used, but sparingly. There appears to be a need to coin a term for the unique relationship between the animals discussed in this paper, but that will be left to those more knowledgeable on the subject than the author.

5 Epizoa

D. Seilacher (2005, personal communication) tentatively identified the majority of the epizoa attached to the ammonite shell as *Placunopsis*, with occasionally some *Ostrea* (Fig. 1). *Placunopsis* were living, attached "guests" that seem to have interacted only with *Quenstedtoceras* (*L.*) *lamberti*, while leaving the other ammonite species in the zone alone. Seilacher (1960) referred to *Placunopsis* as a false oyster, based on the attachment of the right valve rather than the left valve, as in true oysters (such as *Ostrea*). In *Placunopsis*, the lower valve is flat, the upper valve is rounded, and the ornamentation is barely visible. Sometimes the shell structure seems to mimic the ornamentation of *Q.* (*L.*) *lamberti*, perhaps because the ammonite grew over the bivalves while they were still living and growing.

Placunopsis found at the Dubki Quarry seem to have been almost a plague on *Quenstedtoceras* (*L.*) *lamberti*, and are inferred to be the cause of shell deformations

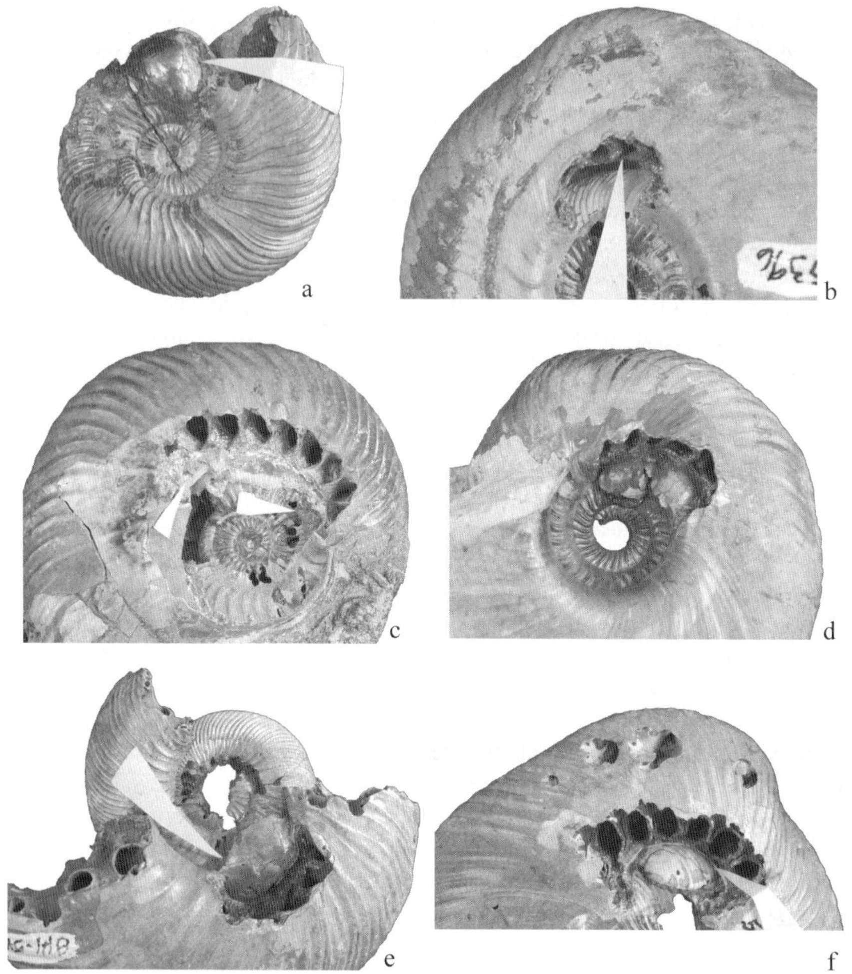

Fig. 16.5 *(a)* Placunopsis *(10 × 11 mm) exposed on* Quenstedtoceras (L.) lamberti *(BHI-5352). (b) Epizoa are missing in exposed void on the venter of* Q. (L.) lamberti *(BHI-5396) (x1). (c) Two small* Placunopsis *(6 mm across) on* Q. (L.) lamberti *(BHI-6000). (d) Two large* Placunopsis *(5 × 7 mm and 6 × 9 mm) on* Q. (L.) lamberti *(BHI-5357). (e)* Placunopsis *(10 × 15 mm) exposed on* Quenstedtoceras (L.) lamberti *(BHI-5470). (f)* Placunopsis *(10.5 × 15 mm) exposed on ventrolateral shoulder of* Q. (L.) lamberti *(BHI-5383).*

in more than 700 specimens in our collections, although the overall percentage is unknown. By comparison, *Ostrea* were found on only eight specimens. While most *Placunopsis* are assumed to be juveniles, some seem to have approached adult size. Attached *Placunopsis* (that were able to be measured on *Q.* (*L.*) *lamberti*) range in size from 2.5 × 3 mm (Fig. 16.4b) to 16 × 18.5 mm (Figs. 16.1, 16.4b, 16.5). Most attached *Placunopsis* are probably smaller than 2 mm, but, if they are, they tend to be destroyed during preparation.

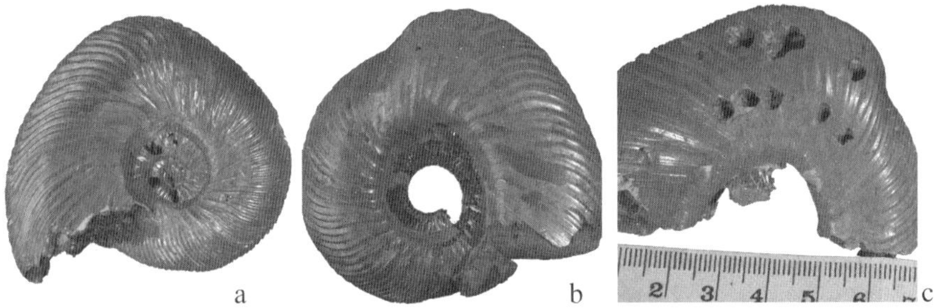

Fig. 16.6 *(a) "Hunchback" deformity on* Quenstedtoceras (L.) lamberti *(BHI-5376) (x0.8). (b) "Hunchback" effect on the venter of* Q. (L.) lamberti *(BHI-5342) (x0.6). (c) Typical "hunchback" on the venter of* Q. (L.) lamberti *(BHI-5383) (x0.7). Note that this is the same specimen as Fig. 5b, but before preparation.*

Fifteen *Quenstedtoceras* (*L.*) *lamberti* have serpulid worm tubes attached to them. Some of the worm tubes appear to have also grown on the ammonites while the ammonites were still alive (Fig. 16.2). Two specimens have large worm tubes that grew on both flanks and across the venter. There are no shell deformities observed resulting from worm-tube attachment, so even though some of the worms grew in a manner that suggests that the ammonites were alive, their attachment to empty ammonite shells could still be a possibility. There are no other types of epizoa detected on *Q*. (*L.*) *lamberti* or any other ammonite genus.

Schindewolf (1936) illustrated four specimens of *Arietites* and *Schlotheimia* from the Jurassic with individual worm tubes (serpulids) growing from one side of the venter to the other. These worms grew larger as the ammonite grew, and their placement on the ammonite shell was determined by how the ammonite shell rotated as the serpulid grew. Landman et al. (1987) observed serpulids, bryozoans, barnacles, and scyphozoans all attached on living *Nautilus*. If present-day serpulids and other epizoa can attach to living *Nautilus*, it is probable that, in the past, serpulids and other epizoa could and did attach to living ammonites (at least in some species).

6 Deformities Caused by Epizoa

The shells of *Quenstedtoceras* (*L.*) *lamberti* contain numerous and bizarre deformities that are attributed to the placement and size of attached epizoa. The pathologies all seem due to *Placunopsis*. In the entire collection of ammonites from Saratov (978 specimens of *Q*. (*L.*) *lamberti*), only 8 specimens have *Ostrea* attached and 15 have serpulids attached. Of the remaining 955 specimens of *Q*. (*L.*) *lamberti*, 655 or 67% of them have deformities known to be caused by *Placunopsis*, and another 101 or 10% are suspected of deformities caused by epizoa. *Placunopsis* attached themselves to the venter, flank, and umbilicus of *Quenstedtoceras*, and sometimes in multiple locations. Within this collection, 4 *Quenstedtoceras* have four *Placunopsis* attached, 12 have three attached, and 38 have two attached. Of these 54 specimens, 17 *Placunopsis* are attached to the flank and venter or flank and umbilicus.

It is known that these epizoa were attached to living ammonites because most of the shell malformations, with the exception of sublethal injuries, were caused by the ammonite growing over the attached organism. In some instances, it also appears that the ammonite grew over the epizoön, but the epizoön somehow dislodged itself, leaving a deformed shell (Fig. 16.5b–f). It is also certain that the epizoa attached when they were still planktonic and then grew. Descriptive names for the location of the deformities will be used in this study, and the causes of the deformities will be discussed. Previously assigned "forma" names are referenced for each deformity, but because there are several types of causes for many of the different "forma," I have chosen not to use this terminology to name the pathology.

6.1 Ventral Attachments

The venter and ventrolateral shoulder of the ammonite are the most common places of attachment of *Placunopsis* on *Quenstedtoceras* (*L.*) *lamberti*. Of 655 known deformities caused by epizoa, 582 or nearly 89% have epizoa attached to the venter or on the ventrolateral shoulder. Obviously, the venter was a more ideal location for attachment without initial discovery, but it was not too favorable for *Placunopsis*. Checa et al. (2002) described ventral deformations on other Jurassic ammonites caused by similar epizoa, and the compensatory growth that was probably undertaken by the ammonites.

6.1.1 Protuberances

Deformity. This deformity is defined as consisting of one or more protuberances on the venter. These protuberances on the ammonite are generally thin, flattened swellings and/or elongated shapes (Figs. 16.3a–c, 16.4a–b). Keupp (2005: Fig. 16.7) illustrated a specimen of *Quenstedtoceras* (*L.*) *lamberti* from the Dubki Quarry with this same deformity. Similar appearing deformities, although caused by bites, were described by Keupp (1976) as *forma inflata* and Kröger (2000) as *forma augata*. Keupp (1976) showed identical multiple ventral protuberances on *Amoeboceras alternans*, which may or may not be due to epicoles. The deformities described as *forma inflata* by Keupp (1996, 2000) and Hengsbach (1996) were not caused by the ammonite growing over an attached epizoön, but rather were the results of bites that had caused a rupture of the mantle, similar to bites seen in scaphites from the US Western Interior (N. L. Larson, 2003).

Cause. This bizarre deformity occurred when a single or several *Placunopsis* attached onto the ammonite venter (Fig. 16.4c). Abnormalities of the ammonite occurred when the shell grew around and over the epizoön attached to the venter, leaving a large, flat, rounded protrusion. The ventral placement of the epizoön caused an unusual deformation to the ammonite and must have been quite terrible for the attached "guest" as well. Because the ammonite grew its shell completely over the epizoön, the epizoön must have died from starvation or suffocation.

Several specimens were taken apart to expose the attached epizoa (Fig. 16.5b–f). While some specimens revealed small and deformed bivalves, most of the bivalves were not very deformed. This additional weight (on the venter) also upset the center of gravity and caused the ammonite to grow off "normal" (Fig. 16.9a), generally resulting in the venter veering off from a straight line, or zigzag as described by Checa et al. (2002).

6.1.2 Hunchbacks

Deformity. A wide, low, broad, sometimes elongated distortion on the venter, which gives the ammonite the appearance of a "hunchback" (Fig. 16.6a–c). Checa et al. (2002) illustrated several deformities caused by epizoa and coined the term zigzag to describe the resulting ammonite growth in relation to the epizoa.

Keupp (2000: 126) illustrated a similar deformity in a specimen of *Pavlovia* sp. cf. *P. iatriensis* from Russia and showed the way that this ammonite shell was

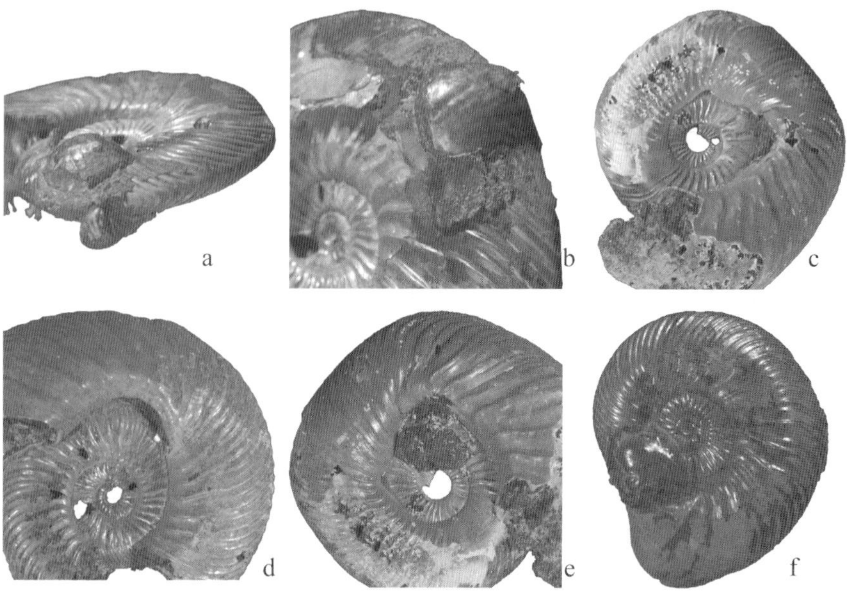

Fig. 16.7 *(a) Two* Placunopsis *(7 × 8.8 mm and 11 × 13 mm) on flank and venter of* Quenstedtoceras (L.) lamberti *(BHI-5362). (b) Close-up of* Placunopsis *(7 × 8.8 mm) (BHI-5362). (c) The effect of a* Placunopsis *on the venter of* Q. (L.) lamberti; *note the place of attachment to the right of the umbilicus (BHI-5382) (x0.6). (d)* Q. (L.) lamberti *with the "hunchback" effect,* Placunopsis *missing (BHI-5370) (x0.8). (e)* Q. (L.) lamberti *with the "hunchback" deformity,* Placunopsis *(10 × 11 mm) on flank and the venter (BHI-5307). (f)* Q. (L.) lamberti *with the "hunchback" deformity,* Placunopsis *(9 × 11 mm) still attached (BHI-5349).*

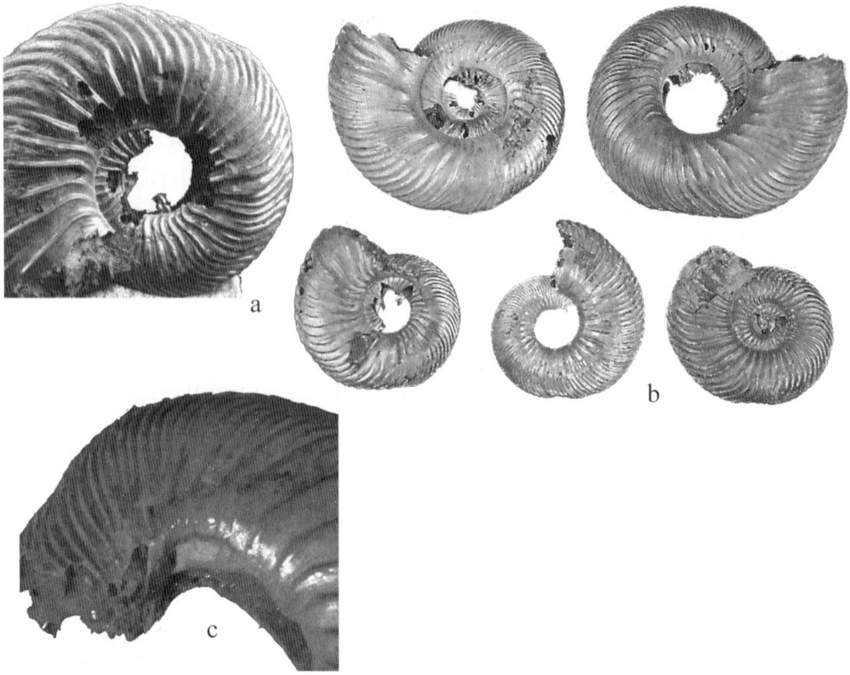

Fig. 16.8 *(a) Ventral "depression" on* Quenstedtoceras (L.) lamberti *(BHI-5454) (x1). (b) Several specimens of* Q. (L.) lamberti *showing the ventral "depression" (x0.8). (c) Note the ventral "depression" on* Q. (L.) lamberti *and the place where the* Placunopsis *rested on the venter to cause the deformity (BHI-5378) (x1.5).*

deformed as it grew over an attached epizoön. Kröger (2000) described this deformation as *forma augata* and similar specimens from the Dubki Quarry were figured by Keupp (2005: Fig. 16.3).

Cause. This is the result of a larger epizoön attached to the venter, and perhaps, partially to the flank (Figs. 163.5b, 16.5d, 16.5f, 16.7a–f). The attached *Placunopsis* seemed to have grown in size before the ammonite grew over it, thus making this deformation quite different from the previous description. The placement and large size of the pelecypod resulted in a deformity that resembles *kyphosis*, which according to Webster's New World Dictionary (Guralnik, 1986) means "a hump, to bend or arch, an abnormal curvature of the spine resulting in a hump or humpback." This could not have been advantageous to either life form, the epizoön dying either before or after the ammonite shell grew over it, and the ammonite shell becoming disfigured in the process.

6.1.3 Depressions

Deformity. A deformity appearing as a depression or dip on the venter (Fig. 16.8a–b). Landman and Waage (1986) illustrated several specimens of *Hoploscaphites nicolletii* with similar ventral depressions. They referred to this as a "stretch pathology" related to the growth of the ammonite as it rapidly reached maturity. The depressions in *Quenstedtoceras* are not the same, because these dips or depressions occur far back on the phragmocone, whereas in *H. nicolletii*, they tend to occur on the shaft of the body chamber. This was not related to a bite or a disease, even though it is similar in appearance to what was figured by Keupp (1977) as *forma aegra aptycha*.

Cause. This deformity is the result of a very small bivalve attached to the venter (Fig. 16.8c), as described in the previous descriptions but with a slightly different distortion. The ammonite added extra shell and grew evenly over the small animal attached to the venter. The continued growth of the ammonite shell then rebounded, leaving a slight depression on the ammonite venter.

6.2 Flank Attachments

Of 655 known deformities caused by epizoa, 61 or about 10% have epizoa attached to the flank. This number is somewhat significantly lower than ventral attachments yet still a much larger place for attachments than the umbilical placements.

Deformity. Bivalves attached to the flanks caused the ammonite to suffer a crooked and twisted venter that deviates from the center (Fig. 16.9a–c). This is similar in appearance to the human deformity called "scoliosis," which means "a lateral curvature of the spine." Ammonite shell distortions, mainly the result of

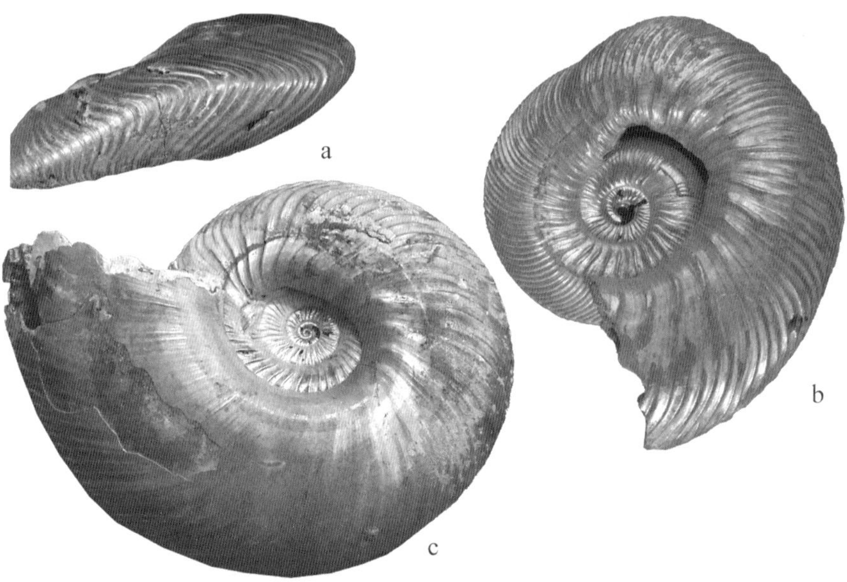

Fig. 16.9 *(a) Curvature of the venter on* Quenstedtoceras (L.) lamberti *(BHI-5384) (x1). (b) Severely deformed* Q. (L.) lamberti, *resulting from attached epizoa (BHI-5338) (x1). (c) Another severely deformed and twisted* Q. (L.) lamberti *(BHI-5466) (x0.8).*

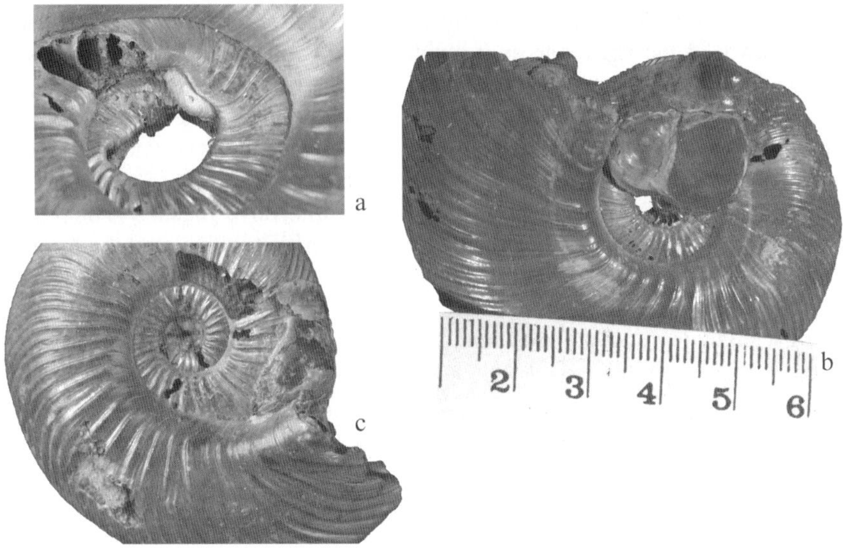

Fig. 16.10 *(a)* Placunopsis *(?) (7.5 mm across) near umbilicus of* Quenstedtoceras (L.) lamberti *(BHI-5613). (b)* Placunopsis *on flank and venter of* Q. (L.) lamberti *(BHI-5340). (c)* Placunopsis *(9.5 × 13.7 mm) on flank and venter of* Q. (L.) lamberti *(BHI-5305).*

healed bites, resembling the deformities seen from the Dubki Quarry have been described as *forma undatecarinata* by Heller (1958), and illustrated by Hengsbach (1979: Fig. 16.8b), and as *forma aegra undatispirata* by Keupp and Ilg (1992) and Keupp (1995, 1996, 2000). Landman and Waage (1986) referred to a similarly twisted venter as "Morton's syndrome," although they did not believe the deformity was caused by an injury. Those ammonites described as having "Morton's syndrome" most likely suffered from unsuccessful predation early in their life.

Cause. Checa et al. (2002) described this deformity as "zigzag." This deformity occurred when the epizoön attached itself to the flank of the ammonite (Fig. 16.10a–c). Depending upon the size and number of animals that attached to the shell, this regularly resulted in grotesque and monstrous deformities. Sometimes there were many of these pelecypods, and commonly they grew quite large on the flanks of the ammonites. In attempting to cover the epizoa, the ammonite had to deal with the ever enlarging and uneven weight distribution caused by the size of these epizoa and its own malformed shell growth. As a result, the ammonite became twisted, with an extremely crooked venter that bent to one side and sometimes back again (Fig. 16.10b, c).

6.3 Umbilical Attachments

Of 655 known deformities caused by epizoa, only 12 or a little more than 1% have epizoa attached on or near the umbilicus. In one-half of these specimens there are more than one *Placunopsis* attached to the ammonite. As evidenced by the low

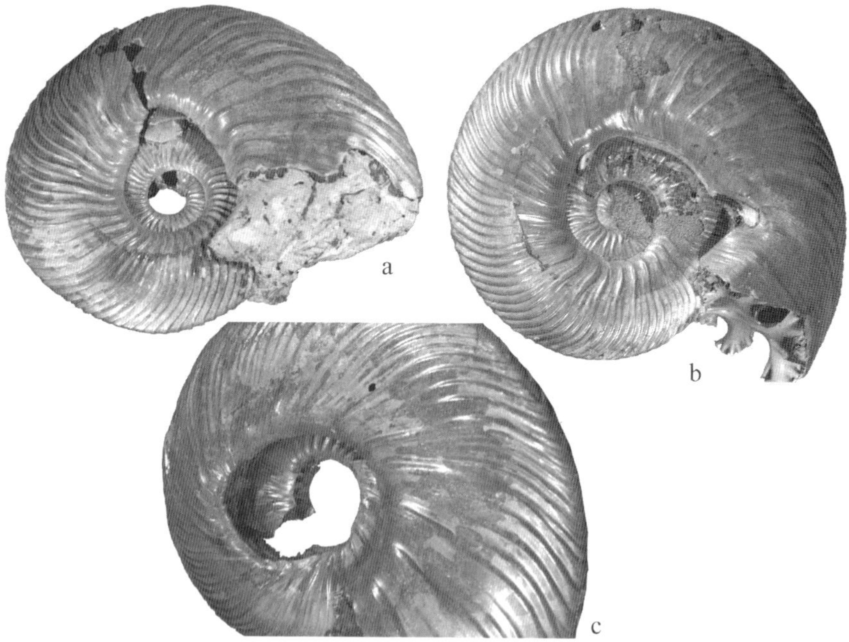

Fig. 16.11 *(a)* Placunopsis *(9.5 mm) near umbilicus of* Quenstedtoceras (L.) lamberti *causing "hunchback" and "depression" deformations (BHI-5346). (b)* Placunopsis *near the umbilicus and on the venter of* Q. (L.) lamberti *caused the deep depression in the dorsum (BHI-5353) (x1). (c)* Q. (L.) lamberti *with the epizoön gone; note the deformation near the umbilicus (BHI-5397) (x1.2).*

numbers, this is the most unusual place for attachment, yet the most easily to see and distinguish.

Deformity. One or more epizoa attached near the umbilicus of the ammonite (Fig. 16.11a–c). This phenomenon has also been described as occurring in *Pavlovia* (well illustrated by Keupp, 1996), although there are no illustrations with the epizoön attached. Keupp and Ilg (1992) referred to this deformity as *forma aegra undatispirata*.

Cause. Newly hatched *Placunopsis* attached themselves in or around the umbilicus of the ammonite and then continued to grow (Fig. 16.12a–f). It may have been impossible for the ammonite to remove epizoa from this location. This point of attachment, of all places, was probably the best for the *Placunopsis* and the least disfiguring for the ammonite. The bivalve was able to survive for a longer time, and, if there were multiple epizoa (Fig. 16.10b), they may have been able to spawn and colonize new areas of the ocean; this could account for different sizes of *Placunopsis* on the ammonites. *Placunopsis* may have been able to "infect" other ammonites after they released their spat (as in living bivalves) into the currents and thus onto other ammonites.

The umbilical attachment site was not too bad for the ammonite. It had an easier time compensating for the additional weight of the epizoön, and its shell growth did not become as contorted as in some of the other attachment places. As seen in Fig. 16.12b–f, there is still considerable distortion around the umbilicus, some on the flanks, and occasionally on the venter, as a result of epizoa attached near the umbilicus.

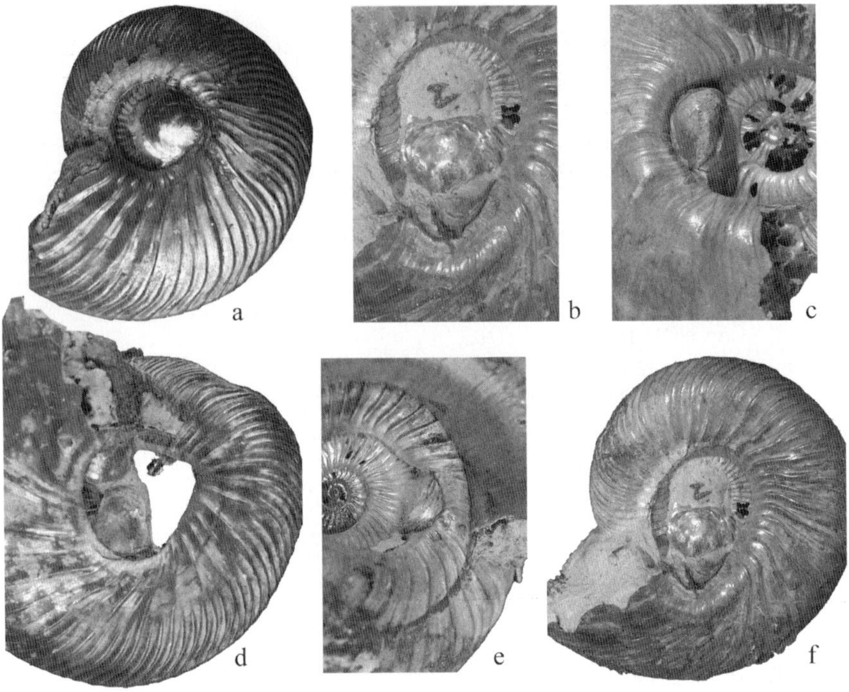

Fig. 16.12 *(a)* Placunopsis *(8.3 × 9.7 mm) attached on the umbilicus of* Quenstedtoceras (L.) lamberti *(BHI-5360). (b)* Placunopsis *(10.5 mm across) attached on the umbilicus of* Q. (L.) lamberti *(BHI-5343). (c)* Placunopsis *(9.7 mm across) attached near the umbilicus of* Q. (L.) lamberti *(BHI-5488). (d) Two* Placunopsis *(large one 7.1 × 9.5 mm) attached near the umbilicus of* Q. (L.) lamberti *(BHI-5467). (e)* Placunopsis *(5 mm across) attached near the umbilicus of* Q. (L.) lamberti *(BHI-5466). (f) Two* Placunopsis *(large one 10 × 10.7 mm) attached near the umbilicus of* Q. (L.) lamberti *causing deformities to the dorsum (BHI-5343).*

7 Healed Shell Fractures

Nonlethal injuries are observed in specimens of all ammonite species from the Dubki Quarry. These injuries have been interpreted as originating as bites or some other form of damage to the ammonite shell or mantle. These pathologies occur as irregularities of the shell in the form of ruptures, scars, wrinkles, folds, scratches, and displaced ribs. They prove that not only did the ammonites have to contend with the infestation of epizoa, but they also had to survive attacks from a variety of predators such as fish, reptiles, and other cephalopods (animals commonly cited as modern cephalopod predators). The injuries in this fauna are all consistent with healed paleopathologies as seen in many publications (for example, Landman and Waage, 1986; Bond and Saunders, 1989; Keupp, 1976, 1996, 2000; Hengsbach 1996; and others).

Hengsbach (1996) listed a fairly complete summary of most of the previous work on pathological ammonites. His use of the German *forma* and *aegra* names to describe the healed ammonite pathologies was not new, but rather a comprehensive overview of the previous work done by so many paleopathologists. The terms *forma* meaning "form" and *aegra* meaning "sick," though appropriate, are very confusing when used in conjunction with the healed wounds found on pathological ammonites. I have chosen to use other names for these injuries, while maintaining the original references for the *forma* names.

Kröger (2002) showed several examples of nonlethal injuries in ammonites. He noted six different types of breakage and repair to the ammonite shell and assigned names from medical terminology. He documented the high incidence of injuries in the genus *Quenstedtoceras*, among many other genera. But because all of the specimens utilized in his paper were from the collection of H. Keupp, who selectively collects such specimens, there was naturally a high incidence of pathological ammonites. Kröger's study dealt with the percentage of different types of sublethal predation, and the most common breakage to the shell. The most common breakage or injury is damage to the aperture of the ammonite. Only a small percentage of shell damage appears as a deeper injury to the flank or venter.

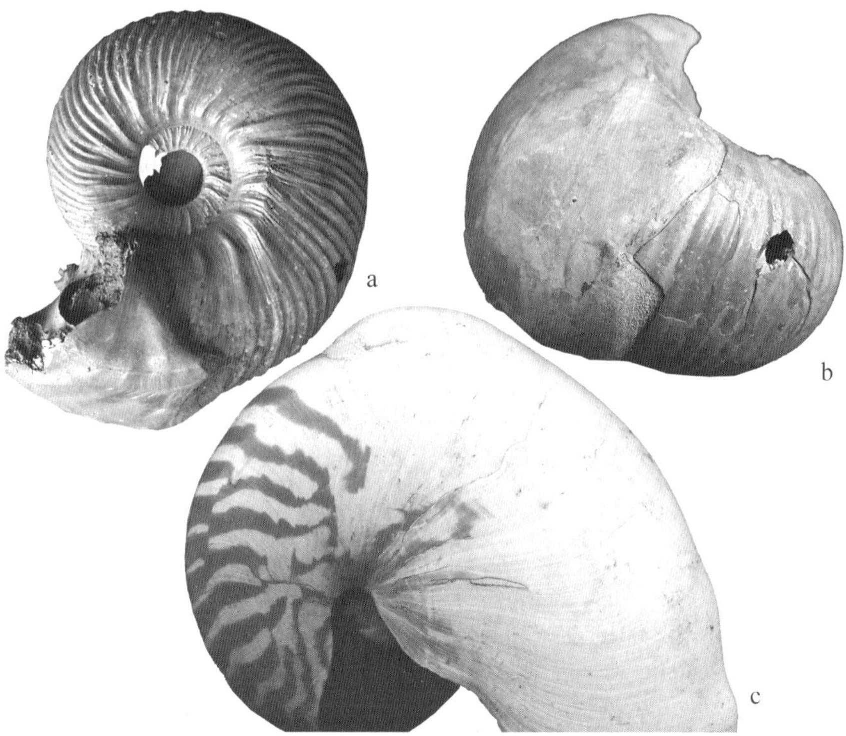

Fig. 16.13 *(a)* Quenstedtoceras (L.) lamberti *showing large-repaired rupture (BHI-5424) (x0.8). (b)* Eboraciceras *showing large-repaired rupture and shell regrowth (BHI-5499) (x0.8). (c) Recent* Nautilus *showing a healed rupture (BHI-5603) (x0.35).*

The percentage of sublethal injuries (as well as deformities caused by epizoa) in the fauna from the Dubki Quarry is difficult to calculate. The specimens used in this study were chosen out of perhaps ten to twenty thousand, or more. Bond and Saunders (1989) reported a 15% incidence of shell injury and repair among Late Mississippian ammonites from the Imo Formation of northwest Arkansas. P. L. Larson (1984) reported that the number of pathological specimens in the family Scaphitidae from the Fox Hills Formation ranged from 15% (*Hoploscaphites nicolletii* Range Zone) to 46.7% (*Jeletzkytes nebrascensis* Range Zone). Landman and Waage (1986) noticed a 10% incidence of shell abnormalities from the *Hoploscaphites nicolletii* Zone and 25–40% from the *Jeletzkytes nebrascensis* Zone based on their ammonite collection. Judging from the thousands of specimens seen and reported from the Saratov locality, it is estimated that perhaps only 10% of the Saratov fauna had any sort of deformities resulting from healed fractures or attached bivalves.

There are several different pathologies that occurred in all ammonite species, as a result of survived predatory attacks. The following list includes some of the different forms of scarring in the ammonites that have been found at the Dubki Quarry.

7.1 Ruptures

Deformity. This deformity is observed as a small or large unornamented swelling of the shell, emanating from one small area (Fig. 16.13a, b). The protrusion is generally round to oblong, and extends away from the rest of the shell. It was described as *forma inflata* by Keupp (1976) and illustrated by Keupp (1995, 1996, 2000), Hengsbach (1996), and N. L. Larson (2003).

Cause. This deformed, bulbous protrusion from the shell is probably the result of a bite through some portion of the body chamber (from fish, reptile, crustacean, or cephalopod) and a subsequent rupture of the mantle through the broken shell (N. L Larson, 2003). As is common in all animals that receive an injury, the ammonite would most likely have tried to immediately repair the rupture or wound. This repair usually resulted in a lack of ornamentation, or a smooth, rounded shell at the point of injury. A similar type of deformity has also been observed in extant *Nautilus* (Fig. 16.13c).

7.2 Rib Displacement

Deformity. A displacement to the sculpture of the ammonite shell, resulting in the conspicuous displacement of the ribs (Fig. 16.14a–c), classified by Hengsbach (1979) as *forma aegra syncosta*, and illustrated by Keupp (1973), Landman and Waage (1986), and Bond and Saunders (1989).

Cause. Most likely, the result of a bite that caused fracturing and displacement of the shell. This deformity has also been observed in present-day *Nautilus* (see Ward, 1987).

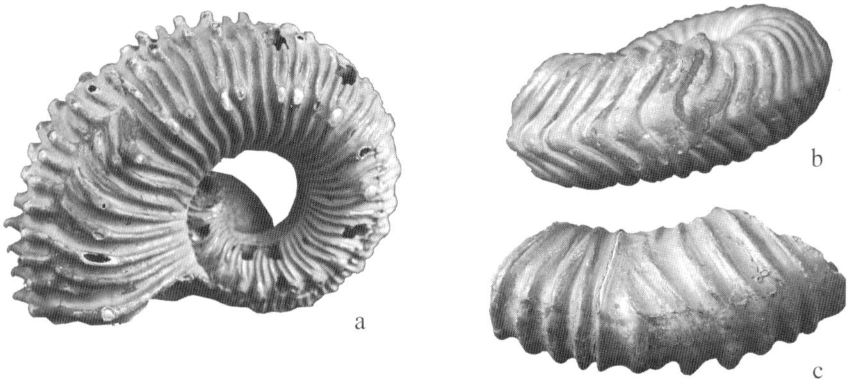

Fig. 16.14 *(a)* Kosmoceras, *injury showing rib displacement (BHI-5599) (x1). (b)* Prorsiceras *with an injury, showing displacement on the venter (BHI-5602) (x1.2). (c)* Rurciceras *(?) showing damage to the venter (BHI-5601) (x1).*

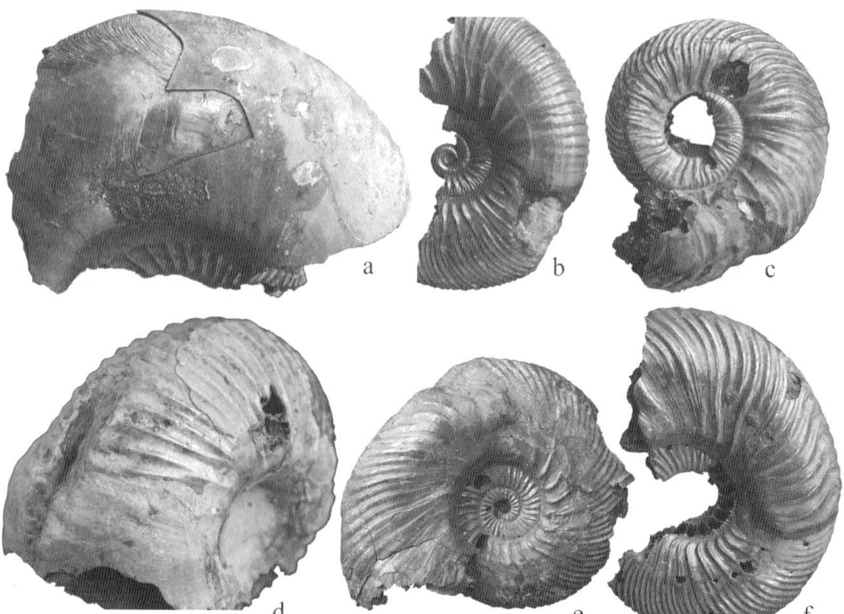

Fig. 16.15 *(a)* Quenstedtoceras (L.) lamberti *with repaired injury (BHI-5415) (x1). (b)* Grossouvria *with repaired bite (BHI-5427) (x1). (c)* Q. (L.) lamberti *showing deformed whorl from a bite (BHI-5426) (x1). (d)* Eboraciceras *showing a deep groove and healed shell (BHI-5608) (x0.8). (e)* Q. (L.) lamberti *with displaced shell (BHI-5432) (x0.7). (f)* Q. (L.) lamberti *showing healed injury as rib repair (BHI-5434) (x0.9).*

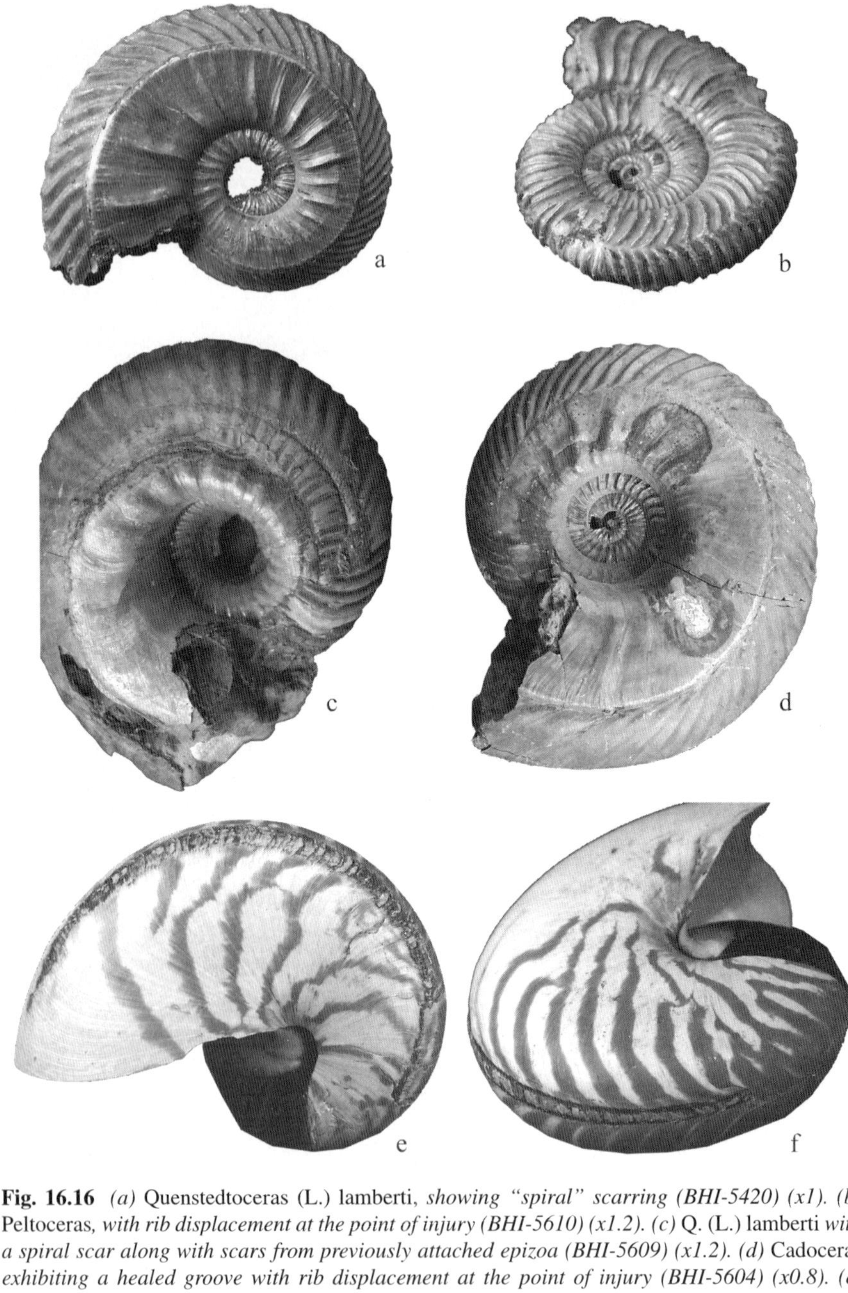

Fig. 16.16 *(a)* Quenstedtoceras (L.) lamberti, *showing "spiral" scarring (BHI-5420) (x1). (b)* Peltoceras, *with rib displacement at the point of injury (BHI-5610) (x1.2). (c)* Q. (L.) lamberti *with a spiral scar along with scars from previously attached epizoa (BHI-5609) (x1.2). (d)* Cadoceras *exhibiting a healed groove with rib displacement at the point of injury (BHI-5604) (x0.8). (e) Recent* Nautilus *showing similar scarring from a healed bite (BHI-5611) (x0.3). (f) Recent* Nautilus *showing "spiral" scarring, and the point of injury (BHI-5612) (x0.2).*

7.3 Scars

Deformity. There are two types of deformities seen on either the flank or the venter, which are the results of a deep bite. One is characterized by the appearance of a large healed injury, regularly seen as a deep depression, remodeled and sculpted, devoid of normal ornamentation, and frequently with portions of the shell completely missing (Fig. 16.15a–f). Similar scarring was described as *forma aegra aptycha* by Keupp (1977) and illustrated by Bond and Saunders (1989).

The second deformity is typified by a long, shallow to deep depression that follows the spiral growth of the ammonite shell (Fig. 16.16a–d). This type of injury was classified by Hölder (1956) as *forma aegra verticata* and figured by Keupp (1979, 1985). This type of scarring was also labeled by P. L. Larson (1984) and Landman and Waage (1986) as a "spiral furrow."

Cause. Both of these depressed scars are the result of severe bites or injuries to the body chamber and mantle during life. The first deformity resulted in the loss of some of the mantle, and as the ammonite attempted to heal the bite, it grew a smooth shell covering the bitten region. Because there was flesh and shell lost in the bite, there is a depression left that never grew back to the original form. If there is no ornament, this indicates that the injury happened somewhere behind the mantle margin, because only the lip of the mantle can create ribbing and ornament. If there is ornament, this indicates that the injury happened near or at the aperture, and that the ammonite managed to create new shell material with its damaged mantle.

The depressed line of scarring is the result of an injury to the lip (or edge) of the mantle in a still-growing ammonite. This type of injury resulted in a damaged and disfigured mantle that would have continued to produce a scarred shell throughout the lifetime of the ammonite, because some portion of the mantle was severely damaged. This type of deformity also has been observed in extant *Nautilus* (Fig. 16.16e, f).

Fig. 16.17 *(a)* Quenstedtoceras (L.) lamberti *with small "nibble" (BHI-5614) (x2). (b)* Grossouvria *(?) with scarring on the venter (BHI-5605) (x1).*

7.4 Scratches

Deformity. This obscure deformity is commonly observed as a scratch on the shell of the ammonite (Fig. 16.17a–b). It may appear in the form of shallow and minute scars that parallel the growth of the ammonite, are restricted to one small area, or are at different angles to the growth. This type of injury has been referred to as "parvus-type" by Kröger (2002), and similar, nonlethal deformities were illustrated by Bond and Saunders (1989).

Cause. Injuries resulting from perhaps a bite or a nibble from another ammonite or an interaction with some other animal. Similar shell damage has been observed to have taken place between breeding pairs of present-day *Nautilus* (see Ward, 1987).

8 Distorted Shapes of Unknown Origin

Deformity. Webster's New World Dictionary (Guralnik, 1986) describes "anamorphism," as "an abnormal change of form which gives the appearance of a different species." The overall shape of the ammonite has become distorted with more shell growth on one side of the ammonite than on the other (Fig. 16.18a–c), making it look quite different when comparing both sides. The venter is commonly shifted to one side, so that one side is generally rounded with both sets of ventral tubercles, whereas the other side is commonly flat. This distortion, which can be viewed by

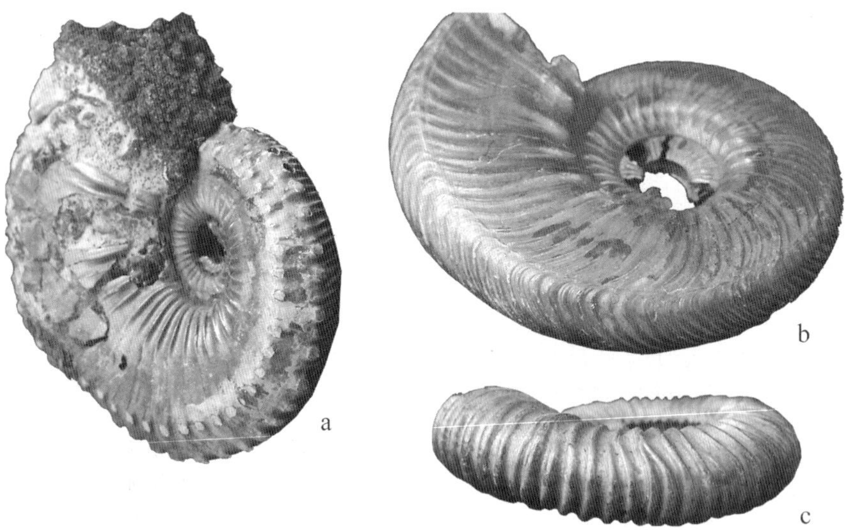

Fig. 16.18 *(a)* Kosmoceras *with nonsymetrical venter (BHI-5607) (x1.2). (b) Trochospiral* Quenstedtoceras (L.) lamberti *"flattened" on one side, round on the other (BHI-5436) (x1). (c)* Rurciceras *(?) with a bent and curved venter (BHI-5606) (x1).*

comparing one side of the ammonite to the other, was described as "forma cacoptycha" by Lange (1941) and illustrated by Keupp (1977, 1984, 2000), and Hengsbach (1996). Kröger (2002) described similar deformities in other ammonites calling them "Harpoceras-type", while Landman and Waage (1986) described this deformity as simply "local asymmetry." Checa et al. (2002) called the deformity "trochospiral growth," attributing the distortions to attachments of epizoa, which caused the ammonites to grow off center.

Cause. It is unknown, in any of the ammonites from this site, whether this distortion is the result of an attached, small organism (such as an epizoön) on the outer flank near the venter in an early growth stage of the ammonite, or the result of a nonlethal bite that damaged the mantle early in life, causing the ammonite to grow in a crooked or asymmetrical manner. Both possibilities are an option because this distortion is commonly observed in *Quenstedtoceras* (*L.*) *lamberti* from the Saratov site. Checa et al. (2002) attributed similar distortions to the attachment of epizoa during an early stage of life, causing tilting or trochospiral growth. Landman and Waage (1986) noted similar distortions in Maastrichtian scaphites from the Western Interior, which have never been reported to have deformities caused by epizoa but have a high percentage of healed bites (P. L. Larson, 1984; Landman and Waage, 1986; N. L. Larson, 2003).

This type of trochospiral growth is known to occur in ammonites of nearly all species, including those from the Late Cretaceous families Scaphitidae, Placenticeratidae, and Sphenodiscidae, and within these Late Cretaceous genera, the cause appears to be from a nonlethal bite earlier in life. This is theorized because attachments of epizoa on scaphites are still unknown. Keupp (1976: Fig. 16.4) noted a similar distortion in *Pleuroceras* and interpreted it as a healed bite.

9 Discussion

Allen (1937) described articulate brachiopods attached to gastropods and scallops and postulated that the molluscs were being used as carriers for the migration of brachiopods into different areas of the sea. Logan et al. (1975) reported on unusual attachments of brachiopods on the surfaces of present-day scallops. These brachiopods were also apparently using the scallops for transportation to different portions of the sea. It is postulated that *Placunopsis* may have inadvertently colonized other areas of the Callovian Sea by attaching themselves to *Quenstedtoceras* (*L.*) *lamberti* and moving with the shoal. The bivalves likely attached themselves to *Q.* (*L.*) *lamberti* while still in the larval stage and grew in place. *Placunopsis* could later release their hold on the ammonites or release their spat while still on the ammonites. This could introduce new settlements of *Placunopsis* into many different areas. Judging by the percentage of distorted ammonites that grew over the epizoa, it appears that perhaps only a small percentage of these *Placunopsis* were successful in further colonization of the sea.

Whether feeding on the waste of the ammonite, or just hitching a ride, as the pelecypod grew, it became so firmly attached, that the ammonite could not dislodge it from the shell once the pelecypod was finally large enough to be a nuisance. Because of the need to increase the size of the shell, both from the additional weight of its growing body and from the increasing weight of the pelecypod, it became necessary to grow over and around the epizoön. In the process, the ammonite shell commonly became grotesquely deformed. Occasionally, the pelecypod is missing from under the ammonite shell (as seen in Fig. 16.5b), even though there is a hole where it used to be. Was the pelecypod able to dislodge itself and move on, or was it somehow dislodged or removed later? Whatever the case, most pelecypods were buried during the construction of the ammonite shell and were unable to escape. The pelecypods must have succumbed to a slow death, unable to escape while the ammonite covered them. The ammonites, which were permanently disfigured, must have had a difficult time swimming properly and because of their grotesque deformities, they may have been less desirable sexually.

There is abundant literature written on pathological ammonites. Yet throughout all of this literature, this widespread ammonite–epizoa relationship is unusual. Hopefully, one day the significance of this relationship may be further explained.

10 Conclusions

The "guest" (*Placunopsis*) was probably planktonic when it attached to the "host" (*Quenstedtoceras* (*L.*) *lamberti*), and in many cases, both were alive at the time of their union. Neither organism needed the other, nor did they live parasitically with each other. For the most part, the "guest" was completely or mostly buried under the shell of the "host." This relationship was not beneficial to either organism, unless the bivalve had reached maturation and was able to spawn. It appears that this may have happened, though it cannot be proved. In this example, the bivalve spat, or offspring, would have been introduced to other parts of the ocean; thus, the "guest" could have benefited. The ammonite host seems never to have benefited; usually, its shell was grotesquely deformed, and its ability to swim and reproduce may have been greatly compromised.

Epizoa did not commonly attach themselves to living ammonites. Davis et al. (1999) pointed out that even in large collections, the attachment of epizoa on ammonites is rare. So why did these ammonites become the hosts of these epizoa? R. A. Davis (2005, personal communication) pointed out to the author that "the ammonoids provided a hard substrate to which the young pelecypods could and did attach." These ammonites happened to be in the wrong place at the wrong time, and something in their anatomy made them suitable for the *Placunopsis* to attach to without being removed. The ribbing on *Quenstedtoceras* (*L.*) *lamberti* is similar to that of other species of ammonites that are found in the same zone. The size of the ammonites did not seem to matter; *Placunopsis* infestation occurred on all parts and on all sizes of *Q.* (*L.*) *lamberti*, yet on no other species.

Why was the pelecypod successful in attaching itself to *Quenstedtoceras* (*L.*) *lamberti*, while leaving the other species of ammonites in the sea alone? The ribbed shell of *Quenstedtoceras* is consistent with the ribbed shells of other genera found in the deposit. There appears to be no noticeable size difference or any other physical shell difference that would favor one species while leaving the others alone. It must be that either the swimming patterns of *Q.* (*L.*) *lamberti* were much different from other ammonite species, or they were unable to clean themselves as efficiently as other ammonites could. The arms of *Q.* (*L.*) *lamberti* could have been much shorter or quite different from those of other species, because they were apparently unable to remove the intruders. Because *Q.* (*L.*) *lamberti* are found in much greater abundance than any of the other ammonite species, they must have lived in much larger shoals, so that there were more of them "hanging around" when the planktonic *Placunopsis* were in the currents looking for a place to attach. *Q.* (*L.*) *lamberti* just happened to be available when they were needed the most.

Of the 655 specimens of *Quenstedtoceras* (*L.*) *lamberti* that were confirmed to have *Placunopsis* attachments, nearly 89% of the *Placunopsis* were attached on or near the venter of the ammonites. A total of 10% had attachments midflank and only a little more than 1% had umbilical attachments. That figure is significant. It seems that *Q.* (*L.*) *lamberti* were unable to remove many epizoa from their venter. Perhaps the venter was out of their line of sight, or maybe even out-of-reach for them. It could have had something to do with their sharp, ribbed keel as well.

The extreme deformities in the ammonite shells indicate the plasticity and resilience of the ammonite shell. The ammonites were able to adapt and recover from a vast variety of problems, including bites and settlements of growing epizoa attached to their shell. As with all known species of ammonites, *Quenstedtoceras* was able to survive many different forms of trauma, such as bites. This resilience is consistent with living cephalopods, as extant species are able to survive a wide array of communal problems, along with friendly and predatory attacks.

It should be possible to calculate the growth rate of *Quenstedtoceras* (*L.*) *lamberti* based on the age and growth rate of the attached epizoa. But because *Placunopsis* were small bivalves throughout their life, the age of the bivalves and how fast these ammonites grew is still not known. It is still not known if the growth of one or two ammonite whorls would have taken several months to a year or more. Further research into *Placunopsis* could help determine more about their growth, which would then lead to a better understanding of the growth and lifespan of *Q.* (*L.*) *lamberti*. We may never know for sure why in the nine described genera of ammonites from the Dubki Quarry, only *Q.* (*L.*) *lamberti* was used by *Placunopsis* for settlement and why it was unable to remove them.

Acknowledgments

I thank Sergey Baskakov and Igor Shumilkin who provided me with much-needed advice and information about the locality. Thanks to Roy Young for showing me the first collection of these pathological abnormalities and taking me to meet with Sergey. Special thanks go to Black Hills

Institute of Geological Research Inc., who purchased all of the specimens for the sole purpose of research. All of the specimens figured in this paper reside in the collection of the Black Hills Museum of Natural History, Hill City, South Dakota. I thank Richard Davis and James Grier for some of the terminology, even if I did not quite find a quick, easy solution to my word problem. Thanks to Steve Jorgensen for editing and comments, also thanks to Robert A. Farrar, Neil H. Landman, Christian Klug, and Royal H. Mapes for their extensive comments, insight, and editorial help. Thanks to Howard Feldman who determined that the epizoa were not brachiopods, but actually pelecypods. A very special thanks to Dolf Seilacher, who was able to identify the pelecypods and finally settle a long debate on what these animals were. He also gave the author other advice and much needed help on this project.

References

Aleksejev, V., and A. Repin. 1986. New data on the Callovian deposits of Malinov Ravine, Saratov area of the Volga Region. *In* M. S. Mesezhnikov (editor), *Jurassic Deposits of the Russian Platform; A Collection of Scientific Works*, pp. 130–137. Geologorazved: Institute of Lenningrad.

Allen, R. S. 1937. On a neglected factor in brachiopod migration. *Records of the Canterbury Museum* **4**: 157–165.

Bond, P. N., and W. B. Saunders. 1989. Sublethal injury and shell repair in Upper Mississippian ammonoids. *Paleobiology* **15**(4): 414–428.

Checa, A. G., T. Okamoto, and H. Keupp. 2002. Abnormalities as natural experiments: a morphogenetic model for coiling regulation in planispiral ammonites. *Paleobiology* **28**(1): 127–138.

Cope, C. W. 1968. Epizoic oysters on Kimmeridgian ammonites. *Palaeontology* **11**(1): 19–20.

Davis, R. A. 1998 (unpublished). Humpty Dumpty's glossary of epizoa and suchlike. Hand-out distributed at the Symposium on "*Ecology and Evolution of Encrusting and Boring Organisms*," Annual Meeting of the North-Central Section, Geological Society of America, 1998, 39pp.

Davis, R. A., R. H. Mapes, and S. M. Klofak. 1999. Epizoa on externally shelled cephalopods. *In* A. Y. Rozanov, and A. A. Shevyrev (editors), *Fossil Cephalopods: Recent advances in their Study*, pp. 32–51. Moscow: Palaeontological Institute, Russian Academy of Sciences.

Guralnik, D. B. 1986. *Webster's New World Dictionary*. New York: Prentice-Hall.

Grier, J. W., and T. Burk. 1992. *Biology of Animal Behavior*, 2nd Edition. St. Louis, Missouri: Mosby Year Book.

Heller, F. 1958. Gehäusemißbildungen bei Amaltheiden, ein neuer Fund aus dem fränkischen Jura. *Geologische Blaetter fuer Nordost-Bayern und angrenzende Gebiete* **8**(2): 66–71.

Hengsbach, R. 1979. Weitere Anomalien an Amaltheen-Gehäusen (Ammonoidea; Lias). *Senckenbergiana lethaea* **60**(1/3): 243–251.

Hengsbach, R. 1996. Ammonoid pathology. *In* N. H. Landman, K. Tanabe, and R. A. Davis (editors), *Ammonoid Paleobiology*, pp. 581–602. New York and London: Plenum Press.

Hölder, H. 1956. Über Anomalien an Jurassischen Ammoniten. *Palaeontographica A* **102**: 18–48.

Hölder, H. 1973. Miscelleana cephalopodica. *Münsterländer Forschungshefte Geologie Paläontologie* **29**: 39–76.

Kase, T., P. A. Johnston, A. Seilacher, and J. B. Boyce. 1998. Alleged mosasaur bite marks on Late Cretaceous ammonites are limpet (patellogastropod) home scars. *Geology* **26**(10): 947–950.

Keupp, H. 1973. Der Wert anomaler Perisphincten (Ammonoidea) für die Deutung der Parabelgenese. *Geologische Blaetter fuer Nordost-Bayern und angrenzende Gebiete* **23**(1): 20–35.

Keupp, H. 1976. Neue Beispiele für den Regenerationsmechanismus bei verletzten und kranken Ammoniten. *Paläontologische Zeitschrift* **50**(1/2): 70–77.

Keupp, H. 1977. Paläopathologische Normen bei Amaltheiden (Ammonoidea) des Fränkischen Lias. *Sonderdruch aus Jahrbuch der Coburger Landesstiftung* **1977**: 263–280.

Keupp, H. 1979. Nabelkanten-Präferenz der forma *verticata* Hölder 1956 bie Dactylioceraten (Ammonoidea, Toarcien). *Paläontologische Zeitschrift* **53**(3/4): 214–219.

Keupp, H. 1984. Pathologische Ammoniten: Kuriositäten oder paläobiologische Dokumente? Part 1. *Fossilien* **6**: 258–262, 267–275.

Keupp, H. 1985. Pathologische Ammoniten: Kuriositäten oder paläobiologische Dokumente? Part 2. *Fossilien* **1**: 23–35.

Keupp, H. 1995 Volumenvergrößernde Anomalien bei Jura-Ammoniten. *Fossilien* **1**: 54–59.

Keupp, H. 1996. Paläopathologische Analyse einer Ammoniten-Vergesellschaftung aus dem Ober-jura Westsibiriens. *Fossilien* **1**: 45–54.

Keupp, H. 2000. *Ammoniten: Paläobiologische Erfolgsspiralen*. Stuttgart: Jan Thorbecke Verlag.

Keupp, H. 2005. Gehörnte Hörner Kranke Ammoniten aus Russland. *Fossilien* **1**: 31–36.

Keupp, H., and A. Ilg. 1992. Paläopathologie der Ammonitenfauna aus dem Obercallovium der Normandie und ihre palökologische Interpretation. *Berliner Geowissenschaftliche Abhandlungen* **30**: 171–189.

Keupp, H., M. Röper, and A. Seilacher. 1999. Paläobiologische Aspecte von syn vivo-besiedelten Ammonoideen im Plattenkalk des Ober-Kimmeridgiums von Brunn in Ostbayern. *Berliner Geowissenschaftliche Abhandlungen* **30**: 121–145.

Kröger, B. 2000. Scalenverletzungen an Jurassischen Ammoniten - ihre paläobiologische und paläoökologische Aussagefähigkeit. *Berliner Geowissenschaftliche Abhandlungen* **33**: 1–97.

Kröger, B. 2002. Antipredatory traits of the ammonoid shell – Indications from Jurassic ammonoids with sublethal injuries. *Paläontologische Zeitschrift* **76**(2): 223–234.

Landman, N. H., and K. M. Waage. 1986. Shell abnormalities in scaphitid ammonites. *Lethaia* **19**: 211–224.

Landman, N. H., W. B. Saunders, J. E. Winston, and P. J. Harries. 1987. Incidence and kinds of epizoans on the shells of live *Nautilus*. *In* W. B. Saunders, and N. H. Landman (editors), *Nautilus: The Biology and Paleobiology of a Living Fossil*, pp. 163–178. New York and London: Plenum Press.

Lange, W. 1941. Die Ammonitenfauna der *Psiloceras*-Stufe Norddeutschlands. *Palaeontographica A* **93**: 1–192.

Larson, N. L. 2003. Predation and pathologies in the Late Cretaceous ammonite family Scaphitidae. *Mid-America Paleontology Society Digest* **26**(3): 1–30.

Larson, P. L. 1984 (unpublished). *Shell Deformities in Scaphitids of the Upper Cretaceous Fox Hills Formation: A Statistical Analysis*. Hill City, South Dakota: Black Hills Institute of Geological Research.

Logan, A. J., P. A. Noble, and G. R. Webb. 1975. An unusual attachment of a recent brachiopod, Bay of Fundy, Canada. *Journal of Paleontology* **49**(3): 557–558.

Maeda, H., and A. Seilacher. 1996. Ammonoid taphonomy. *In* N. H. Landman, K. Tanabe, and R. A. Davis (editors), *Ammonoid Paleobiology*, pp. 544–578. New York and London: Plenum Press.

Meischner, D. 1968. *Placunopsis* as an epizoan of *Ceratites* (Palaeoecology, Mollusca). *Lethaia* **1**: 156–174.

Meledina, S. V. 1988. Callovian. *In* G. Y. Krymholts, M. S. Mesezhnikov, and G. E. G. Westermann (editors), *The Jurassic Ammonite Zones of the Soviet Union*. Geological Society of America Special Paper **223**: 33–39.

Oxford English Dictionary. 1989. 2nd edition. Oxford: Oxford University Press. OED Online. http://dictionary.oed.com/cgi/entry/50245036, retrieved 14 March, 2007

Parker, S. B. (editor). 1994. *Dictionary of Scientific and Technical Terms*. 5th edition. New York: McGraw-Hill.

Reineck, H. E., and I. B. Singh. 1980. *Depositional Sedimentary Environments, with Reference to Terrigenous Clastics*, 2nd Edition. New York and Berlin: Springer.

Seilacher, A. 1960. Epizoans as a key to ammoniod ecology. *Journal of Paleontology* **34**(1): 189–193.

Seilacher, A. 1982. Ammonite shells as habitats in the Posidonia Shale of Holzmaden – floats or benthic islands? *Neues Jahrbuch fuer Geologie und Paläontologie Monatshefte*. **1982**: 98–114.

Seilacher, A., and H. Keupp. 2000. Wie sind Ammoniten geschwommen? *Fossilien* **5**: 310–313.

Schindewolf, O. H. 1936. Über Epöken auf Cephalopoden-Gehäusen. *Paläontologische Zeitschrift* **16**: 15–31.

Sowerby, J. 1812–1829. *The Mineral Conchology of Great Britain*. London: Meredith.

Ward, P. 1987. *The Natural History of Nautilus*. Boston: Allen & Unwin.

Chapter 17
Biogeography of Kutch Ammonites During the Latest Jurassic (Tithonian) and a Global Paleobiogeographic Overview

Subhendu Bardhan,[1] Sabyasachi Shome,[2] and Pinaki Roy[1]

[1]Department of Geological Sciences, Jadavpur University, Kolkata 700032, India, s_bardhan01@yahoo.co.uk;

[2]Geological Survey of India, 15A&B Kyd Street, Kolkata 700016, India

Keywords: biogeography, Jurassic, ammonites, India, Kutch

1 Introduction

The Late Tithonian ammonite-bearing horizons have very limited geographic occurrence in Kutch, being restricted only to the western part of the mainland. A 15 m thick sequence consisting mainly of oolite-shale alternations and coarse grained sandstone yields the terminal Tithonian faunal assemblages. The previous comprehensive report comes from Spath's work (1927–1933) where he described some Early Tithonian ammonites from farther east within the mainland, but the typical Upper Tithonian assemblages are restricted only to the westernmost part, i. e., towards the paleoshoreline (Fig. 17.1).

Spath described altogether seven genera of Late Tithonian age and these include mostly species of the genus *Virgatosphinctes* Uhlig. Stratigraphic knowledge of the Kutch ammonites was very poor during Spath's time and the collections were made mostly by other workers. Our repeated field investigations and systematic collections revealed that the Upper Tithonian assemblages are restricted mainly to the top part of the Umia Member (Mitra et al., 1979) (Fig. 17.2). The stratigraphic distribution of all species and genera described by Spath (1927–1933) has been firmly established on the basis of additional collection. In addition, our recent collection has unearthed many other genera of various

N. H. Landman et al. (eds.), *Cephalopods Present and Past: New Insights and Fresh Perspectives*, 375–395.

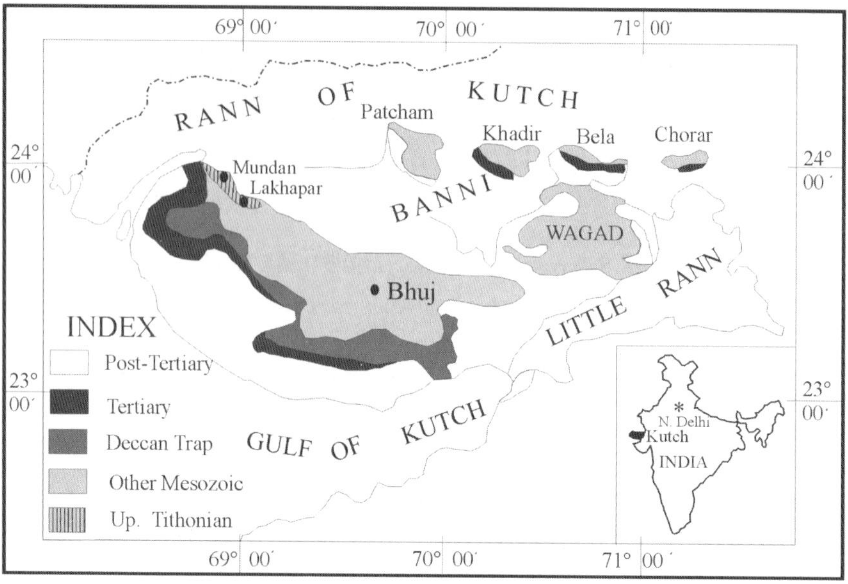

Fig. 17.1 *Geological map of Kutch. Note small aerial extent of the Upper Tithonian beds.*

provinces and different levels within the Upper Tithonian. These include *Durangites* Burckhardt, *Corongoceras* Spath, *Tithopeltoceras* Arkell, *Himalayites* Uhlig, and *Pterolytoceras* Spath (Shome et al., 2004; Shome et al., 2005; Shome and Roy, 2006). *Durangites* and *Micracanthoceras micracanthum* (Oppel) are the zonal indices of tropical Europe (see Cecca, 1999). It now appears that the Upper Tithonian assemblage in Kutch is represented by at least ammonites of three Mediterranean zones (cf. Tavera et al., 1986; Cecca, 1999).

The state-of-the-art of the Upper Tithonian ammonite assemblages of the world has been reevaluated in the light of new data and the nature of endemism and diversity patterns have been studied in the present endeavor. The Jurassic–Cretaceous boundary arguably marks a mass extinction event and ammonites evidently show a high extinction signal. We believe it is critical to understand the biogeography of the Late Tithonian ammonites in order to evaluate the patterns and extent of the end-Jurassic extinction event.

2 Upper Tithonian Assemblages of Different Faunal Provinces

The paleobiogeographic history of Mesozoic ammonites is marked by the repeated alternation of high degrees of provincialism and cosmopolitanism. The peak of ammonite provincialism occurred during the Late Jurassic and Early Cretaceous (Gordon, 1976). Different names were given for the Upper Jurassic stages in different regions because of this high endemism, i.e., Tithonian in the Tethyan and Indo-

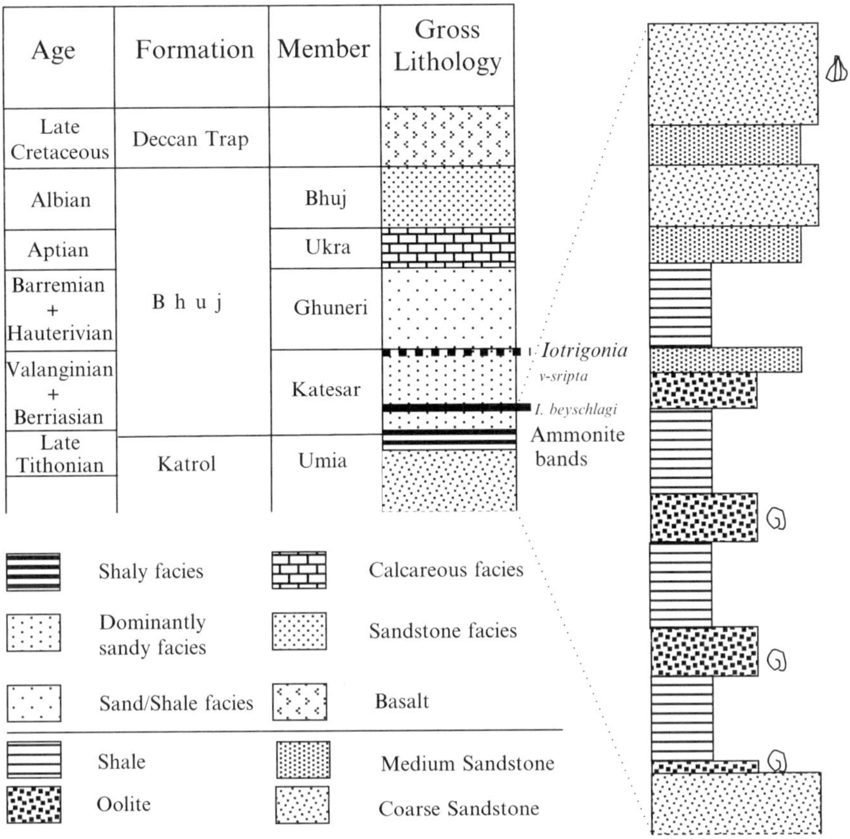

Fig. 17.2 *Stratigraphic section showing the position of the Late Tithonian ammonites in Kutch (vertical thickness not to scale).*

Pacific realms, Portlandian in southern England and Paris, and Volgian in Russia. Even within provinces, subprovinces can still be recognized, e.g., in the Mediterranean (see Cecca, 1999).

Because of the endemism of Tithonian ammonites, interprovincial correlation is a challenge and forms the focus of recent paleobiogeographic research (see references below). This has greatly improved our understanding of Tithonian ammonites, their paleobiogeography, oceanographic barriers, and migrational routes. Paleobiogeographic terminology is still in a state of flux and suffers from nomenclatural chaos because of the subjectivity involved (see for details Westermann, 2000a, b; Cecca and Westermann, 2003) and inherent complexity of the paleobiogeographical patterns (Dommergues, 2005, personal communication). We here try to follow the paleobiogeographic classification used by the leading workers (Riccardi, 1991; Enay and Cariou, 1997, 1999; Cecca, 1999, 2000; Zakharov and Rogov, 2003) of Tithonian ammonites and include the standardized terminology as suggested by Westermann (2000a, b; 2005, personal communication) and Cecca and Westermann (2003) within parentheses. Following is a discussion of recent research results on the occurrences of Late Tithonian genera ("leiostracans" excluded) in different non-Boreal provinces/regions. A summary of results is presented in Table 17.1.

Table 17.1 *List of Late Tithonian ammonite genera in different non-Boreal provinces/realms.* x = *present;* * = *genus survived the Jurassic–Cretaceous boundary.*

	Indo-Madagascar	Mediterranean	Andean	Caribbean	Austral (Indo-Pacific)
Anavirgatites	x	–	–	–	–
Andalusphinctes	–	x	–	–	–
Aspidoceras	x	x	x	x	x
Aulacosphinctes	x	x	x	x	x
Aulacosphinctoides	x	–	–	–	x
Baeticoceras	–	x	–	–	–
*Berriasella**	x	x	x	x	x
Blanfordiceras	x	–	x	–	x
*Bochianites**	–	x	–	x	–
Corongoceras	x	x	x	x	x
Curdubiceras	–	x	–	–	–
*Dalmasiceras**	–	x	–	–	–
Danubisphinctes	–	x	–	–	–
Dickersonia	–	–	x	x	–
Djurjuriceras	x	x	x	–	–
Durangites	x	x	–	x	–
*Haploceras**	x	x	–	x	–
Hemisimoceras	x	x	–	–	–
Hemispiticeras	–	–	x	–	–
Hildoglochiceras	x	–	–	–	–
Himalayites	x	x	x	–	x
Kossmatia	–	x	–	x	x
Lamencia	–	x	–	–	–
Lytogyroceras	–	x	–	–	–
*Lytohoplites**	x	–	x	x	x
*Malbosiceras**	–	x	–	–	–
Mazaplites	–	–	–	x	–
Micracanthoceras	x	x	x	x	–
Moravisphinctes	–	x	–	–	–
Nebrodites	x	–	–	–	–
*Negreliceras**	–	x	x	x	–
*Neolissoceras**	x	x	–	–	–
Neoperisphinctes	–	x	–	–	–
Oloriziceras	–	x	–	–	–
Oxylenticeras	–	x	–	–	–
Paraulacosphinctes	–	x	–	–	–
Paraboliceras	–	–	–	–	x
Parapalliceras	–	x	–	–	–

(continued)

Table 17.1 (continued)

Parabilliceratoides	–	–	–	–	x
Parodontoceras	–	–	x	x	–
Pectinatites	–	–	x	–	–
Phanerostephanus	x	–	–	–	–
Proniceras	x	x	–	–	x
Protacanthodiscus	x	x	x	–	–
Protancyloceras	x	–	–	–	–
Pseudoargentiniceras	–	x	–	–	–
Pseudodiscosphinctes	–	x	–	–	–
Pseudoinvoluticeras	x	–	x	–	–
Pseudolissoceras	–	–	x	x	x
Salinites	–	–	–	x	–
Schaiveria	–	x	–	–	–
Simoceras	–	–	x	–	–
Simolytoceras	–	x	–	–	–
Simosphinctes	–	x	–	–	–
Simplisphinctes	–	x	–	–	–
Spiticeras[*]	x	–	x	–	x
Suarites	–	–	–	x	–
Subalpinites[*]	–	x	–	–	–
Subdichotomoceras	x	x	–	–	x
Substeueroceras	–	x	x	x	x
Substreblites[*]	–	x	–	–	–
Subthurmannia[*]	–	x	–	–	–
Tithopeltoceras	x	x	–	–	–
Umiaites	x	–	–	–	–
Vinalesites	–	–	x	x	–
Virgatosphinctes	x	–	x	–	x
Wichmanniceras	–	–	x	–	–
Windhauseniceras	–	–	x	–	–
Zittelia	–	x	–	–	–

The Indo-Madagascan Province was well established since the Middle Jurassic (Arkell, 1956). During the Tithonian, the most fossiliferous regions are Kutch (Spath, 1927–1933), Baluchistan (Fatmi, 1972; Fatmi and Zeiss, 1994), and Madagascar (Collignon, 1960). The other areas included Somalia and Tanzania, from which only a few genera have been reported (see Enay and Cariou, 1997; Cecca, 1999). The Himalayan faunas, which show "sub-austral" affinity (Indo-Pacific of Westermann, 2000a, b) (see Enay and Cariou, 1997), are not included in this province. Previously it was believed (e.g., Cecca, 1999) that the Indo-Madagascan Late Tithonian ammonites were less diverse. The recent discoveries of several new ammonite genera from Kutch now reveal that this province is the second most diverse (see Table 17.1). Kutch, northwest Pakistan, and Madagascar together include 27 genera (Table 17.2) under four families: Olcostephanidae, Neocomitidae, Himalayitidae, and Aspidoceratidae.

Like other provinces, the Indo-Madagascan Province initially also shows well-marked endemism. The Early Tithonian is characterized mainly by many species of

Table 17.2 *Late Tithonian ammonite genera present in three different regions of the Indo–Madagascan Province. Symbols are same as in Table 1.*

	Kutch	Madagascar	NW Pakistan
Anavirgatites	–	x	–
Aspidoceras	–	–	x
Aulacosphinctes	x	x	x
Aulacosphinctoid	x	x	x
*Berriasella**	–	x	–
Blandfordiceras	–	–	x
Corongoceras	x	x	–
Djurdjuriceras	–	x	–
Durangites	x	–	–
*Haploceras**	x	–	x
Hemisimoceras	–	x	–
Hildoglochiceras	–	x	x
Himalayites	x	x	x
*Lytohoplites**	–	x	–
Micracanthoceras	x	x	x
Nebrodites	x	–	–
*Neolissoceras**	–	x	x
Phanerostephanus	–	x	–
Proniceras	–	x	x
Protacanthodiscus	–	–	x
*Protancyloceras**	–	x	–
*Spiticeras**	–	x	x
Subdichotomoceras	–	–	x
Tithopeltoceras	x	–	–
Umiaites	x	–	–
Virgatosphinctes	x	x	x

Virgatosphinctes (see Spath, 1927–1933). Endemism is so pronounced that even some leiostracans (e.g., *Pterolytoceras*) were exclusive to this province and Spiti Himalayas (Shome and Roy, 2006). Endemism became somewhat blurred, however, due to the invasion of many cosmopolitan forms during the Late Tithonian marine transgression (Haq et al., 1987; Hallam, 1992). The subfamily Virgatosphinctinae continued, but younger forms included a new stock having large body size (authors' personal observation).

In the Mediterranean Province, Late Tithonian ammonite assemblages are known only from a few areas in southern Spain, Italy, and France (see Cecca, 1999). Like Kutch, this province also experienced considerable reduction of the habitat due to regression, especially in the southern European platform and epicontinental basins (Fourcade et al., 1991; Cecca, 1999). However, southern Spain is highly fossiliferous and includes 33 genera (see Tavera, 1985). This is also the highest level of ammonite diversity in any region of the world during the Late Tithonian.

The Mediterranean assemblage was dominated by genera belonging to mainly three families. "Perisphinctidae" showed renewed diversity (Tavera, 1985) and included 11 genera. The "Berriasellidae" were also equally diverse with 11 genera (Tavera, 1985). Additionally, Simoceratidae, which is typical of the Mediterranean facies during the Early Tithonian, persisted. Olcostephanidae and Opellidae were represented by sparse genera.

Leanza (1981) and Riccardi (1991) have analyzed the Upper Tithonian ammonite assemblages of the Andean Province. Three main fossiliferous areas are north-western South America, central-western South America, and southern South America. Their faunas are fairly homogeneous, but the similarity coefficient index varies from place to place (Riccardi, 1991). Each area includes Andean endemic genera which help in establishing intraprovincial correlation. Altogether, 25 genera have been recognized in this province. Assemblages in southern South America are slightly distinct in the sense that they show closer affinity with "Austral" New Zealand and "Ethiopean" (=Indo-Madagascan, described here) Madagascar. Even at the species level, this affinity is evident especially during the Late Tithonian (see below).

The Caribbean Province was first proposed by Westermann (1984) and Taylor et al. (1984) who used the term for the Middle Jurassic assemblages. On the basis of ammonite assemblages, Myczyńsky and Pszczolkowski (1994) suggested that this province was also well established during Late Jurassic and Early Cretaceous times. The province includes faunas of Cuba and Mexico.

The Tithonian ammonites of Cuba are moderately diverse. Myczyńsky and Pszczolkowski (1994) reported 13 genera from eight families. They are dominated by the cosmopolitan forms and some endemic taxa at the species level (see Cecca, 1999). The assemblages show mixed provincial characters dominated by Mediterranean faunas. However, geographically wide ranging genera like *Corongoceras*, *Himalayites*, *Durangites*, and *Aulacosphinctes* Uhlig, which are now described from Kutch (Shome et al., 2004), are also present. Myczyńsky (1989) mentioned many species of *Virgatosphinctes* including *V. denseplicatus*, which is a typical Indo-Madagascan form. However, Cecca (1999), and previously Enay (1972) and Callomon (in Hillebrandt et al., 1992), doubted the generic affiliation of this Cuban form.

In Mexico, ammonite assemblages show a fair degree of homogeneity (Imlay, 1939; Verma and Westermann, 1973; Olóriz et al., 1999) with the Cuban taxa. Verma and Westermann (1973) described altogether nine genera, including *Kossmatia* Uhlig, a genus not found in Cuba but typical of the "Austral Province" (Indo-Pacific Realm) (Enay and Cariou, 1997).

The Austral Realm (Indo-Pacific) covers east peri-Gondwanan areas including Antarctica (Thomson, 1980, 1982; Riccardi, 1991), New Zealand (Stevens, 1997), and Papua New Guinea (Indonesia) (Riccardi, 1991) and extends up to the Himalayas (see Enay and Cariou, 1997; Westermann, 2000b). This is a less diverse region as far as the Late Tithonian ammonites (14 genera) are concerned. Other than the Himalayas, Tithonian ammonites are poorly known from the rest of the areas. Recently, Enay and Cariou (1997, 1999) reported seven genera under three families, i.e., Olcostephanidae, Neocomitidae, and Himalayitidae from Spiti in India and the Nepal Himalayas. They mentioned that Himalayan forms are characterized by endemic taxa, which show some degree of Austral affinity. The family Neocomitidae dominates the assemblage and includes three genera, *Berriasella* Uhlig, *Blanfordiceras* Cossmann, and *Corongoceras*.

Panboreal Superrealm ammonites are high latitude faunas living in the areas north of 45–50° N (Westermann, 2000a; Zakharov and Rogov, 2003). This belt was well established since the Middle Jurassic and correlation, therefore, between the Volgian and Tithonian is still a problem. However, during the Late Jurassic, Boreal–Tethyan ammonites show latitudinal shifts time and again, and thermally anomalous assemblages have been found in many Boreal areas indicating a northward migration of several Tethyan genera (see Zakharov and Rogov, 2003; and references therein).

The composition of Boreal ammonite assemblages during the latest Jurassic differs greatly from the contemporary Kutch faunas, which are essentially Tethyan in affinity. We follow Zakharov and Rogov (2003) in dividing the Panboreal Superrealm into four realms, Boreal West Europe (Northwestern Europe), Boreal East Europe (Volga Basin and Russian Platform), Boreal Pacific (Chukotka–Canadian Province and Primorje), and Boreal Eastern Pacific (British Columbia and California). We present the boreal data in a later section, to show diversity and extinction intensity. The genera characteristic of these regions are listed in Table 17.3 (Arkell, 1956; Arkell et al., 1957; Wright et al., 1996; Cecca, 1999; and Zakharov and Rogov, 2003; and references therein).

3 Affinity of Kutch Assemblage

It is already mentioned that the Upper Tithonian assemblages of Kutch can be correlated with the three biozones of the Late Tithonian of the Mediterranean Province. The lower assemblage in Kutch, however, is quite distinct and includes mostly virgatosphinctinid genera, especially the *Virgatosphinctes denseplicatus* group of species (Spath, 1927–1933; Krishna et al., 1996). *Virgatosphinctes denseplicatus* is typically restricted to peri-Gondwanan regions. In the upper part, *Virgatosphinctes* spp. continues, but they show evolutionary changes, becoming large and coarsely and distantly ornate. This kind of evolution has also been noticed in Spiti and Nepal (see Enay and Cariou, 1997), Baluchistan (Fatmi, 1972), and also in Madagascar (Collignon, 1960).

The Kutch assemblage also contains other faunas which show complex paleobiogeographic affinities. At the generic level, it shows affinity to various provinces. For example, *Tithopeltoceras*, which is only found in a restricted belt (30° N) within the Mediterranean Province has recently been described from Kutch (Shome et al., 2005). In addition, *Durangites*, which is found dominantly in the Mediterranean Province (Tavera, 1985; Cecca, 1999) and Mexico (Imlay, 1939; Verma and Westermann, 1973), has also been recorded from Kutch (Shome et al., 2004), while the austral *Himalayites*, is now found to have extended up to the Kutch basin (authors' personal observation). *Corongoceras* and *Micracanthoceras*

Table 17.3 *Late Tithonian ammonite genera present in different Boreal realms. Symbols are same as in Table 1.*

	Boreal West Europe	Boreal East Europe	Boreal Pacific	Boreal Eastern Pacific
Aulacosphinctes	–	x	x	–
*Berriasella**	x	x	x	x
*Bochianites**	x	x	–	x
Chetaites	–	–	x	–
Craspedites	–	x	–	–
*Cyrtosiceras**	–	x	–	–
*Dalmasiceras**	–	x	x	x
Danubisphinctes	x	–	–	–
Dorsoplanites	–	x	x	–
Durangites	–	x	–	–
Glochiceras	–	x	–	–
*Graniericeras**	–	x	–	–
*Haploceras**	x	x	–	–
Kachpurites	–	x	–	–
Kossmatia	–	–	–	x
Lamencia	–	x	–	–
Lomonossovella	–	x	–	–
*Negreliceras**	–	–	–	x
Notostephanus	–	–	–	x
Praechetaites	–	x	–	–
Proniceras	–	–	x	x
*Protancyloceras**	–	x	–	–
Pseudovirgatites	x	–	–	–
*Spiticeras**	x	x	–	x
*Subcraspedites**	–	–	x	–
Substeueroceras	–	–	–	x
*Substreblites**	–	–	–	x
Sutneria	–	x	–	–
Titanites	–	–	–	x
Zaraiskites	x	–	–	–

Spath are cosmopolitan genera and they also have been found in Kutch (Spath, 1927–1933; Shome et al., 2004). *Pterolytoceras*, a leiostracan, has recently been recovered from Kutch (Shome and Roy, 2006). This genus was earlier reported from Tithonian beds of Madagascar, northwestern Pakistan, and Spiti Himalayas (Collignon, 1960; Fatmi, 1972; Krishna et al., 1982). It is found in shallow water sediments with other genera in Kutch. The restricted paleobiogeographic distribution of *Pterolytoceras* also suggests its shallow water habitat (see Shome and Roy, 2006). It is endemic, short lived, and restricted only to the Indo-Madagascan Province.

The high faunal diversity of Kutch is compatible with the sea level rise during the Late Tithonian (see Haq et al., 1987; Hallam, 1992), which facilitated some

Himalayatidae also participated in this "bioevent par excellence." *Blanfordiceras* and *Spiticeras*, the two typical genera of the Himalayas, are also found in Baluchistan, Madagascar, and in the Andean Province. But they are still absent in Kutch, which we believe is perhaps due to collection failure. *Himalayites*, another typical Himalayan form is now known to extend up to Kutch.

Spath (1927–1933) described a new endemic genus *Umiates* Spath from Kutch and reported two species, *U. rajnathi* Spath and *U. minor* Spath. We have recently collected additional specimens and studied Spath's type material kept in the repository of the Geological Survey of India in Kolkata. His two species appear to be conspecific and *Umiates* specimens are large, septate, and macroconchs. Detailed taxonomic analysis (to be published elsewhere) reveals that *Umiates* is very close to microconchiate *Proniceras* Burckhardt and they may form a dimorphic pair. During the Late Tithonian, *Proniceras* had a paleobiogeographic distribution in the Subbetic Mediterranean (Tavera, 1985), Mexico (Imlay, 1939), Baluchistan (Fatmi, 1972; Fatmi and Zeiss, 1994), Nepal Himalayas (Enay and Cariou, 1997), and Madagascar (Collignon, 1960). Thus, like many other forms, it also showed southward migration during the Late Tithonian.

Tithopeltoceras was reported only from the Mediterranean Province until recently (Shome et al., 2005). It is a "strange" genus (Enay, 1973; Olóriz and Tavera, 1979; Tavera, 1985), even within this province. It has restricted occurrences in the Micracanthum Zone of southern Spain, central Italy, and the Balearic islands and thus is distributed in a very narrow zone coinciding with the 30°N paleolatitude. This narrow latitudinal distribution was perhaps constrained by temperature. *Tithopeltoceras* has now been found in Kutch and significantly, the paleoposition of Kutch during the Tithonian was also 30°, but south of the equator (Shome et al., 2005). Other genera of Himalayatinae had a wide biogeographic spread (Olóriz and Tavera, 1989; Cecca, 1999). *Tithopeltoceras* perhaps being stenothermal was restricted mainly in the subtropic zone, on either side of the equator.

5 Paleobiogeography of Mass Extinction

Study of the diversity, distribution patterns, and endemism of the Late Tithonian ammonites is especially important because soon, at the end-Tithonian, the majority of them died. Arguably, the end of the Jurassic marks a mass extinction event (Raup and Sepkoski, 1986; but for an opposite view, see Hallam, 1986; and Hallam and Wignall, 1997). Hallam (1986) claimed that family level extinction data of ammonites are not impressive, but he admitted that taxonomic resolution down to lower level along with stratigraphic precision can only help in understanding the extent of extinction. Since then, floods of information on ammonite diversity, paleobiogeography, and refinements of stratigraphic ranges across the Jurassic–Cretaceous boundary have emerged (e.g., see Fatmi and Zeiss, 1994; Tavera et al., 1986; Riccardi, 1991; Enay and Cariou, 1997; Cecca, 1999; Olóriz et al., 1999;

Zakharov and Rogov, 2003; and many others) and a new treatise on Cretaceous ammonites has appeared (Wright et al., 1996). Literature scanning reveals that out of 84 ammonite genera (excluding leiostracans) of the Late Tithonian, 67 did not pass the Jurassic–Cretaceous boundary, thus showing a significant extinction signal (80%). Hallam (1986) tried to show that the end-Jurassic extinction event is of moderate scale and regional in nature. But many other workers also have investigated the extinctions of other organisms during the end-Tithonian and found significant extinction peaks. For example, Kelly (1984–92) analyzed bivalve data of England and Zakharov and Yanine (1975) also demonstrated extinction on the basis of Central Russian bivalves. In both areas, bivalve diversity was reduced significantly. Ager (1975, in Hallam, 1986), Davis (1975, in Hallam, 1986) and Sandy (1988) attempted to assess brachiopod mortality across the Jurassic–Cretaceous boundary, while others studied foraminifers (e.g., Jenkins and Murray, 1981), coccoliths (Hamilton, 1982) and ostracods (Ascoli et al., 1984; Whatley, 1988). All of them noted a major turn over of taxa at the boundary. More significantly, like the K–T boundary extinction, the Jurassic–Cretaceous extinction event also affected terrestrial and taxonomically different biota. Bakker (1993) recognized near-total extinction of plesiosaur reptiles and Benton (1995) claimed greater mortality of continental faunas at family level than of marine groups at the end of the Tithonian. It would be interesting to know, therefore, the biogeographic distribution patterns of the Late Tithonian ammonites which may, in turn, help in understanding the nature and causes of the extinction (also see Raup and Jablonski, 1993).

Late Tithonian ammonite genera have wide geographic distribution. Longitudinally, they are found from the boreal eastern Pacific to Austral Indonesia and the latitudinal distribution ranges from the Arctic regions in the north to Antarctica in the southern hemisphere. There are at least 60 major fossil occurrences in the world from where different assemblages of Late Tithonian ammonites have been recorded (see Cecca, 1999; Zakharov and Rogov, 2003). Since the number of genera in many localities is few and the Late Jurassic ammonites overall show strong endemism, we have plotted diversity data on a provincial, subprovincial, and regional basis (Fig. 17.3). Geographic and stratigraphic distributions have been drawn from the literature cited above and are also subjected to an updating and evaluation procedure. Stratigraphic ranges of some genera have been shown to be different in different literature, but we follow the data provided by the new Treatise (Wright et al., 1996).

The distribution of ammonite genera shows a broad latitudinal gradient with diversity decreasing towards at least the South Polar Region (see also Tables 17.1–3). Diversity is maximum in the subtropics on either side of the equator. The Mediterranean diversity is represented by 44 genera and in the south, the Indo-Madagascan Province and the Himalayas include 29 genera (see Fig. 17.3 and Tables 17.1–2). Similar trends of different taxa also existed during other mass extinction events such as the K–T (see Stanley, 1984, 1988; Jablonski, 1986; Zinsmeister et al., 1989). High latitude Tithonian ammonites were more diverse in some provinces in the northern hemisphere than in the southern hemisphere. This is due to latitudinal shifts towards the north shown by many Tethyan genera during the Late Tithonian (see Zakharov

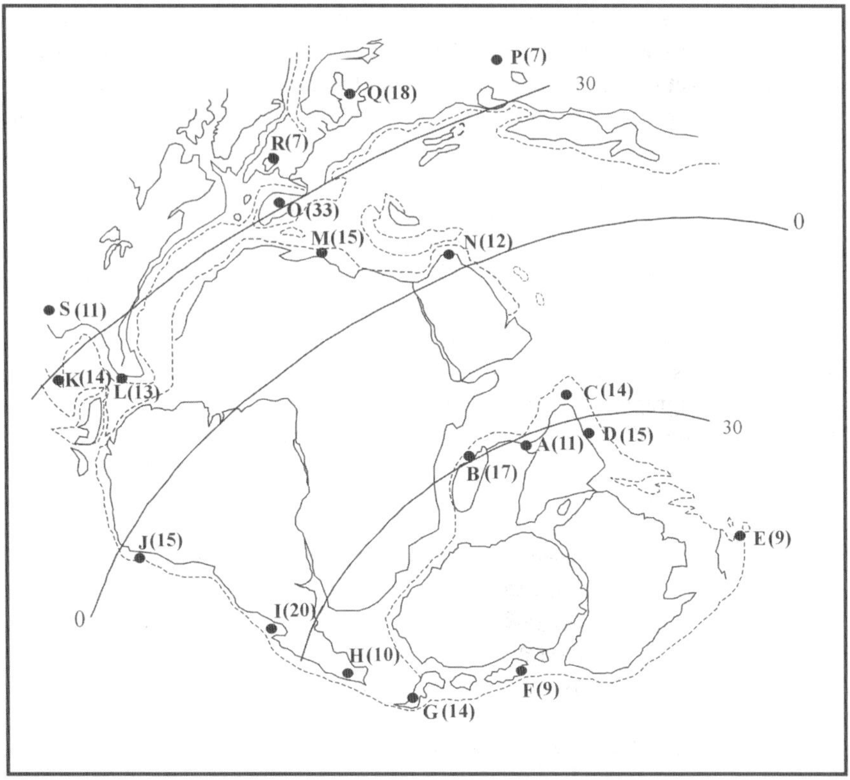

Fig. 17.3 *Diverisity of ammonite genera in different regions during the Late Tithonian. The base map is modified after Enay and Cariou (1997). A = Kutch; B = Madagascar; C = Northwest Pakistan; D = Himalayas; E = Indonesia; F = New Zealand; G = Antarctica; H = Southern South America; I = Central-West South America; J = Northwestern South America; K= Mexico; L= Cuba; M = North Africa; N = Iraq; O = Southern Spain; P = Boreal Pacific, shown at Primorje; Q = Boreal East Europe, shown at Crimea; R = Boreal West Europe, shown at England; S = Boreal Eastern Pacific, California. Sources are mentioned in the text.*

and Rogov, 2003; and references therein). Climatic equability during the entire Jurassic was envisaged by Hallam (1969). Since the Arctic taxa are not found south of 30° N, Zakharov and Rogov (2003) believed in the existence of a temperature gradient during the Late Jurassic – Early Cretaceous period. Jeletzky (1984) also suggested a cooling event during the terminal Tithonian on the eastern coast of the Paleopacific. In addition to fossil data, recent general circulation model simulations of the Jurassic climate also reveal latitudinal thermal gradients at certain times and a semi-arid climate in many tropical areas especially during the Late Jurassic – Early Cretaceous (Page, 2005). Our study on *Tithopeltoceras* also shows latitudinal control on ammonite distribution (Shome et al., 2005). Surprisingly, high latitude southern hemisphere Austral faunas are marked by low endemism (12%).

The Majority of Late Tithonian ammonite genera became extinct at the Jurassic–Cretaceous boundary (80% as mentioned earlier). We intend here to study the extinction intensity of ammonites in different areas and to know whether there is any geographic control on extinction intensity. During the K–T extinction, some workers believe (e.g., Copper, 1977; Stanley, 1988; Banerjee and Boyajion, 1996; but see Raup and Jablonski, 1993, for a different view), that tropical areas were more severely affected than temperate or polar regions. This is envisaged because tropical taxa show greater endemism than cold water forms (Jablonski, 1986). We have followed Raup and Jablonski (1993) in choosing the extinction metric for our analysis. According to them "extinction was quantified as the proportion of genera found in an assemblage, or local group of assemblages, that suffered global extinction in the final stage" This is adopted here mainly because of the rarity of truly continuous stratigraphic sections across the Jurassic–Cretaceous boundary, and direct comparison of assemblages immediately below and above the boundary would result in only a limited analysis. However, unlike Raup and Jablonski (1993), we have included only Late Tithonian genera and thus have avoided genera which became extinct during the Early Tithonian (cf. Hallam, 1986). Otherwise, extinction

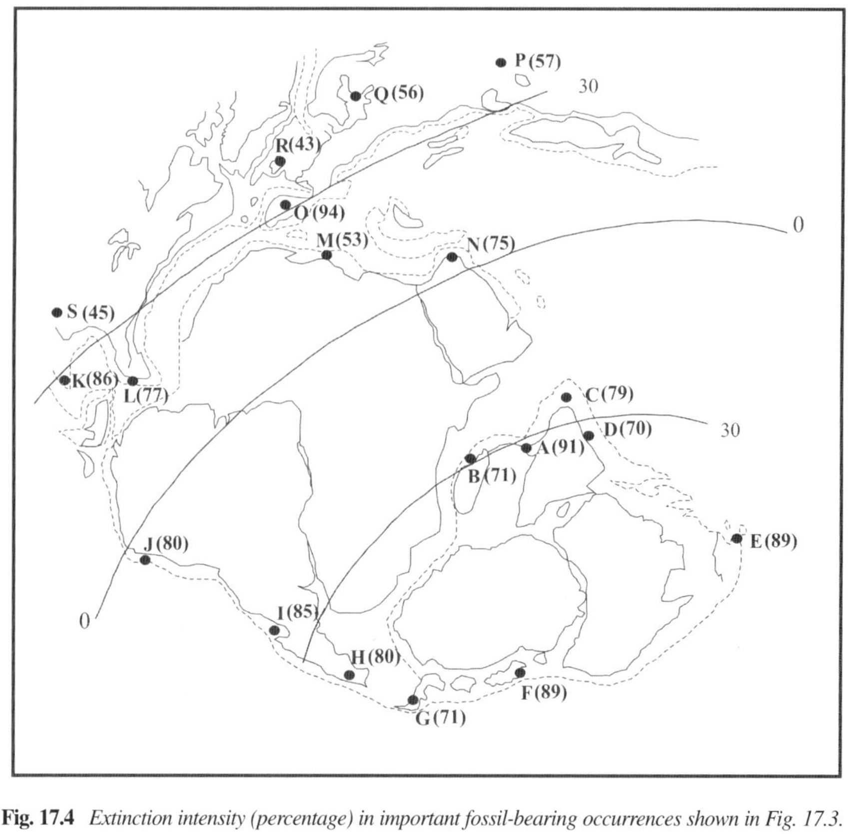

Fig. 17.4 *Extinction intensity (percentage) in important fossil-bearing occurrences shown in Fig. 17.3.*

values at the Jurassic–Cretaceous boundary would be artificially elevated. Extinction percentages of different provinces/regions are plotted in Fig. 17.4.

Extinction appears to be high and uniform throughout the world except in the Panboreal Superrealm. Here, the mean extinction is 50%. This is mainly for two reasons. First, some Boreal realms have low diversity (see Fig. 17.3). Second, the Panboreal Superrealm was invaded by many Tethyan genera during the Late Tithonian (see Table 17.3), some of which survived the Jurassic–Cretaceous boundary crisis (for details see below). In the boreal west Pacific Realm, diversity is less; of seven genera, three are Tethyan forms which crossed the Jurassic–Cretaceous boundary, but surviving genera are inevitably pooled in the global data, thus decreasing the regional extinction percentage. The majority of Boreal endemics (12 out of 15, i.e., 80%) as well as all immigrant genera died out in these areas during the J–K transition. If the Tethyan genera are removed from the count, the "sanctuaries" of extinction in northern higher latitudes disappear and the mean extinction becomes 86%. Thus, extinction patterns show no gradient across either latitude or longitude. The mean extinction is 73% and the median is 77%. This indicates a quite high extinction signal all over the world.

Kutch and the tropical Mediterranean southern Spain have higher extinction percentages (91% and 94%, respectively) and can be considered as "hot spots" of extinction. In Kutch, although the lowermost Cretaceous includes marine strata, the basin became extremely shallow (Biswas, 1977). Few gastropod and bivalve genera survived the Jurassic–Cretaceous boundary, but they also show species level turnover (Das, 2002; authors' personal observation). All ammonite genera including leiostracans suffered regional extermination (Bardhan et al., 1989). Admittedly, the present extinction metric has the disadvantage that it fails to locate a local hot spot because the genera that died in that area may have survived elsewhere. In Kutch it so happened that all the genera also went extinct globally, so that this local hot spot was not hidden by this effect. This is perhaps due to the fact that the Kutch Sea harbored shallow water faunas which suffered most during the extinction.

The Mediterranean Province has an unusual number of endemics. A total of 14 of the 33 genera in Spain are endemic (42%) and none of them crossed the J–K boundary. This has greatly increased the extinction value. Kutch and southern Spain were located in the subtropical latitudes, but we do not believe that these areas were worst hit during the extinction. The high latitude Austral faunas (in spite of low endemism) have higher extinction values than other tropical areas.

Analysis of 17 genera surviving the extinction indicates that they belong to six families out of nine which persisted up to the Late Tithonian. Neocomitidae were most diverse (14 genera at the J–K boundary of which seven genera entered into the Cretaceous). Unfortunately, we do not have at present the species level database of all genera and, therefore, cannot tell whether any selectivity of survival of species-rich genera existed. However, three successful genera, i.e., *Berriasella*, *Lytohoplites*, and *Spiticeras* were diverse during the Late Tithonian (see Collignon, 1960; Tavera, 1985).

The surviving tropical genera quickly diversified immediately after the boundary in the earliest Berriasian and belonged to two important families (Neocomitidae and Olcostephanidae) of the Early Cretaceous (see Wright et al., 1996, Table 17.2). It is tempting to suggest that like background extinvction, cosmopolitan genera are resistant to mass extinction (see Jablonski, 1989) at the Jurassic–Cretaceous boundary. However, 13 of the 19 cosmopolitan genera (with at least distribution in more than two provinces, see Table 17.1) went extinct. One selective element, however, becomes apparent; of 15 Tethyan genera which show climatic shifts towards northern high latitudes (see Table 17.3), eight survived the J–K boundary crisis. This perhaps suggests their ability to maintain range expansion in times of environmental stress. We have excluded leiostracan data from our analysis. Leiostracans have a higher rate of survival (only 30% became extinct) and were present in the Early Cretaceous both in the tropical and extratropical areas (Arkell et al., 1957; Wright et al., 1996; and Shome and Roy, 2006). Leiostracans, which are believed to be offshore dwellers (not necessarily deeper water, see Kennedy and Cobban, 1976; Bardhan et al., 1993), have lower extinction rates throughout the Jurassic (House and Senior, 1981).

The Late Tithonian ammonites are in general less diverse or poorly known from the whole Mediterranean Tethys (Cecca, 1999). This is due to large-scale destruction of the southern European platform and epicontinental habitat attributed to marine regression (Fourcade et al., 1991). But in southern Spain the situation was quite different. Here the ammonites were the most diverse in the world (33 genera, see Table 1, and Tavera et al., 1986) and belong to the Ammonitico Rosso facies, which represented the pelagic, epioceanic environments including submarine highs (see Cecca, 1992; Cecca, 1999). This area is one of the rare sites where continuous sections across the Jurassic–Cretaceous boundary and the earliest Cretaceous ammonite assemblages are found (Tavera, 1985; Cecca, 1998). *Berriasella,* which suffered regional extinction elsewhere, crossed the boundary here and diversified quickly afterwards (see Tavera, 1985; Tvera et al., 1986). There is a general agreement that the Jurassic–Cretaceous boundary interval witnessed a major sea level lowering and regression took place in all provinces (e.g., Cecca, 1999; Zakharov and Rogov, 2003). This may be one of the causul factors of large-scale extinction of the Jurassic ammonites and other groups. A few genera survived, perhaps due to chance, with areas like Subbetic Spain serving as an "island refugia" (see Kauffman and Erwin, 1995) where *Berriasella* and several other Mediterranean genera survived.

In conclusion, analysis of the biogeography of the end-Jurassic ammonites reveals that extinction is homogeneous throughout the world. It shows no selectivity by latitude (except a few extinction "hot spots") and geographic range (both endemic and pandemic groups suffered equally). Shallow water forms suffered most, while pelagic leiostracans have significantly lower extinction intensity, but this is due to low extinction rate in this group during the entire Jurassic. The Jurassic–Cretaceous ammonite extinction pattern is consistent with that observed in bivalves during the K–T mass extinction event (see Raup and Jablonski, 1993; Jablonski and Raup, 1995).

Acknowledgments

We are thankful to G. S. Roy for extending logistic support during fieldwork. The help rendered by the local administration and people of Lakhapar is also acknowledged. The Department of Science and Technology provided financial support for SB. SS wishes to thank the Director

General, Geological Survey of India for permitting him to do the work. Jim Haggart, Gerd Westermann, Jean-Louis Dommergues, and Neil Landman critically read the manuscript and provided valuable suggestions. Financial support was also provided to the third author by the Centre for Scientific and Industrial Research.

References

Arkell, W. J. 1956. *Jurassic Geology of the World*. Edinburgh and London: Oliver and Boyd.

Arkell, W. J., B. Kummel, and C. W. Wright. 1957. *Mesozoic Ammonoidea*, *In* R. C. Moore (editor), *Treatise on Invertebrate Paleontology, Part L, Mollusca 4*, pp. L80–L437. Lawrence, KS, and New York: University of Kansas Press and Geological Society of America.

Ascoli, P., C. W. Poag, and J. Remane. 1984. Microfossil zonation across the Jurassic-Cretaceous boundary on the Atlantic margins of North America. *Geological Association of Canada Special Paper* **21**: 31–48.

Bakker, R. T. 1993. Plesiosaur extinction cycles-events that mark the beginning, middle and end of the Cretaceous. *Geological Association of Canada Special Paper* **21**: 641–664.

Banerjee, A., and G. Boyajion. 1996. Changing biologic selectivity of extinction in the Foraminifera over the past 150 m.y. *Geology* **24**: 607–610.

Bardhan, S., S. Shome, P. K. Bose, and G. Ghose. 1989. Faunal crisis and marine regression across the Jurassic-Cretaceous boundary in Kutch, India. *Mesozoic Research* **2**(1): 1–10.

Bardhan, S., S. K. Jana, and K. Datta. 1993. Preserved color pattern of a phylloceratid ammonoid from the Jurassic Chari Formation, Kutch, India, and its functional significance. *Journal of Paleontology* **67**: 140–143.

Benton, M. J. 1995. Diversification and extinction in the history of life. *Science* **268**: 52–58.

Biswas, S. K. 1977. Mesozoic rock stratigraphy of Kutch. *Quarterly Journal of the Geological Mineralogical and Metallurgical Society of India* **49**:1–52.

Campbell, C. A., and J. W. Valentine. 1977. Comparability of modern and ancient marine faunal provinces. *Paleobiology* **3**: 49–57.

Cecca, F. 1992. Ammonite habitats in the Early Tithonian Western Tethys. *Lethaia* **25**(3): 257–267.

Cecca, F. 1998. Early Cretaceous (pre-Aptian) ammonites of the Mediterranean Tethys: palaeoecology and palaeobiogeography, *Palaeogeography, Palaeoclimatology, and Palaeoecology* **138**: 305–323.

Cecca, F. 1999. Palaeobiogeography of Tethyan ammonites during the Tithonian (latest Jurassic). *Palaeogeography, Palaeoclimatology, and Palaeoecology* **147**: 1–37.

Cecca, F. 2002. *Palaeobiogeography of Marine Fossil Invertebrates*. London and New York: Taylor & Francis.

Cecca, F., and G. E. G. Westermann. 2003. Towards a guide to palaeobiogeographic classification. *Palaeogeography, Palaeoclimatology, and Palaeoecology* **201**: 179–181.

Collignon, M. 1960. *Atlas des fossiles caractéristiques de Madagascar*. Fascicule VI (Tithonique). *Réport. Malagache Survey. Geólogie*. Tananarive, pls. 134–175.

Copper, P. 1977. Paleolatitudes in the Devonian of Brazil and the Frasnian-Famenian mass extinction. *Palaeogeography, Palaeoclimatology, and Palaeoecology* **22**: 1–60.

Cox, L. R. 1940. The Jurassic lamellibranch fauna of Kachh (Cutch). *Palaeontologica Indica, Geological Survey of India IX* **3**(3): 1–157.

Das, S. S. 2002. Two new pleurotomariid (Gastropoda) species, including the largest *Bathrotomaria*, from the Berriasian (Early Cretaceous) of Kutch, western India. *Cretaceous Research* **23**: 99–109.

Das, S. S. 2003. New assemblage of the Mesozoic gastropod faunas of Kutch, western India – a study of systematics, palaeobiogeography and evolution. Unpublished Ph.D. thesis, Jadavpur University, 161p.

Enay, R. 1972. Paléobiogéographie des ammonites du Jurassique terminal (Tithonique/ Volgien/ Portlandien) et mobilité continentale. *Geobios* **5**(4): 355–407.

Enay, R. 1973. Upper Jurassic (Tithonian) ammonites. *In* A. Hallam (editor), *Atlas of Palaeobiogeography*, pp. 297–307. Amsterdam: Elsevier.

Enay, R., and E. Cariou. 1997. Ammonite faunas and palaeobiogeography of the Himalayan belt during the Jurassic: initiation of Late Jurassic austral ammonite fauna. *Palaeogeography, Palaeoclimatology, and Palaeoecology* **134**: 1–38.

Enay, R., and E. Cariou. 1999. Jurassic ammonite faunas from Nepal and their bearing on the palaeobiogeography of the Himalayan belt. *Journal of Asian Earth Science* **17**: 829–848.

Enay, R., and J. R. Geyssant. 1975. Faunes tithoniques des chaîns bétiques (Espagne méridionale). *Mémoires du Bureau de Recherches Géologiques et Minières* **86**: 39–55.

Fatmi, A. N. 1972. Stratigraphy of Jurassic and Lower Cretaceous rocks and Jurassic ammonites from northern areas of West Pakistan. *Bulletin British Museum (NH) Geology* **20**: 297–380.

Fatmi, A. N., and A. Zeiss. 1994. New Upper Jurassic and Lower Cretaceous (Berriasian) ammonite faunas from the Sember Formation ("Belemnite shales") of Southern Baluchistan, Pakistan. *Geobios Mémoire Spécial* **17**: 175–185.

Fourcade, E., J. Azéma, F. Cecca, M. Bonneau, B. Peybernès, and J. Decourt. 1991. Essai de reconstitution cartographique de la paléogéographie et des paléoenvironments de Téthys au Tithonique supérieur (138–135 Ma). *Bulletin de la Société géologique de France* **162**(6): 1197–1208.

Fürsich, F. T., and D. K. Pandey. 2003. Sequence stratigraphic significance of sedimentary cycles and shell concentrations in the Upper Jurassic-Lower Cretaceous of Kachchh, western India. *Palaeogeography, Palaeoclimatology, and Palaeoecology* **193**: 285–309.

Gordon, W. A. 1976. Ammonoid provincialism in space and time. *Journal of Paleontology* **50**(3): 521–535.

Hallam, A. 1969. Faunal Realms and facies in the Jurassic. *Palaeontology* **12**(1): 1–18.

Hallam, A. 1986. The Pliensbachian and Tithonian extinction events. *Nature* **319**: 765–768.

Hallam, A. 1992. *Phanerozoic Sea-level Changes*. New York: Columbia University Press.

Hallam, A., and P. B. Wignall. 1997. *Mass Extinctions and Their Aftermath*. Oxford: Oxford University Press.

Hamilton, G. B. 1982. Triassic and Jurassic calcareous nanofossils. *In* A. R. Lord (editor), *A Stratigraphic Index of Calcareous Nanofossils*, pp. 17–39. Chichester: Ellis Horwood.

Haq, B. U., J. Hardenbol, and P. R. Vail. 1987. Chronology of fluctuating sea levels since the Triassic (250 million years ago to present). *Science* **235**: 1156–1167.

Hillebrandt, Av., P. Smith, G. E. G. Westermann, and J. H. Callomon. 1992. Ammonite zones of the circum-Pacific region. *In* G. E. G. Westermann (editor), *The Jurassic of the Circum-Pacific*, pp. 247–272. Cambridge: Cambridge University Press.

House, M. R., and J. R. Senior. 1981. *The Ammonoidea. The Evolution, Classification, Mode of Life and Geological Usefulness of a Major Fossil Group. Systematics Association Special Volume 18*. London: Academic Press.

Imlay, R. W. 1939. Upper Jurassic ammonites from Mexico. *Geological Society of America Bulletin* **50**: 1–78.

Jablonski, D. 1986. Causes and consequences of mass extinctions: a comparative approach. *In* D. K. Elliot (editor), *Dynamics of Extinction*, pp. 183–230. New York: Wiley.

Jablonski, D. 1989. The biology of mass extinction: a palaeontological view. *Philosophical Transactions of the Royal Society of London B* **325**: 357–368.

Jablonski, D., and D. M. Raup. 1995. Selectivity of end-Cretaceous marine bivalve extinctions. *Science* **268**: 389–391.

Jeletzky, J. A. 1984. Jurassic-Cretaceous boundary beds of western and Arctic Canada and the problem of the Tithonian-Berriasian stages in the Boreal Realm. *In* G. E. G. Westermann (editor), *Jurassic-Cretaceous Biochronology and Biogeography of North America. Geological Association of Canada Special Paper* **25**: 175–255.

Jenkins, D. G, and J. W. Murray (editors). 1981. *Stratigraphic Atlas of Fossil Foraminifera*. Chichester: Ellis Horwood.

Kauffman, E. G., and D. H. Erwin. 1995. Surviving mass extinctions. *Geotimes* **40**: 14–17.

Kelly, S. R. A. 1984–92. Bivalvia of the Spilsby Sandstone and Sandringham Sands (Late Jurassic-Early Cretaceous) of southern England. *Monograph of the Palaeontological Society*, 123pp., pls 27.

Kennedy, W. J., and W. A. Cobban. 1976. Aspects of ammonite biology, biogeography, and biostatigraphy. *Special Papers in Palaeontology* **17**: 1–94.

Krishna, J., S. Kumar, and I. B. Singh. 1982. Ammonoid stratigraphy of the Spiti Shale (Upper Jurassic), Tethys Himalaya, India. *Neues Jahrbuch für Geologie und Paläontologie, Monatschafte* **10**: 580–592.

Krishna, J., B. Pandey, and D. B. Pathak. 1996. Ammonoid chronology in the Tithonian of Kachchh (India). *Geological Research Forum* **1–2**: 205–214.

Leanza, H. A. 1981. The Jurassic-Cretaceous boundary beds in west central Argentina and their ammonite zones. *Neues Jahrbuch für Geologie und Paläontologie, Abhandlungen* **161**(1): 62–92.

Matsumoto, T., and H. Sakai. 1983. On some Jurassic ammonites from Muktinath, Nepal. *Memoir Faculty of Science, Kyushu University Geology* **25**(1): 75–91.

Mitra, K. C., S. Bardhan, and D. Bhattacharya. 1979. A study of Mesozoic stratigraphy of Kutch, Gujarat with special reference to rock-stratigraphy of Keera dome. *Bulletin of the Indian Geological Association* **12**: 129–143.

Myczyńsky, R. 1989. Ammonite biostratigraphy of the Tithonian of western Cuba. *Annales Societatis Geologorum Poloniae* **59**: 43–125.

Myczyńsky, R. and A. Pszczolkowski. 1994. Tithonian stratigraphy and microfacies in the Sierra del Rossario, Western Cuba. *Studia Geologica Polonica* **105**: 91–108.

Olóriz, F., and J. M. Tavera. 1979. Consideraciones sobre el género *Tithopeltoceras* Arkell (1953) en las Cordilleras Béticas (Zona subbética, sector central). *Estudios Geológicos* **35**:137–147.

Olóriz, F., and J. M. Tavera. 1989. The significance of Mediterranean ammonites with regard to the traditional Jurassic-Cretaceous boundary. *Cretaceous Research* **10**: 221–237.

Olóriz, F., A. B. Villasenor, A. Gonzalez, and G. E. G. Westermann. 1999. Ammonite biostratigraphy and correlation in the Upper Jurassic – Lower Cretaceous Lacaja Formation of north – central Mexico (Sierra Catorce, San Luis Potosí). *In* F. Olóriz, and T. Rodriguez (editors), *Advancing Research on Living and Fossil Cephalopods*, pp. 463–491. New York: Kluwer Academic/Plenum.

Page, K. N. 2005. Jurassic. *In* R.C. Selley, L. R. M. Cocks, and I. R. Plimer (editors), *Encyclopedia of Geology*, pp. 352–360. London: Elsevier.

Raup, D. M., and J. J. Sepkoski, Jr. 1986. Periodic extinction of families and genera. *Science* **231**: 833–836.

Raup, D. M., and D. Jablonski. 1993. Geography of end-Cretaceous marine bivalve extinctions. *Science* **260**: 971–973.

Riccardi, A. C. 1991. Jurassic and Cretaceous marine connections between the southeast Pacific and Tethys. *Palaeogeography, Palaeoclimatology, and Palaeoecology* **87**: 155–189.

Sandy, M. R. 1988. Tithonian brachiopods. *Mémoires de la Société Géologique de France, Nouvelle* Série **154**: 71–74.

Shome, S., S. De, P. Roy, and S. S. Das. 2004. Ammonites as biological stopwatch and biogeographical blackbox – a case study from the Jurassic-Cretaceous boundary (150 Ma) of Kutch, Gujarat. *Current Science* **86**(1): 197–202.

Shome, S., S. Bardhan, and S. De. 2005. Record of *Tithopeltoceras* (Ammonoidea) from the Tithonian of Kutch, India and its stratigraphic and paleobiogeographic significance. *Journal of Paleontology* **79**(3): 619–624.

Shome, S., and P. Roy. 2006. New record of *Pterolytoceras* Spath, 1927 from the Upper Jurassic (Late Tithonian) of Kutch, western India and its palaeobiogeographic significance. *Indian Minerals* **59**(1/2): 57–64.

Spath, L. F. 1927–33. Revision of the Jurassic cephalopod fauna of Kachh (Cutch). *Palaeontologia Indica, Geological Survey of India, New Series* **9**(2): 1–945.

Stanley, S. M. 1984. Temperature and biotic crises in the marine realm. *Geology* **12**: 205–208.

Stanley, S. M. 1988. Paleozoic mass extinctions: shared patterns suggest global cooling as a common cause. *American Journal of Science* **288**: 334–352.

Stevens, G. R. 1997. The Late Jurassic ammonite fauna of New Zealand. *Monograph of Institute of Geology and Nuclear Science* **18**: 1–216.

Tavera, J. M. 1985. *Les ammonites del Tithonico superior-Berriasense de la Zona Subbética (Cordilleras Béticas)*. Ph.D. thesis of University of Granada, 381pp.

Tavera, J. M., A. Checa, F. Olóriz, and M. Company. 1986. Mediterranean ammonites and the Jurassic-Cretaceous boundary in southern Spain (Subbetic Zone). *Acta Geologica Hungarica* **29**(1–2): 151–159.

Taylor, D. G. H., J. H. Callomon, R. Hall, P. L. Smith, H. W. Tipper, and G. E. G. Westermann. 1984. Jurassic ammonite biogeography of western North America; the tectonic implications. Circum – Pacific Jurassic Research Group Project **171**(2). *Geological Association of Canada Special Paper* **27**: 13–16.

Thomson, M. R. A. 1980. Mesozoic ammonite faunas of Antarctica and the break-up of Gondwana. *5th Gondwana Symposium Proceedings*: 269–275.

Thomson, M. R. A. 1982. Late Jurassic fossils from Low Island, South Shetland Islands. *British Antarctic Survey Bulletin* **56**: 25–35.

Verma, H. M., and G. E. G. Westermann. 1973. The Tithonian (Jurassic) ammonite fauna and stratigraphy of Sierra Cartoce, Sar Luis Potosi, Mexico. *Bulletin of American Paleontology* **63**(277): 107–320.

Westermann, G. E. G. 1984. Jurassic – Cretaceous biochronology and paleogeography of North America. *Geological Association of Canada Special Paper* **27**:105.

Westermann, G. E. G. 1990. New developments in ecology of Jurassic-Cretaceous ammonoids. *In* G. Pallini, F. Cecca, S. Cresta, and M. Santantonio (editors), *Atti del Secondo Convegno Internazionale Fossili, Evolutione, Ambiente*, pp. 459–478. Pergola.

Westermann, G. E. G. 2000a. Biochore classification and nomenclature in paleobiogeography: an attempt at order. *Palaeogeography, Palaeoclimatology, and Palaeoecology* **158**: 1–13.

Westermann, G. E. G. 2000b. Marine faunal realms of the Mesozoic: review and revision under the new guidelines for biogeographic classification and nomenclature. *Palaeogeography, Palaeoclimatology, and Palaeoecology* **163**: 49–68.

Whatley, R. C. 1988. Patterns and rates of evolution among Mesozoic Ostracoda. *In* T. Hanai, N. Ikeya, and K. Ishizaki (editors), *Evolutionary Biology of Ostracoda*, pp. 1021–1040. Tokyo: Kodansha.

Wright, C. W., J. H. Callomon, and M. K. Howarth. 1996. Cretaceous Ammonoidea. *In* R. L. Kaesler (editor), *Treatise on Invertebrate Paleontology, Part L, Mollusca 4, Revised, Volume 4*: 1–362. Boulder, Colorado, and Lawrence, KS: The Geological Society of America and the University of Kansas Press.

Zakharov, V. A., and B. T. Yanine. 1975. Les bivalves á la fin du Jurassique et au début du Crétacé. *Mémoires du Bureau de Recherches Géologiques et Minières* **86**: 221–228.

Zakharov, V. A., and M. A. Rogov. 2003. Boreal-Tethyan mollusk migrations at the Jurassic-Cretaceous boundary, time and biogeographic ecotone position in the Northern Hemisphere. *Stratigraphy and Geological Correlation* **11**(2): 152–171. [English Translation]

Zinsmeister, W. J., R. M. Feldmann, M. O. Woodburne, and D. H. Elliot. 1989. Latest Cretaceous/earliest Tertiary transition on Seymour Island. *Journal of Paleontology* **63**: 731–738.

Chapter 18
Ammonite Touch Marks in Upper Cretaceous (Cenomanian-Santonian) Deposits of the Western Interior Seaway

Neil H. Landman[1] and William A. Cobban[2]

[1]Division of Paleontology (Invertebrates), American Museum of Natural History, 79th Street and Central Park West, New York, NY 10024, USA, landman@amnh.org;

[2]70 Estes Street, Lakewood, CO 80226, USA

Keywords: ammonites, touch marks, taphonomy, Upper Cretaceous, Western Interior

1 Introduction

Over the course of collecting in the last 50 years in the Upper Cretaceous US Western Interior, one of us (WAC) has assembled a collection of "trace fossils" left by ammonites as they touched the sea floor. These impressions provide clues to the taphonomic history of ammonite shells. They also serve as biostratigraphic markers in the absence of ammonite fossils themselves. In the following pages, we illustrate some of these impressions in association with the species that probably produced them.

Ammonite impressions are part of a larger category known as tool marks. The terminology of tool and scour marks has been thoroughly reviewed by Dzułynski and Sanders (1962). We use the term "touch mark" in a general sense for the impression made by the ammonite. What we actually observe of course, is the cast or negative of the impression on the underside of the overlying bed. Twenhofel (1939: 565) referred to such negatives as "counterparts."

Originally, ammonite touch marks were interpreted as trace fossils produced by vertebrates, e.g., tetrapod claw scratches, or ripple effects caused by fish swimming just above the bottom (as reviewed by Maeda and Seilacher, 1996). Rothpletz (1909) published one of the first correct interpretations, based on his analysis of ammonite

N. H. Landman et al. (eds.), *Cephalopods Present and Past: New Insights and Fresh Perspectives*, 396–422.

touch marks from the Jurassic Solnhofen Limestone of Germany. Seilacher (1963) did an extensive study on the marks from this formation, and in a series of memorable figures, illustrated how these marks were produced by waterlogged ammonite shells (perisphinctids and aspidoceratids), in various states of preservation, as they bounced, rolled, and swayed on the seafloor. Gaillard (1977), in a thorough analysis using quantitative data, documented ammonite touch marks in Jurassic strata of the Champagnole region, France. He interpreted the marks as having been produced by the spines of *Euaspidoceras*, as these shells dragged along the bottom. More recently, Summesberger et al. (1999) illustrated roll marks, attributed to perisphinctids, from the Lipica Formation of Slovenia. Each mark ends in a paintbrush-like structure, which the authors interpreted as the impression of the ammonite hyponome.

To better interpret these touch marks, several workers tried to reproduce them experimentally. Dzułynski and Sanders (1962) rolled ammonites on modeling clay, creating marks very similar to those observed in nature. Barthel et al. (1990), in their study of the Solnhofen Limestone, reproduced the marks in this formation by rolling ammonites on wet mud.

Illustrated specimens are reposited in the American Museum of Natural History (AMNH), New York, New York, and the US National Museum (USNM), Washington, DC.

2 Localities

The ammonite touch marks that we describe occur in Upper Cretaceous (middle Cenomanian-middle Santonian) deposits of the US Western Interior. There are 27 localities ranging from Montana to New Mexico, representing 14 ammonite range zones (Figs. 18.1–3). Most of the ammonite touch marks are newly documented, although some have previously been reported (Dzułynski and Sanders, 1962: pl. 19B; Mudge, 1972: A68; Scott et al., 1986: Fig. 12f, g).

The ammonite impressions occur in the Dakota Sandstone of western Colorado, the Juana Lopez Member of the Mancos Shale of New Mexico, the Tropic Shale of Utah, the Lincoln and Hartland Members of the Greenhorn Limestone of Colorado, the Turner Sandy Member of the Carlile Shale of South Dakota, the Floweree and Ferdig Members of the Marias River Shale of Montana, the Cody Shale of Wyoming, and the Niobrara Formation of New Mexico. Ammonite touch marks have not been observed from any strata above the middle Santonian.

The localities of the ammonite touch marks are shown in Figs. 18.1–3 and listed below:

1. US Geological (USGS) Mesozoic loc. D2611. South Fork of Sun River, southwest of junction with Bear Creek, sec. 34, T22N, R10W, Lewis and Clark County, Montana. Approximately 110 ft (33 m) above base of Ferdig Member of Marias River Shale (Mudge, 1972).

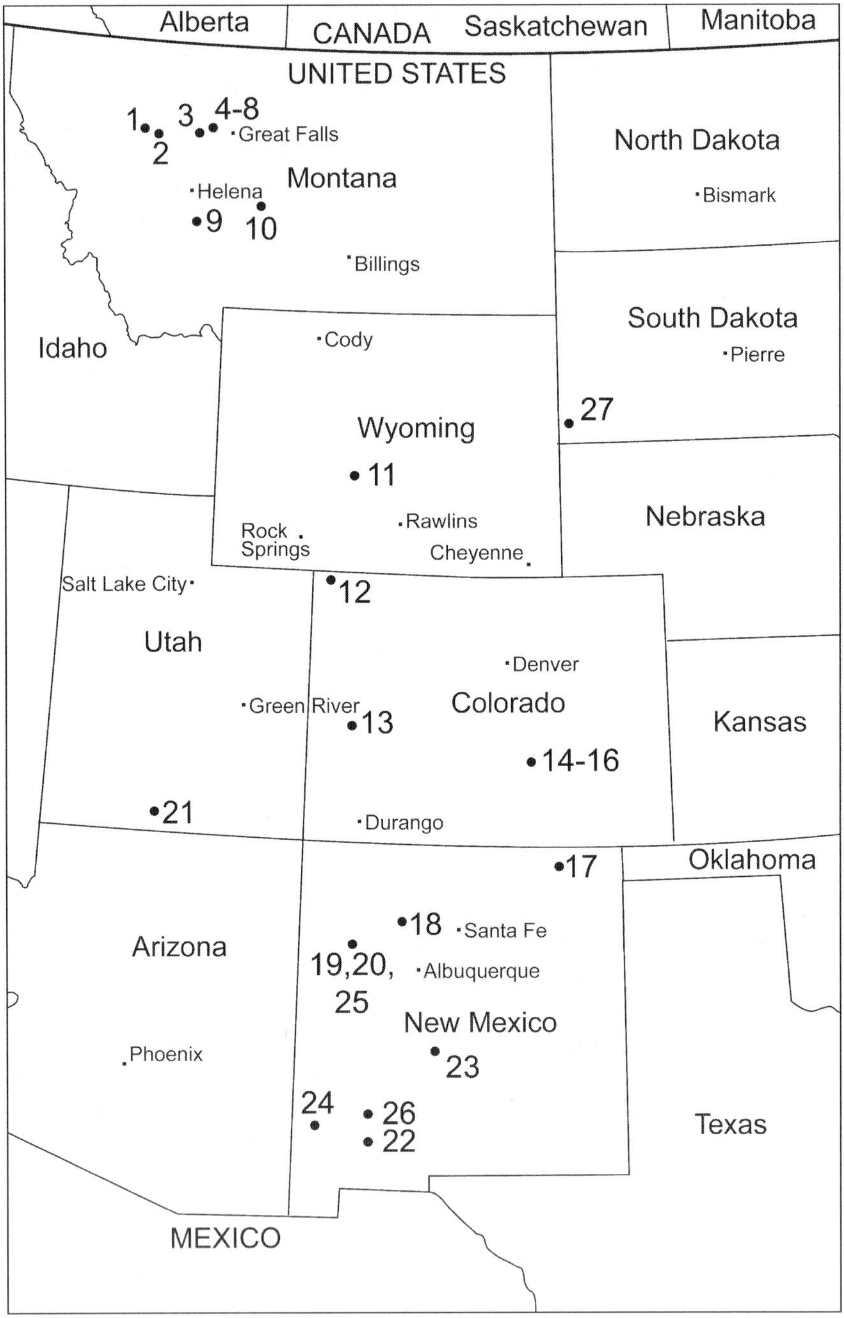

Fig. 18.1 *Map of localities mentioned in the text.*

STAGES AND SUBSTAGES		WESTERN INTERIOR AMMONITE TAXON RANGE ZONES	WESTERN INTERIOR INOCERAMID INTERVAL ZONES	LOCALITIES OF TRACES
Santonian	Upper	*Desmoscaphites bassleri*	*Sphenoceramus lundbreckensis*	
		Desmoscaphites erdmanni		
		Clioscaphites choteauensis	*Cordiceramus muelleri*	
	Middle	*Clioscaphites vermiformis*	*Cordiceramus bueltenensis*	←11?
			Cordiceramus cordiformis	
	Lower	*Clioscaphites saxitonianus*	*Cladoceramus undulatoplicatus*	
Coniacian	Upper	*Scaphites depressus*	*Magadiceramus crenelatus*	
			Magadiceramus subquadratus	←17,18
	Middle	*Scaphites ventricosus*	*Volviceramus involutus* *Volviceramus koeneni*	
	Lower	*Scaphites preventricosus*	*Cremnoceramus crassus crassus*	
			Cremnoceramus crassus inconstans	
			Cremnoceramus deformis dobrogensis	
			Cremnoceramus deformis erectus	
Turonian	Upper	*Scaphites mariasensis*	*Cremnoceramus waltersdorfensis*	
		Prionocyclus germari	*Mytiloides scupini*	←1,2,4-10
		Scaphites nigricollensis	*Mytiloides incertus*	
		Scaphites whitfieldi	*Inoceramus dakotensis*	←20
			Inoceramus perplexus	
	Middle	*Scaphites ferronensis*	*Inoceramus dimidius*	←27
		Scaphites warreni		
		Prionocyclus macombi	*Inoceramus* aff. *dimidius*	
		Prionocyclus hyatti	*Inoceramus howelli*	←12
		Collignoniceras praecox	*Inoceramus* n. sp.	
		Collignoniceras woollgari	*Mytiloides hercynicus*	←19,21, 25,26
			Mytiloides subhercynicus	
	Lower	*Mammites nodosoides*	*Mytiloides mytiloides*	
		Vascoceras birchbyi	*Mytiloides kossmati*	←22
		Pseudaspidoceras flexuosum		
		Watinoceras devonense	*Mytiloides puebloensis*	
Cenomanian (Part)	Upper	*Nigericeras scotti*	*Mytiloides hattini*	
		Neocardioceras juddii	*Inoceramus pictus*	
		Burroceras clydense		
		Euomphaloceras septemseriatum		
		Vascoceras diartianum		
		Dunveganoceras conditum	*Inoceramus ginterensis*	←13,15
		Dunveganoceras albertense		←3
		Dunveganoceras problematicum		←14
		Dunveganoceras pondi	*Inoceramus prefragilis*	←16
	Middle	*Plesiacanthoceras wyomingense*		
		Acanthoceras amphibolum	*Inoceramus rutherfordi*	←24
		Acanthoceras bellense	*Inoceramus arvanus*	←23
		Acanthoceras muldoonense	*Inoceramus eulessanus*	
		Acanthoceras granerosense		
		Conlinoceras tarrantense		

Fig. 18.2 *Biostratigraphic distribution of ammonite touch marks. Numbers on the right side correspond to localities plotted in Fig. 18.1 and mentioned in the text. The chart is reproduced from Cobban et al. (2006).*

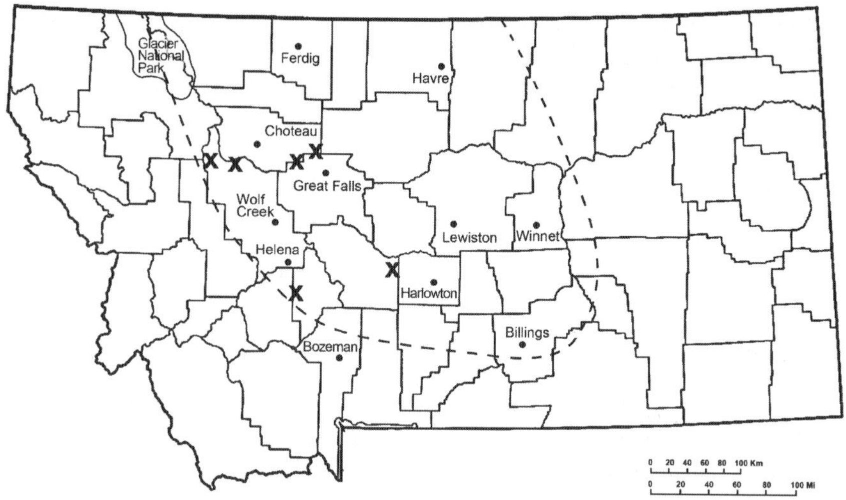

Fig. 18.3 *Map of Montana showing approximate limits of the facies containing scaphitid touch marks in the Ferdig Member of the Marias River Shale. The Xs indicate localities of known touch marks (a single X may represent several closely spaced localities).*

2. USGS Mesozoic loc. D1493. Barr Creek, SW¼ sec. 8, T21N, R8W, Lewis and Clark County, Montana. Ferdig Member of Marias River Shale.
3. USGS Mesozoic loc. D12278. Cone Hill, N½ sec. 13, T22N, R1W, Teton County, Montana. Middle part of Floweree Member of Marias River Shale.
4. USGS Mesozoic loc. D2013. NE¼ sec. 22, T23N, R2E, Teton County, Montana. Bed "N" of Ferdig Member of Marias River Shale (Erdmann et al., 1947; Cobban et al., 1976: 45, 47).
5 USGS Mesozoic loc. D2014. NW¼ sec. 23, T23N, R2E, Teton County, Montana. Bed "N" of Ferdig Member of Marias River Shale (Erdmann et al., 1947; Cobban et al., 1976: 45, 47).
6. USGS Mesozoic loc. D2019. NW¼ sec. 23, T23N, R2E, Teton County, Montana. Approximately 32 ft (9.7 m) above bed "N" of Ferdig Member of Marias River Shale (Erdmann et al., 1947; Cobban et al., 1976: 45, 47).
7. USGS Mesozoic loc. D2254. SW¼ sec. 34, T23N, R1W, Teton County, Montana. Approximately 5.5 ft (1.5 m) above bed "N" of Ferdig Member of Marias River Shale (Erdmann et al., 1947; Cobban et al., 1976: 45, 47).
8. USGS Mesozoic loc. D14366. Northern part of Great Falls 7½ minute quadrangle, T23N, R's 2 and 3E, Teton and Chouteau Counties, Montana. Upper part of Ferdig Member of Marias River Shale.
9. USGS Mesozoic loc. D14085. Indian Creek west of Townsend near middle of sec. 6, T6N, R1E, Broadwater County, Montana. Holter Sandstone Member of Marias River Shale (Groff, 1963).

10. USGS Mesozoic loc. D4593. NW¼ sec. 9, T9N, R11E, Meagher County, Montana. Carlile Shale.
11. USGS Mesozoic loc. D14088. NW¼ sec. 24, T29N, R96W, Fremont County, Wyoming. Cody Shale.
12. USGS Mesozoic loc. D9280. North of Vermillion Creek, NW¼ sec. 30, T10N, R100W, Moffat County, Colorado. Basal 10 ft (3 m) of Frontier Sandstone Member of Mancos Shale.
13. USGS Mesozoic loc. D14083. Highway 50, approximately 7 miles (11 km) west of Delta, Delta County, Colorado. Near top of Dakota Formation.
14. USGS Mesozoic loc. D1307. West of Pueblo, near west line of sec. 30, T20S, R65W, Pueblo County, Colorado. Hartland Shale Member of Greenhorn Limestone.
15. USGS Mesozoic loc. D13241. West of Pueblo, NW¼ sec. 25, T20S, R66W, Pueblo County, Colorado. Hartland Shale Member of Greenhorn Limestone.
16. USGS Mesozoic loc. D13831. Near center of W½ sec. 25, T20S, R66W, Pueblo County, Colorado. Upper part of Lincoln Member of Greenhorn Limestone.
17. USGS Mesozoic loc. D11309. SE¼ sec. 23, T29N, R25E, Colfax County, New Mexico. Sandy Member of Niobrara Formation.
18. USGS Mesozoic loc. D13967. SE¼ sec. 8, T17N, R1W, Sandoval County, New Mexico. El Vado Sandstone Member of Mancos Shale.
19. USGS Mesozoic loc. D10590. Sec. 17, T15N, R12W, McKinley County, New Mexico. Mancos Shale.
20. USGS Mesozoic loc. D13101. SW¼ sec. 17, T15N, R11W, McKinley County, New Mexico. Near top of Juana Lopez Member of Mancos Shale.
21. USGS Mesozoic loc. D13760. NW¼ sec. 14, T40S, R1W, Kane County, Utah. Approximately 175 ft (53 m) below top of Tropic Shale.
22. USGS Mesozoic loc. D11531. Cookes Range, NE¼ sec. 13, T21S, R9W, Luna County, New Mexico. Lower part of sandy unit overlying Bridge Creek Limestone Member of Mancos Shale.
23. USGS Mesozoic loc. D10128. Carthage area, NE¼ sec. 8, T5S, R2E, Socorro County, New Mexico. Thin-bedded siltstone beds beneath Marker bentonite bed of Mancos Shale.
24. USGS Mesozoic loc. D12744. Riley Canyon, NW¼ sec. 21, T18S, R20W, Hidalgo County, New Mexico. Mancos Shale.
25. USGS Mesozoic loc. D5770. NW¼ sec. 28, T15N, R12W, McKinley County, New Mexico. Slightly above Bridge Creek Limestone Member of Mancos Shale.
26. USGS Mesozoic loc. D11022. SW¼ NE¼ sec. 13, T14S, R4W, Sierra County, New Mexico. Shaly siltstone unit 30 feet (9 m) above top of *Sciponoceras gracile* Zone, Mancos Shale.
27. USGS Mesozoic loc. D12874. NE¼ NW¼ sec. 36, T7S, R6E, Fall River County, South Dakota. Float from 5–15 ft (1.5–4.6 m) above the base of the Turner Sandy Member of the Carlile Shale.

3 Description of Ammonite Touch Marks

Careful examination of the ammonite touch marks reveals the species that probably produced them (Figs. 18.4–21, arranged in taxonomic order). The main evidence for this determination is the impression of the ornamental features (ribs, tubercles, and keels), which form a distinctive pattern. We compared this pattern to specimens in our collections, bearing in mind that the pattern of ornamentation commonly changes during ontogeny. We examined those ammonites known to occur in the same strata as the impressions and in age-equivalent strata from elsewhere.

Most of the touch marks were produced by acanthoceratid ammonites including species of *Collignoniceras*, *Calycoceras*, and *Tarrantoceras*. These ammonites were widespread in the Late Cretaceous Western Interior Seaway. Other touch marks in our collection were produced by scaphitid ammonites.

The majority of touch marks are impressions of the venter or ventrolateral edge of the shell, as inferred from the kind and distribution of ornamental features. For example, the ventral impression of scaphitid ammonites consists of a series of 5–10 straight ridges resembling scratch marks (Fig. 18.5A), while the ventral impression of *Tarrantoceras* consists of three parallel rows of small bumps (Fig. 18.12A). The ventrolateral impression of *Collignoniceras woollgari* (Mantell, 1822) consists of a series of clavate ridges (Fig. 18.17A), while the ventrolateral impression of *Fagesia catinus* (Mantell, 1822) shows a break in the shape of the ridges, corresponding to a change in the shape of the whorl section as it passes from the flanks to the venter (Fig. 18.14).

The ammonite touch marks show a broad range in size. Those attributed to *Dunveganoceras conditum* (Haas, 1951) (Fig. 18.8) and *Calycoceras* aff. *C. canitaurinum* (Haas, 1949) (Fig. 18.10) are the largest, whereas those attributed to *Scaphites corvensis* Cobban, 1952 (Fig. 18.6B) are the smallest. In addition, the impressions vary in how faithfully they reproduce the original morphology. For example, the impression of the shell flanks attributed to *Prionocyclus novimexicanus* (Marcou, 1858) is very faithful (Fig. 18.19), whereas the ventrolateral impression attributed to *Protexanites bourgeoisianus* (d'Orbigny, 1850) is distorted, with twisted ribs (Fig. 18.16A). The impressions also vary in their degree of relief. For example, the touch mark attributed to *Calycoceras* aff. *C. canitaurinum* (Fig. 18.10) is very prominent, whereas that attributed to *Pseudaspidoceras flexuosum* Powell, 1963, is very faint (Fig. 18.13).

The touch marks of scaphitid ammonites are extremely abundant (Figs. 18.4, 18.5, 18.6A–D). For example, there are nine sets of impressions on a single slab 10 cm on each side (Fig. 18.5A). All of the impressions on this slab reflect the venter and show no preferred orientation.

The most spectacular ammonite impressions are roll marks or "tire tracks." Figure 18.11 depicts a "tire track" approximately 10 cm long and 4 cm wide, probably produced by an acanthoceratid ammonite like *Tarrantoceras* rolling on the seafloor. The track consists of five parallel rows of bumps and forms a slightly arcuate pattern. Other roll marks in our collection were produced by *Collignoniceras woollgari*. In Fig. 18.17C, the roll mark is slightly arcuate and represents the

Fig. 18.4 *Scaphitid touch marks, USNM 534416, from the upper part of the Ferdig Member of the Marias River Shale in northcentral Montana (loc. 4, Fig. 18.1). The touch marks (arrows) and steinkerns can be assigned to* Scaphites corvensis *Cobban, 1952. Similar impressions are also present in the Ferdig Member elsewhere in northcentral (locs. 5–8), northwestern (locs. 1, 2), westcentral (loc. 9), and southcentral Montana (loc. 10). X1.*

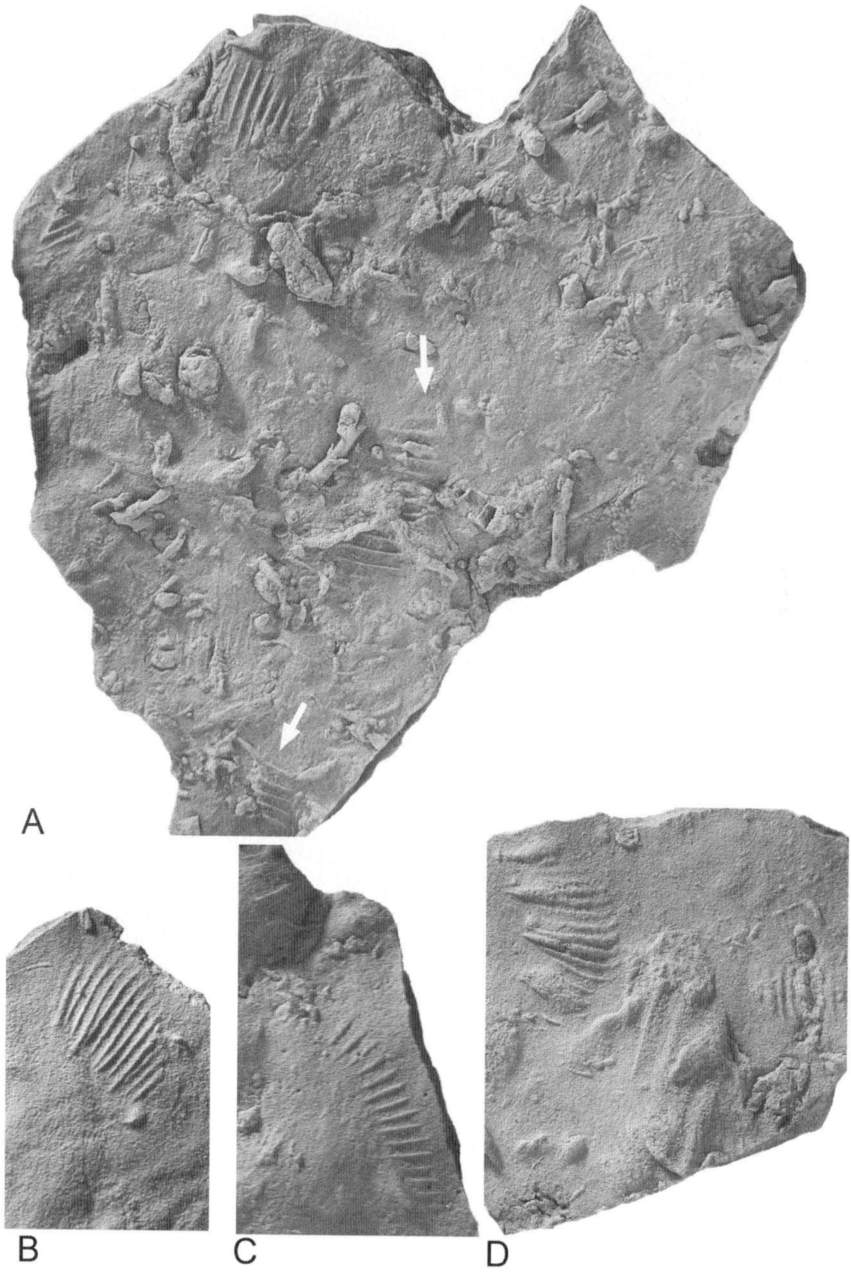

Fig. 18.5 *Scaphitid touch marks (arrows) attributed to* Scaphites corvensis *Cobban, 1952, Ferdig Member of the Marias River Shale in northcentral Montana (loc. 6, Fig. 18.1). A. USNM 534417. B. USNM 534418. C. USNM 534419. D. USNM 534420. All figures X1.*

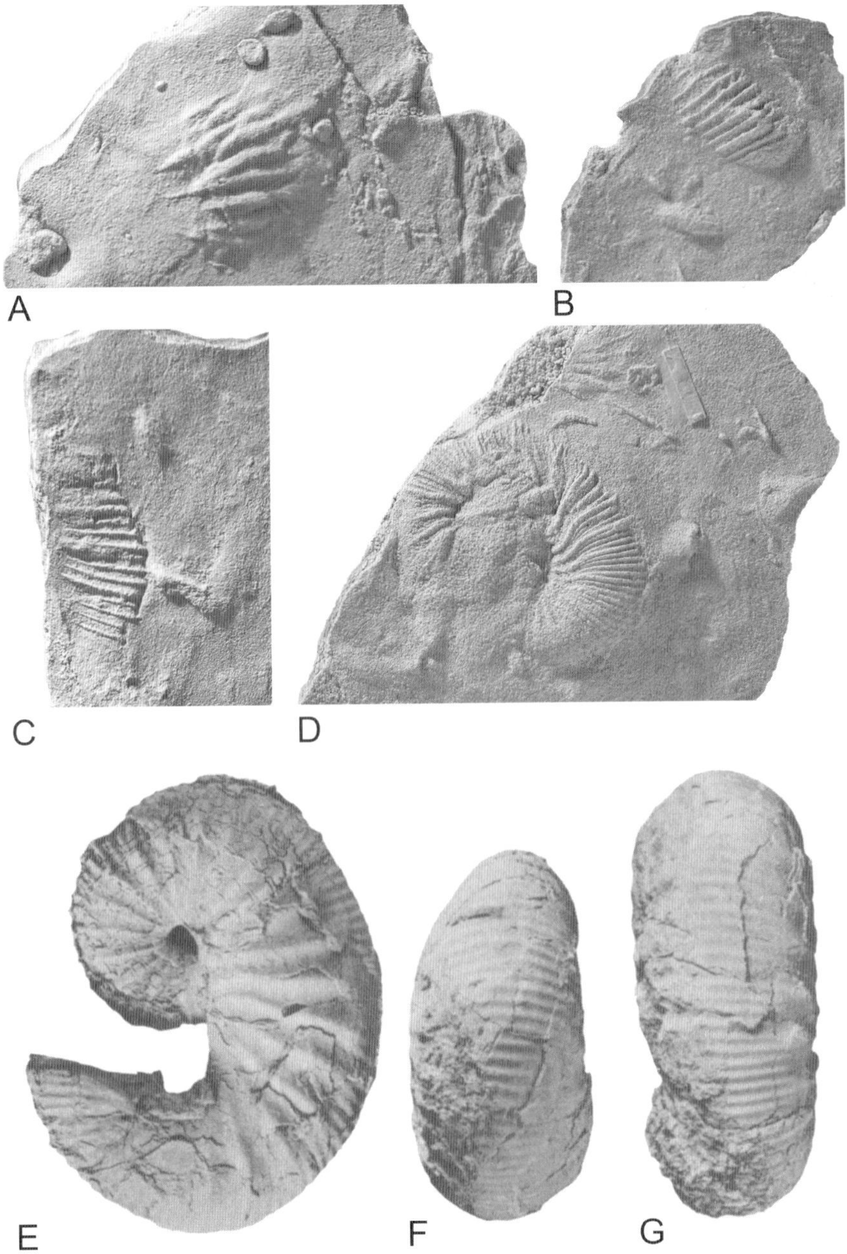

Fig. 18.6 *A–D. Scaphitid touch marks attributed to* Scaphites corvensis *Cobban, 1952, Ferdig Member of the Marias River Shale in northcentral Montana (Fig. 18.1). A. USNM 534421 (loc. 8, Fig. 18.1). B. USNM 534422. C. USNM 534423 (loc. 7, Fig. 18.1). D. USNM 534424 (loc. 7, Fig. 18.1). E–G. Holotype of* Scaphites corvensis *Cobban, 1952, USNM 106755, USGS Mesozoic loc. 20939, Cody Shale, Montana. E. Left lateral view. F. Ventral view of the hook. G. Ventral view of the shaft. All figures X1.*

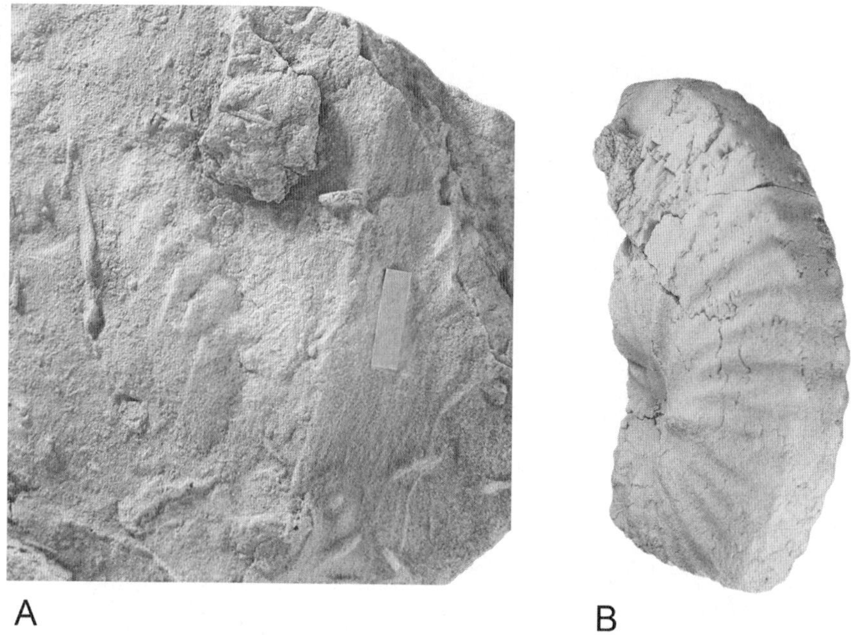

Fig. 18.7 *A. Touch mark, USNM 534425, from a sandy bed in the Floweree Member of the Marias River Shale, Montana (loc. 3, Fig. 18.1), assigned to* Metoicoceras mosbyense *Cobban, 1953, because this is the only species of* Metoicoceras *known from this unit. B. A specimen of* M. mosbyense, *USNM 534426, photographed obliquely for comparison, to show the flattened venter bordered by ventrolateral clavi, USGS Mesozoic loc. 21487, Mosby Sandstone, eastcentral Montana. Both figures X1.*

impression of the ventrolateral tubercles and serrated keel. In Fig. 18.17E, the impression of the tubercles and keel is smaller, reflecting a specimen at an earlier stage of ontogeny. These roll marks perfectly match those produced experimentally by Dzułynski and Sanders (1962), using specimens of *C. woollgari*.

Many ammonite touch marks are also associated with scour and tool marks of unknown origin. These marks may have formed before, after, or at the same time as the ammonite touch marks. Such marks are present, for example, in the Ferdig Member of the Marias River Shale (Figs. 18.5, 6A–D) and the Turner Sandy Member of the Carlile Shale (Fig. 18.21). Other structures may represent burrows that formed at the interface between the mud and the overlying sediments (Fig. 18.5A).

4 Discussion

Ammonite touch marks record the impact of the shells as they bounced and rolled on a firm mud bottom in relatively shallow water. The impressions are preserved on the basal surfaces of siltstones and fine sandstones. The waterlogged shells were

Fig. 18.8 *Touch mark, USNM 534427, showing five straight, narrow, closely spaced ribs, from the upper part of the Dakota Formation near Delta in westcentral Colorado (loc. 13, Fig. 18.1). This impression was made by a large ammonite such as the middle late Cenomanian species* Dunveganoceras conditum *Haas, 1951 (Fig. 18.9B), although this species has not been recorded in Colorado west of Pueblo.*

probably resuspended and carried along by bottom currents associated with storms. As the storms subsided, the ammonites were "the first particles to touch bottom, so that their markings could be immediately cast by the sand, in whose suspension they had been transported" (D. Seilacher, 2005, personal communication).

This scenario requires a substrate firm enough to retain impressions of the shells. The firmness of the bottom may have been due to the cohesive properties of the sediments, perhaps related to the presence of bacterial films or mucus. The

Fig. 18.9 *A. Touch mark, USNM 534428, from the Hartland Shale Member of the Greenhorn Limestone near Pueblo, Colorado (loc. 15, Fig. 18.1), showing fairly closely spaced ribs bearing inner and outer ventrolateral tubercles, suggesting the outer whorl of the phragmocone of an adult specimen of* Dunveganoceras conditum *Haas, 1951, which is the only ammonite like it known from this stratigraphic unit. X1. B. Paratype of* D. conditum, *AMNH 27686, left lateral view, illustrated for comparison, Frontier Formation, T39N R83W, Wyoming. X 0.33.*

ammonite impressions were subsequently covered by the sediments entrained in the current. Afterward, the bottom must not have experienced any significant erosion that would have removed the impressions, or any large-scale bioturbation that would have obliterated them.

Fig. 18.10 *A. Touch mark, USNM 534429, showing broad, thick, alternating ribs, attributed to* Calycoceras *aff.* C. canitaurinum *(Haas, 1949) as defined by Cobban (1988a), from the Hartland Shale Member of the Greenhorn Limestone near Pueblo, Colorado (loc. 14, Fig. 18.1). Ammonites in the underlying Lincoln Member include* C. canitaurinum. *X1.* B. Calycoceras *aff.* C. canitaurinum, *USNM 376912, right lateral view, illustrated for comparison, USGS Mesozoic loc. 23154, Frontier Formation, Wyoming (Cobban, 1988a: pl. 5). In central Wyoming,* Calycoceras *aff.* C. canitaurinum *occurs in the Frontier Formation just above* C. canitaurinum, *as at Pueblo. X 0.33.*

Fig. 18.11 *Roll mark collected as float, USNM 534430, from the Lincoln Member of the Greenhorn Limestone near Pueblo, Colorado (loc. 16, Fig. 18.1). Several species of acanthoceratid ammonites could have made this type of roll mark. The only acanthoceratid ammonites recorded from the Lincoln Member in the Pueblo area are* Acanthoceras amphibolum *Morrow, 1935,* Calycoceras canitaurinum *(Haas, 1949), and* Tarrantoceras *sp. (Cobban and Scott, 1972; Sageman and Johnson, 1985; see fig. 18.12B, C). Of these,* Tarrantoceras *is the most likely candidate, but none of the specimens of Lincoln age is known to be large enough to have produced a roll mark of this size. X1.*

The ammonite impressions reflect the imprint of a curved surface on a flat bottom and as a result, do not always faithfully record the ornamentation. In addition, depending on the fragmentation of the shell, the strength and direction of the current, the firmness of the bottom, and the nature of the contact (resting, dragging, skipping), the impressions can be indistinct or distorted.

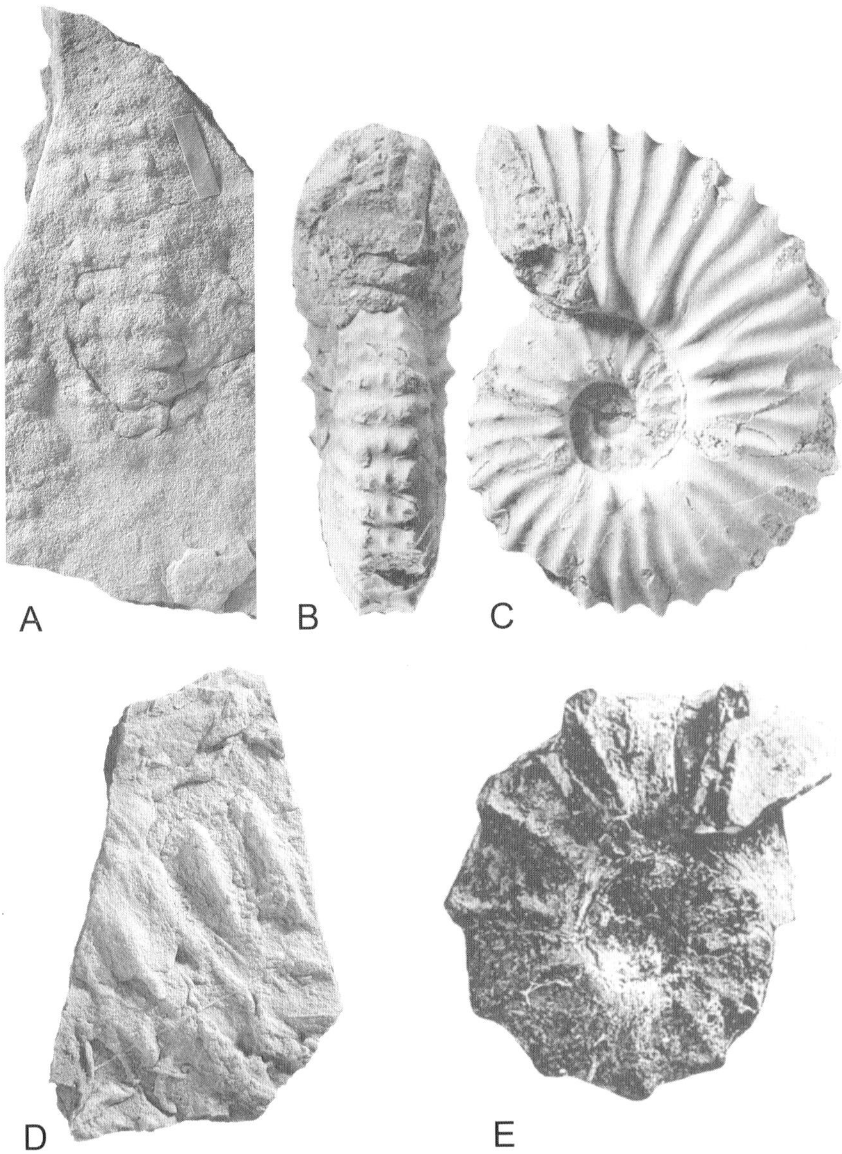

Fig. 18.12 *A. Roll mark of an acanthoceratid ammonite, USNM 534431, such as* Tarrantoceras *Stephenson, 1955 (loc. 24, fig. 18.1). B, C. Holotype of* Tarrantoceras sellardsi *(Adkins, l928), USNM 400760, illustrated for comparison, USGS Mesozoic loc. D12626, Tarrant Formation, Texas (Cobban, 1988b: pl.1, Figs. 6, 7). B. Ventral view. C. Right lateral view. D. Touch mark, USNM 534432, attributed to* Acanthoceras alvaradoense *Moreman, 1942 (loc. 23, Fig. 18.1). This ammonite occurs a little below the Marker bentonite bed at other localities in New Mexico. E. Acanthoceras* alvaradoense, *left lateral view, illustrated for comparison, Tarrant Formation, Texas (Moreman, l942: pl. 32, Fig. 6). All figures X1.*

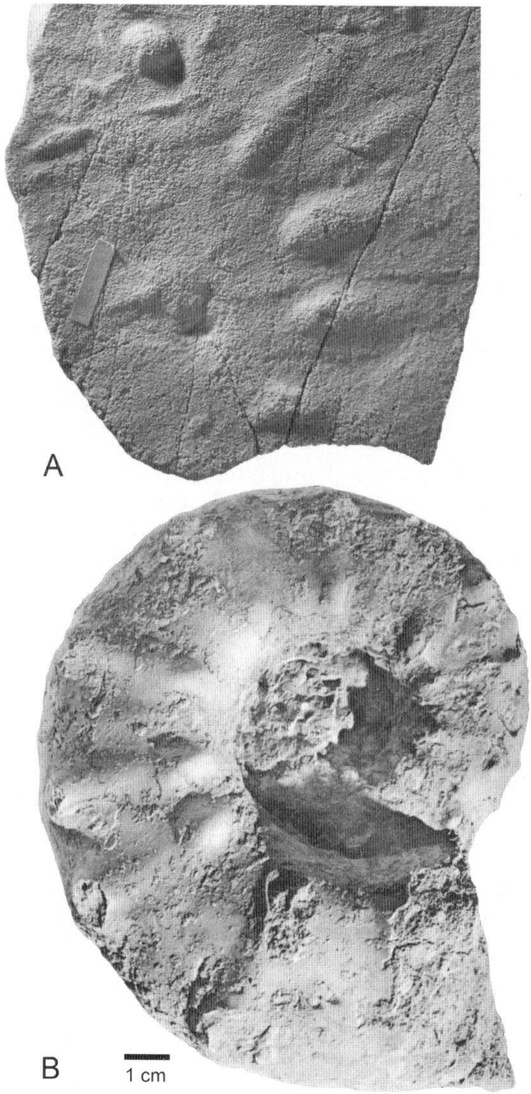

Fig. 18.13 *A. Touch mark, USNM 534433, attributed to* Pseudaspidoceras flexuosum *Powell, 1963, from the* P. flexuosum *Zone in the Mancos Shale of southwestern New Mexico (loc. 22, Fig. 18.1; Cobban et al., 1989: 64). The wide umbilicus and prominent umbilical tubercles are characteristic of this species. X1. B. Holotype of* Pseudaspidoceras paganum, *Reyment, 1954, C. 47422, right lateral view, illustrated for comparison, lower Turonian, Pindiga, Bauchi Province, Nigeria (Reyment, l959: pl.4, Fig. 1). X0.75.*

Ammonite touch marks can be used to derive information on current direction. The ventral touch marks of the scaphitid ammonites show no preferred orientation (Fig. 18.5A), perhaps implying variable current directions. In contrast, roll marks reveal the current track, but it is difficult to determine up from down current. The longer the shell rolled, the longer the roll mark, and the easier to follow the trace.

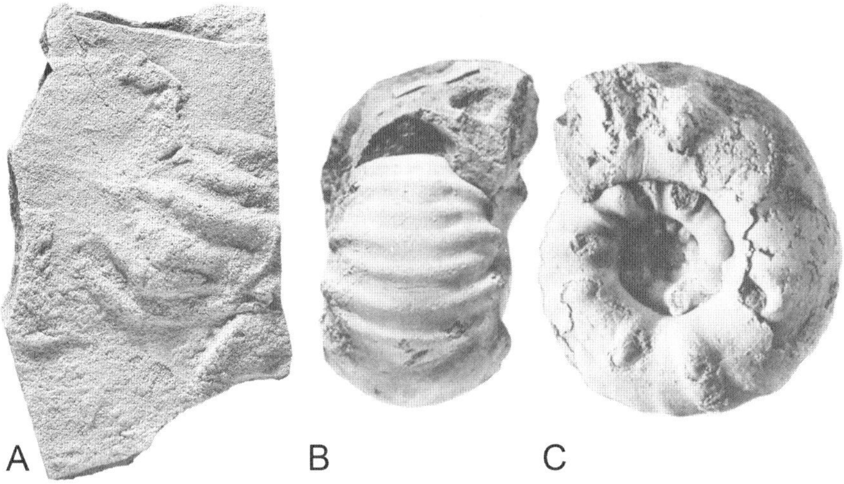

Fig. 18.14 *A. Touch mark, USNM 534434, of the phragmocone of* Fagesia catinus *(Mantell, 1822) (loc. 22, fig. 18.1). Specimens of this Turonian species have been found in calcareous concretions at this stratigraphic level. B, C. A similarly sized specimen of* F. catinus, *USNM 425388, illustrated for comparison, USGS Mesozoic loc. D11009, Colorado Formation, southwestern New Mexico (Cobban et al., 1989: Fig. 92GG, HH). B. Ventral view. C. Right lateral view. All figures X1.*

Fig. 18.15 *Touch mark attributed to* Protexanites bourgeoisianus *(d'Orbigny, 1850), USNM 534435, from the El Vado Member of the Mancos Shale of northwestern New Mexico (loc. 18, Fig. 18.1), because there is no other ammonite like it from this stratigraphic unit (see Fig. 18.16B). X1.*

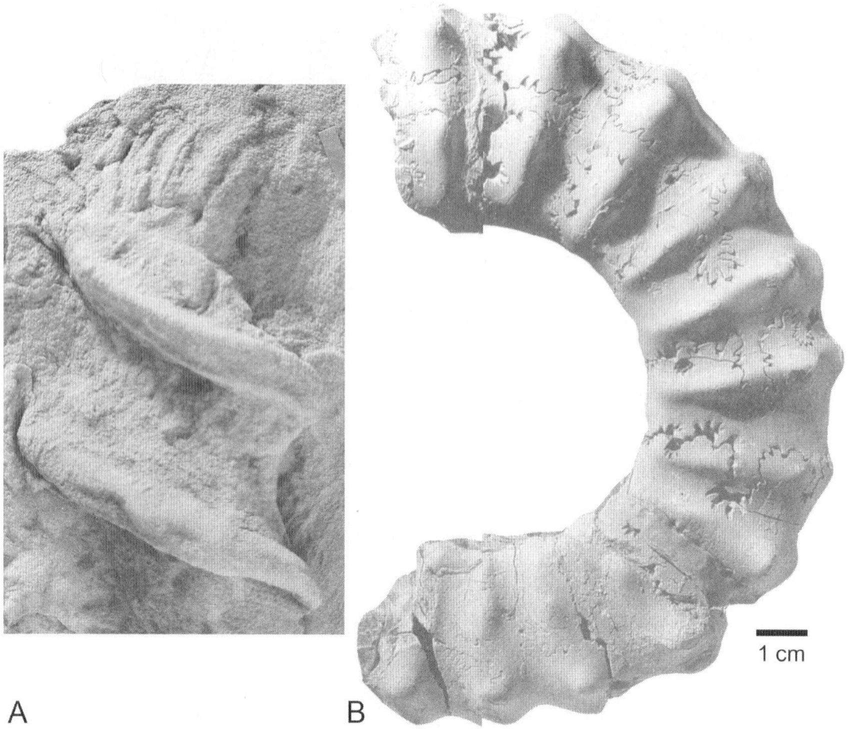

A B

Fig. 18.16 *A. Touch mark attributed to* Protexanites bourgeoisianus *(d'Orbigny, 1850), USNM 534436, from the sandy member of the Niobrara Formation of northeastern New Mexico (loc. 17, Fig. 18.1), a stratigraphic unit that has yielded an upper Coniacian assemblage of the* Scaphites depressus *Zone. The impression is slightly distorted and the ventrolateral tubercles are missing. Note the ribbing of a juvenile nearby, much like the specimen illustrated by Kennedy and Cobban (1991: pl. 8, Fig. 1). X1. B. A specimen of* P. bourgeoisianus, *USNM 433795, right lateral view, illustrated for comparison, USGS Mesozoic loc. 23100, Cody Shale, Wyoming (Kennedy and Cobban, 1991: Fig. 20). X0.75.*

The shape of the component parts of the roll mark may help indicate the initial point of contact between the shell and the seafloor, i.e., the up current end.

The touch marks in the Ferdig Member of the Marias River Shale are remarkable because of their broad geographic distribution, covering an area of several hundred square kilometers in northcentral Montana (Fig. 18.3). They are confined to the upper Turonian *Scaphites corvensis* Zone of the Ferdig Member of the Marias River Shale. This unit consists of silty, noncalcareous shale with lenses of very fine grained sandstone (Mudge, 1972; Cobban et al., 1976; Lemke, 1977).

P.E. Cloud, in an unpublished report from the US Geological Survey (1959), characterized the paleoenvironment of this unit as follows: "a generally quiet marine water body deep enough to be spared strong wave or current action, shallow enough to be within reach of some standing waves, sediments anaerobic enough to

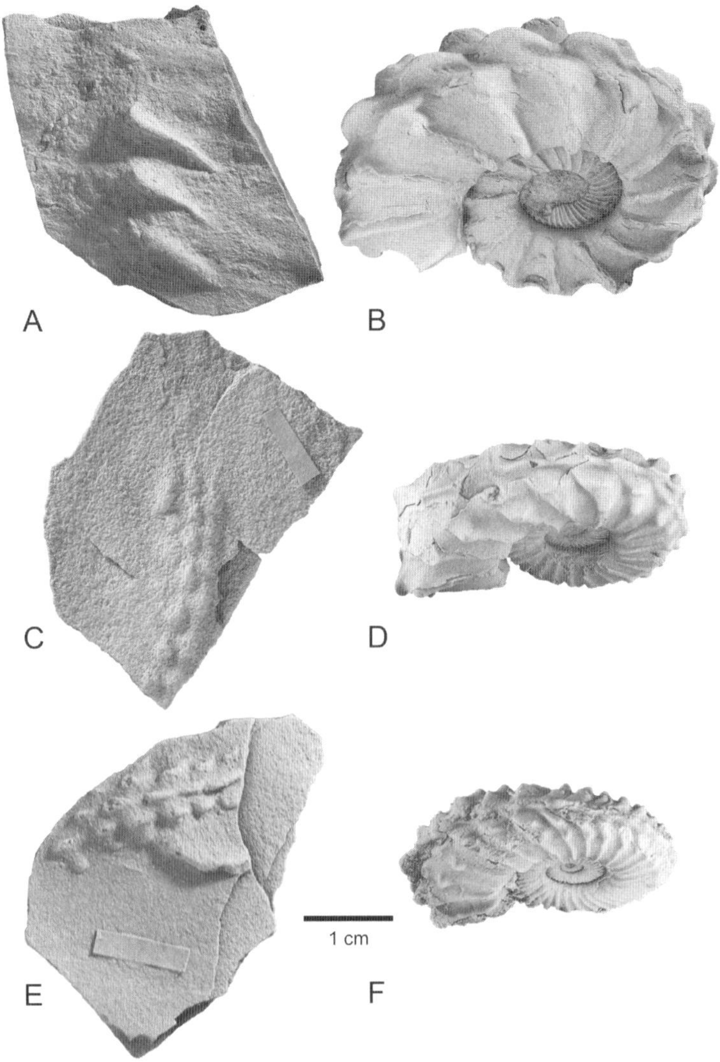

Fig. 18.17 *Touch marks attributed to* Collignoniceras woollgari *(Mantell, 1822). A. USNM 534437, showing three sharply defined narrow ribs bearing inner ventrolateral tubercles, from part of the Mancos Shale of northwestern New Mexico (loc. 19, Fig. 18.1) that contains this species in limestone concretions. X1. B. A similarly sized specimen of* C. woollgari, *USNM 534438, photographed obliquely for comparison, USGS Mesozoic loc. D9896, Carlile Shale, Black Hills area, Wyoming. X1. C. Roll mark, USNM 534439, from the Tropic Shale of southern Utah (loc. 21, Fig. 18.1), that contains this species in limestone concretions. X1. D. A similarly sized specimen of* C. woollgari, *USNM 534440, photographed obliquely for comparison, USGS Mesozoic loc. D3754, Carlile Shale, Black Hills area, South Dakota. E. Roll mark, USNM 534441, from the Mancos Shale, New Mexico (loc. 25, Fig. 18.1). X1.5. F. A similarly sized specimen of* C. woollgari, *USNM 534442, photographed obliquely for comparison, Carlile Shale, Black Hills area, South Dakota or Wyoming. X1.5.*

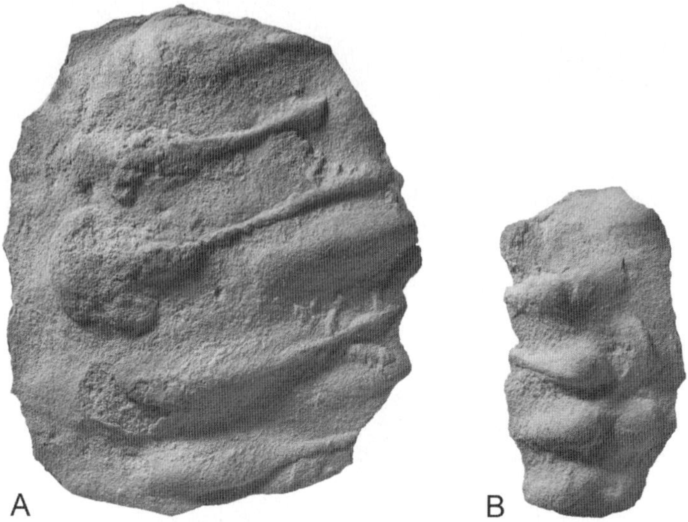

Fig. 18.18 *Touch marks attributed to* Collignoniceras woollgari *(Mantell, 1822), USGS Mesozoic loc. D11022, southwestern New Mexico (loc. 26, Fig. 18.1). A. USNM 534443. B. USNM 534444. Both figures X1.*

limit burrowing activity that would destroy lamination but rich in nutrients for surface feeders, in an area far enough away from a river system for its normal sediments to consist of silt from pulsatory system overflow but near enough to be within reach of heavy and rapid settling-out from occasional flood deliveries."

Most of the touch marks in the Ferdig Member reflect impressions of the venter of the body chamber near the point of recurvature. This implies that the shells were in a nearly vertical orientation with the phragmocone on top. Other impressions reflect the ventrolateral region where the ribs bifurcate. Some of these marks are relatively long, indicating that a broad area of the shell surface contacted the bottom (Fig. 18.5D). Fragments of actual shells, i.e., steinkerns, are also present at this site (Fig. 18.4), and probably represent the very specimens that made the touch marks.

It is possible that the touch marks in the Ferdig Member record a sequence in the taphonomic history of the ammonite shells. For example, the ventral impressions may have been produced during life. Scaphitid ammonites with hook-shaped body chambers were oriented with the phragmocone on top and the hook-like body chamber on the bottom (Landman, 1987). It is conceivable that these animals hovered just above the bottom and occasionally touched it. The ventrolateral and flank impressions may have been produced after death, during progressive stages of shell fragmentation and waterlogging. The steinkerns represent the final disposition of the empty shells.

On the other hand, these ammonite touch marks are also associated with tool marks of unknown origin. It is perhaps more parsimonious to argue that all of these impressions simply reflect transport of waterlogged shells and other debris during

Fig. 18.19 *A. Touch mark, USNM 534445, attributed to* Prionocyclus novimexicanus *(Marcou, 1858) (loc. 20, Fig. 18.1). This unusual and rare preservation clearly shows an ammonite that briefly rested on its side. X1. B. A similarly sized specimen of* P. novimexicanus, *USNM 498417, right lateral view, illustrated for comparison, USGS Mesozoic loc. 21191, Carlile Shale, Butte County, South Dakota (Kennedy et al., 2001: Fig. 99F). X0.85.*

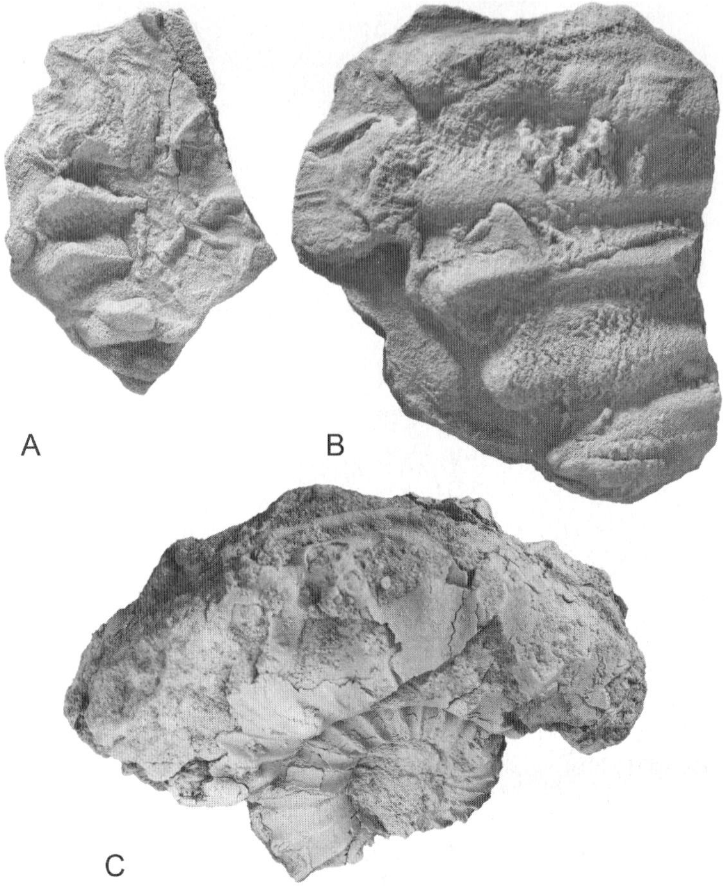

Fig.18.20 *A, B. Touch marks from the middle Turonian part of the Mancos Shale (loc. 12, Fig. 18.1) of northwestern Colorado attributed to* Prionocyclus hyatti *(Stanton, 1894) because this is the only prionocyclid ammonite known from the basal 3 m of the Frontier Sandstone Member in this area. A. USNM 534446. B. USNM 534447. C. A similarly sized specimen of* P. hyatti, *USNM 534448, photographed obliquely for comparison, USGS Mesozoic loc. D14365, Semilla Sandstone, New Mexico. All figures X1.*

periodic storm activity. The extent of waterlogging may have influenced how easily the shells were resuspended, and the kinds of impressions that they made.

5 Conclusions

Touch marks provide a window into the taphonomic history of ammonite shells. If the shells still retained some buoyancy, they would have behaved like lightweight materials, bouncing and skipping on the seafloor. If they were completely waterlogged, they could have been resuspended by bottom currents.

Fig. 18.21 *Touch mark attributed to* Prionocyclus wyomingensis *Meek, 1876, USNM 534486, from the Turner Sandy Member of the Carlile Shale of southwestern South Dakota (loc. 27, Fig. 18.1). X1.*

The formation of touch marks depended on a number of factors related to the ammonites and the paleoenvironment including: (1) the shape of the ammonite shells – inflated or compressed; (2) the kind of ornamentation – keels, spines, ribs, tubercles, or clavi; (3) the state of preservation of the shells – whole or fragmented; (4) the degree of waterlogging of the shells, which is directly related to the amount of residual buoyancy; (5) the strength and direction of the currents; (6) the nature of the contact – dragging, bouncing, or rolling; and (7) the suitability of the bottom sediments to take an impression. After the formation of the impressions, they were immediately cast by overlying silts and fine sands. Subsequently, the bottom could not have experienced major episodes of erosion and bioturbation or the marks would have been completely erased.

Conditions for the formation and preservation of ammonite touch marks seem rare, but the broad geographic distribution of the facies containing scaphitid touch marks in the Ferdig Member of the Marias River Shale of Montana implies that these conditions occasionally prevailed over wide areas. The absence of ammonite touch marks in strata in the US Western Interior above the middle Santonian, on the other hand, suggests that such conditions were not present in this region during the later part of the Cretaceous Period.

Acknowledgments

We thank Dolf Seilacher (Yale University, New Haven, Connecticut), Neal L. Larson (Black Hills Museum of Natural History, Hill City, South Dakota) and Royal H. Mapes (Ohio University, Athens, Ohio) for reviewing an earlier draft of this manuscript and making many valuable suggestions. We thank the US Geological Survey for permission to study their collections. Stephen Thurston prepared the photographs and Stephanie Crooms word-processed the manuscript.

References

Adkins, W. A. 1928. Handbook of Texas Cretaceous fossils. *University of Texas Bulletin* **2838**: 1–385.

Barthel, K. W., N. H. M. Swinburne, and S. Conway Morris. 1990. *Solnhofen: A Story in Mesozoic Paleontology*. Cambridge: Cambridge University Press.

Cobban, W. A. 1952. Scaphitid cephalopods of the Colorado group. *U.S. Geological Survey Professional Paper* **239**: 1–42 (1951 imprint).

Cobban, W. A. 1953. Cenomanian ammonite fauna from the Mosby sandstone of central Montana. *U.S. Geological Survey Professional Paper* **243-D**: D45–D55.

Cobban, W. A. 1988a. Some acanthoceratid ammonites from upper Cenomanian (Upper Cretaceous) rocks of Wyoming. *U.S. Geological Survey Professional Paper* **1353**: 1–17.

Cobban, W. A. 1988b. *Tarrantoceras* Stephenson and related ammonoid genera from Cenomanian (Upper Cretaceous) rocks in Texas and the Western Interior of the United States. *U.S. Geological Survey Professional Paper* **1473**: 1–30.

Cobban, W. A., C. E. Erdmann, R. W. Lemke, and E. K. Maughan. 1976. Type sections and stratigraphy of the members of the Blackleaf and Marias River Formations (Cretaceous) of the Sweetgrass Arch, Montana. *U.S. Geological Survey Professional Paper* **974**: 1–63.

Cobban, W. A., S. C. Hook, and W. J. Kennedy. 1989. Upper Cretaceous rocks and ammonite faunas of southwestern New Mexico. *New Mexico Bureau of Mines and Mineral Resources Memoir* **45**: 1–137.

Cobban, W. A., and G. R. Scott. 1972. Stratigraphy and ammonite fauna from the Graneros Shale and Greenhorn Limestone near Pueblo, Colorado. *U.S. Geological Survey Professional Paper* **645**: 1–108.

Cobban, W. A., I. Walaszczyk, J. D. Obradovich, and K. C. McKinney. 2006. A USGS zonal table for the Upper Cretaceous middle Cenomanian-Maastrichtian of the Western Interior of the United States based on ammonites, inoceramids, and radiometric ages. *U.S. Geological Survey Open- File Report* **2006–1250**, 45p.

Dzułynski, S., and J. E. Sanders. 1962. Current marks on firm mud bottoms. *Transactions of the Connecticut Academy of Arts and Sciences* **42**: 57–96.

Erdmann, C. E., J. T. Gist, J. W. Nordquist, and G. W. Beer. 1947. Map of the areal and structural geology of T. 35 N., R. 3 W., Toole County, Montana, showing oil pools in West Kevin district, Kevin-Sunburst oil field. *U.S. Geological Survey*, January 1947, scale: 1 inch equals 1 mile.

Gaillard, C. 1977. Cannelures d'érosion et figures d'impact dues à des coquilles d'ammonites à épines (Oxfordien supérieur du Jura français). *Eclogae Geologicae Helvetiae* **70**(3): 701–715.

Groff, S. L. 1963. Stratigraphic correlations for Montana and adjacent areas. *Montana Bureau of Mines and Geology Special Publication* **31** (chart).

Haas, O. 1949. Acanthoceratid Ammonoidea from near Greybull, Wyoming. *Bulletin of the American Museum of Natural History* **93**(1): 1–39.

Haas, O. 1951. Supplementary notes on the ammonite genus *Dunveganoceras. American Museum Novitates* **1490**: 1–21.

Kennedy, W. J., and W. A. Cobban. 1991. Coniacian ammonite faunas from the United States Western Interior. *Special Papers in Palaeontology* **45**: 1–96.

Kennedy, W. J., W. A. Cobban, and N. H. Landman. 2001. A revision of the Turonian members of the ammonite subfamily Collignoniceratinae from the United States Western Interior and Gulf Coast. *Bulletin of the American Museum of Natural History* **267**: 1–148.

Landman, N. H. 1987. Ontogeny of Upper Cretaceous (Turonian-Santonian) scaphitid ammonites from the Western Interior of North America: Systematics, developmental patterns, and life history. *Bulletin of the American Museum of Natural History* **185**(2): 118–241.

Lemke, R. W. 1977. Geologic map of the Great Falls quadrangle, Montana. *U.S. Geological Survey Geologic Quadrangle Map* GQ-1414, scale 1:62,500.

Mantell, G. 1822. The fossils of the South Downs, or illustrations of the geology of Sussex. London: Lupton Relfe.

Maeda, H., and A. Seilacher. 1996. Ammonoid taphonomy. *In* N. H. Landman, K. Tanabe, and R. A. Davis (editors), *Ammonoid Paleobiology*, pp. 543–578. New York: Plenum Press.

Marcou, J. 1858. Geology of North America; with two reports on the prairies of Arkansas and Texas, the Rocky Mountains of New Mexico, and the Sierra Nevada of California. Zurich: Zürcher and Furrer.

Meek, F. B. 1876. A report on the invertebrate Cretaceous and Tertiary fossils of the upper Missouri country. *U. S. Geological Survey of the Territories (Hayden)* **9**: 1–629.

Moreman, W. L. 1942. Paleontology of the Eagle Ford Group of north and central Texas. *Journal of Paleontology* **16**(2): 192–220.

Mudge, M. R. 1972. Pre-Quaternary rocks in the Sun River Canyon area, northwestern Montana. *U.S. Geological Survey Professional Paper* **663-A**: 1–142.

Orbigny, A. d'. 1850–52. Prodome de Paléontologie stratigraphique universelle des animaux mollusques et rayonnés 2. Paris: Masson.

Powell, J. D. 1963. Cenomanian-Turonian (Cretaceous) ammonites from Trans-Pecos Texas and northeastern Chihuahua, Mexico. *Journal of Paleontology* **37**(2): 309–322.

Reyment, R. A. 1954. Some new Upper Cretaceous ammonites from Nigeria. *Colonial Geology and Mineral Resources* **4**(3): 248–270.

Rothpletz, A. 1909. Ueber die Einbettung der Ammoniten in die Solnhofener Schichten. Abhandlungen der Mathematisch-Physikalischen Klasse der Königlich Bayerischen Akademie der Wisssenschaften **24**(2): 311–337.

Sageman, B. B., and C. C. Johnson. 1985. Stratigraphy and paleobiology of the Lincoln Limestone Member, Greenhorn Limestone, Rock Canyon anticline, Colorado. *SEPM Field Trip Guidebook 4*, 1985 Midyear Meeting, pp. 100–109. Golden, Colorado.

Scott, G. K., W. A. Cobban, and E. A. Merewether. 1986. Stratigraphy of the Upper Cretaceous Niobrara Formation in the Raton Basin, New Mexico. *New Mexico Bureau of Mines and Mineral Resources Bulletin* **115**: 1–34.

Seilacher, A. 1963. Umlagerung und Rolltransport von Cephalopoden-Gehäusen. *Neues Jahrbuch für Geologie und Paläontologie Monatshefte* **1963**: 593–615.

Stanton, T. W. 1894. The Colorado formation and its invertebrate fauna. *U.S. Geological Survey Bulletin* **106**: 1–288 [1893 imprint].

Stephenson, L. W. 1955. Owl Creek (Upper Cretaceous) fossils from Crowley's Ridge, southeastern Missouri. *U.S. Geological Survey Professional Paper* **274**: 97–140.

Summesberger, H., B. Jurkovšek, and T. Kolar-Jurkovšek. 1999. Rollmarks of soft parts and a possible crop content of Late Cretaceous ammonites from the Slovenian Karst. *In* F. Olóriz, and F.J. Rodriquez-Tovarz (editors), *Advancing Research on Living and Fossil Cephalopods*, pp. 335–344. New York: Kluwer Academic/Plenum.

Twenhofel, W. H. 1939. *Principles of Sedimentation*. New York: McGraw-Hill.

Chapter 19
Some Data on the Distribution and Biology of the Boreal Clubhook Squid *Moroteuthis robusta* (Verrill, 1876) (Onychoteuthidae, Teuthida) in the Northwest Pacific

Alexei M. Orlov

Russian Federal Research Institute of Fisheries and Oceanography (VNIRO), 17, V. Krasnoselskaya, Moscow 107140, Russia, orlov@vniro.ru

Keywords: boreal clubhook squid, *Moroteuthis robusta*, distribution, mantle length, body mass, Kurile Islands, Kamchatka

1 Introduction

The boreal clubhook squid *Moroteuthis robusta* (Verrill, 1876) is one of the largest living squids (Clarke, 1966; Nesis, 1982). This species is considered as a Pacific boreal squid inhabiting the coastal waters of the North Pacific between Japan and California including the Bering Sea (Nesis, 1982). The boreal clubhook squid is an important link in trophic chains of the North Pacific ecosystems. This species feeds on slowly moving benthos or pelagic organisms (Anderson, 1996). It is a food item for sperm whales *Physeter catodon*, fur seals *Callorhynus ursinus* and Pacific sleeper shark *Somniosus pacificus* (Kawakami, 1976; Clarke and MacLeod, 1980; Fiscus, 1993; Orlov, 1999; Flinn et al., 2002).

The boreal clubhook squid was considered to be cosmopolitan and occurring in all oceans (Akimushkin, 1963). In subsequent publications, the range of this squid was limited to only the North Pacific, including the high seas area of this ocean (Nesis, 1982; Roper et al., 1984; Pearcy, 1991; Ishida et al., 1999). Several papers (Hochberg and Fields, 1980; Nesis, 1982; Bizikov, 1996) point out that the boreal clubhook squid inhabits the near-bottom waters of the outer shelf and bathyal zone. Recent studies

N. H. Landman et al. (eds.), *Cephalopods Present and Past: New Insights and Fresh Perspectives*, 423–433.

showed that this squid may often occur in the upper 50 m layer of the Bering Sea (Radchenko, 1992; Sinclair et al., 1999), southern Sea of Okhotsk and adjacent waters (Shuntov et al., 1993; Radchenko et al., 1997; Nagasawa et al., 1998), and Pacific waters off the Kurile Islands (Ivanov, 1998; Ivanov and Sukhanov, 2002). There is a rare occurrence of its capture with bottom trawling in deepwater fisheries (rockfish, halibut, etc.), with bottom longlines when fishing for Pacific halibut and sablefish, or with salmon nets (Anderson, 2002; Hines, 2002; Miller, 2003). In Taiwanese waters this squid is considered as a potential fisheries resource (Mao-Sen Su et al., 2004). Once it was even recorded at fish markets in Seattle (Anderson, 2002). In some areas of the North Pacific, the biomass of clubhook squid may be rather high, reaching 20,200 metric tons in the southern Sea of Okhotsk and adjacent waters (Shuntov et al., 1993; Radchenko et al., 1997) and 27,700 metric tons in the Pacific waters off the Kurile Islands (Ivanov and Sukhanov, 2002).

Despite the long period of study of the North Pacific water area, with its numerous fisheries, frequent occurrence and relatively high abundance of clubhook squid, the data on the biology of this animal, accumulated up to the present time, are extremely scarce. One may consider that enough knowledge is available only on the structure of the gladius and statolith (Bizikov and Arkhipkin, 1997) and the design of beaks (Iverson and Pinkas, 1971), some features of growth (Tsuchiya and Okutani, 1991), chemical composition of tissues (Hayashi et al., 1990), and parasites (McLean et al., 1987; Shinn and McLean, 1989).

The purpose of this paper is to provide some data on distribution, length, and weight frequencies, and the relation between the length of the mantle and body mass of the boreal clubhook squid in the Pacific off the northern Kurile Islands and southeastern Kamchatka.

2 Material and Methods

The outer shelf and mesobenthal, within 80–850 m, were surveyed with bottom trawls from the Japanese trawlers (*Tora maru 58*, *Tomi Maru 82*, and *Tomi Maru 53*) on the Pacific side of the northern Kurile Islands and southeastern Kamchatka in 1993–2002. That work was made pursuant to the joint program of Russian Federal (VNIRO, Moscow), Sakhalin (SakhNIRO, Yuzhno-Sakhalinsk), and Kamchatka (KamchatNIRO, Petropavlovsk-Kamchatsky) Research Institutes of Fisheries and Oceanography aimed to study and assess the stocks of the groundfish species.

The species were identified using identification sheets and descriptions prepared by Nesis (1982). The vertical distribution, and the distribution dependent on bottom temperatures were analyzed by the frequency of occurrence in catches. They were calculated as percentages of hauls with squid within a specific range of depths or temperatures from the total number of tows when squid were taken. The pattern of vertical distribution of 81 individuals of *Moroteuthis robusta* was examined, as well as the distribution of 38 individuals as a function of bottom temperature. The length of the mantle measured along the dorsal side (DML) of 63 squids was used for size frequency analysis. The body mass frequencies were analyzed in a subsample of 41 individuals. Body mass of squids was determined just after capture in the fish factory and included weights of head, tentacles, and viscera. Water and fish were removed from the mantle cavity before weighing. Larger specimens were dissected and the total squid body mass was determined as a sum of the dissected parts.

3 Results

Throughout the entire period of studies in 1993–2002 near the northern Kurile Islands and southeastern Kamchatka in the Pacific, the boreal clubhook squid was found in 61 bottom trawl catches, the frequency of occurrence being less than 1%. On the whole, the share of squid within any catch-yielding hauls was 1.4% by weight. The catch per haul varied from 1 to 4 specimens; respective averages were 1.33 individuals (ind.) and 14.38 kg. As recalculated by 1 h hauls, we obtained the mean values of 0.88 ind. and 6.27 kg. The species occurred all across the area surveyed from 52°N in the north to 47°50'N in the south (Fig. 19.1). However, the

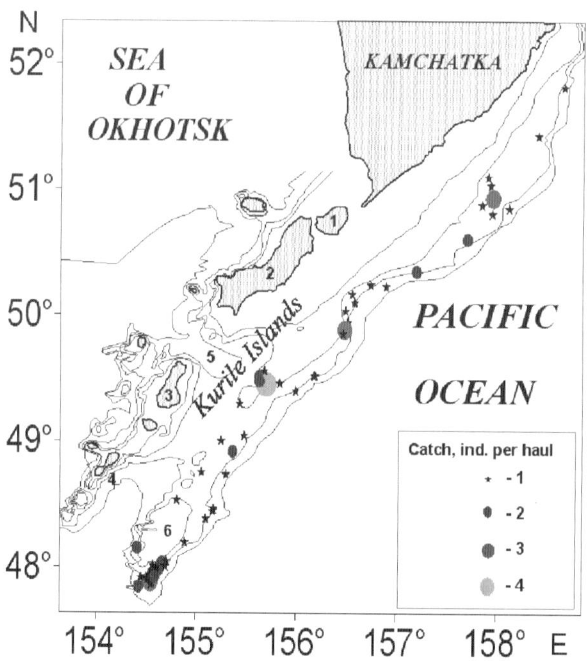

Fig. 19.1 *Spatial distribution and relative abundance (ind. per haul) of the boreal clubhook squid in the Pacific waters off northern Kurile Islands and southeastern Kamchatka, 1993–2002 (1 – Shumshu Isl., 2 – Paramushir Isl., 3 – Onekotan Isl., 4 – Shiashkotan Isl., 5 – Fourth Kurile Strait, 6 – underwater plateau; thin lines are isobaths 100, 200, 500, and 1000 m).*

maximum concentration was recorded against the Fourth Kurile Strait, though it was on the southern part of the slope of the underwater plateau southeast of Shiashkotan Island that *Moroteuthis robusta* was most frequent.

In the Pacific waters of the northern Kurile Islands and southeastern Kamchatka, the boreal clubhook squid occurred in catches taken from 198–620 m (average depth of 399.4 m). Meanwhile, the majority of squids (66.7%) were recorded within the 351–550 m range of depths (Fig. 19.2).

The boreal clubhook squid seen in the Pacific waters off the northern Kurile Islands and southeastern Kamchatka occurred at bottom temperatures of 0.45–4.15°C (average 3.09°C); most individuals (72.9%) were recorded within 3.1–4.0 °C (Fig.19.3).

In waters of the Pacific near the northern Kurile Islands and southeastern Kamchatka, the species under review was represented by individuals having 43–184 cm mantles (average 91.76 cm). No single size group prevailed (Fig. 19.4). The squids having 46–55 cm (15.9%) and 111–135 cm (28.6%) mantle were most frequent.

In our catches the species' body mass was within 1.1–36.4 kg (average 11.91 kg). Over half of those individuals (51.2%) weighed less than 10 kg each (Fig. 19.5).

Given below is the relation obtained from 41 measurements ($R^2 = 0.9266$):

$W = 1.839 \times 10^{-4} \times DML^{2.4290}$, where

W is total body mass (kg), and

DML is mantle length measured along the dorsal side (cm).

The values calculated by the formula given agree well with the empirical data (Fig. 19.6).

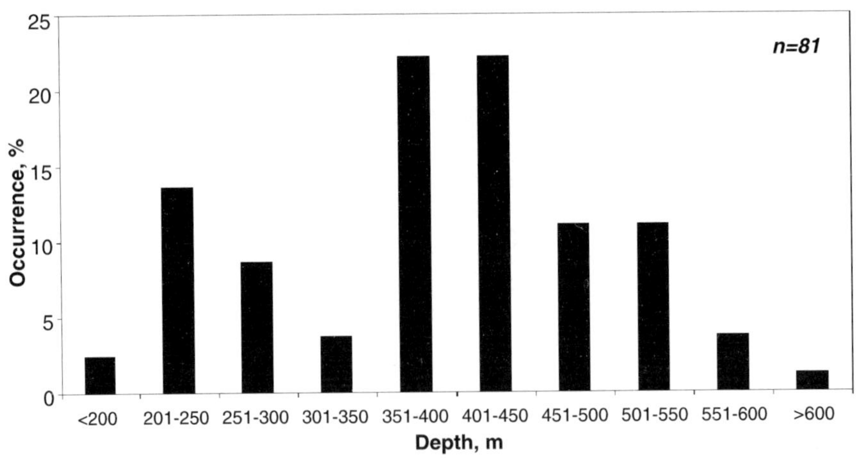

Fig. 19.2 *Vertical distribution of the boreal clubhook squid in the Pacific waters off the northern Kurile Islands and southeastern Kamchatka, 1993–2002 (n = number of observations).*

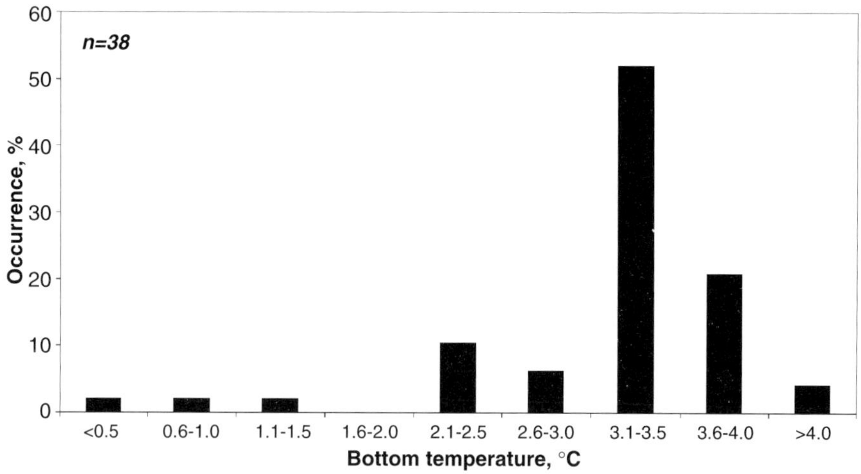

Fig. 19.3 *Distribution of the boreal clubhook squid in the Pacific waters off the northern Kurile Islands and southeastern Kamchatka depending on bottom temperature, 1993–2002 (n = number of observations).*

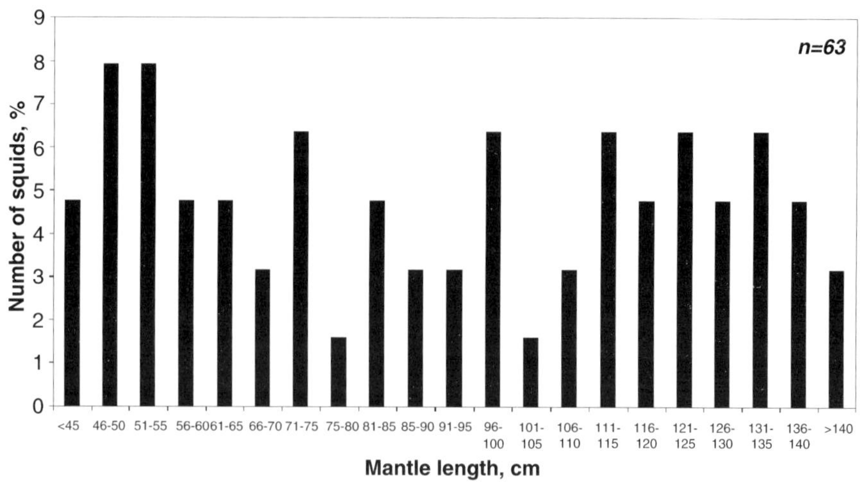

Fig. 19.4 *Length frequencies of the boreal clubhook squid in the Pacific waters off the northern Kurile Islands and southeastern Kamchatka, 1993–2002 (n = number of squids measured).*

4 Discussion

Comparison of our data with those of Ivanov (1998) show that the boreal clubhook squid is less frequently caught near the bottom than within the upper (0–50 m) pelagic layer (<1% vs. 17.1% respectively). On the one hand, these data are in

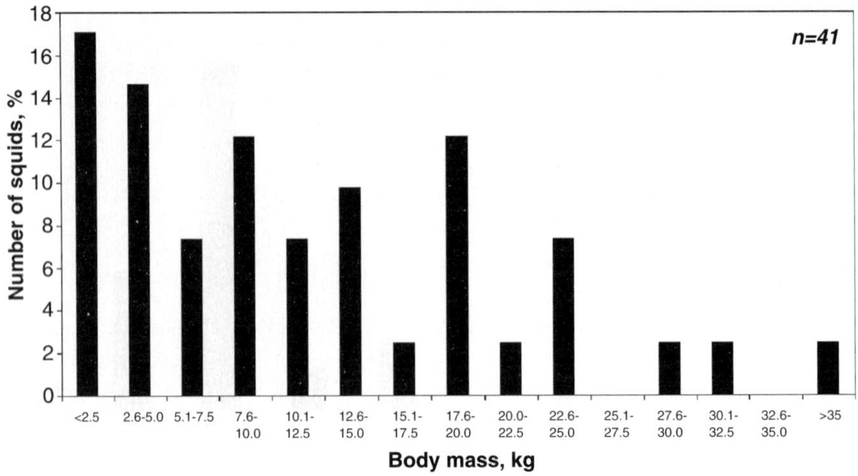

Fig. 19.5 *Weight frequencies of the boreal clubhook squid in the Pacific waters off the northern Kurile Islands and southeastern Kamchatka, 1993–2002 (n = number of squids weighed).*

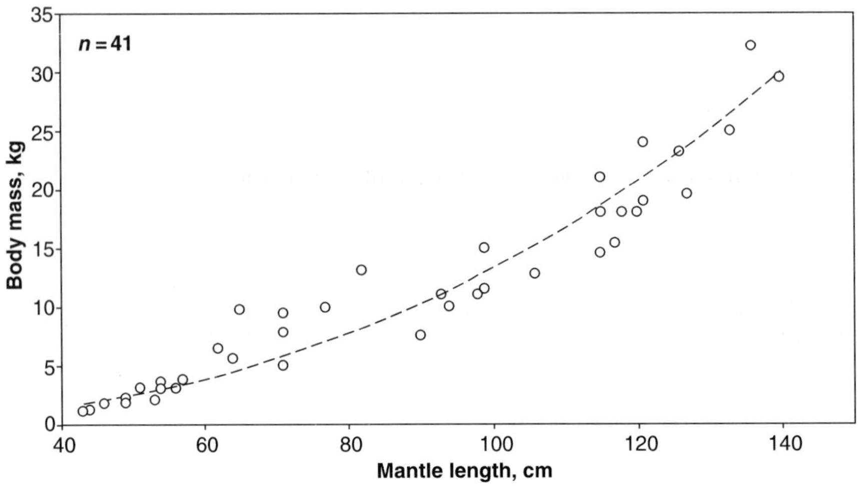

Fig. 19.6 *Length–weight relationship in the boreal clubhook squid in the Pacific waters off the northern Kurile Islands and southeastern Kamchatka, 1993– 2002 (n = number of squids weighed and measured).*

contradiction with the suggestion (Hochberg and Fields, 1980; Nesis, 1982; Bizikov, 1996) that this squid primarily inhabits near-bottom layers of the outer shelf and bathyal zone. On the other hand, this may not reflect real species' abundance or distribution patterns near the bottom and in upper ocean layers but is the result of different recovery capability of bottom trawls (vertical opening 5–7 m) and large midwater trawls (vertical and horizontal openings 50–80 m).

The data available on the species' frequency of occurrence make it possible to assume that this species is much less abundant in the eastern part of the North Pacific. One proof of this is that, as some data indicate, only six individuals of the boreal clubhook squid were caught during the last 29 years in the waters off the coast of Washington, which yield over 100,000 metric tons of fish annually (Miller, 2003). According to other sources (Anderson, 2002), somewhat over 50 squid of this species were found in the Puget Sound area (Washington) between 1949 and 2002. In our case, over 60 individuals of this species were recorded only in bottom trawl catches during the decade of research.

The data available on the depths of the species' habitat are scarce, and they are somewhat contradictory. There is a view that this species is common at great depths (Meachum, 2002). Nesis (1982) pointed out that this squid inhabits the near-bottom layers of the lower sublittoral and upper bathyal zones. The only paper suggesting a concrete vertical range for the species of 100–600 m is Hochberg and Fields (1980). A lot of papers indicated frequent and even abundant captures of this squid in the epipelagic layer (Pearcy, 1991; Radchenko, 1992; Shuntov et al., 1993; Radchenko et al., 1997; Nagasawa et al., 1998; Ivanov, 1998; Sinclair et al., 1999; Ishida et al., 1999; Ivanov and Sukhanov, 2002). There have been records of capturing the boreal clubhook squid in the Bering Sea from depths of 300–450 m (Bizikov and Arkhipkin, 1997; Miller, 2003) whereas it was caught at 700–910 m off the US West Coast (Anon., 2002; Hines, 2002). The pattern of its vertical distribution in the area studied is similar to that of the abundant northwest Pacific near-bottom Commander squid *Berryteuthis magister* (both species have similar ranges and inhabit biotopes of the same sort); according to various sources, the latter squid's concentration density is greatest in the range of 150–600 m (Railko, 1983; Nesis, 1989; Fedorets et al., 2000).

The temperatures in the species' habitats are virtually unknown. Anderson (2002) pointed out that the water temperature around Puget Sound fluctuates from 6°C to 13°C, while it is about 7°C at the depth of 250 m off the coast of Washington. In temperate waters, boreal clubhook squid occurs in the ocean at surface temperatures ranging from 6°C to 25°C with maximum frequency from 6°C to 9°C (Pearcy, 1991). In terms of thermal conditions, the species' habitat is close to that of the Commander squid whose major concentrations tend to be found at bottom temperatures exceeding 3.65°C (Verkhunov, 1996).

The boreal clubhook squid ranks second by size among the squids occurring in the North Pacific, the first being *Architeuthis japonicus* whose mantle can be up to 6 m of the total length of over 20 m (Roper et al., 1984). The maximum length data on the clubhook squid are quite contradictory. Akimushkin (1963) suggests that the maximum size is 6.8 m (total length with tentacles) and 2.47 m (mantle length). The maximum length data provided by Miller (2003) are close to that (6.1 m). Other authors put forward the values of 2 m (Roper et al., 1984) or 2.5 m (Nesis, 1982) as maximum lengths of the mantle, or over 4 m as an overall length (Anderson, 1996). Several papers (Armstrong, 2001; Anon., 2002; Hines, 2002) refer to information that the boreal clubhook squid can reach the length of 914 cm (30 ft.) which appears unlikely since there have been no documented reports of finding squids of this size. Our data agree with those of Hochberg and Fields (1980) that the squids of this species with a mantle of about 1 m were most frequently caught. Akimushkin (1963) reported that 1.5–2 m squids are not rare. Of the species found in bottom trawl catches in the western Bering Sea, 11 individuals had 97.5–141.5 cm long mantles (Bizikov and Arkhipkin, 1997). Some dead squids, 152–335 cm long have sporadically been found on the US West Coast (Armstrong, 2001; Meachum, 2002; Miller, 2003), though it is unclear which length of the dead specimens was measured: total length or that of the mantle. Consequently, the shortage of data prevents a current judgment regarding similarity or disparity in the size composition of the boreal clubhook squid in various parts of its range. In the western Bering Sea, squids captured were approximately of the same size (97.5–141.5 cm long mantles), age, sex, and stage of maturity (immature females) (Bizikov and Arkhipkin, 1997). This fact, according to the opinion of those authors demonstrates that the western Bering Sea serves as a feeding ground for immature females of the boreal clubhook squid. Unfortunately, we have no data on sex composition of our squid catches. The longevity of this species is unknown though it is suggested that its life span exceeds one year (Bizikov and Arkhipkin, 1997). The size composition (very small number of squids with 75–80 cm and 100–105 cm DML) of our catches may reflect the complex age structure of the population with the existence of two or three age groups. Juvenile boreal clubhook squids are known mostly from waters off Japan (Tsuchiya and Okutani, 1991). Specimens with mantle less than 95 cm do not occur in the western Bering Sea (Bizikov and Arkhipkin, 1997) but are very abundant (52.4%) in the Pacific off the northern Kurile Islands and southeastern Kamchatka (our data). Therefore, it is possible to suggest that the tropical and subtropical waters from Taiwan and Japan to California are the spawning grounds of the boreal clubhook squid. The temperate and boreal waters of the North Pacific (including Bering Sea) serve as feeding grounds for this species. It is likely that squids perform feeding migrations gradually with growth and return to spawning grounds while mature; the life span may take about 2–3 years.

The body mass data of the species in question are contradictory as well. Roper et al. (1984) suggested 50 kg as the maximum weight. According to other sources (Miller, 2003), the weight of the boreal clubhook squid may reach 181–272 kg, though findings of squids having such a body mass have not been documented. The recorded weight of the individuals near the US West Coast was 36.3–45.4 kg (Armstrong, 2001; Meachum, 2002; Miller, 2003), which agrees with the previously published data (Roper et al., 1984).

The mantle length-body mass relationship in the species under review has not been previously investigated. The exponential coefficient in the equation obtained is close to that of the Commander squid (2.5312–2.8028 depending on the sex and state of maturity) of the western Bering Sea (Bizikov and Golub, 1996), which may be evidence of the similarity in the mechanisms of growth in both species.

5 Conclusions

The range of the boreal clubhook squid in the North Pacific includes coastal waters (outer shelves and continental slopes) and high seas from Taiwan and Japan to California in the south to the Bering Sea in the north. It is likely that the abundance of the boreal clubhook squid is higher in the northwestern Pacific than in the northeastern part of the ocean.

The known vertical range of this squid is from upper ocean layers to 910 m off the bottom. It seems that this species is more abundant within epipelagic ocean layers than near the bottom at outer shelves and continental slopes. Near the bottom, the majority of squids occupy the 350–500 m depth range.

The bottom temperature at which boreal clubhook squid is caught, ranges from 6°C to 25°C (modal 6°C –9°C) within upper oceanic layers and from 0.5°C to 4.2°C (modal 3°C –4°C) off the bottom.

The boreal clubhook squid is one of the largest living squids, with the mantle reaching 2–2.5 m and body mass about 50 kg. The majority of individuals have a mantle length about 1 m and body mass less than 10 kg.

Spawning grounds of boreal clubhook squid are most probably the tropical and subtropical waters from Taiwan and Japan to California. Temperate and boreal waters of the North Pacific (including the Bering Sea) serve as feeding grounds for this species. It is likely that squids perform feeding migrations gradually with growth and return to spawning grounds while mature. The life span of the species may take about 2–3 years.

References

Akimushkin, I. I. 1963. *Cephalopod Mollusks of the USSR Seas*. Moscow: USSR Academy of Science [In Russian].

Anderson, R. C. 1996. Records of *Moroteuthis robusta* (Cephalopoda: Onychoteuthidae) in Puget Sound (Washington State, USA). *Of Sea and Shore* **19**: 111–113.

Anderson, R. C. 2002. Records of the robust clubhook squid *Moroteuthis robusta* (Cephalopoda: Onychoteuthidae) in Puget Sound (Washington State, USA). *World Wide Web Electronic Publication*. http://www.dal.ca/~ceph/TCP/Mrobusta.html. 25 November 2002.

Anonymous, 2002. Squid in college. *Newsletter of the School of Aquatic & Fishery Sciences University of Washington,* Spring 2002. Seattle, WA: University of Washington.

Armstrong, B. 2001. Of Sea Vipers and Leviathans. *Alaskan Southeaster*, September 2001.

Bizikov, V. A. 1996. *Atlas of Morphology and Anatomy of Squid Gladius*. Moscow: VNIRO Publishing [In Russian].

Bizikov, V. A., and A. I. Arkhipkin. 1997. Morphology and microstructure of the gladius and statolith from the boreal Pacific giant squid *Moroteuthis robusta* (Oegopsida; Onychoteuthidae). *Journal of Zoology London* **241**: 475–492.

Bizikov, V. A., and A. N. Golub. 1996. Size-mass relationship of *B. magister. In* A. A. Jelizarov (editor), *Commercial Aspects of Biology of Commander Squid Berryteuthis magister and of Fishes of Slope Communities in the Western Part of the Bering Sea*, pp. 38–40. Moscow: VNIRO Publishing [In Russian].

Clarke, M. R. 1966. Review of systematics and ecology of oceanic squids. *Advances in Marine Biology* **4**: 91–300.

Clarke, M. R., and N. MacLeod. 1980. Cephalopod remains from sperm whales caught off western Canada. *Marine Biology* **59**: 241–246.

Fedorets, Y. A., P. P. Railko, V. D. Didenko, et al. 2000. *Guide to Search Schoolings and Fishery on Commander Squid off the Kuril Islands*. Vladivostok: TINRO-Center [In Russian].

Fiscus, C. H. 1993. Catalogue of cephalopods at the National Marine Mammal Laboratory. *United States Department of Commerce, NOAA Technical Memorandum NMFS-AFSC* **16**: 1–183.

Flinn, D., A. Trites, E. J. Gregr, and R. I. Perry. 2002. Diets of fin, sei, and sperm whales in British Columbia; an analysis of commercial whaling records, 1963–1967. *Marine Mammal Science* **18**: 663–679.

Hayashi, K., H. Kishimura, and Y. Sakurai. 1990. Level and composition of diacyl glyceryl ethers in the different tissues and stomach contents of giant squid *Moroteuthis robusta. Bulletin of the Japanese Society of Scientific Fisheries* **56**: 1635–1990.

Hines, S. 2002. Squid gets education instead of dinner. *University week: University of Washington* **19** (25 April): np.

Hochberg, F. G., and W. G. Fields. 1980. Cephalopoda: the squids and octopuses. *In* R. H. Morris, D. P. Abbott, and E. C. Haderlie (editors), *Intertidal Invertebrates of California*, pp. 429–444. Stanford, CA: Stanford University Press.

Ishida, Y., T. Azumaya, and M. Fukuwaka. 1999. Summer distribution of fishes and squids caught by surface gillnets in the western North Pacific Ocean. *Bulletin of Hokkaido National Fisheries Research Institute* **63**: 1–18.

Ivanov, O. A. 1998. Epipelagic community of fishes and squids in the Kuril Island area of the Pacific Ocean in 1986–1995. *Izvesiya TINRO* **124**: 3–54 [In Russian].

Ivanov, O. A., and V. V. Sukhanov. 2002. *Structure of Nektonic Communities of Kuril Islands waters*. Vladivostok: TINRO-Center [In Russian].

Iverson, I. K. L., and L. Pinkas. 1971. A pictorial guide to beaks of certain eastern Pacific cephalopods. *California Department of Fish and Game Fishery Bulletin* **152**: 83–105.

Kawakami, T. 1976. Squids found in the stomach of sperm whales in the northwestern Pacific. *Scientific Reports of Whales Research Institute Tokyo* **3**: 145–151.

McLean, N., F. G. Hochberg, and G. L. Shinn. 1987. Giant protistan parasites on the gills of cephalopods. *Disease of Aquatic Organisms* **3**: 119–125.

Mao-Sen Su, Don-Chung Liu, Shih-Chin Chen, and Ya-Ke Hsu. 2004. *Introduction to Fisheries Research Institute*. Keelung, Taiwan: Fisheries Research Institute.

Meachum, L. 2002. Research is squid's life after death. *The Sun*, 17 July.

Miller, B. 2003. Graduate student catches giant squid in Bering Sea. *The Daily Mississippian*, 5 September.

Nagasawa, K., Y. Ueno, J. Sakai, and J. Mori. 1998. Autumn distribution of epipelagic fishes and squids in the Okhotsk Sea and western North Pacific Ocean off the Kuril Islands and southeast Hokkaido. *Bulletin of National Research Institute of Far Seas Fisheries* **35**: 113–130.

Nesis, K. N. 1982. *Brief Guide to Cephalopods of the World Oceans*. Moscow: Lyogkaya I Pishchevaya Promyshlennost [In Russian].

Nesis, K. N. 1989. Teuthofauna of the Sea of Okhotsk. Biology of squids *Berryteuthis magister* and *Gonatopsis borealis* (Gonatidae). *Zoologichesky Zhurnal* **68**: 45–56 [In Russian].

Orlov, A. M. 1999. Capture of especially large Pacific sleeper shark *Somniosus pacificus* (Squalidae) and some notes on its ecology in the northwestern Pacific. *Voprosy Ikhtiologii* **39**: 558–563 [In Russian].

Pearcy, W. G. 1991. Biology of the transition region. *In* J. A. Wetherall (editor), Biology, oceanography, and fisheries of the North Pacific Transition Zone and Subarctic Frontal Zone. *NOAA Technical Report NMFS* **105**: 39–56.

Radchenko, V. I. 1992. The role of squids in the Bering Sea pelagic ecosystem. *Okeanologiya* **32**: 1093–1101 [In Russian].

Radchenko, V. I., I. V. Melnikov, A. F. Volkov. 1997. Environmental conditions and composition of plankton and nekton in epipelagic layer of the southern Sea of Okhotsk and adjacent Pacific waters in summer. *Biologiya Morya* **23**: 15–25 [In Russian].

Railko, P. P. 1983. Biology and distribution of the Commander squid *Berryteuthis magister* in the area of Kuril Islands. *In* Y. I. Skorobogatov, and K. N. Nesis (editors), *Taxonomy and ecology of Cephalopoda,* pp. 97–98. Leningrad: Zoological Institute [In Russian].

Roper, C. F. E., M. J. Sweeny, and C. E. Nauen. 1984. Cephalopods of the world. *FAO Fisheries Synopsis,* **125**: 1–277.

Shinn, G. L., and N. McLean. 1989. *Hochbergia moroteuthensis* gen. et sp. nov., a giant protistan parasite from the giant squid *Moroteuthis robusta* (Mollusca: Cephalopoda). *Disease of Aquatic Organisms* **6**: 197–200.

Shuntov, V. P., V. I. Radchenko, V. I. Chuchukalo, et al. 1993. Structure of planktonic and nektonic communities of the upper epipelagic zone in the Sakhalin-Kuriles region in the period of anadromous migrations of salmons. *Biologiya Morya* **4**: 32–43 [In Russian].

Sinclair, E. H., A. A. Balanov, T. Kubodera, et al. 1999. Distribution and ecology of mesopelagic fishes and cephalopods. *In* T. R. Loughlin, and K. Ohtani (editors), *Dynamics of the Bering Sea*, pp. 485–508. Fairbanks, AK: University of Alaska Sea Grant.

Tsuchiya, K., and T. Okutani. 1991. Growth stages of *Moroteuthis robusta* (Verill, 1881) with the re-evaluation of the genus. *Bulletin of Marine Sciences* **49**: 137–147.

Verkhunov, A. V. 1996. Role of oceanographic factors in formation of aggregations of *B. magister* and demersal fishes. *In* A. A. Jelizarov (editor), *Commercial Aspects of Biology of Commander Squid Berryteuthis magister and of Fishes of Slope Communities in the Western Part of the Bering Sea*, pp. 150–155. Moscow: VNIRO Publishing [In Russian].

Verrill, A. E. 1876. Note on gigantic cephalopods: A correction. *American Journal of Science and Arts* **12**: 236–237.

Chapter 20
Habitat Ecology of *Enteroctopus dofleini* from Middens and Live Prey Surveys in Prince William Sound, Alaska

D. Scheel, A. Lauster, and T. L. S. Vincent

Environmental Sciences, Alaska Pacific University, Anchorage, AK 99508, dscheel@alaskapacific.edu

Keywords: octopus, Alaska, ecology, habitat, population

1 Introduction

The population ecology of mobile predators is often regulated by a complex mix of factors. Understanding the mechanisms that regulate the abundance and distribution of organisms remains a central goal of marine ecology (Estes and Peterson, 1998). Recent studies have emphasized the success of using multiple mechanisms to understand control of population dynamics, particularly in marine species (National Science Foundation, 1998; Connolly and Roughgarden, 1998; Karlson, 2002). At a fundamental level, population change may be represented as births (recruitment) minus deaths plus immigration minus emigration. The latter two may be combined as habitat selection. In this paper, we focus on habitat selection in the giant Pacific Octopus *Enteroctopus dofleini* (Wülker, 1910), whether habitat selection can be

N. H. Landman et al. (eds.), *Cephalopods Present and Past: New Insights and Fresh Perspectives*, 434–458.

linked to diet and prey availability, and the potential role of habitat selection in local population fluctuations.

The dynamics of habitat association in octopus, including *Enteroctopus dofleini*, are complex and likely important to fisheries and marine communities (e.g., Garstang, 1900; Griffiths and Hockey, 1987; Smith and Herrnkind, 1992; Mather, 1993). However, habitat selection by octopuses has received little rigorous attention (Anderson, 1997). Limited research on octopus distribution suggests that many coastal benthic species are characterized by substantial density fluctuations at a particular site (Garstang, 1900; Rees and Lumby, 1954; Mather, 1982), and that this is also the case for *E. dofleini* (see Hartwick and Thorarinsson, 1978; Hartwick et al., 1984a). While both ontogenetic and migratory mechanisms are suggested to account for this, the emphasis for *E. dofleini* has been on migration (Hartwick, 1983; Hochberg, 1998). Octopuses may exercise substantial choice in matters of diet and shelter (Iribarne, 1990; Laidig et al., 1995; Anderson, 1997) and often modify shelters to their satisfaction (Legac, 1969; Mather, 1994; D. Scheel, personal observation, 1998).

A number of physical habitat characteristics may influence octopus density and distribution (Anderson, 1997; Scheel, 2002). The availability of shelters influences octopus distribution and may limit abundance (Aronson, 1986), although not always in obvious ways (Frazer and Lindberg, 1994) nor for populations in rocky or reef habitats (Ambrose, 1988; Anderson, 1997). Dens do not appear to be limiting for *Enteroctopus dofleini* in the eastern North Pacific (Hartwick, 1983; Hartwick et al., 1984b), although this may not be the case in the western North Pacific where a successful commercial lair-pot fishery for this species suggests that dens are in demand (Hartwick, 1983), at least during the migratory movements.

Physical habitat characteristics may have a large influence on *Enteroctopus dofleini* ecology, as suggested by differences between eastern and western North Pacific movement patterns (for review of fisheries biology, see Hartwick and Barriga, 1997; Gillespie et al., 1998). In waters off Vancouver Island (eastern North Pacific), Mather et al., (1985) found that *E. dofleini* had relatively small home ranges over a 15-day period; and Hartwick et al., (1984b) found that 59% of tagged individuals occupied a single den for over a month. Hartwick et al., (1988) used lair-pot sampling at adjacent onshore and offshore sites to capture, tag, and release 191 individuals. On recapture of 81, none had moved between the two sites. This site fidelity in rocky habitats contrasts with data from Japan, where tagged *E. dofleini* were recovered in a commercial fishery by lair-pot trap-lining on soft substrate fishing grounds (Mottet, 1975; Nagasawa et al., 1993). Thirty-two percent of recovered octopuses (85 of 264 tags) made relatively long-distance movements of 5–50 km over periods of 6–18 months (Sato, 1994, 1996); the remainder moved less than 5 km. No evidence from the eastern Pacific indicates this scale of movement, despite at least a handful of studies that had the potential to detect it. Rigby (2003, 2004) found that the timing of movements in Japanese waters corresponded to onshore temperatures exceeding the bounds of optimal physiological functioning of

E. dofleini (7°C–9.5°C). This suggests that differences in seasonal temperature fluctuations between the eastern and western Pacific may account for differences in behavior, and underscores the importance of habitat selection to octopus ecology and fisheries. However, further data are required to warrant this conclusion.

In the eastern Pacific, neither den availability nor seasonal temperature changes appear to drive the distribution of *Enteroctopus dofleini*; and it is reasonable to consider whether biological factors such as the distribution of prey, predators, or vegetative cover may drive patterns of distribution and abundance. Few studies have obtained quantitative data about predation on octopuses, although several studies documented the presence of potential predators (including fish and marine mammal predators) or indirect evidence of partial predation (i.e., scars or partial arm loss). Hartwick et al. (1988) report a higher incidence of such scars on *Enteroctopus dofleini* from deeper habitats (37–110 m), as well as the occurrence of octopuses in the stomachs of fish predators (Lingcod *Ophiodon elongatus* and Spiny dogfish *Squalus acanthias*) from those depths. Octopuses feed on a diverse array of prey species and are often considered to be generalist predators. Field studies have found that octopus diets, including those of *E. dofleini*, may be influenced by both food preferences and prey availability (Hartwick et al., 1981; Ambrose, 1984). Thus, prey availability is also likely to influence octopus distribution (Anderson, 1997).

In the eastern north Pacific, juvenile *Enteroctopus dofleini* (to approximately 15 kg) found in shallow water occupy dens under boulders and in rock crevices, and feed on a variety of benthic prey, including primarily crabs and bivalves (Hartwick et al., 1981; Vincent et al., 1998). During this life stage, they appear to be central-place foragers (*sensu* Curio, 1976) occupying home ranges of less than 250 m^2 and switching dens frequently (Hartwick et al., 1984b; Mather et al., 1985). Prey are often brought back to the den and consumed there; the remains of hard-shelled prey accumulate at the mouth of the den as a midden. Diet as characterized by midden piles collected from dens has been a principle avenue of inquiry (Hartwick et al., 1981; Mather et al., 1985; Cosgrove, 1987; Anderson, 1994; Vincent et al., 1998; Dodge and Scheel, 1999), although very little work has been done to examine what portion of diet is represented in midden piles or to understand the foraging ecology of juvenile *E. dofleini* (for exceptions, see Mather et al., 1985; Vincent et al., 1998). In this part of their range, *E. dofleini* appear to be dependent on rocky habitat for denning, although they apparently forage in other habitats such as eel grass beds and sand flats as well as in rocky areas (Hartwick et al., 1981).

In intertidal habitats in Prince William Sound, Alaska, Scheel (2002) found *Enteroctopus dofleini* on soft substrates (mud, sand, or gravel) rather than hard (bedrock, rock outcrops, large cobble fields) and where boulders were present rather than absent. Octopus numbers increased fivefold in areas adjacent to dense kelp versus those adjacent to areas with sparse kelp (Scheel, 2002). Diet composition appeared to be variable with small changes in depth or location (Vincent et al., 1998), possibly because of concomitant changes in prey populations. In a small sample of intertidal prey abundance, octopus counts were higher on those transects where live crabs were more abundant, suggesting that selection of habitats within the intertidal may be influenced by prey abundance (Vincent et al., 1998).

This paper examines prey availability and quality (energy value, handling time) as aspects of habitat quality associated with local variation in counts of *Enteroctopus dofleini* in Prince William Sound, Alaska, and describes efforts to address the role of habitat selection and foraging behavior in octopus ecology.

2 Methods

2.1 Study Sites and Population Trends

Observations were made along the south-central Alaskan coast during May, June, and July from 1995–1996, 1998, and 2001–2004. Most work was conducted in Prince William Sound, AK. The majority of sites around Prince William Sound were visited once in 1995, 1996, or 1998. In addition, several sites were visited repeatedly: most notably Gibbon's Anchorage on Green Island (60°16°N, 147°27°W) plus a nearby cove, and Ellamar near the village of Tatitlek (60°53°N, 146°42°W), a site at which octopuses are harvested, plus a site on nearby Busby Island where no known harvest occurs. A description of habitats surveyed and octopus counts for the years 1995, 1996, and 1998 was previously published (Scheel, 2002).

Population data are based on beach walk transects conducted at these sites. See Scheel (2002) for detailed methods. Transect data for 2001–2004 were collected at the Green Island, Ellamar, and Busby Island sites mentioned above. Differences in octopus density (count per 1,000 m^2 surveyed area) between years and among sites were analyzed with nonparametric median tests and Kruskal–Wallace one-way analyses of variance by ranks (SPSS 12.0 2003), as counts and densities were nonnormal (zero was the most frequent value).

2.2 Prey: Middens, Live Surveys, and Energetics

In conjunction with transect sampling, remains of recent prey items (estimated to be <7 days old based on algae growth on inner surfaces and signs of weathering) were collected from all middens found on each transect, identified to species, measured, and counted (minimum number of individuals represented). Octopus bite or drill marks were recorded for each prey item (Dodge and Scheel, 1999; Vincent et al., 1998). Surveys of live prey were also conducted. Live prey surveys consisted of 0.5×0.5 m quadrats in 1996 and 1.0×1.0 m quadrats in 2001–2004, located on beach walk transects, at the start and at 1–9 randomly chosen distances from the start along the transect length. Within each quadrat, rocks, and kelp were moved and all epibenthic mobile individuals of potential prey were identified to species and counted. Rocks were replaced to their original orientation and position to minimize disruption to the habitat; rocks too large to replace precisely were sampled to the best of our ability without moving. Carapace size or shell width were recorded for bivalves, gastropods, chitons, and crabs. Specimens were collected, frozen, and held at –40°C for subsequent analysis of energy content. Analyzed samples were thawed, dissected, the soft tissue dried in a drying oven, and pelletized. Pellets were combusted using a Parr Microbomb calorimeter to determine energy content. Analyses in this paper focus on the crabs *Cancer oregonensis*, *Telmessus cheiragonus*, *Lophopanopeus bellus*, and *Pugettia gracilis* (major hard-bodied prey, Vincent et al., 1998), as well as *Cancer productus* and the bivalve *Protothaca staminea*, the next most frequent hard-bodied prey that commonly occurred in octopus middens in Prince William Sound intertidal habitats. These six taxa are hereafter referred to as "major hard-bodied prey species." Two additional crab species, *Hapalogaster mertensii* and *Cryptolithodes sitchensis*, were also included in some analyses to contrast species comprising a minor portion of prey remains but commonly occurring in live prey surveys.

2.3 Relocation Experiment

Three octopuses in 1996 and four in 2003 were captured, tagged, and released; and their subsequent selection of habitats monitored. All tagged and released octopuses were obtained within the intertidal zone during beach walk surveys. Sonic-tagged animals were given unique identifiers (names) for record-keeping. The 1996 animals were released where captured (control for handling artifacts, *sensu* Chapman, 1986); those in 2003 were relocated away from the preferred octopus habitat and released either in shallow water (relocation control) or deep water (experimental treatment). Octopuses caught were weighed, sexed, and measured (interocular and mantle length). Midden piles were collected from the dens of these animals for analysis of diet. Octopus 4 kg or larger were transferred to a research vessel for anesthesia and tagging and were placed in holding tanks with flow-through seawater. Cold-water anesthesia was used (Anderson, 1996; Andrews and Tansey, 1981). Octopus and seawater were gradually chilled to 3°C, at which temperature they became nonresponsive, and held at 2–3°C during tag implantation. Tags were either attached through arm muscle on cable-tie loops (1996) or implanted through a small incision in the dermal layer only, above the web, on the third left arm of the octopus (2003). The implant wound was blotted dry, pinched closed, and sealed with Nexaband, a veterinary adhesive. Respiration rates, color, and posture returned to preanesthesia condition shortly after the water and octopus were rewarmed to ambient temperature. It should be noted here that all implanted tags (2003) were lost within 24 h of release of the octopus (see Results). Loss of a tag (regardless of attachment method) was suspected after a period of at least 24 h without any tag movement; in several cases (2 of 3 in 1996; 1 of 4 in 2003), tags were confirmed lost by recovering of the tag from a recently occupied octopus den. Due to loss of tags in 2003 within 24 h of implant, subcutaneous implantation by the method described here cannot be recommended for future studies.

Capture, release, and relocation habitats were surveyed via SCUBA surveys for all sites, except one below 40 m depth. That site was surveyed by camcorder lowered to the site. In each case a 1 m^2 plot was sampled. Variables collected were depth, slope, substrate, kelp and *Zostera* cover, available prey, and presence or absence of boulders (Scheel, 2002). A habitat quality index (HQI) was calculated as the sum of a depth index (shallow, <10 m = 1; deeper = 0), substrate (soft, silt to gravel = 1, or hard, cobble to bedrock = 0), boulders (present = 1, absent = 0), kelp cover (by percent cover 0, 0.25, 0.5, etc.), and presence of major hard-bodied prey species (Vincent et al., 1998; 0.2 per species constrained to a maximum prey score of 1.0). Two versions of this index were calculated, one including all five habitat characters (HQI_5, which thus varies from one to five) and one without prey data (HQI_4, which varies from one to four). Only HQI_4 were calculated for 1996 surveys, as prey data were not collected at that time. For either index, increasing scores indicate habitat characteristics associated with greater octopus counts (Vincent et al., 1998; Scheel, 2002).

We calculated HQI_4s for all transects (1995–2004). Transects with missing data were omitted. The distribution of HQI_4 scores was examined to confirm whether HQIs actually correlate with octopus count at a site.

3 Results

Population trends are based on 82 beach walk transect surveys described in Scheel (2002, only transects in Prince William Sound included) plus an additional 25 beach walk transects conducted in the period 2001–2004 using the same methods. On these transects, we recorded 90 octopuses in a total of 333,000 m^2 area surveyed. Prey of octopuses are described here from a total of 249 middens (containing 2,731 items) and 88 live prey quadrats (containing 1,098 items). All samples were collected during beach walk transects in Prince William Sound and Kachemak Bay in 1996, 1998, and 2001–2004.

3.1 Population Trends

Octopus densities were lowest in 1995 and 2004, while density rose in the intervening years (Fig. 20.1; nonparametric medians test: N = 107, chi-square = 15.537, df = 5, asymptotic sig. = 0.008; Kruskal–Wallace ANOVA: N = 107, chi-square = 11.284, df = 5, asymptotic sig. = 0.046). Overall beach walk densities in 2002 (N = 7, average density = 0.85 per 1,000 m^2) were nearly four times those found in 1995 (N = 30, average density = 0.22 per 1,000 m^2), and have declined since (Fig. 20.1). An apparent low at the start of the study in 1995, and a subsequent rise and then fall again by 2004 can be found in all Green Island sites taken together (Kruskal–Wallace ANOVA: N = 37, chi-square = 1.897, df = 5, not significant), in

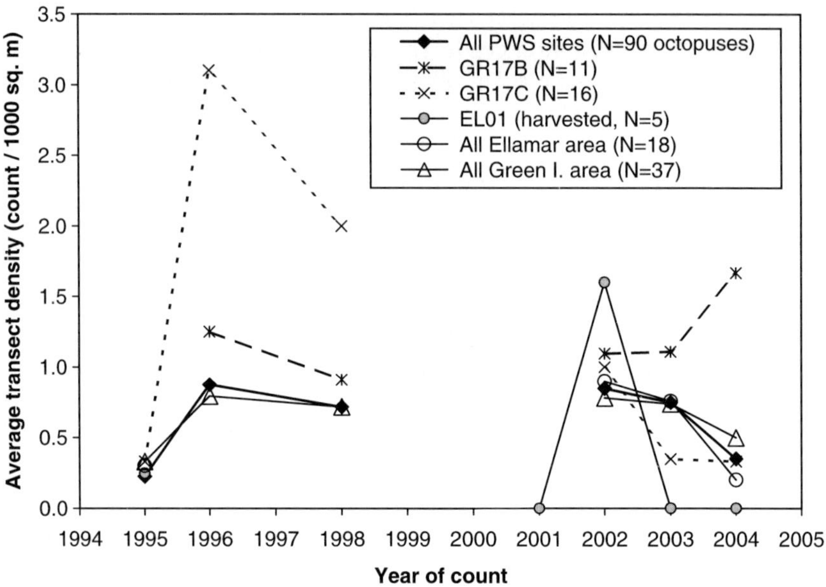

Fig. 20.1 *Octopus counts over all sites (heavy line) and select individual sites. PWS- Prince William Sound; GR17B, GR17C -Green Island sites; EL01 -harvested Ellamar site. Sampling effort was not equal at each site.*

the most frequently sampled Green Island sites taken individually (Fig. 20.1; GR17C and in contrast GR17B), and in all Ellamar areas sites (including Busby Island) taken together (Kruskal–Wallace ANOVA: N = 37, chi-square = 4.152, df = 5, not significant; see Fig. 20.1). Year to year differences at each location did not approach significance due to the lower sample size when the data were subdivided. One site showing a contrasting pattern is the harvested EL01 site (Ellamar), which although low in 1995 shows an apparent decline to near-zero octopus densities in the period 2001–2004. Only one octopus was found (in 2002) on beach walk transects during this period. However, this site is regularly harvested by local residents, who are reported to harvest the beach in the days or months preceding our beach walk transects. Exact harvest rates are not known, but human residency at Ellamar has increased over the decade of this study.

3.2 Prey: Middens, Live Surveys, and Energetics

A sample of 26 prey items representing eight species of common crabs and bivalves collected during beach walk transects were analyzed for energy content. Energy density of prey varied from 3,000 to 5,800 cal/g dry weight. However, variation was not a function of prey species (Fig. 20.2, top; ANOVA: N = 26, df = 25, f = 0.250, p = 0.95).

Table 20.1 *Sizes (carapace or valve width) of prey species commonly represented in both live prey surveys and octopus middens.*

Species	N_M (N_L)[1]	Mean width (cm)		HSD[2]
		Middens	Live	
Hapalogaster mertensii	1 (126)	1.4	1.3	a
Pugettia gracilis	107 (121)	2.0	1.6	ab
Lophopanopeus bellus	63 (122)	2.2	1.5	ab
Cancer oregonensis	219 (121)	2.3	1.6	bc
Chlamys sp.	61 (0)	2.3	–	c
Protothaca staminea	110 (0)	2.8	–	d
Cryptolithodes sp.	17 (8)	4.1	4.5	e
Telmessus cheiragonus	224 (41)	4.6	2.9	e
Cancer productus	102 (4)	5.8	3.5	f

[1] N_M = number of prey items from middens (N_L = number from live prey surveys).
[2] Tukey HSD homogeneous subsets on midden and live prey samples combined. Species marked with different letters were significantly different in size distribution.

Species were, therefore, lumped for analysis of energy density by size. No correlation existed between energy density and either dry weight of whole body soft tissue (meat per prey item) or prey size (carapace or valve width; see Fig. 20.2, top; ANOVA: N = 26, df = 25, f = 0.150, p = 0.861). However, the amount of meat per prey item increased exponentially with prey size (Fig. 20.2, bottom; exponential curve fitting: *Cryptolithodes sitchensis* omitted, N = 23, adjusted R square = 0.784, df = 22, F = 80.88, p < 0.00001). This was not true, however, of *C. sitchensis* (Fig. 20.2, bottom), a crab species with an unusually wide and flat carapace that covers both the body and legs of the crab. Whole-body energy content per prey item may be expressed as cal/g dry-weight (Fig. 20.2, top) times g dry-weight per item (Fig. 20.2, bottom) and, thus, rises rapidly with prey size. With the exception of *C. sitchensis*, whole body energy of prey items is not correlated with species of prey except as prey species vary in maximum body size.

A total of 249 middens containing 2,731 items were recorded, of which 2,260 items (83%) were major hard-body prey (see methods; Vincent et al., 1998), *Cryptolithodes* sp. or *Hapalogaster mertensii.* The sample of live individuals of potential prey consisted of N = 88 quadrats in which were recorded 1,098 items; of these, 26 quadrats (30%) contained at least one major hard-body prey, *Cryptolithodes* sp. or *H. mertensii* individual, and a total of 566 individuals of these species were recorded in quadrats.

The prey representation from the middens (Table 20.1, N = 904 items) was shifted toward larger sizes, relative to that measured during live prey quadrats (N = 543, Fig. 20.3 top). Average carapace sizes differed among species (Table 20.1; ANOVA: F = 76.8, df = 8, p < 0.001). The largest crabs were *Cancer productus* and *Telmessus cheiragonus* (*Cryptolithodes* sp. also had a large carapace but contains very little meat for its size, see above); while the smallest were *Hapalogaster mertensii* (a species avoided by octopuses, Vincent et al., 1998), *Pugettia gracilis*, and *Lophopanopeus bellus* (Table 20.1). For species represented

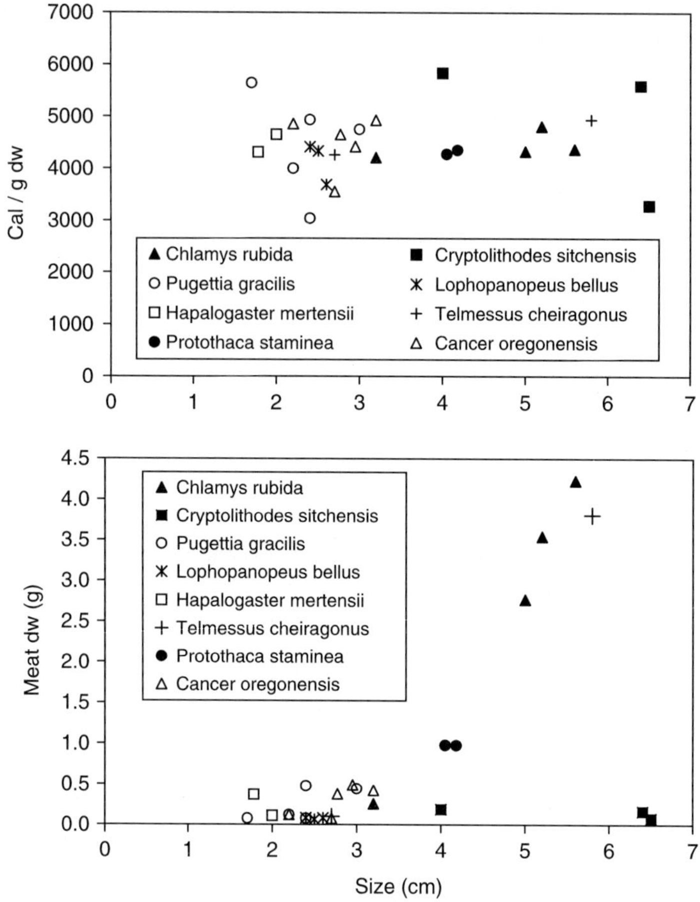

Fig. 20.2 *Energy density (per gram, top panel) and whole-body energy (bottom) of select species of octopus prey.*

in both middens and live prey samples, mean size in middens was larger than mean size in live prey samples (Table 20.1, ANOVA, N = 1,447 prey items of seven taxa; difference by species, between live prey and middens: $F = 7.978$, $df = 6$, $p < 0.001$). For three of four crab species represented by more than 40 individuals in both live prey and midden samples (Table 20.1), the size of remains from middens increased over the study period while the average size in live prey quadrats decreased (Fig. 20.3, middle panel and *Cancer oregonensis* in bottom panel). Size differences among years for *P. gracilis* did not show this pattern; and those of other species were not considered due to only minor representation either in the middens or in the live prey surveys.

Methods by which octopus open prey may leave marks on hard remains of prey. The size of carapace or valve remains of the most common species in

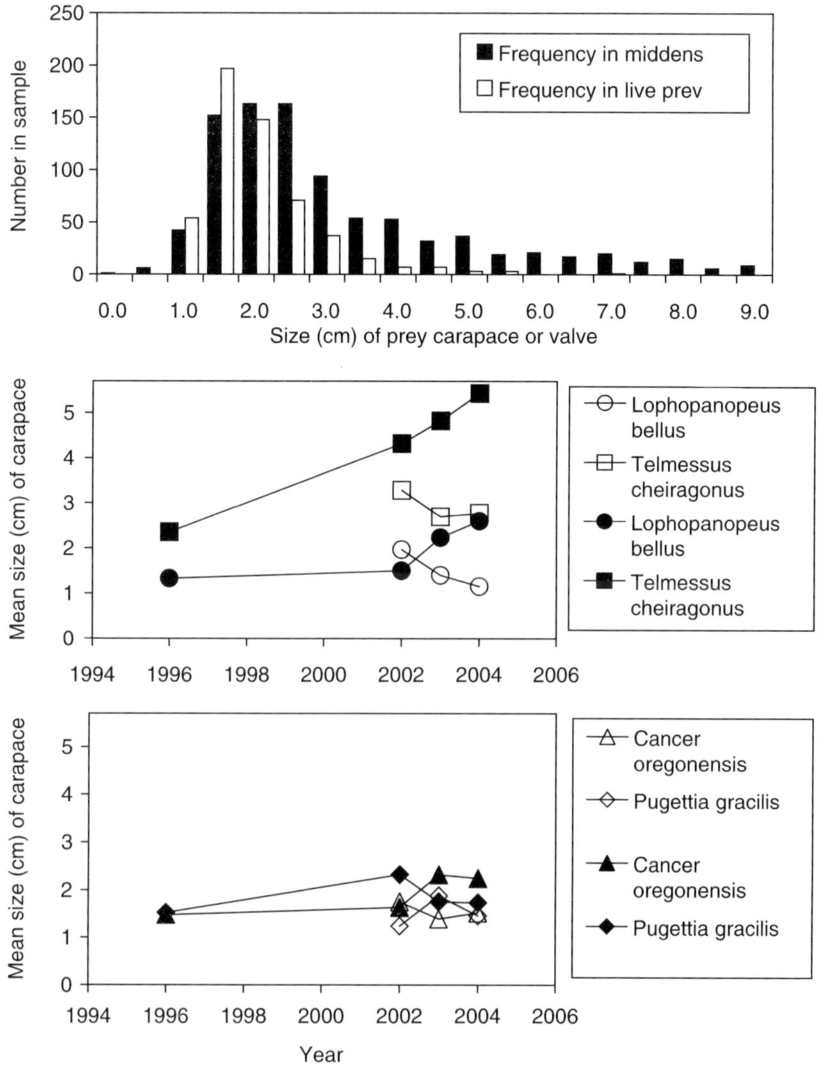

Fig. 20.3 *Number of prey by size (top), and mean size of prey (middle and bottom) in middens (solid symbols) and live prey surveys (open symbols).*

middens varied among prey species and by type of mark found (Table 20.2; N = 824 prey items from middens [615 unmarked, 183 drilled, and 26 bitten] of six species; there was a significant interaction effect between species and method of opening prey; see Table 20.2 for results for individual species.). *Telmessus* crabs were bitten or drilled on the carapace least often (these crabs were often bitten on the legs, however) and two species of *Cancer* crabs and the clam *Protothaca staminea* were drilled or bitten most often (Table 20.2). Furthermore, for *Cancer*

Table 20.2 *Proportion of carapaces and bivalve shells in middens bearing drill or bite marks for select species (top), and for* P. staminea *fed to named captive octopuses (bottom). Drilled remains were significantly larger than unmarked remains of* P. staminea *in both middens and prey fed to captive octopuses*[a]*, and for* C. productus *carapaces in middens.*

		Number (Proportion)				
Species	N	Drilled	Unmarked	Bite	F[b]	P
Telmessus cheiragonus	224	5 (2%)	213 (95%)	6 (3%)[c]	1.283	0.279
Pugettia gracilis	107	23 (21%)	84 (79%)	0 (–)	0.036	0.850
Lophopanopeus bellus	63	14 (22%)	49 (78%)	0 (–)	0.371	0.545
Protothaca staminea	110	32 (29%)	69 (63%)	9 (8%)	7.531	0.001[d]
Cancer productus	101	30 (30%)	67 (66%)	4 (4%)	4.158	0.018[e]
Cancer oregonensis	219	79 (36%)	133 (61%)	7 (3%)	0.882	0.415
Ellamar	57	37 (65%)	6 (11%)	14 (25%)		
Wesley Sands	28	12 (43%)	14 (50%)	2 (7%)		
Inky Waters	44	21 (48%)	21 (48%)	2 (5%)		
Ocho	130	79 (61%)	40 (31%)	11 (8%)		

[a]Tukey HSD comparisons, drill-bite, mean difference = 0.03 cm, p = 0.975; drill-unmarked, mean difference = 0.3 cm, p = 0.005; bite-unmarked, mean difference = 0.2 cm, p = 0.162.

[b]df = 2 except where no bites were recorded, in which cases, df = 1.

[c]Although *T. cheiragonus* remains seldom bear bite marks on the carapace, 35 of 148 legs (25%) of this species found in middens in 1996–2004 bore bite marks (Dodge and Scheel, 1999). Because only one leg may need to be bitten to subdue each crab, this may reflect the percentage of individual *Telmessus* that are opened by biting. *Cancer productus* and possibly other crabs may also be bitten on the leg, apparently with less frequency (Dodge and Scheel, 1999).

[d]Drilled valves (mean size = 3.3 cm) larger than unmarked (mean size = 2.6 cm) and bitten valves (2.5 cm). Tukey HSD comparisons, drill-bite, mean difference = 0.8 cm, p = 0.049; drill-unmarked, mean difference = 0.7 cm, p = 0.001; bite-unmarked, mean difference = 0.08, p = 0.960.

[e]Drilled carapaces (mean size = 6.6 cm) larger than unmarked (5.3 cm); while bitten (6.0 cm) were not significantly different from either drilled or unmarked. Tukey HSD comparisons, drill-bite, mean difference = 0.5 cm, p = 0.882; drill-unmarked, mean difference = 1.3 cm, p = 0.014; bite-unmarked, mean difference = 0.7 cm, p = 0.748.

productus and *P. staminea*, drilled individuals were significantly larger than those unmarked; and for *P. staminea*, drilled individuals were also significantly larger than bitten. These differences were not seen for the other species. Remains of *P. staminea* prey of four captive octopuses were analyzed in the same way (Table 20.2), and showed similar trends although octopuses were fed food of slightly different average sizes.

Midden compositions were examined for the years 1996, 1998, and 2002–2004 for dens (N = 102 middens) in the Green Island and Ellamar areas only (where octopus count transects were repeated most often). Midden contents were analyzed as summed prey size (an index of the energy content represented by each midden), and count and species diversity of items per midden. Summed prey sizes but neither number of items nor species diversity (Shannon index) per midden were significantly lower in 1996 and 1998 than in other years (Kruskal–Wallace ANOVA by years, Table 20.3). There were no significant differences

Table 20.3 *Kruskal Wallis Test for differences in midden content, with Year and Region as grouping variables: (a) Midden content differed between years in the size of prey (sum of prey length), but not in number of items per midden nor diversity (Shannon Index, SI). (b) Mean value of summed prey size and number of items [and items of major hard-bodied prey (MHB) only] per midden by year. Size of prey items was lower in 1996 and 1998 than in later years.*

(a)	Years Chi-Square	df	Asymp. Sig.	Regions Chi-Square	df	Asymp Sig.
Sum prey length	16.096	4	0.003	1.068	1	0.301
Items per midden	8.372	4	0.079	0.110	1	0.741
SI	7.263	4	0.123	1.461	1	0.227

(b)	Year	Sum of prey size	Number of items	Number of MHB items
	1996	16.8	6.4	3.7
	1998	7.1	4.7	3.6
	2002	21.6	5.6	4.3
	2003	19.8	6.7	4.9
	2004	19.8	5.2	3.9
	Total	17.9	5.9	4.2

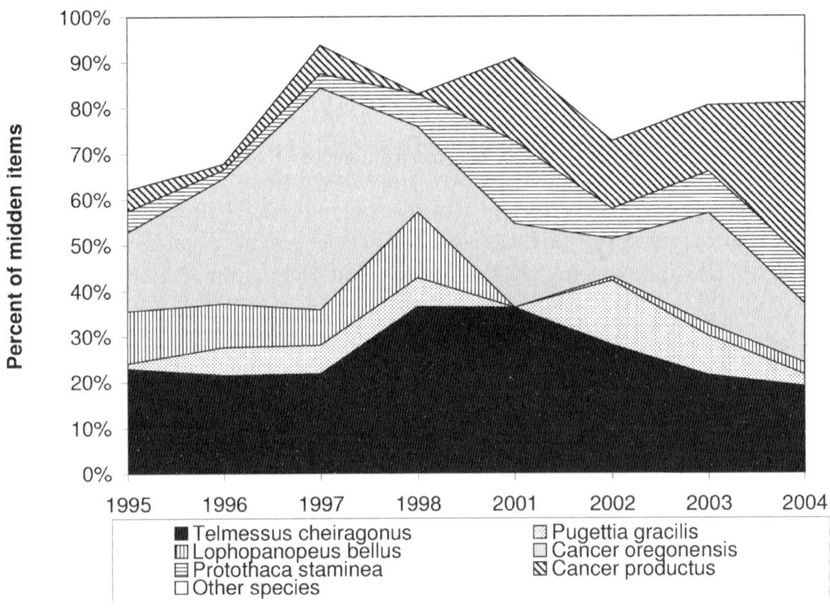

Fig. 20.4 *Over-all species composition of remains found in the midden.*

between the Green Island and Ellamar areas. At Green Island sites, the low summed size of midden items in 1996 and 1998 appears to be due to the scarcity (absence in 1998) of the large crab *Cancer productus* (mean carapace width = 5.8 cm, N = 102) in that year (see Table 20.1, Fig. 20.4). This crab increased to

18% of midden items in 2001 and reached a high of 34% of all midden items in 2004. In contrast, the other large crab found in middens, *Telmessus cheiragonus* (Table 20.1, Fig. 20.4), gradually declined in abundance in middens from a high of 37% in 1998 to a low of 19% in 2004; and *Cancer oregonensis*, a small crab, also declined in representation in the middens since 1997 (Fig. 20.4). Over the same period and on the same beaches, the overall abundance of five species of crabs in live prey samples increased (Fig. 20.5; MANOVA: $F = 6.176$, $df = 20$, $p < 0.001$; differences by sampling region were not significant: $F = 0.535$, $df = 10$, $p < 0.864$). These increases were due to increased counts of *C. oregonensis* (between subject effects, $F = 12.563$, $df = 4$, $p < 0.001$), *Pugettia gracilis* ($F = 12.360$, $df = 4$, $p < 0.001$), and included the first occurrences of *C. productus* in these samples in 2004. Significant variation in *Lophopanopeus bellus* counts ($F = 4.214$, $df = 4$, $p = 0.003$) offset some of these increases in 2003, which was nevertheless the year in which crabs were at their greatest abundance. Counts of *T. cheiragonus* did not change significantly over these years.

3.3 Relocation Experiment

Tag retention, determined as the time postrelease before the tag stopped moving for all subsequent relocations, was 17–51 days in 1996, but less than 24 h in 2003 due to different attachment methods. Thus, results in 2003 reflect very short-term movements, and do not necessarily indicate final choice of habitat. Immediately on release, octopuses either readily reoccupied their previous dens (1996, released where captured), sought nearby shelter under kelp if it was available (2003, shallow releases), or fled (2003, deep releases, no cover in vicinity). Two octopuses, possibly seeking shelter, approached divers immediately on release.

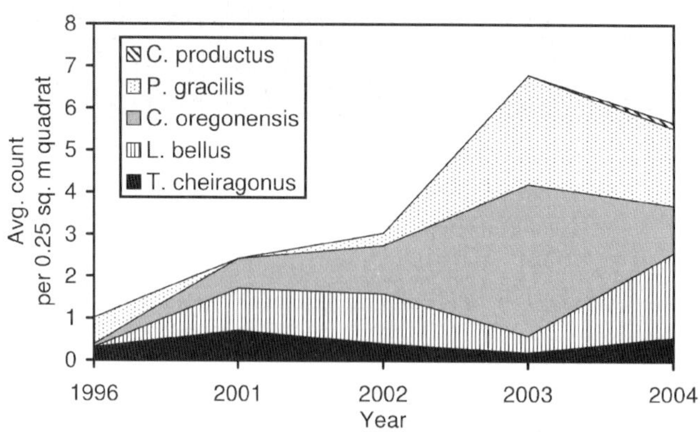

Fig. 20.5 *Individuals per quadrat of select crab species in live prey surveys, demonstrating the increase in prey availability over the course of this study.*

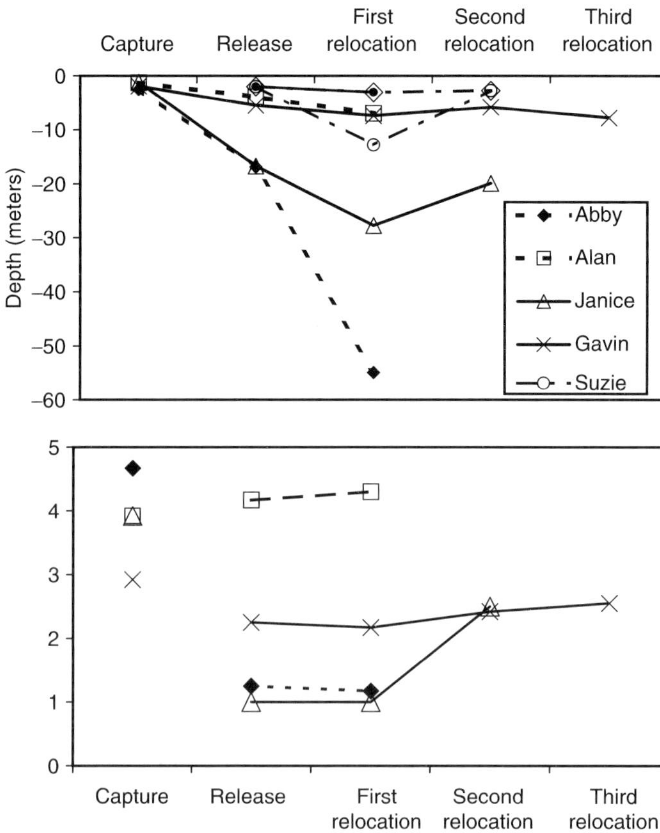

Fig. 20.6 *Captured and released octopuses moved deeper and selected higher quality habitats during relocaton.*

All octopuses moved deeper by 2–40 m following release (Table 20.4), although not all remained deeper at their final relocation. The 1996 controls (released where captured) moved 3–13 m deeper, and subsequently returned to shallow water (<3 m). Two moved within hours of release; the third (Petunia) remained in her release location for three days before briefly moving deeper and then returning to her intertidal capture location. Habitat quality indices (without prey, HQI_4s) for the 1996 releases did not increase above capture/release sites as the octopuses moved (Table 20.4). The 2003 controls (relocated shallow) each moved deeper by <4 m; while the 2003 treatment (relocated deep) each moved deeper by >10 m (Fig. 20.6, top). However, Gavin (relocated shallow) was released in a location where substantially deeper water was not directly accessible. While the distance that the octopuses moved was not correlated with the time postrelease until tag loss, Alan (relocated shallow) moved least and was released into the highest HQI, suggesting a link between movement and

Table 20.4 *Octopuses in the relocation experiment. Depths relative to mean lower low water.*

Octopus	Weight (kg)	Sex	Capture depth (m)	Release depth (m)	Final relocation depth (m)	Capture HQI_4	Release HQI_4	Final Relocation HQI_4	Lowest relocation HQI_4
Petunia[1]	6.3	F	+0.1	+0.1	−0.3	3.5	=capture	3.5	
Beth[1]	7.1	F	−1.0	−1.0	−2.7	3.5	=capture	3.0	
Suzi[1]	7.4	F	−1.0	−1.0	−2.7	3.5	=capture	3.0	
Alan[2]	4.0	M	−1.3	−3.9	−7.0	3.8	4.0	4.0	=final[3]
Gavin[2]	5.3	M	−1.9	−5.5	−7.8	2.3	2.3	2.5	2.0
Abby[2]	7.6	F	−2.5	−16.9	−54.9	4.0	1.3	1.0	=final[3]
Janice[2]	5.1	F	−1.2	−16.7	−19.9	3.8	1.0	2.0	1.0

[1]1996 octopus, released at capture site (control for handling artifacts). Lowest HQIs at relocation sites not available as only release and final relocation sites were surveyed.

[2]2003 octopus, relocated and released either shallow (<6 m, relocation control) or deep (>10 m, treatment). Total distance moved within the first 24 h postrelease was 22 m for Alan, and 297, 344, and 563 m respectively for Gavin, Abby, and Janice.

[3]Alan and Abby were only tracked to a single relocation site, that was thus the lowest, highest, and final relocation HQI.

habitat quality. Indeed, Alan was the only octopus to be released in an area with HQIs greater than his capture location.

For relocated (2003) octopuses, mean postrelease change in habitat quality index was 0.64, with only one value (for Abby) being negative (Fig. 20.6, bottom). Due to small sample size, habitat quality was not significantly different between capture and release sites; between release and relocation sites; nor between release and final relocation sites (HQI_5 paired sample t-test, release to capture: $t = 1.919$, $df = 3$, $p = 0.076$, $n = 4$; release to first relocation: $t = 0.151$, $df = 3$, $p = 0.445$, $n = 4$; release to final relocation: $t = -1.305$, $df = 3$, $p = 0.142$, $n = 4$).

Across all 128 transects conducted 1995–2004, those in habitats with HQI_4 greater than or equal to 2.75 had significantly greater octopus density than those in habitats with lower HQI_4s (Independent samples t-test: $HQI \geq 2.75$, $N = 85$, avg density = 1.0 octopus per 1,000 sq. m surveyed; $HQI < 2.75$, $N = 41$, avg density = 0.4 octopus/1,000 sq. m, $t = 2.823$, $df = 121$, $p = 0.006$, $N = 2$ transects dropped due to missing data; equal variances not assumed). Thus, there were significantly more octopuses on transects with higher habitat quality. HQI_4 did not, however, correlate with midden contents, which as indicators of octopus foraging success might also be expected to indicate habitat quality. No relationships were apparent between HQI_4 or octopus density and either a Shannon Index (SI) of prey diversity in midden piles, the number of items in a midden, or summed size of all items in a midden. While surprising, this lack of correlation would be consistent with time- or risk-minimizing, if risk-adverse octopuses invested more foraging effort in poor habitats and spent more time in shelter where habitats were richer.

4 Discussion

Because foraging is so clearly related to an animal's fitness, natural selection should result in the evolution of animals that are effective foragers. It may or may not follow that animal population fluctuations are influenced by foraging success and food availability. A wide body of work explores these two relationships, including consumer–resource competition, foraging, and habitat selection theories for the relationship of feeding to fitness, and trophic theory for the relation of resource (prey) availability to population size.

Foragers may show functional (Holling, 1959), aggregative (relating numbers of foragers to the abundance of their food, e.g., Piatt, 1990) and population responses to food availability. Functional responses determine the diet of individual predators, while aggregative and population responses together determine local populations of foragers. Foraging theory and habitat selection theory together have been used to address whether foragers maximize foraging functions of fitness by aggregating in particular habitats (Vincent et al., 1996). Here we examine whether considerations of classical foraging theory (Schoener, 1971; Pulliam, 1974; Curio, 1976; Stephens and Krebs, 1986), including energetic value of prey, prey handling time, and prey abundance can explain patterns of variation in octopus counts.

Because population density is determined by recruitment, habitat selection, and mortality, it is very difficult to tease apart historical reasons for density changes without data on all three of these aspects of octopus ecology. We had available data on population counts, habitat characteristics, and trends in prey availability only; and thus were restricted to a correlative retrospective that should be more useful in framing specific hypotheses about octopus habitat selection than in explicitly testing them. Within these limitations, however, we found that octopuses have diets reflecting rate-maximizing foraging strategies, but that population trends did not reflect prey availability.

4.1 Population Trends

Within the sample years surveyed in this study, average octopus densities have varied nearly fourfold from lows of 0.22–0.33 octopus/1,000 m^2 in 1995 and 2004 to a high of 0.72–0.85 octopus/1,000 m^2 in 1998 and 2002 (Fig. 20.1). There was considerably more variation among sites (Fig. 20.1).

Can the sources of variation in a population be assigned? The Ellamar study site (EL01, Fig. 20.1) was near a village, while other sites were not near human population centers. Reports from residents at Ellamar were that harvest from this site occurred every year and had increased over the course of the study (although no records were kept); and octopuses were recorded at Ellamar in our samples only in 1995 and 2002. No known harvest occurs at our other study sites which are remote from human residences. In general, it is very difficult to accumulate data that directly measure recruitment or harvest and other mortality. However, at widely separated sites in the Sound (Fig. 20.1, All Ellamar area vs. All Green I. Area) population trends were similar, indicating that the sources of variation driving the trends were regional rather than local. Local sources might include recovery of sea otter populations from the 1989 *Exxon Valdez* oil spill or from hunting by Native Alaskans in the vicinity of the Ellamar study area. Regional sources might include climatic and oceanographic drivers, or human harvests that drive changes in predators or prey of octopuses over the scale of the entire Sound.

Our analyses in this paper were, therefore, focused on determining whether population fluctuations could be accounted for by changes in habitat characteristics that influence habitat selection. In this study, higher octopus densities were found at sites with higher habitat quality indices (HQI_4: included measures for depth, substrate, boulders, kelp cover, but not presence of prey). During this study, depth, substrate type, and boulders at sites have not changed, and therefore cannot account for fluctuations in octopus counts. However, kelp cover may change, or may correlate with other biological parameters such as prey availability.

Using foraging and habitat selection theory, the effect of prey availability on octopus counts can be considered in at least three ways. With or without an aggregative response, if octopuses acted as foraging time- or risk-minimizers (Schoener, 1971), where fitness was maximized when foraging time spent to obtain a given energy requirement was minimized, then no relationship between prey availability and octopus counts would be expected (other than at a threshold point). In this case, midden content, however, would be stable in the face of varying prey abundance as octopuses expended more effort foraging in habitats with low foraging success but minimized effort in habitats with higher success rates. Alternatively, if octopuses acted as rate maximizing foragers and habitat selectors, population counts as well as midden contents would parallel prey availability (Schoener, 1971). If octopuses were rate maximizing foragers, but population counts were limited by differential recruitment or mortality rather than resource abundance, then, once again, no relationship between prey availability and octopus count would be found. However, in this case, midden content, although not octopus counts, would be expected to mirror differences in prey availability. (Note that this list of alternatives is not exhaustive.) Finally, foraging theory makes predictions about how prey selectivity, handling time, and encounter rates are related to foraging success, and our data on octopus diet can be considered in light of these three parameters.

4.2 Prey: Middens, Live Surveys, and Energetics

What then can be discerned regarding these predictions from our analyses of prey and middens? First, although octopuses have been referred to as generalist (unselective) predators (Mather, 1993; Smith, 2003; but see Ambrose, 1984), our data indicate that they were selective of certain prey species and prey sizes, possibly in a manner consistent with foraging theory for rate-maximizing foragers. Octopus middens lacked certain prey species and contained excesses of others, relative to live prey surveys (compare relative abundance in middens and live surveys, Table 20.1), indicating either octopus selection of preferred prey or differential susceptibility of prey to octopus predation. Availability of *Cancer oregonensis*, *Pugettia gracilis*, and possibly *Cancer productus* increased over the course of the study (Fig. 20.5). In the middens, however, *P. gracilis* did not increase in representation while *C. oregonensis* declined, and *C. productus* increased out of proportion to its occurrence in live prey surveys (Fig. 20.4). These patterns indicate selective predation rather than differential susceptibility of prey. Second, octopuses selected larger individuals of all prey species (Table 1) and for several did so increasingly over the duration of the study (Fig. 20.3). Because prey species are equivalent in energetic content (Fig. 20.2, upper panel) and larger prey have greater energy content (Fig. 20.2, lower panel), this resulted in greater foraging success (higher summed prey size in middens) in later years of the study (Table 20.3). Higher foraging success was the result of choosing larger individuals in later years (Fig. 20.3, lower panels) as well as greater availability and use of larger species (Figs. 20.4, 5) and not catching more individuals (Table 20.3). Third, octopus methods for opening prey (drill, bite, or pull, Hartwick et al., 1978; Steer and Semmens, 2003, with pull generally represented by unmarked shells although, for details, see Mather and Nixon, 1995; Grisley et al., 1996; Dodge and Scheel, 1999) depended on the size and species of prey (Table 20.2), with larger prey more likely to be drilled or bitten. Drilling and biting take considerably longer than pulling to open prey (Steer and Semmens, 2003).

As gain rate is a complex function of prey availability (encounter rate), handling time, and energetic content, it is not possible to determine whether these patterns represent rate-maximization foraging without further data. Predictions for time- or risk-minimizing are that foragers should minimize variance in gain rates in productive habitats but maximize it in unproductive ones; while for rate-maximizing foragers predictions are that diet breadth should be narrower in productive habitats and greater in unproductive ones. In this study, early years were less productive than later years (Fig. 5). The increase in use of larger species as their abundance increased, and the selection of larger prey items, represented a narrowing of diet breadth, and was therefore consistent with rate-maximizing foraging.

4.3 Relocation Experiment

Tagging has been a challenge in octopus studies. Mather et al. (1985) tagged four *Enteroctopus dofleini*; their attachment method, using metal hooks, allowed 4–15 days of observation. Mather et al. (1985) documented that octopuses make frequent short foraging trips and establish home ranges of 250 m^2; however, they did not provide habitat data. Octopuses have sensitive skin, and have been witnessed to tear out tags (G. Ivey, personal communication to AL, 2004): Ivey implanted tags dermally and had an average retention of 27 h. She has had tag retention as long as 14 days using tags attached by cable tie through the septum (G. Ivey, personal communication to DS, 2004), a method similar to that reported in this study for the 1996 releases. Attachment through the mantle (Rigby, 2004) has resulted in tag retention exceeding 21 days. The tags attempted in this study were quite large; use of smaller tags implanted inside the mantle (Rigby, 2004; G. Ivey, personal communication to DS, 2004; in squid, Webber and O'Dor, 1986; O'Dor et al., 1988) appears to be the method of choice for active instrumentation of cephalopods.

All instrumented octopuses in this study, regardless of treatment, moved deeper on release. Do these data support any interpretation other than that octopuses increase movement rates (at random with respect to habitat quality) or move to deeper water on handling? As we released five of seven octopuses in shallow water (<6 m depth, Table 20.4) one could argue that by chance alone they would move to deeper water. However, the data do not entirely support this. Petunia (released where captured) remained three days at her release site prior to moving, while Alan (relocated shallow) was released into higher quality habitat than his capture site and moved little at all (his tag was found in an unoccupied den with a midden, indicating he denned and foraged in the area). Furthermore, the greatest increase in depth came from the two treatment octopuses, which by this argument had the most opportunity to move shallow, but did not.

Alternatively, perhaps handled octopuses prefer deeper water as a response to stress (i.e., to avoid visual predators), the opposite pattern to that predicted by the hypothesis of active habitat selection for high-HQI sites. This explains the greater depth attained by octopuses relocated to deep water, and the fact that all 1996 octopuses (released where captured) initially moved at least slightly deeper. However, the 1996 and 2003 controls released in high HQI habitat showed little movement or depth change (Beth, Petunia, Alan, Table 20.4), while Suzi returned from deeper water to a site with a higher HQI. While this suggests that initial movement to deep water may be a result of handling, it also seems that relocated octopuses do not remain in deeper habitats for long.

Finally, it appears that octopuses show more movement when released into low HQI habitat. Indeed, for the 2003 octopuses, release HQIs are inversely related to the total distance moved by each octopus (Table 20.4, footnote 2) and not correlated with tag retention time. Due to short tag retention times on all 2003 releases, this study was unable to determine whether octopuses stayed for long at any relocation site. However, these results, albeit preliminary, indicate that the postrelease movements were affected by habitat quality [as reflected in release depth, Scheel (2002) and in HQI] at least as much or more than by chance or by handling artifacts. Octopuses seem more likely to leave a low-HQI site than to leave a high-HQI site. Further tracking and habitat data are needed to confirm this conclusion.

4.4 Habitat Selection and Foraging Models – Status and Need

The above sections led us to suggest that, first, octopus middens are consistent with octopuses acting as rate-maximizing foragers; and second, octopus response to relocation depends on habitat quality. With these initial conclusions in hand, can we now correlate prey availability and the abundance of octopuses as a rate-maximizing predator showing aggregative response to prey?

This straightforward prospect does not hold up to scrutiny. While HQI_4 and octopus density were correlated with each other, neither showed significant correlation with any aspect of midden content, including diversity of items or measures of foraging success (number of items, summed length of items). Characters used in calculating HQI may be related to cover for octopuses (boulders and substrate), food availability (prey presence or absence), or both (kelp, depth, substrate), but other interpretations are possible. For example, kelp may serve as cover, as a food resource aggregating prey populations, or cover to potential prey, or may alter octopus and prey larval recruitment (Gaines and Roughgarden, 1987). While the correlation of these characters with octopus counts has previously been established (Scheel, 2002), little data exist to test mechanisms causing the correlation.

Nor did octopus population trends mirror prey population trends as measured in live prey surveys over the course of this study. Octopus densities increased and then declined over time (Fig. 20.1), while major prey densities increased throughout the study (Fig. 20.5). Furthermore, of the three predictions formulated above, the data allow us to reject both rate-maximizing forager alternatives: midden data was stable in the face of changing prey availability. Nor did octopus count parallel habitat productivity as reflected in live prey abundance. We are thus left with the alternative of octopuses as risk- or time-minimizing foragers. This behavior is consistent with the importance of shelter in octopus ecology. However, a more sophisticated approach to formulating hypotheses about foraging and habitat selection is needed.

Patch (habitat) selection models include the trade-off of foraging gain rate against (1) metabolic, (2) predation, and (3) missed–opportunity costs of foraging (Charnov, 1976; Brown, 1988). While foraging, octopuses may select prey as rate-maximizing foragers (as suggested by their selectivity for larger prey items and species). Metabolic costs are unlikely to vary among habitats. However, movement among foraging patches (Charnov, 1976) may be minimized due to high predation risk while in poor HQI_4 patches, which are characterized by low shelter and greater depth. Octopus life history reflects specialization on risk-minimization strategies (Hanlon and Messenger, 1996). Octopuses display diverse cryptic, deimatic, and protean escape behaviors[1]. Other risk-minimizing strategies include use of dens for shelter, and transport of captured prey to the den for handling and consumption. Few data are available to examine perceived predation risk among octopuses, but within- and between-patch travel rates while foraging in habitats with varying cover could be used to examine active management of risk by octopuses. Missed opportunities costs potentially include those for reproductive activities, social activities, grooming and physiological maintenance, and den construction or maintenance. Octopuses increasingly forego foraging at the onset of maturity and reproductive activities, so that trade-off is managed by life history and unlikely to affect the submature juvenile octopuses in this study. Octopuses are solitary, and, hence, social

[1] Deimatic displays function to threaten, startle or frighten, from Deimos, Greek god of terror; while protean behaviors involve changing shape, from Proteus, a shape-changing sea-god.

activities are not important. Finally, grooming and den maintenance are both performed at the den and seem to occupy relatively little time, although the role of sleep in octopus is still uncertain. These trade-offs are promising avenues for further research, but at this time seem likely to be of minor importance to octopuses relative to foraging success and risk from predators.

Patch selection by octopuses should be modeled as a function of trade-off between foraging success within depletable patches (gain rate) and predation risk. The need for this type of model is emphasized by the following considerations. First, the lack of correlation between octopus counts or HQI_4, and any aspect of midden content suggests that prey abundance be excluded as a determinant of habitat selection. Second, if so, of the habitat characteristics measured (depth, substrate, boulders, kelp cover), only kelp cover varies from year-to-year. Third, we do not have data to exclude the hypothesis that kelp cover may directly affect predation rates upon octopuses; however, kelp effects on prey densities or perception of predation risk were not suggested by our analyses (i.e., measured prey density did not significantly correlate with octopus counts; nor did foraging success rate measured from midden contents, which is expected to correlate with foraging effort).

Testing of this proposed model would require detailed data on within- and between-patch movement rates, and characteristics of habitat through which octopuses forage, as well as perceived predation risk estimated as proportion of time spent in and associated with a den; prey encounter rate, measured as counts

Fig. 20.7 *Shadow during field testing in Prince William Sound. Shadow is a prototype submersible vehicle designed to remotely videotape a sonic-tagged benthic marine animal.*

of representative prey encountered per minute foraging; average energetic rate of gain, measured as energetic content of captured prey per minute foraging over the full duration of each foraging bout; and giving-up harvest rate (also a measure of perceived risk), measured as energetic content of captured prey per minute foraging at the end of each foraging bout. Giving-up harvest rate corresponds to a rate that just balances metabolic costs, predation costs, and missed–opportunity costs of foraging in a patch (Charnov, 1976; Brown, 1988).

Such data have been hard to collect in foraging studies, even for terrestrial organisms, and new techniques are required to obtain such data for mobile benthic marine organisms such as octopuses. Shallow, tropical water videography has shown some promise for octopuses (Forsythe and Hanlon, 1997). One approach under development by the authors for cold or deep water is the use of acoustic-positioning telemetry, already adapted for cephalopod studies (e.g., O'Dor et al., 1988; Rigby, 2004), to simultaneously track both a target animal and a submersible vehicle for collection of behavioral and habitat data as the target animal forages and seeks shelter (Shadow, Fig. 20.7). Telemetry has been successfully used for positioning information in cephalopod studies, and the addition of underwater data would substantially increase the utility of such studies in interpreting movement data.

Acknowledgments

This material is based upon work supported by the National Science Foundation under Grant No. 0115882 to DLS (2001–2003), by Alaska Pacific University through donations from the At-Sea Processors Association under grants to AL and DLS (2000–2003), by the West Coast and Polar Regions Undersea Research Center (University of Alaska, Fairbanks, National Undersea Research Program, NOAA) under a 1998 grant to DLS and TLSV, and by the Exxon Valdez Oil Spill Trustee Council under grants to DLS (1995–1997).

References

Authorities for species retrieved 25 Feb 2004 from the Integrated Taxonomic Information System (ITIS) on-line database at http://www.itis.usda.gov.

Ambrose, R. F. 1984. Food preferences, prey availability, and the diet of *Octopus bimaculatus* Verrill. *Journal of Experimental Marine Biology and Ecology* **77**: 29–44.

Ambrose, R. F. 1988. Population dynamics of *Octopus bimaculatus*: influence of life history patterns, synchronous reproduction and recruitment. *Malacologia* **29**(1): 23–39.

Anderson, R. C. 1994. Octopus bites clam. *The Festivus* **25**(5): 58–59.

Anderson, R. C. 1996. Sedating and euthanizing octopuses. *Drum and Croaker* **27**: 7–8.

Anderson, T. 1997. Habitat selection and shelter use by *Octopus tetricus*. *Marine Ecology Progress Series* **150**: 149–155.

Andrews, P. L. R., and E. M. Tansey. 1981. The effects of some anaesthetic agents in *Octopus vulgaris*. *Comparative Biochemistry and Physiology* 70C: 241–247.

Aronson, R. B. 1986. Life history and den ecology of *Octopus briareus* Robson in a marine lake. *Journal of Experimental Marine Biology and Ecology* **95**: 37–56.

Brown, J. S. 1988. Patch use as an indicator of habitat preference, predation risk, and competition. *Behavioral Ecology and Sociobiology* **22**: 37–47.

Chapman, M. G. 1986. Assessment of some controls in experimental transplants of intertidal gastropods. *Journal of Experimental Marine Biology and Ecology* **103**: 181–201.

Charnov, E. L. 1976. Optimal foraging, the marginal value theorem. *Theoretical Population Biology* **9**: 129–136.

Connolly, S. R., and J. Roughgarden. 1998. A latitudinal gradient in northeast Pacific intertidal community structure: evidence for an oceanographically based synthesis of marine community theory. *American Naturalist* **151**(4): 311–326.

Cosgrove, J. A. 1987. Aspects of the natural history of *Octopus dofleini*, the Giant Pacific Octopus. M.Sc. thesis, University of Victoria.

Curio, E. 1976. *The Ethology of Predation*. New York: Springer.

Dodge, R., and Scheel, D. 1999. Remains of the prey: recognizing the middens of *Octopus dofleini*. *The Veliger* **42**(3): 260–266.

Estes, J. A., and C. H. Peterson. 1998. The dynamics of marine benthic/demersal ecosystems. Whitepaper, NSF Workshop on OEUVRE -Ocean Ecology: Understanding and vision for research. On the web at www.joss.ucar.edu/joss_psg/project/oce_workshop/oeuvre/estes_peterson.html, retrieved 19 Jul 2004.

Forsythe, J. W., and R. T. Hanlon. 1997. Foraging and associated behavior by *Octopus cyanea* Gray, 1849 on a coral atoll, French Polynesia. *Journal of Experimental Marine Biology and Ecology* **209**:15–31.

Frazer, T. K., and W. J. Lindberg. 1994. Refuge spacing similarly affects reef-associated species from three phyla. *Bulletin of Marine Science* **55**(2–3): 388–400.

Gaines, S. D., and J. Roughgarden. 1987. Fish in offshore kelp forests affect recruitment to intertidal barnacle populations. *Science* **235**: 479–481.

Garstang, W. 1900. The plague of octopus on the south coast, and its effect on the crab and lobster fisheries. *Journal of the Marine Biological Association of the United Kingdom* **6**: 260–273.

Gillespie, G. E., G. Parker, and J. Morrison. 1998. A review of octopus fisheries biology and British Columbia octopus fisheries. *Canadian Stock Assessment Secretariat Research Document 98/87*. Ottawa, Canada: Fisheries and Oceans Canada.

Griffiths, C. L., and P. A. R. Hockey. 1987. A model describing the interactive roles of predation, competition and tidal elevation in structuring mussel populations. *South African Journal of Marine Science* **5**: 547–556.

Grisley, M. S., P. R. Boyle, and L. N. Key. 1996. Eye puncture as a route of entry for saliva during predation on crabs by the octopus *Eledone cirrhosa* (Lamarck). *Journal of Experimental Marine Biology and Ecology* **202**: 225–237.

Hartwick, E. B. 1983. *Octopus dofleini*. *In* P. R. Boyle (editor), *Cephalopod Life Cycles*. Volume I, pp. 277–291. London: Academic Press.

Hartwick, E. B., R. F. Ambrose, and S. M. C. Robinson. 1984a. Dynamics of shallow-water populations of *Octopus dofleini*. *Marine Biology* **82**: 65–72.

Hartwick, E. B., R. F. Ambrose, and S. M. C. Robinson. 1984b. Den utilization and the movements of tagged *Octopus dofleini*. *Marine Behaviour and Physiology* **11**: 95–110.

Hartwick, E. B., and I. Barriga. 1997. *Octopus dofleini*: biology and fisheries in Canada. *In* M. A. Lang, and F. G. Hochberg (editors), *Proceedings of the workshop on The Fishery and Market Potential of Octopus in California*, pp. 45–56. Washington, DC: Smithsonian Institution.

Hartwick, E. B., P. A. Breen, and L. Tulloch. 1978. A removal experiment with *Octopus dofleini* (Wulker). *Journal of the Fisheries Research Board of Canada* **35**: 1492–1495.

Hartwick, E. B., S. M. C. Robinson, R. F. Ambrose, D. Trotter, and M. Walsh. 1988. Inshore-offshore comparison of *Octopus dofleini* with special reference to abundance, growth and physical condition during winter. *Malacologia* **29**(1): 57–68.

Hartwick, B., and G. Thorarinsson. 1978. Den associates of the giant Pacific octopus, *Octopus dofleini* (Wulker). *Ophelia* **17**: 163–166.

Hartwick, B., L. Tulloch, and S. MacDonald. 1981. Feeding and growth of *Octopus dofleini* (Wulker). *The Veliger* **24**(2): 129–138.

Hochberg, F. G. 1998. Class Cephalopoda. *In* P. V. Scott, and J. A. Blake (editors), *Taxonomic atlas of the benthic fauna of the Santa Barbara basin and the western Santa Barbara channel.* Volume 8, The Mollusca, Part I: The Aplacophora, Polyplacophora, Scaphopoda, Bivalvia, and Cephalopoda. Santa Barbara, California: Santa Barbara Museum of Natural History.

Holling, C. S. 1959. Some characteristics of simple types of predation and parasitism. *The Canadian Entomologist* **91**(7): 385–398.

Iribarne, O. O. 1990. Use of shelter by the small Patagonian octopus *Octopus tehuelchus*: availability, selection and effects on fecundity. *Marine Ecology Progress Series* **66**: 251–258.

Karlson, R. H. 2002. *Dynamics of Coral Communities. Population and Community Biology Series.* Volume 23. New York: Kluwer Academic Publishers.

Laidig, T. E., P. B. Adams, C. H. Baxter, and J. L. Butler. 1995. Feeding on euphausiids by *Octopus rubescens. California Fish and Game* **81**(2): 77–79.

Legac, M.1969. Some observations on the building up of hiding-places by the *Octopus vulgaris* Lam. in the Channel regions of the north-eastern Adriatic. *Thalassia Jugoslavica* **5**: 193–199.

Mather, J. A. 1982. Factors affecting the spatial distribution of natural populations of *Octopus joubini* Robson. *Animal Behaviour* **30**: 1166–1170.

Mather, J. A. 1993. Octopuses as predators: implications of management. *In* T. Okutani, R. K. O'Dor, and T. Kubodera (editors), *Recent Advances in Cephalopod Fisheries Biology*, Volume 1, pp. 275–282. Tokyo: Tokai University Press.

Mather, J. A. 1994. 'Home' choice and modification by juvenile *Octopus vulgaris* (Mollusca: Cephalopoda): specialized intelligence and tool use? *Journal of Zoology London* **233**: 359–368.

Mather, J. A., and M. Nixon. 1995. *Octopus vulgaris* (Cephalopoda) drills the chelae of crabs in Bermuda. *Journal of Molluscan Studies* **61**: 405–506.

Mather, J. A., S. Resler, and J. Cosgrove. 1985. Activity and movement patterns of *Octopus dofleini. Marine Behaviour and Physiology* **11**: 301–314.

Mottet, M. G. 1975. The fishery biology of *Octopus dofleini* (Wulker). Technical Report No. 16, Management and Research Division, Washington Department of Fisheries.

Nagasawa, K., S. Takayanagi, and T. Takami. 1993. Cephalopod tagging and marking in Japan: a review. *In* T. Okutani, R. K. O'Dor and T. Kubodera (editors), *Recent Advances in Cephalopod Fisheries Biology*. Volume 1, pp. 313–329. Tokyo: Tokai University Press.

National Science Foundation. 1998. OEUVRE Ocean Ecology: Understanding and vision for research. Report on the 1998 workshop OEUVRE -Ocean Ecology: Understanding and vision for research. Biological Oceanography, National Science Foundation. On the web at www.joss.ucar.edu/joss_psg/project/oce_workshop/oeuvre, retrieved 19 Jul 2004.

O'Dor, R. K., D. M. Webber, and F. M. Boegeli. 1988. A multiple buoy acoustic-radio telemetry system for automated positioning and telemetry of physical and physiological data. *In* C. J. Amlaner Jr. (editor), *Proceedings of the Tenth International Symposium on Biotelemetry*, pp. 444–452. Fayetteville: The University of Arkansas Press.

Piatt, J. F. 1990. The aggregative response of common murres and Atlantic puffins to schools of capelin. *Avian Biology* **14**: 36–51.

Pulliam, H. R. 1974. On the theory of optimal diets. *American Naturalist* **108**: 59–74.

Rees, W. J., and J. R. Lumby. 1954. The abundance of octopus in the English channel. *Journal of the Marine Biological Association of the United Kingdom* **33**: 515–536.

Rigby, R. P. 2004. Ecology of immature octopus, *Enteroctopus dofleini*: growth, movement and behavior. Ph.D. dissertation. Hokkaido University, Hokkaido, Japan.

Rigby, R. P., and Y. Sakurai. 2004. Temperature and feeding related to growth efficiency of immature octopuses *Enteroctopus dofleini. Suisanzoshoku* **52**(1): 29–36.

Sato, K. 1994. Saving the mizudako. *Fisheries Research* **13**(6) (Suppl.): 82–89. [In Japanese]

Sato, K. 1996. Survey of sexual maturation in *Octopus dofleini* in the coastal waters off Cape Shiriya, Shimokita Peninsula, Aomori Prefecture. *Nippon Suisan Gakkaishi* **62**(3): 355–360. [In Japanese, with English abstract]

Scheel, D. 2002. Characteristics of habitats used by *Enteroctopus dofleini* in Prince William Sound and Cook Inlet, Alaska. P.S.Z.N. *Marine Ecology* **23**(3): 185–206.

Schoener, T. W. 1971. Theory of feeding strategies. *Annual Review of Ecology and Systematics* **2**: 369–404.

Smith, C. D. 2003. Diet of *Octopus vulgaris* in False Bay, South Africa. *Marine Biology* **143**: 1127–1133.

Smith, K. N., and W. F. Herrnkind. 1992. Predation on early juvenile spiny lobsters *Panulirus argus* (Latreille): influence of size and shelter. *Journal of Experimental Marine Biology and Ecology* **157**: 3–18.

Steer, M. A., and J. M. Semmens. 2003. Pulling or drilling, does size or species matter? An experimental study of prey handling in *Octopus dierythraeus* (Norman, 1992). *Journal of Experimental Marine Biology and Ecology* **290**: 165–178.

Stephens, D. W. and J. R. Krebs. 1986. Foraging theory. *In* J. R. Krebs and T. Clutton-Brock (editors), Monographs in behavior and ecology. Princeton, New Jersey: Princeton University Press.

Vincent, T. L. S., D. Scheel, J. S. Brown, and T. L. Vincent. 1996. Trade-offs and coexistence in consumer-resource models: it all depends on what and where you eat. *American Naturalist* **148**(6): 1038–1058.

Vincent, T. L. S., D. Scheel, and K. Hough. 1998. Aspects of the diet and foraging behavior of *Octopus dofleini* in its northernmost range. *Marine Ecology* **19**(1): 13–29.

Webber, D. M., and R. K. O'Dor. 1986. Monitoring the metabolic rate and activity of free-swimming squid with telemetered jet pressure. *Journal of Experimental Biology* **126**: 205–224.

Wülker, G. 1910. Ueber Japanische Cephalopoden. Beitrage zur kenntnis der systematic und anatomie der dibranchiaten. *Abhandlungen der mathematische- physikalische Klasse der Koeniglich Bayerischen Akademie der Wissenschaften* **3** (Suppl. 1): 1–77.

Index

Like most indices, this one is nowhere near as thorough as it might have been; the tremendous number of items that might have been included would have made the index completely unwieldy.

In general, people whose works are cited in a given chapter are not indexed; those mentioned in the given chapter are listed in the end of that chapter. For items that appear many times throughout a chapter or section of a chapter, only the first occurrence in that chapter or section is indexed, with the page number followed by "ff". Material in any appendices or in the References section of a given chapter is referred to in the text of that chapter, so it has not been entered into the index. No concerted attempt has been made to organize taxa into a Linnaean hierarchy, primarily because there is no one such hierarchy that meets with universal approval amongst cephalopod workers. In the index, "phylogeny" encompasses patterns through geologic time, whereas "organic evolution" comprises the processes. Some entries are in the format "285-295"; this does not necessarily mean that the discussion extends for the eleven full pages, but that the topic appears on each of the eleven pages.

Indices are idiosyncratic constructs of their compilers. We hope that the following one is useful to you nonetheless.

D

E

F

P

Q

R

S

U

V

W

X

Y

Z